Practical Handbook of

GENETIC ALGORITHMS

Applications
Volume I

Edited by Lance Chambers

CRC Press
Boca Raton New York London Tokyo

Library of Congress Cataloging-in-Publication Data

Chambers, Lance.
Practical handbook of genetic algorithms: applications, volume I / Lance Chambers.
p. cm.
Includes bibliographical references (p. -) and index.
ISBN 0-8493-2519-6 (v. 1 : alk. paper)
1. Genetic algorithms. I. Title.
QA402.5.C44 1995
005.1--dc20
95-17139
CIP

Direct all inquiries to CRC Press, Inc., 2000 Corporate Blvd., N.W., Boca Raton, Florida 33431.

International Standard Book Number 0-8493-2519-6
Library of Congress Card Number 95-17139
Printed in the United States of America 1 2 3 4 5 6 7 8 9 0
Printed on acid-free paper

I dedicate this work to Tony Noakes, a great friend and mentor. I will miss you.

The Editor

Lance Chambers is Manager: Data and Analysis at the Department of Transport in Western Australia where he has been working for over seven years in various areas but currently concentrating on answering transport policy questions and the solution of transport policy problems.

He has also worked in academia, consulting and industry with a number of companies over a number of decades and as a full- and part-time lecturer for Curtin University of Technology teaching both graduate and post-graduate courses in the quantitative disciplines. Being unable to bear the internal politics of any of these types of organisation he decided to leave and become a public servant.

Lance graduated from the Western Australian Institute of Technology (since renamed Curtin University of Technology) majoring in Marketing and Operations Research. From there he proceeded to travel to the USA to undertake further studies at Purdue University, West Laffayette, Indiana. Working in the areas of Finance (with Prof. Lewellen) and Operations Management (Prof. De Kluyver) he finalised his work in 1984 and returned to Australia.

Since joining the Department of Transport Lance has been their most prolific researcher, publisher of journal articles, generator of conference papers and all-round iconoclast. He is, proudly, a member of no societies but is an active participant in the Australian transport scene (academic and consulting) where his particular strengths are recognised and employed to assist in the creation of change. This arrangement suits him very well since he can get in, have his say, and get out.

Lance is currently working in the area of the application of GAs to assist in solving highly complex policy problems in the transport arena and he is assisted in the work by Prof MAP Taylor of the University of South Australia, the Western Australian Department of Transport and the Australian Commonwealth Government.

TABLE OF CONTENTS

Preface

Genetic Algorithms

In the real world, the process of natural selection controls evolution. Organisms most suited for their environment tend to live long enough to reproduce, whereas less-suited organisms often die before producing young or produce fewer and/or weaker young. In artificial-life simulations and in the applications of genetic algorithms, we can study the process of evolution by creating an artificial world, populating it with pseudo-organisms and giving those organisms a goal to achieve. Using genetic algorithms, we can then use a very crude form of evolution to teach the organisms to achieve the stated goal. If the goal, given to these organisms, is the formulation of a problem we confront in our own world we can, by observing the behaviours and characteristics of the successful organisms discover solutions to our own problems.

What's a genetic algorithm? An algorithm is simply a series of steps for solving a problem. A *genetic algorithm,* then, is a problem-solving method that uses genetics as its model of problem solving. In short, genetic algorithms apply the rules of reproduction, gene crossover, and mutation to these pseudo-organisms so those organisms can pass beneficial and survival-enhancing traits[1] to new generations.

Artificial Genotypes

In the real world, an organism's characteristics are encoded in its DNA. Genetic algorithms store the characteristics of artificial organisms in an *electronic genotype,* which mimics the DNA of natural life. This electronic genotype is nothing more than a long string of bits (a bit is the smallest piece of data a computer can process. It can be only one of two values: 0 or 1) A bit in the genotype string can be 'on' (has the value 1), or can be 'off' (has the value of 0). The existance of a certain charaterisitic can be indicated by whether a particular bit is set to on or off (e.g. a gene that is turned on could indicate that the genetype [physical manifestation of the gene (the person, animal, etc.)] has blue eyes and if it is off the genetype has brown eyes). Because some behaviors or characteristics might be too complex to be represented by a single on or off, a number of bits are sometimes used to determine how those characteristic or behaviour are manifest.

Mutation as an Evolutionary Factor

Mutation is another factor that affects evolution in the real world. When creatures reproduce, there's no guarantee the resulting genotype will be reproduced perfectly. Often, reproduction introduces unpredictable changes to the genotype,

---

[1] A survival-enhancing trait is one that helps the organism improve its ability to reach the goal set for the organism. In the standard genetic algorithm (GA) it is survival of the pseudo-species that is important not the survival of an individual. Therefore very good (fit) parents can die off – but as long as they have passed on their good traits we have a viable population of genes that can continue to enhance the average fitness of the population of genes.

creating new characteristics that may or may not be advantageous. Those creatures with positive mutations tend to live to reproduce. Those with negative mutations die off.

In genetic algorithms, mutation is introduced by changing a randomly determined bit in the genotype. The programmer programs this change to occur according to a percentage chance. For example, a programmer might determine that mutation is to occur once for every 100 matings, which means about one percent of all offspring will contain a mutated gene.

The differences between a-life and genetic algorithms in application are, I believe, minor. The genetic algorithm has no need to feed its population for the phenotype to remain alive nor do two genes need to physically meet in their computer world to mate. Other than that the differences have to do with the 'world' the phenotype lives in and what represents success in this world. An example, and not an extreme one, is a world that represents the design of an aircraft and success being measured in costs of construction, flight stability, seating capacity, operational complexity, etc. Therefore measures of success are not necessarily simple, as in survival, number of times bred or the destruction of enemies, but are rather more complex, often having a multitude of success factors and the world within which the phenotype lives being an electronic construct of a mathematical world that represents a problem for which we are seeking a solution.

Genetic algorithms are also only of value where it is possible to test, often many hundreds of thousands of times, the efficacy of a suggested solution. Therefore we don't want genetic algorithms to be used to test how well a human-life-dependent process works, unless we can electronically simulate the process very accurately. In the operating theater of a hospital I wouldn't want a genetic algorithm to make suggestions to my surgeon on how to operate on me. GA: "Doctor try cutting out the heart and throwing it on the floor." Surgeon proceeds to do so. GA: "How did that work?" Surgeon: "The patient is dead!" GA: "Okay. Bad gene. I'll give it a low fitness value so that it has a low probability of breeding and being passed on as a viable method of solving the problem on hand." Okay for those who come after me, they won't have their heart cut out and thrown on the floor. But the GA might suggest some other equally deadly test in its search for a good way to cure a headache. However, if we can simulate the process accurately we can then proceed with the simulation and allow the GA to 'kill' the patient as many times as it likes in its search for a solution. Then once the GA has come up with a good solution we can implement that solution in the real world and go around curing headaches in thier millions.

What would the objective function (measure of success) be for the GA exampled above? One measure would be that the patient survives the cure, another might be that the solution is cheap to implement, that the patient can administer the cure themselves, that there are no side effects that make the cure worse than the sickness, etc. The 'best' phenotype may be something as simple as, "Take an Aspro and call me in the morning." My experiences with GAs have led me to believe that if you can structure the problem sufficiently accurately and if there exists a solution to your problem, then a GA will find it. I have been continually impressed and surprised by the ability of GAs to elicit useful and usable solutions to highly complex problems.

Nature was smart enough to develop the process of biological evolution and what we have out of that process is one of the most fascinating mechanism for devloping lifeforms that can live in the most delicate, fearsome, dry, wet, hot, cold, acid, caustic environments on earth. Each possible niche that could hold life on the world holds life. GAs have been working in nature for millions of years and all we have done is mimic the effectiveness of natural processes to solve some of our more intractable problems. If evolution and genetic process can solve all those many millions of life based problems why can we not use the same processes to solve our far less complex problems? So we have taken what nature has to offer and have used it with thanks.

As you come to understand GAs and their processes you will come to understand the power of the genetic based problem solving paradigms that lie behind them. Many problems that we face are generally considered intractable. These are problems that we have no simple method of solving. It is not possible to move easily from one starting point directly towards a solution. The reasons are because as one changes one factor in the environment in question other factors change as a consequence (the problems can be very complex mathematically). We have all heard the screech that comes from a speaker if a microphone is held too close. The solution to that problem, of positive feedback, is to move the microphone away. However; what would you do if you were in a room filled with tens of thousands of microphones, with tens of thousands of speakers, and an ear shattering screech? You job is to eliminate the screech without turning off the power. As you moved one microphone you would change the level of sound but in the process would be moving the microphone closer to another speaker that is screeching so the speaker connected to that microphone would increase its sound level which would make the microphones close to that speaker send signals to their speakers to increase volume and so it would go on, in an ever increasing cycle until the speakers all exploded.

Now let a GA loose in an electronically simulated environment that duplicates the above scenario. Objective: to eliminate the sound. I believe that within minutes, on a decent computer, you will have a solution on where to move the microphones, and in what order, to eliminate the sound, or at least get it down as far as it can go. The solution to a highly intractable problem becomes easy to lay hands on.

There is a chance that GAs can help in understanding the process of learning. How do we, as humans, learn throughout life? We learn, often by trail and error, GAs learn that way. We also learn by observing the effects of actions on others (Harry does something dumb on the swing and falls on his head. We tend to learn from that). GAs also learn that way. They observe the efficacy of individual phenotypes and manipulate the probability of the genes responsible for the phenotype of breeding, based upon that efficacy. Therefore the whole population of phenotypes 'learn' about the behaviour of others in the population by having particular characteristics not being bred into the next population.[2]

I hope that as you read this book you will also come to be as fascinated by GAs as I am.

---

2 It is probably not right to say they learn but rather that certain characteristics are forgotten (unlearned and bred out of the gene pool?).

Chapter 0

J.E. Everett
Department of Information Management and Marketing
The University of Western Australia
Nedlands, Western Australia 6009

jeverett@ecel.uwa.edu.au

Model Building, Model Testing and Model Fitting

0-8493-2519-6/95/$0.00 + $.50

Abstract
One major area of usefulness for genetic algorithms lies in testing and fitting quantitative models. Sensible discussion of this use of genetic algorithms depends upon a clear view of the nature of quantitative model building and testing. In this chapter we consider the formulation of such models, and the various approaches that might be taken to fit model parameters. A hierarchy of optimisation methods is discussed, ranging from analytical methods, through various types of hill climbing, randomised search and genetic algorithms. A number of examples illustrate that modelling problems do not fall neatly into this clear-cut hierarchy. For this reason, a judicious selection of hybrid methods, selected according to the model context, is preferred to any pure method alone in designing efficient and effective methods for fitting parameters to quantitative models.

0.1 Uses of Genetic Algorithms

0.1.1 Optimising or Improving the Performance of Operating Systems

Genetic algorithms can be useful for two largely distinct purposes. One purpose is the selection of parameters to optimise the performance of a system. Usually we are concerned with a real or realistic operating system, such as a gas distribution pipeline system, traffic lights, travelling salesmen, allocation of funds to projects, scheduling, handling and blending of materials and so forth. Such operating systems typically depend upon decision parameters which can be chosen (perhaps within constraints) by the system designer or operator. Appropriate or inappropriate choice of decision parameters will cause the system to perform better or worse, as measured by some relevant objective or fitness function. In realistic systems, the interactions between the parameters are not generally amenable to analytical treatment, and the researcher has to resort to appropriate search techniques. Most published work on genetic algorithms has been concerned with this use to optimise operating systems (or at least to improve them by approaching the optimum).

0.1.2 Testing and Fitting Quantitative Models

The second potential use for genetic algorithms has been less discussed, but lies in the field of testing and fitting quantitative models. Scientific research into a problem area classically consists of the iterative process of building explanatory or descriptive models, collecting data, testing the models and, when discrepancies are found, modifying the models and then repeating the process until the problem is solved, or the researcher retires, dies or runs out of funds, and interest passes on to a new problem area.

In using genetic algorithms to test and fit quantitative parameters, we are again searching for parameters which will optimise a fitness function. However, in contrast to the situation where we were trying to maximise the performance of an operating system, we are now trying to find parameters that minimise the misfit between the model and the data. The fitness function, perhaps more appropriately referred to as the 'misfit function', will be some appropriate function of the difference between the observed data values and the data values that would be predicted from the model. Optimising involves finding parameter values for the model that minimise the misfit function. In some applications, it is conventional to refer to the misfit function as the 'loss' or 'stress' function. For the purposes of this chapter, 'fitness', 'misfit', 'loss' and 'stress' can be considered as synonymous.

0.1.3 Maximising versus Minimising

We might be tempted to distinguish the two major areas of potential for genetic algorithms – optimising operating systems and fitting quantitative models – as corresponding to the difference between maximising an operating system's performance measure and minimising the misfit between a model and a set of observed data. This distinction, while useful, must not be pressed too far, since maximising and minimising can always be interchanged. Maximising an operating system's performance is equivalent to minimising its shortfall from some unattainable ideal. Conversely, minimising a misfit function is equivalent to maximising the negative of the fitness function.

0.1.4 Purpose of this Chapter

The use of genetic algorithms to optimise or improve the performance of operating systems is discussed in many of the chapters of this book. The purpose of the present chapter is to concentrate on the second use of genetic algorithms: the fitting and testing of quantitative models. An example of such an application, using a genetic algorithm to fit multidimensional scaling models, appears in Chapter 9.

It is important to consider the strengths and limitations of the genetic algorithm method for model fitting. To understand whether genetic algorithms are appropriate for a particular problem, it is necessary first to consider the various types of quantitative model and appropriate ways of fitting and testing them. In so doing, we will see that there is not a one-to-one correspondence between problem types and methods of solution. Indeed, a particular problem may contain elements of a range of model types, and therefore may be more appropriately tackled by a hybrid method, which may incorporate genetic algorithms with other methods, rather than by a single pure method.

0.2 Quantitative Models

0.2.1 Parameters

Quantitative models generally include one or more parameters. As a simple example, if we have a model that says that children's weights are linearly related to their heights, then the model contains two parameters: an intercept (the weight of a hypothetical child of zero height) and a slope (the increase in weight for each unit increase in height). Such a model can be tested by searching for parameter values that fit real data to the model. Using the children's weight and height model, if we could find no values of the intercept and slope parameters that adequately fit a set of real data to the model, we would be forced to abandon or to modify the model.

In cases where parameters could be found that adequately fit the data to the model, then the values of the parameters are likely to be of use in several ways. The parameter values will aid attempts to use the model as a summary way of describing reality, to make predictions about further as yet unobserved data, and perhaps even to give explicative power to the model.

0.2.2 Revising the Model or Revising the Data?

If an unacceptable mismatch occurs between a fondly treasured model and a set of data, then it may be justifiable, before abandoning or modifying the model, to question the validity or relevance of the data. Cynics might accuse some practitioners, notably a few economists and psychologists, of having a tendency to take this too far, to the extent of ignoring, discrediting or discounting any data that do not fit received models. However, in all sciences, the more established is a model, the greater the body of data evidence that is required to overthrow it.

0.2.3 Hierarchic or Stepwise Model Building - The Role of Theory

Generally, following the principal of Occam's razor, it is advisable to start with models that are too simplistic (which usually means, they contain less parameters than are required). If a simplistic model is shown to inadequately fit observed data, then we reject the model in favour of a more complicated model with more parameters. In the height and weight example this might, for instance, be achieved by adding a quadratic term to the model equation predicting weight from height.

In building successively more complex models it is, where possible, preferable to base the models upon some theory. If we have a theory that says children's height would be expected to vary with household income, then we are more justified in including the variable in the model. If, as often happens, a variable is included because it helps the model fit the data, but without any prior theoretical justification, then that may be interesting exploratory model building, but more analysis of more data and explanatory development of the theory will be needed to place much credence on the result.

As we add parameters to a model, in a stepwise or hierarchic process of increasing complexity, we need to be able to test whether each new parameter added has improved the model sufficiently to warrant its inclusion. We also need some means of judging when the model has been made complex enough: that is, when the model fits the data acceptably well.

In these questions as to whether added parameters are justified, and whether a model adequately fits a data set, the concepts of significance and meaningfulness become important.

0.2.4 Significance and Meaningfulness

It is important to distinguish statistical significance from statistical meaningfulness. The explanatory power of a parameter can be statistically significant but not meaningful, or it can be meaningful without being significant, or it can be neither or both significant and meaningful.

In model building, we require any parameters we include to be statistically significant and to be meaningful. If a parameter is statistically significant, then that means the data set as extreme as we have got would be highly unlikely if the parameter were absent or zero. If a parameter is meaningful, then it explains a useful proportion of whatever it is that our model is setting out to explain.

The difference between significance and meaningfulness is best illustrated by an example. Consider a sample of 1000 people from each of two large communities.

Their heights have all been measured. The average height of one sample was 1 cm more than the other. The standard deviation of height for each sample was 10 cm. We would be justified in saying that there was a significant difference in height between the two communities, because if there really were no difference between the population, the probability of getting such a sampling difference would be about 0.1%. Accordingly, we are forced to believe that the two communities really do differ in height. However, the difference between the communities' average heights is very small compared with the variability within each community. One way to put it is to say that the difference between the communities explains only one percent of the variance in height. Another way of looking at it is to compare two individuals chosen at random, one from each community. The individual from the taller community will have a 46% chance of being shorter than the individual from the other community, instead of the 50% chance if we had not known about the difference. All in all, it would be fair to say that the difference between the two communities' heights, while significant, is not meaningful. Following Occam's razor, if we were building a model to predict height, we might not in this case consider it worthwhile to include community membership as a meaningfully predictive parameter.

Conversely, it can happen that a parameter appears to have great explicative power, but our evidence is insufficient to be significant. Using the same example, if we had sampled just one member from each community and found they differed in height by 15 cm, that would be a meaningful pointer to further data gathering, but could not be considered significant evidence in its own right. In this case, we would have to collect more data before we could be sure that the apparently meaningful effect was not just a chance happening.

Before a new model, or an amplification of an existing model by adding further parameters, can be considered worth adopting we need to demonstrate that its explanatory power (its power to reduce the misfit function) is both meaningful and significant.

In deciding whether a model is adequate, we need to examine the residual misfit:

- If the misfit is neither meaningful nor significant, we can rest content that we have a good model;

- If the misfit is significant but not meaningful, then we have an adequate working model;

- If the misfit is both significant and meaningful, the model needs further development;

- If the misfit is meaningful but not significant, we need to test further against more data.

The distinction between significance and meaningfulness provides one very strong reason for the use of quantitative methods both for improving operating systems and for building and testing models. The human brain operating in qualitative mode has a tendency to build a model upon anecdotal evidence, and subsequently to accept evidence that supports the model and reject or fail to notice evidence that does not support the model. A disciplined, carefully designed and well-documented quantitative approach can help us avoid this pitfall.

0.3 Analytical Optimisation

Many model fitting problems can be solved analytically, without recourse to iterative techniques such as genetic algorithms. In some cases, the analytical solubility is obvious. In other cases, the analytical solution may be more obscure and require analytical skills unavailable to the researcher.

If the analytical solution lies beyond the powers of the researcher, or if the problem is likely to become non-analytical as we look at more and more realistic data sets, then the researcher might be justified in using iterative methods even when they are not strictly needed.

However, the opposite case is also quite common: a little thought may reveal the problem to be susceptible to analytical treatment. As we shall see, it can happen that parts of a more intractable problem can be solved analytically, reducing the number of parameters that have to be solved by iterative search. A hybrid approach including partly analytical methods can then reduce the complexity of an iterative solution.

0.3.1 An Example - Linear Regression

Linear regression models provide an example of problems that can be solved analytically.

Consider a set of 'n' data points $\{x_i , y_i \}$ to which we wish to fit the linear model:

$$y = a + bx \qquad (0.1)$$

The model has two parameters 'a' (the intercept) and 'b' (the slope), as shown in Figure 0.1.

The misfit function to be minimised is the mean squared error F(a,b):

$$F(a,b) = \Sigma(a + bx_i - y_i)^2 /n \qquad (0.2)$$

Differentiation of F with respect to a and b shows F is minimised when:

$$b = (\Sigma y_i \, \Sigma x_i - n\Sigma y_i \, \Sigma x_i) / ((\Sigma x_i)^2 - n\Sigma x_i {}^2) \qquad (0.3)$$

$$a = (\Sigma y_i - b\Sigma x_i)/n \qquad (0.4)$$

It is important that the misfit function be statistically appropriate. For example, if we had reason to believe that scatter around the straight line should increase with x, then use of the misfit function defined in equation (0.2) would lead to points of large x being given too much weight relative to points of small x. In this case, with misfit function to be minimised would be F/x. Sometimes the appropriate misfit function can be optimised analytically, other times it cannot, even though the model may itself be quite simple.

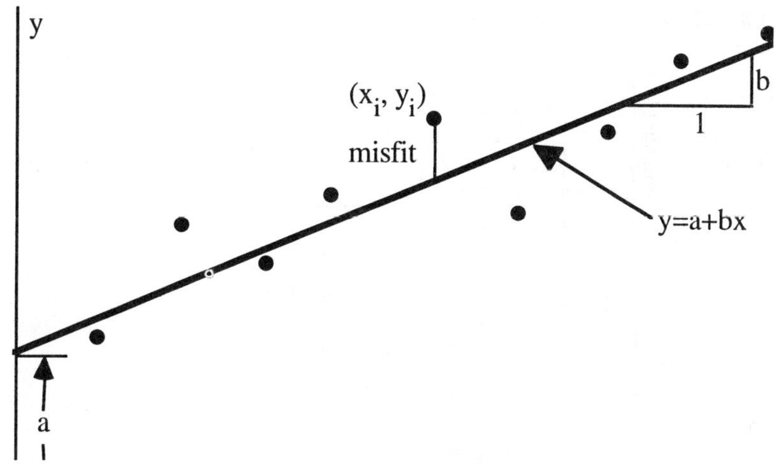

Figure 0.1: Simple Linear Regression

More complicated linear regression models can be formed with multiple independent variables:

$$y = a + b_1x_1 + b_2x_2 + b_3x_3 + b_4x_4 \ldots\ldots \quad (0.5)$$

Analytical solution of these multiple regression models is described in any standard statistics textbook, together with a variety of other analytically soluble models. However, many models and their misfit functions cannot be expressed in an analytically soluble form. In these cases we will need to consider iterative methods of solution.

0.4 Iterative Hill Climbing Techniques

There are many situations where we need to find the global optimum, or a close approximation to the global optimum, of a multidimensional function which we cannot optimise analytically. For many years, various "hill-climbing" techniques have been used to search iteratively toward an optimum. The term "hill-climbing" should strictly be applied only to maximising problems, with techniques for minimising being identified as "valley-descending". However, a simple reversal of sign converts a minimising problem into a maximising one, so it is customary to use the "hill-climbing" term to cover both situations.

A very common optimising problem occurs when we try to fit some data to a model. The model may include a number of parameters, and we want to choose the parameters to minimise some function which represents the "misfit" between the data and the model. The values of the parameters can be thought of as coordinates in a multidimensional space, and the process of seeking an optimum involves some form of systematic search through this multidimensional space.

0.4.1 Iterative Incremental Stepping Method

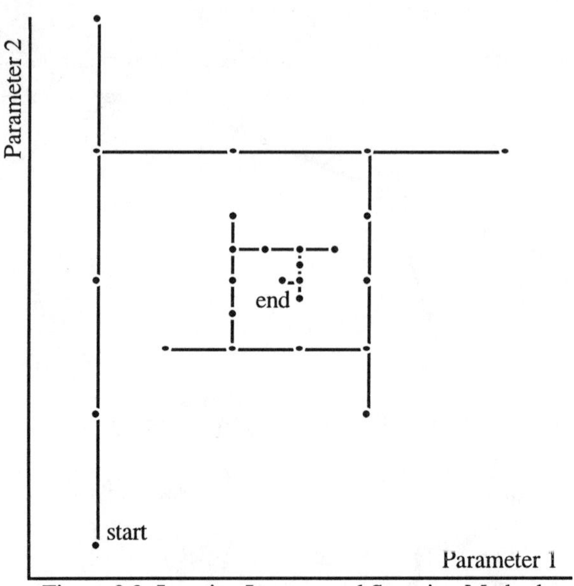

Figure 0.2: Iterative Incremental Stepping Method

The simplest, moderately efficient way of searching for an optimum in a multidimensional space is by the iterative incremental stepping method, illustrated in Figure 0.2.

In this simplest form of hill-climbing, we start with a guess as to the coordinates of the optimum. We then change one coordinate by a suitably chosen (or guessed) increment. If the function gets better, we keep moving in the same direction by the same increment. If the function gets worse, we undo the last increment and start changing one of the other coordinates. This process continues through all the coordinates until all the coordinates have been tested. We then halve the increment, reverse its sign, and start again. The process continues until the increments have been halved enough times that the parameters have been determined with the desired accuracy.

0.4.2 An Example - Fitting the Continents Together

A good example of this simple iterative approach is the computer fit of the continents around the Atlantic, which I carried out in 1963. This study provided the first direct quantitative evidence for continental drift (Bullard, Everett and Smith, 1965). It had long been observed that the continents of Europe, Africa and North and South America looked as if they fit together. I digitised the spherical coordinates of the contours around the continents and used a computer to fit the jigsaw together. The continental edges were fit by shifting one to overlay the other as closely as possible. This shifting, on the surface of a sphere, was equivalent to rotating one continental edge by a chosen angle around a pole of chosen latitude and longitude. There were thus three coordinates to choose to minimise the measure of misfit:

• The angle of rotation;

• The latitude and longitude of the pole of rotation.

The three coordinates were as shown in Figure 0.3, in which point P_i on one continental edge is rotated to point P_i' close to the other continental edge.

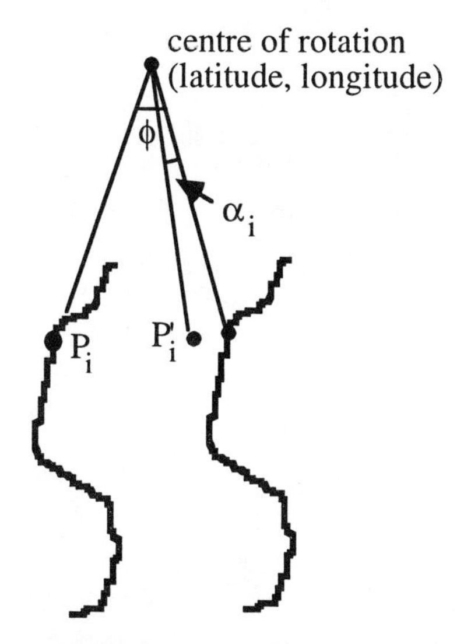

Figure 0.3: Fitting Contours on the Opposite Sides of an Ocean

The misfit function, to be minimised, was the mean squared underlap or overlap between the two continental edges after rotation.

If the underlap or overlap is expressed as an angle of misfit α_i , then the misfit function to be minimised is:

$$F = \Sigma\alpha_i^2 / n \tag{0.6}$$

It can easily be shown that F is minimised if α, the angle of rotation is chosen so that:

$$\Sigma\alpha_i = 0 \tag{0.7}$$

So, for any given centre of rotation, the problem can be optimised analytically for the third parameter, the angle of rotation, by simply making the average overlap zero.

Minimising the misfit can therefore be carried out using the iterative incremental stepping method, as shown above in Figure 0.2, with the two parameters being the latitude and longitude of the centre of rotation. For each centre of rotation being evaluated, the optimum angle of rotation is found analytically to make the average misfit zero.

There was in fact a fourth parameter, the depth contour at which the continental edges were digitised. This parameter was treated by repeating the study for a number of contours: first for the coastline (zero depth contour) and then for the 200, 1000, 2000 and 4000 metre contours. Gratifyingly, the minimum misfit function was found to be obtained for contours corresponding to the steepest part of the continental shelf, as shown in Figure 0.4.

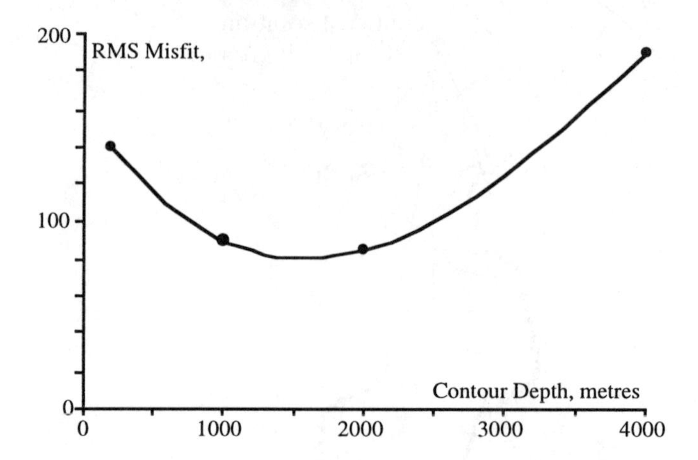

Figure 0.4: Least Misfit for Contours of Steepest Part of Continental Shelf

This result, that the best fit was obtained for the contour line corresponding to the steepest part of the continental shelf, provided good theory-based support for the model. Applying the theory of continental drift, and postulating that the continents around the Atlantic are the remains of a continental block that has been torn apart, we would then indeed expect to find that the steepest part of the continental shelf provides the best definition of the continental edge, and therefore best fits the reconstructed jigsaw.

The resulting map for the continents around the Atlantic is shown in Figure 0.5. Further examples of theory supporting the model are found in such details as the extra overlap in the region of the Niger delta, where more recent material has been washed into the ocean, bulging out that portion of the African coastline.

0.4.3 Other Hill Climbing Methods

A more direct approach to the optimum can be achieved by moving in the direction of steepest descent. If the function to be optimised is not directly differentiable, then the method of steepest decent may not improve the efficiency, because it is necessary to evaluate the function stepped out along each parameter axis in order to determine the direction of steepest descent.

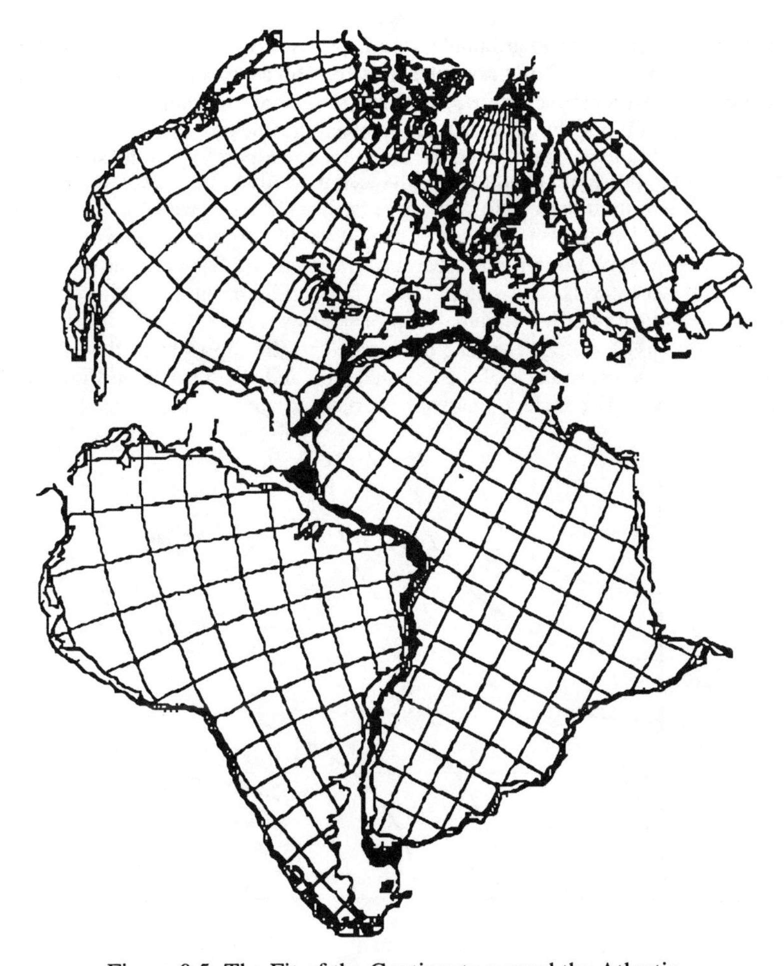

Figure 0.5: The Fit of the Continents around the Atlantic

Another modification that can improve the efficiency of approach to the optimum is to determine the incremental step by a quadratic approximation to the function. The function is computed at its present location, and at two others equal amounts to either side. The increment is then calculated so as to take us to the minimum of the quadratic fit through these three points, or to the reflection of the maximum if the curvature is convex upwards. Repeating the process can lead us to the minimum in fewer steps than would be needed if we used the iterative incremental stepping method. A fuller description of this quadratic approximation method can be found in Chapter 9 (Figure 0.3).

0.4.4 The Danger of Entrapment on Local Optima and Saddle Points

Although the continental drift problem required an iterative solution, the clear graphical nature of its solution suggested that local optima were not a problem of concern. This possibility was in fact checked for by starting the solution at a number of widely differing centres of rotation, and finding that they all gave consistent convergence to the same optimum. When only two parameters require iterative solution, it is usually not difficult to establish graphically whether local optima are a problem. If the problem requires iteration on more than two parameters, then it may be very difficult to check for local optima.

While iterating along each parameter, it is also possible to become entrapped at a point which is a minimum with respect to each parameter, with the point being not even a local optimum but just a saddle point. Figure 0.6 illustrates this possibility for a problem with two parameters, p and q. The point at the centre, marked with an asterisk, is a saddle point. The saddle point is a minimum with respect to changes in either parameter, p or q. However, it is a maximum along the direction (p+q), going from the bottom left to the top right of the graph. If we explore by changing each of the parameters p and q in turn, as in Figure 0.2, then we will wrongly conclude that we have reached a minimum.

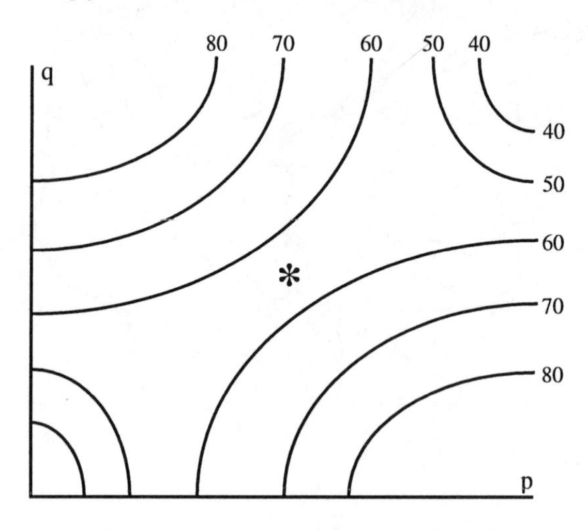

Figure 0.6: Entrapment at a Saddle Point

0.4.5 The Application of Genetic Algorithms to Model Fitting
For problems that have multiple local optima, or even for problems where we do not know whether a single optimum is unique, both the iterative incremental step method and methods of steepest descent lead to a danger of the solution being trapped in a local optimum. The device of restarting the iteration from a wide variety of starting points may provide some safeguard against entrapment in a local minimum, although there are problems where any starting point could lead us to a local optimum before we reached the global optimum. For this type of problem, genetic algorithms offer a preferable means of solution. Genetic algorithms offer the attraction that all parts of the feasible space are potentially available for exploration, so the global minimum should be attained if premature convergence can be avoided.

We will now consider a model building problem where a genetic algorithm can be usefully incorporated into the solution process.

0.5 Assay Continuity in a Gold Prospect

To illustrate the application of a genetic algorithm as one tool in fitting a series of hierarchical models, we will consider an example of real economic significance.

0.5.1 Description of the Problem

In Europe there exists a copper mine that has been mined underground since at least Roman times. The ore body measures nearly a kilometre square by a couple of hundred metres thick. It is now worked out as a copper mine, but only a very small proportion of the ore body has been removed. Over the past forty years, as the body was mined, core samples were taken and assayed for gold and silver as well as copper. About 1500 of these assays are available from locations scattered unevenly through the ore body.

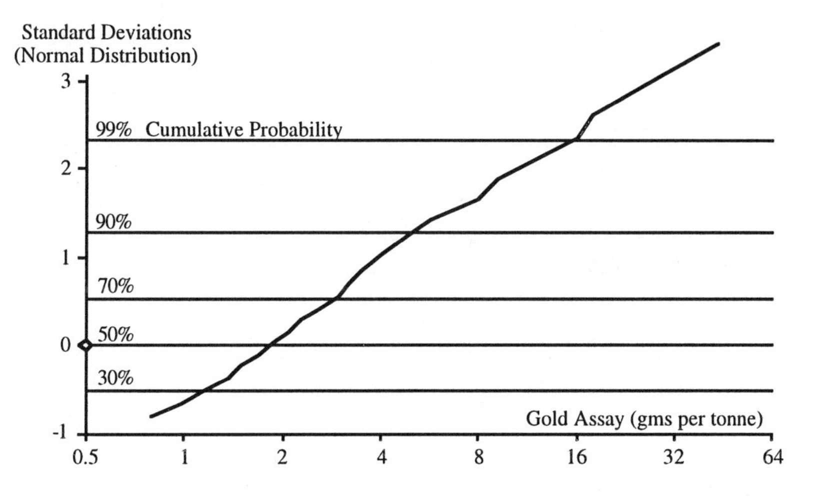

Figure 0.7: Cumulative Distribution of Gold Assays, on Log-Normal Scale

The cumulative distribution of the gold assay values is plotted on the log-normal scale of Figure 0.7. The plot is close to being a straight line, confirming that the assay distribution is close to being log normal, as theory would predict for this type of ore body. If the gold concentration is the result of a large number of random multiplicative effects, then the Central Limit theorem would lead us to expect the logarithms of the gold assays to be normally distributed, as we have found.

The gold assays average about 2.6 gms per tonne, have a median of 1.9 gms per tonne, and correlate only weakly (0.17) with the copper assays. It is therefore reasonable to suppose that most of the gold has been left behind by the copper mining. The concentration of gold is not enough to warrant underground mining, but the prospect would make a very attractive open-cut mine provided the gold assays were representative of the whole body.

To verify which parts of the ore body are worth open-cut mining, extensive further core-sample drilling is needed. Drilling is itself an expensive process, and it would be of great economic value to know how closely the drilling has to be carried out to give an adequate assessment of the gold content between the drill holes. If the holes are drilled too far apart, interpolation between them will not be valid, and expensive mistakes may be made in the mining plan. If the holes are drilled closer together than necessary, then much expensive drilling will have been wasted.

0.5.2 A Model of Data Continuity

Essentially, the problem reduces to estimating a data 'continuity distance' across which the gold assays can be considered to be reasonably well correlated. Assay values that are from locations far distant from one another will be uncorrelated. Assay values will become identical (within measurement error) as the distance between their locations tends to zero. So the expected correlation between a pair of assay values will range from close to one (actually the test/retest repeatability) when they have zero separation, down to zero correlation when they are far distant from each other. The questions remain as to:

- How fast the expected correlation diminishes with distance, and
- What form the diminution with distance takes.

The second question can be answered by considering three points strung out along a straight line, separated by distances r_{12} and r_{23} as shown in Figure 0.8. Let the correlation between the assays at point 1 and point 2, points 2 and 3, and points 1 and 3 be ρ_{12}, ρ_{23}, ρ_{13} respectively.

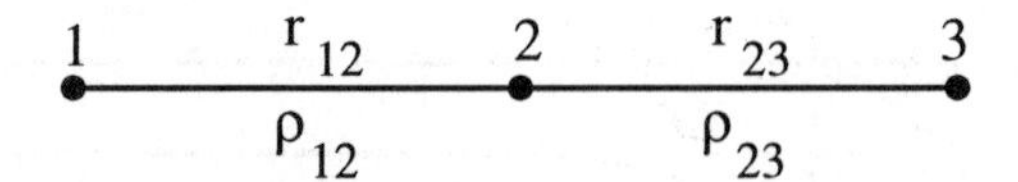

Figure 0.8: Assay Continuity

It can reasonably be argued that, in general, knowledge of the assay at point 2 gives us some information about the assay at point 3. But the assay at point 1 tells us no more about point 3 than we already know from the assay at point 2. We have what is essentially a Markov process. This assumption is valid unless there can be shown to be some predictable cyclic pattern to the assay distribution.

Examples of Markov processes are familiar in marketing studies where, for instance, knowing what brand of toothpaste a customer bought two times back adds nothing to the predictive knowledge gained from knowing the brand they brought last time. Or in the field of finance, for the share market, if we know yesterday's share price, we will gain no further predictive insight into today's price by looking up what the price was the day before yesterday. Similarly, with the gold assays, the assay from point 1 tells us no more about the assay for point 3 than we have already learned if we know the assay at point 2, unless there is some predictable cyclic behaviour in the assays.

Consequently, we can treat ρ_{12} and ρ_{23} as orthogonal, so:

$$\rho_{13} = \rho_{12} \cdot \rho_{23} \tag{0.8}$$

$$\rho(r_{13}) = \rho(r_{12}+r_{23}) = \rho(r_{12}).\rho(r_{23}) \qquad (0.9)$$

To satisfy equation (0.9), and the limiting values $\rho(0) = 1$, $\rho(\infty) = 0$, we can postulate a negative exponential model for the correlation coefficient $\rho(r)$, as a function of the distance r between the two locations being correlated.

MODEL 1 $$\rho(r) = \exp(-kr) \qquad (0.10)$$

Model 1 has a single parameter, 'k,' whose reciprocal represents the distance at which the correlation coefficient falls to (1/e) of its initial value. Thus the value of k answers our first question as to how fast the expected correlation diminishes with distance. However, the model makes two simplifying assumptions, which we may need to relax.

Firstly, Model 1 assumes implicitly that the assay values have perfect accuracy. If the test/retest repeatability is not perfect, we should introduce a second parameter $a = \rho(0)$, where $a < 1$. The model then becomes:

MODEL 2 $$\rho(r) = a.\exp(-kr) \qquad (0.11)$$

The second parameter 'a' corresponds to the correlation that would be expected between repeat samples from the same location, or the test/retest repeatability. This question of the test/retest repeatability explains why we do not include the cross products of assays with themselves to establish the correlation for zero distance: these auto cross products would have an expectation of one, since they are not subject to test/retest error.

The other implicit assumption is that the material is homogeneous along the three directional axes x, y and z. If there is geological structure that pervades the entire body, then this assumption of homogeneity may be invalid. We then need to add more parameters to the model, expanding kr to allow for different rates of fall off (k_a, k_b, k_c), along three orthogonal directions, or major axes, (r_a, r_b, r_c). This modification of the model is still compatible with Figure 0.8 and equation (0.9), but allows for the possibility that the correlation falls off at different rates in different directions.

$$k\,r = \mathrm{sqrt}(k_a^2\, r_a^2 + k_b^2\, r_b^2 + k_c^2\, r_c^2) \qquad (0.12)$$

These three orthogonal directions of the major axes (r_a, r_b, r_c) can be defined by a set of three angles (α, β, γ). Angles α and β define the azimuth (degrees east of north) and inclination (angle upwards from the horizontal) of the first major axis. Angle γ defines the direction (clockwise from vertical) of the second axis, in a plane orthogonal to the first. The direction of the third major axis is then automatically defined, being orthogonal to each of the first two. If two points are separated by distances (x, y, z) along north, east and vertical coordinates, then their separation along the three major axes is given by:

$$a = x.\cos\alpha.\cos\beta + y.\sin\alpha.\cos\beta + z.\sin\beta \qquad (0.13)$$
$$b = -x(\sin\alpha.\sin\gamma+\cos\alpha.\sin\beta.\cos\gamma) + y(\cos\alpha.\sin\gamma-\sin\alpha.\sin\beta.\cos\gamma)+z.\cos\beta.\cos\gamma \qquad (0.14)$$
$$c = x(\sin\alpha.\cos\gamma-\cos\alpha.\sin\beta.\sin\gamma) - y(\cos\alpha.\cos\gamma+\sin\alpha.\sin\beta.\sin\gamma) + z.\cos\beta.\sin\gamma \qquad (0.15)$$

The six parameters (k_a, k_b, k_c) and (α, β, γ) define three-dimensional ellipsoid surfaces of equal assay continuity. The correlation between assays at two separated points is now $\rho(r_a, r_b, r_c)$, a function of (r_a, r_b, r_c), the distance between the points along the directions of the three orthogonal major axes.

Allowing for the possibility of directional inhomogeneity, the model thus becomes:

MODEL 3 $\rho(r_a, r_b, r_c) = a.\exp[-\mathrm{sqrt}(k_a^2 r_a^2 + k_b^2 r_b^2 + k_c^2 r_c^2)]$ (0.16)

In Model 3, the correlation coefficient still falls off exponentially in any direction, but the rate of fall-off depends upon the direction. Along the first major axis, the correlation falls off by a ratio 1/e for an increase of $1/k_a$ in the separation. Along the second and third axes, the correlation falls off by 1/e for increases of $1/k_b$ and $1/k_c$ respectively in the separation.

0.5.2.1 A Model Hierarchy

In going from Model 1 to Model 2 to Model 3, as in equations (10), (11) and (16), we are successively adding parameters:

1) $\rho(r) = \exp(-kr)$

——> 2) $\rho(r) = a.\exp(-kr)$

——> 3) $\rho(r_a, r_b, r_c) = a.\exp[-\mathrm{sqrt}(k_a^2 r_a^2 + k_b^2 r_b^2 + k_c^2 r_c^2)]$

The three models can thus be considered as forming a hierarchy. In this hierarchy, each successive model adds explanatory power at the cost of using up more degrees of freedom in fitting more parameters. Model 1 is a special case of Model 2, and both are special cases of Model 3. As we go successively from Model 1 to Model 2 to Model 3, the goodness of fit (the minimised misfit function) cannot get worse, but may improve. We have to judge whether the improvement of fit achieved by each step is sufficient to justify the added complexity of the model.

0.5.3 Fitting the Data to the Model

As we have seen, the assay data were found to be log-normally distributed to a close approximation. Accordingly, the analysis to be described here was carried out on standardised logarithm values of the assays. Some assays had been reported as zero gold content: these were in reality not zero, but below a reportable threshold. The zero values were replaced by the arbitrary low measure of 0.25 gms/tonne before taking logarithms. The logarithms of the assay values had their mean subtracted and were divided by the standard deviation, to give a standardised variable of zero mean, unit standard deviation and approximately normal distribution. The cross-product between any two of these values could therefore be taken as an estimate of the correlation coefficient. The cross-products provide the raw material for testing Models 1, 2 and 3.

The 1576 assay observations yield over 1,200,000 cross products (not including cross products of assays with themselves). Of these cross products, 362 corresponded to radial distances less than a metre, 1052 between 1 and 2 metres, then steadily increasing numbers for each one-metre shell, up to 1957 in the interval between 15 and 16 metres. The average cross product in each concentric shell can be used to estimate the correlation coefficient at the centre of the shell.

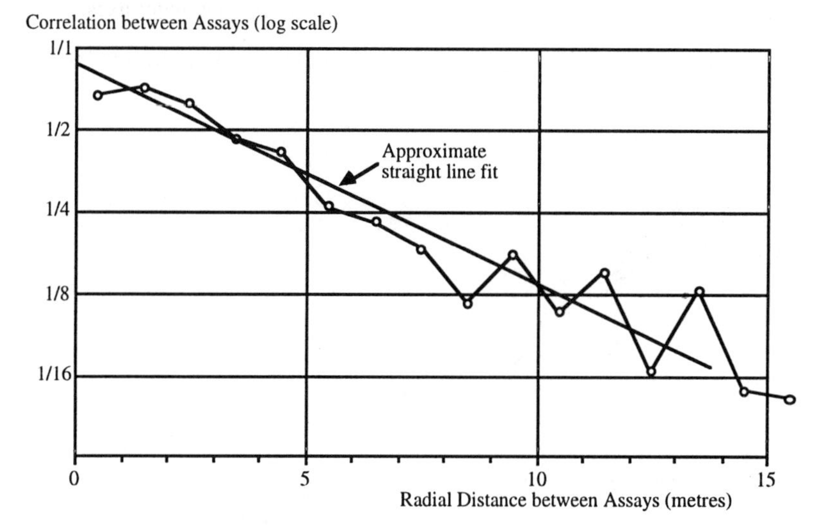

Figure 0.9: Log Correlations as a Function of r, the Inter-Assay Distance

The average cross product for each of these one-metre shells is plotted in Figure 0.9, and provides an estimate of the correlation coefficient for each radial distance. The vertical scale is plotted logarithmically, so the negative exponential Model 1 or 2 (equations 0.10 or 0.11) should yield a negatively sloping straight-line graph. The results appear to fit this model reasonably well.

The apparently increasing scatter as the radial distance increases is an artifact of the logarithmic scale. Figure 0.10 shows the same data plotted with a linear correlation scale. The curved thin line represents a negative exponential fit by eye. It is clear that the scatter in the observed data does not vary greatly with the radial distance. There is also no particular evidence of any cyclic pattern to the assay values. The data provide empirical support for the exponential decay model that we had derived theoretically.

0.5.4 The Appropriate Misfit Function

The cross product of two items selected from a pair of correlated standardised normal distributions has an expected value equal to the correlation between the two distributions. We can accordingly construct a misfit function based upon the difference between the cross product and the modelled correlation coefficient.

So, in using our Models 1, 2 or 3 to fit the correlation coefficient to the observed cross products 'p_i' as a function of actual separation $\underline{r}_i = (x_i, y_i, z_i)$, our objective is to minimise the misfit function:

$$F = \Sigma[p_i - \rho(\underline{r}_i)]^2/n \qquad (0.17)$$

The objective in fitting a model is to find parameters which minimise F. The magnitude of the residual F, and the amount of residual F removed in going through the hierarchy of models will help in judging the meaningfulness and overall explanatory power of the model.

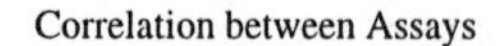

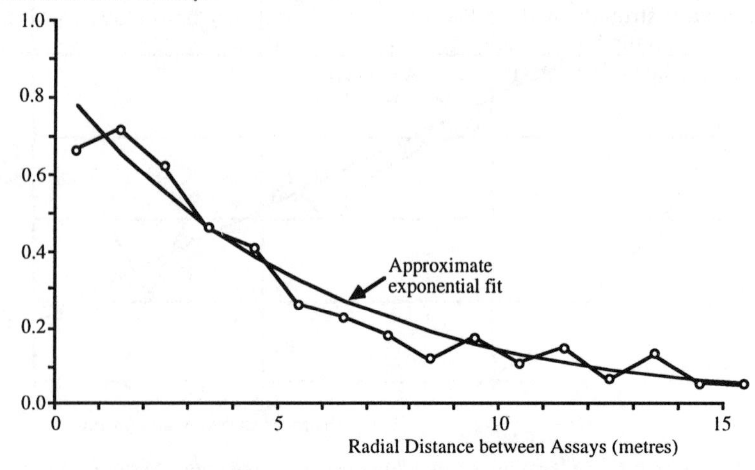

Figure 0.10: Correlations as a Function of r, the Inter-Assay Distance

We have seen that the available 1576 observations could yield more than a million cross products. The model fitting to be described here will be based upon the 4674 cross products that existed for assays separated by less than 5 metres. As discussed above, the cross products were formed from the standardised deviations of the logarithms of the gold assays. Cross products of assays with themselves were of course not used in the analysis.

Using data for separations of less than 5 metres is admittedly rather arbitrary. To use the more than a million available cross products would take too much computer time. An alternative approach would be to sample over a greater separation range. However, given the exponential fall-off model, the parameters will be less sensitive to data for greater separations, so it was decided to carry out these initial investigations with a manageable amount of data by limiting the separation between assays to 5 metres. The theoretical model of Figure 0.8 and equation (0.9) gives us the reassurance that establishing the model for separations less than 5 metres should allow extrapolation to predict the correlations for greater separations.

0.5.5 Fitting Models of One or Two Parameters

We will first consider the models that require only one or two parameters. The misfit function for these models can be easily explored without recourse to a genetic algorithm. The analysis and graphs to be reported here were produced using an Excel spreadsheet.

0.5.5.1 Model 0

If we ignore the variation with 'r', the inter-assay distance, we would just treat the correlation coefficient as a constant, and so could postulate:

$$\textbf{MODEL 0} \qquad \rho(r) = a \qquad (0.18)$$

It is clear that this model is inadequate, since we have already strong evidence that ρ does vary strongly with r. The estimate 'a' will just be the average cross product within the 5 metre radius that we are using data from. If we increased the data radius, the value of the parameter a would decrease.

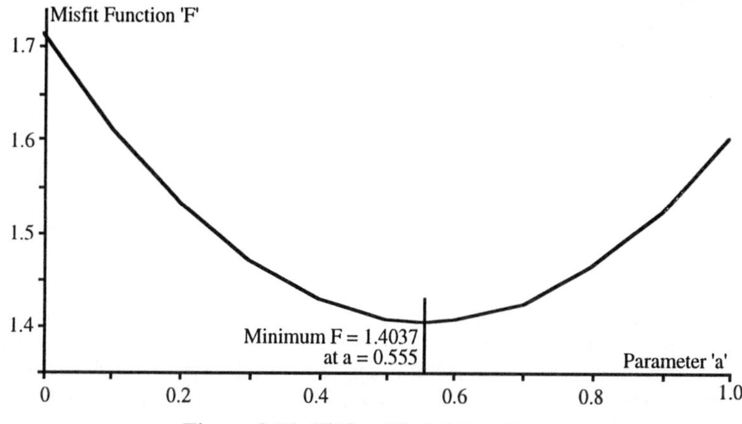

Figure 0.11: Fitting Model 0: $\rho(r) = a$

Although the parameter 'a' can be simply estimated analytically by computing the average value of the cross product, the same answer can also be obtained by iteratively minimising the misfit function F in equation (0.17), using the hill climbing methods described in the earlier sections. The results are shown in Figure 0.11. The misfit function is U-shaped, with a single minimum. This procedure gives us a value of the misfit function, 1.4037, to compare with that obtained for the other models.

It should be pointed out that Model 0 and Model 1 do not share a hierarchy, since each contains only a single parameter and they have different model structures. But Models 2 and 3 can be considered as hierarchical developments from either of them.

0.5.5.2 Model 1

Model 1 again involves only a single parameter 'k', which cannot be solved for analytically. An iterative approach (as described in the earlier sections) is needed to find the value of the parameter which minimises the misfit function. Since only one parameter is involved, we can easily explore the range of this parameter over its feasible range ($k > 0$) and confirm that we do not get trapped in a local minimum, so the model does not require a genetic algorithm. The graph in Figure 0.12 shows the results of this exploration.

Model 1 gives a misfit function of 1.3910, somewhat better than the misfit of 1.4037 for Model 0. This is to be expected, because the exponential decline with distance of Model 1 better accords with our theoretical understanding of the way the correlation coefficient should vary with distance between the assays.

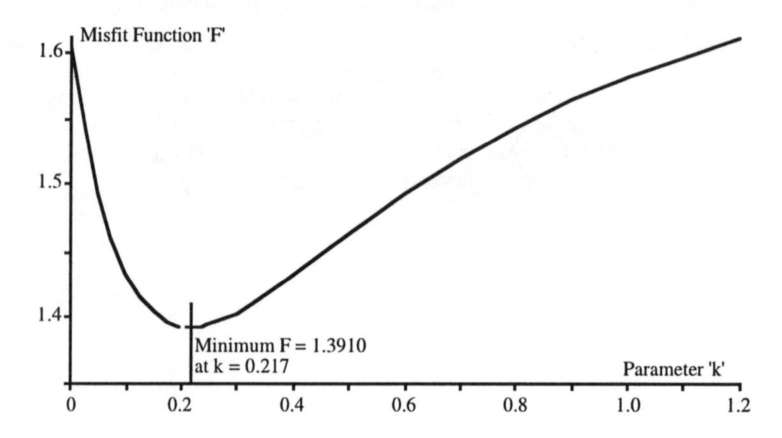

Figure 0.12: Fitting Model 1: $\rho(r) = \exp(-kr)$

0.5.5.3 Model 2

Introducing a second parameter 'a' as a constant multiplier forms Model 2. Since there are still only two parameters, it is easy enough to explore the feasible space iteratively, and establish that there is only the one global minimum for the misfit function F.

The results for Model 2 are summarised in Figure 0.13. The thin-line graphs each show the variation of F with parameter k for a single value of the parameter a. Each is a U-shaped curve with a clear minimum. The locus of these minima is the thick line, which is itself U-shaped. Its minimum is the global minimum, which yields a minimum misfit function F equal to 1.3895, a marked improvement on the 1.3910 of Model 1. For this minimum, the parameter a is equal to 0.87 and parameter k equals 0.168.

It should be pointed out that a hybrid analytical and iterative combination can more efficiently be used to find the optimum fit to Model 2. For this hybrid approach we combine equations (0.11) and (0.17), to give the misfit function for Model 2 as:

$$F = \Sigma[p_i - a.\exp(-kr_i)]^2 /n \qquad (0.19)$$

Setting to zero the differential with respect to 'a',

$$a = \Sigma[p_i.\exp(-kr_i)] / \Sigma[\exp(-2kr_i)] \qquad (0.20)$$

So for any value of k, the optimum value of a can be calculated directly, without iteration. There is therefore need to explore only the one parameter, k, iteratively. This procedure leads directly to the bold envelope curve of Figure 0.13.

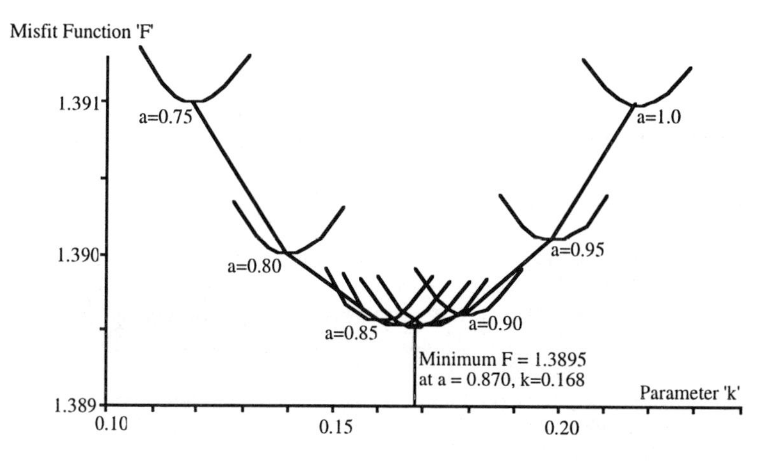

Figure 0.13: Fitting Model 2: $\rho(r) = a.\exp(-kr)$

0.5.5.4 Comparison of Model 0, Model 1 and Model 2

Figure 0.14 summarises the results of the three models analysed so far.

Model 0 is clearly inadequate, because it ignores the relation between correlation and inter-assay distance. Model 2 is preferable to Model 1 because it gives an improved misfit function, and because it allows for the fact that there will be some test/retest inaccuracy in the assays.

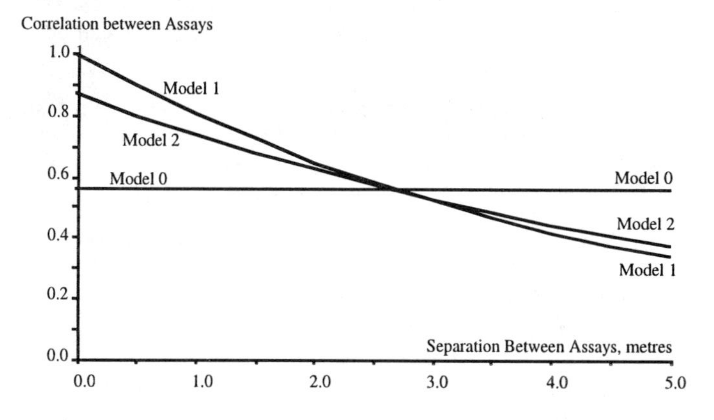

Figure 0.14: Comparing Model 0, Model 1 and Model 2

0.5.5.5 Interpretation of the Parameters

Although none of the assays are from identical locations, the value 0.87 of the intercept parameter 'a' for Model 2 can be interpreted as our best estimate of the correlation that would be found between two assays made of samples taken from identical locations. However, each of these two assays can be considered as the combination of the 'true' assay plus some orthogonal noise. Assuming the noise components of the two assays to be uncorrelated with each other, the accuracy of a single estimate, or the correlation between a single estimate and the 'true' assay is given by $\sqrt{a}$.

The value 0.168 of the exponential slope parameter 'k' tells us how quickly the correlation between two assays dies away as the distance between them decreases. The correlation between two assays will have dropped to one-half at a distance of about three metres, a quarter at six metres, and so on. Given that the data to which the model has been fitted includes distances up to only five metres, it might be objected that conclusions for greater distances are invalid extrapolations "out of the window". This objection would be valid if the model was purely exploratory, without any theory base. However, our multiplicative model of equation (8) is based on theory, and predicts that doubling the inter-assay distance should square the correlation coefficient. With this theoretical base, we are justified in extrapolating the exponential fit beyond the data range, so long as we have no reason to doubt the theoretical model.

0.5.6 Fitting the Non-Homogeneous Model 3

Model 3, as postulated in equation (0.16), replaces the single distance variation parameter 'k' by a set of six parameters, to allow for the fact that structure within the ore body may cause continuity to be greater in some directions than in others.

$$\rho(r_a, r_b, r_c) = a.\exp[-\text{sqrt}(k_a^2 r_a^2 + k_b^2 r_b^2 + k_c^2 r_c^2)] \qquad (0.21)$$

We saw that the model includes seven parameters, the test/retest repeatability 'a'; the three fall-off rates (k_a, k_b, k_c); and three angles (α, β, γ) defining the directions of the major axes, according to equations (0.13) to (0.15). These last six parameters make allowance for possible inhomogeneity, and can be used to define ellipsoid surfaces of equal continuity.

Although we cannot minimise the misfit function analytically with respect to these six parameters, we can again minimise analytically for the multiplying parameter 'a'. For any particular set of values of the parameters (k_a, k_b, k_c); and three angles (α, β, γ), the derivative $\partial F/\partial a$ is set to zero when:

$$a = \frac{\Sigma[p_i.\exp[-\text{sqrt}(k_a^2 r_{ai}^2 + k_b^2 r_{bi}^2 + k_c^2 r_{ci}^2)]}{\Sigma[\exp[-2.\text{sqrt}(k_a^2 r_{ai}^2 + k_b^2 r_{bi}^2 + k_c^2 r_{ci}^2)]} \qquad (0.22)$$

In equation (0.22), the cross products p_i are values for assay pairs with separation (r_{ai}, r_{bi}, r_{ci}). The model can thus be fitted by a hybrid algorithm, with one of the parameters being obtained analytically and the other six by using iterative search or a genetic algorithm.

An iterative search is not so attractive in solving for six parameters, because it now becomes very difficult to ensure against entrapment in a local minimum. Accordingly, a genetic algorithm was developed to solve the problem.

0.5.6.1 The Genetic Algorithm Program

The genetic algorithm was written using the simulation package Extend, with each generation of the genetic algorithm corresponding to one step of the simulation. The use of Extend as an engine for a genetic algorithm is described more fully in Chapter 9. The coding within Extend is in C. The program blocks used for building the genetic algorithm model are provided on disk, in an Extend library titled "GeneticCorrLib". The Extend model itself is the file "GeneticCorr".

The genetic algorithm used the familiar genetic operators of selection, crossover and mutation. Chapter 9 includes a detailed discussion of the application of these operators to a model having real (non-integer) parameters. It will be sufficient here to note that:

- The population size used could be chosen, but in these runs was 20.
- Selection was elite (retention of the single best yet solution) plus tournament (selection of the best out of each randomly chosen pair of parents).
- "Crossover" was effected by random assignment of each parameter value from the two parents to each of two offspring.
- "Random Mutation" consisted of the addition of a normally distributed adjustment to each of the six parameters. The adjustment had zero mean and a preset standard deviation, referred to as the "mutation radius".
- "Projection Mutation" involved projection to a quadratic minimum along a randomly chosen parameter. This operator is discussed fully in Chapter 9.
- In each generation, the "best yet" member was unchanged, but nominated numbers of individuals were subjected to crossover, mutation and projection mutation. In these runs, in each generation, ten individuals (five pairs) were subjected to crossover, five to random mutation, and four to projection mutation. Since the method worked satisfactorily, optimisation of these numbers was not examined.

The solution already obtained for Model 2 was used as a starting solution. For this homogenous solution, the initial values of (k_a , k_b , k_c) were each set equal to 0.168, and three angles (α, β, γ) were each set equal to zero.

The model was run for a preset number of generations (500), and kept track of the misfit function for the "best yet" solution at each generation. It also reported the values of the six iterated parameters for the "best yet" solution of the most recent generation, and calculated the 'a' parameter, according to equation (22).

As an alternative to running the genetic algorithm, the simulation could also be used in "Systematic Projection" mode. Here, as discussed in Chapter 9, each of the parameters in turn is projected to its quadratic optimum. This procedure is repeated in sequence for all six parameters until an apparent minimum is reached. As we have seen earlier, such a systematic downhill projection faces the possibility of entrapment on a local optimum, or even on a saddle point (see Figure 0.6).

0.5.6.2 Results Using Systematic Projection

The results of three runs using systematic projection are graphed in Figure 0.15. For each iteration, the solution was projected to the quadratic minimum along each of the six parameters (k_a , k_b , k_c) and (α, β, γ). The order in which the six parameters were treated was different for each of the three runs. It is clear that at least one of the runs has become trapped on a local optimum (or possibly a saddle point, as in Figure 0.6).

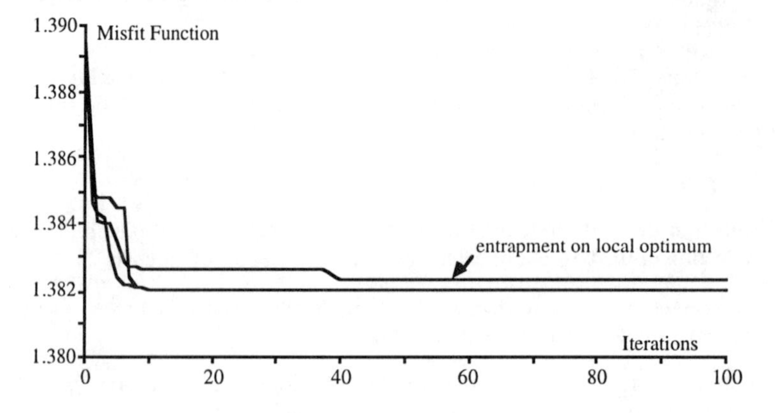

Figure 0.15: Fit of Model 3 Using Systematic Projection

0.5.6.3 Results Using the Genetic Algorithm

Figure 0.16 shows the results for seven runs of the genetic algorithm. Although these took much longer to converge, they all converged to the same solution, suggesting that the genetic algorithm has provided a robust method for fitting the multiple parameters of Model 3.

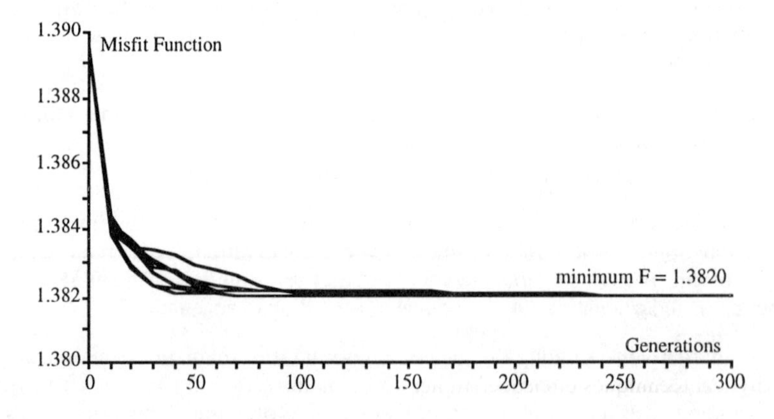

Figure 0.16: Fit of Model 3 Using the Genetic Algorithm

0.5.6.4 Interpretation of the Results

At the converged solution the parameters had the following values:

$$a = 0.88;\ (k_a, k_b, k_c) = (0.307, 0.001, 0.006);\ (\alpha, \beta, \gamma) = (34^\circ, 19^\circ, -43^\circ)$$

The test/retest repeatability of 0.88 is similar to the 0.87 obtained for Model 2.

The results suggest that the fall-off of the correlation coefficient is indeed far from homogeneous with direction. Along the major axis the fall-off rate is 0.307 per metre. This means the correlation decreases to a proportion 1/e in each 3.3 (=1/0.307) metres. The figure corresponds to a halving of the correlation coefficient every 2.3 metres.

Along the other two axes, the fall-off of the correlation coefficient is much slower, of the order of one percent or less per metre.

The results are compatible with the geologically reasonable interpretation that the material has a planar or bedded structure. The correlation would be expected to fall off rapidly perpendicular to the planes, but to remain high if sampled within a plane.

The direction of the major axis is 34° east of north, pointing 19° upwards from the horizontal. The planes are therefore very steeply dipped (71° of horizontal). Vertical drilling may not be the most efficient form of drilling, since more information would be obtained by sampling along the direction of greatest variability, along the major axis. Collecting samples from trenches dug along lines pointing 34° east of north may be a more economical and efficient way of gathering data.

Further analysis of data divided into subsets from different locations within the project would be useful to determine whether the planar structure is indeed uniform, or whether it varies in orientation in different parts of the ore body. If the latter turns out to be the case, then the results we have obtained represent some average over the whole prospect.

0.6 Conclusions

In this paper I have attempted to place genetic algorithms in context by considering some general issues of model building, model testing and model fitting. We have seen how genetic algorithms fit in the top end of a hierarchy of analytical and iterative solution methods.

Just as the models that we wish to fit tend to be hierarchical, with models of increasing complexity being adopted only to the extent that simpler models prove inadequate, so also do analytical, hill climbing and genetic algorithms form a hierarchy of tools. There is generally no point using an iterative method if an analytical one is available to do the job more efficiently. Similarly, genetic algorithms do not replace our standard techniques, but rather supplement them. As we have seen, hybrid approaches can be fruitful. Analytical techniques embedded in iterative solutions reducing the number of parameters needing iterative solution. Solutions obtained by iterative techniques on a simpler model provide useful starting values for parameters in a genetic algorithm.

Thus, genetic algorithms are most usefully viewed, not as a self-contained area of study, but rather as providing a useful set of tools and techniques to combine with methods of older vintage to enlarge the areas of useful modelling.

Reference

Bullard E.C. Everett J.E. & Smith A.G. (1965). The fit of the continents around the Atlantic, *Philosophical Transactions of the Royal Society,* 258, 41-51.

Chapter 1

Geoff Bartlett
1 Dalkieth Road
Dalkieth
Perth
West Australia

Genie: A First GA

0-8493-2519-6/95/$0.00 + $.50

1.1 Introduction

This paper outlines an implementation of a GA in a simulation/optimisation program called Genie. The structure of the GA shown here follows the Genie program in spirit, but most of the examples used have been modified for clarity. For reasons of memory management, code efficiency and the general dictums of real life, the actual implementation is somewhat different.

1.2 Genie

Genie was developed in 1991 as a pilot project, and as such was more of a sketch of a GA than a fully featured implementation. Even so, the result worked remarkably well. Genie is a Trends Integration Procedure (TIP) simulation model allied with a GA It is beyond the scope of this paper to give a detailed description of the TIP model, but some explanation is necessary to put the GA into context. TIP models are growth models. Each factor in the model may have an innate capacity for growth, but more importantly, the factors can be cross linked. This means that change in one factor can cause change in one or more other factors. This cross effects travel through the linkage chains, and can result in recursive effects. The models may have one hundred or more factors.

The purpose of GA's in Genie is to search for starting configurations that will produce a particular desired outcome. To achieve this there are two special types of factors in the model.

The first are the *control factors*. It is the initial values for these factors that the GA will manipulate. Any number of control factors may be nominated. Each control factor must have its domain declared. This is the range of useful values that the factor may take. Without this constraint the range from $-\infty$ to ∞ would have to be searched, making it an impossible task. The set of control factors form the *solution* for the model, and it is this solution set that the GA will try to optimize.

The second class of factors are the *target factors*. At the end of the model run, the final values of these factors are compared against user defined target values. The closer to the target value the better. The deviations from the targets for all of the target factors are aggregated to supply a total fitness for the model. A fitness of zero is a perfect solution.

1.3 Code Examples

Examples of source code are used extensively to illustrate the implementation of the GA The code is in Pascal, which is close enough to what passes for pseudo code most of the time. It should be no problem to anyone with even a passing association with computer languages to understand the intent of the code. The examples are not meant to be put together to create a working GA, but to outline the Genie implementation.

In particular, Genie kept track of many of the entities found in the GA in a way which would overly complicate the code examples. This housekeeping is not essential to the principles of the GA, but is useful in a working implementation. In lieu of details of the housekeeping, where necessary empty procedures have been supplied which explain what is happening.

1.4 Similies and Space

At times it is necessary to try and describe the terrain that the GA is traversing. Three dimensional topographical terms and concepts such as hills, ruggedness, steepness etc. generalize well to n-dimensions and are used where applicable.

1.5 Data Structures

The two distinct elements in the GA are individuals and populations. An individual is a single solution. The population is the set of individuals currently involved in the search process.

1.6 Individuals

An individual is a single solution. The term individual is used to group together the two forms of a solution: the *chromosome*, which is the raw 'genetic' information (*genotype*) that the GA deals with, and the *phenotype* which is the expression of the chromosome in the terms of the model, in this case the factor starting values. A chromosome is subdivided into genes. A gene is the GA's representation of a single factor value for a control factor.

Each factor in the solution set corresponds to a gene in the chromosome.

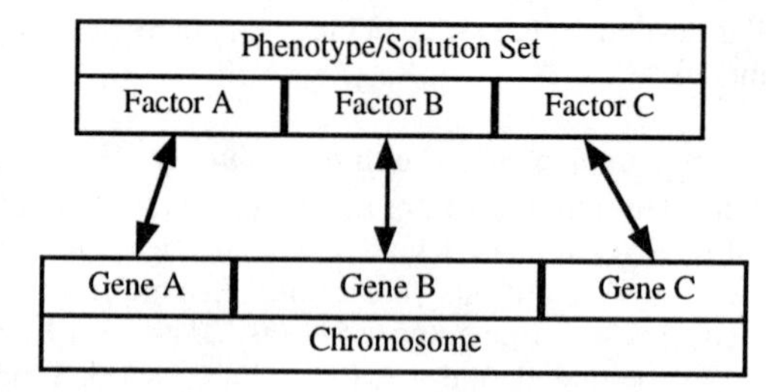

1.7 Genes

A gene is a bit string of arbitrary length. The bit string is a binary representation of the number of intervals from a lower bound. As previously mentioned, control factors must have an upper and lower bound declared. This range is divided into the number of intervals that can be expressed by the gene's bit string. A bit string of length n can represent $(2^n - 1)$ intervals. The size of the interval would be *range*/ (2^n-1). For example:

> Factor A has a range of -0.1 to +0.1.
> The bit string is five bits in length.
> This gives 31 intervals, or a interval size of 0.00645...
>
> The bit string '00000' represents zero intervals, so the gene's value is equal to the lower bound.
>
> $-0.1 + (0 * 0.00645) = -0.1$
>
> The bit string '11111' represents 31 intervals, so the gene's value is equal to the upper bound.
>
> $-0.1 + (32 * 0.00645) \approx +0.1$

The bit string '01101' represents 13 intervals, so the gene's value is equal to -0.01875.

$$-0.1 + (13 * 0.00645) \approx -0.01677$$

The size of the bit string is of paramount importance. If the interval is too coarse, the GA may never be able to find an optima simply because the genes can't express a value close enough to the optima. Too fine a resolution may result in excessive hair splitting in the search.

The structure of each gene is defined in a record of phenotyping parameters. The phenotyping parameters are instructions for mapping between genotype and phenotype, or in other words, encoding a solution set into a chromosome and decoding a chromosome to a solution set. This mapping is necessary to convert solution sets from the model into a form that the GA can work with, and for converting new individuals from the GA into a form that the model can evaluate.

This is the data structure defining a gene.

```
TCrossSites =               (oneSite,twoSites);
TPhenoTypeParams =          RECORD
                  {The model factor associated with this gene}
                            factID:            INTEGER;

                            {The factor range}
                            minValue:          REAL;
                            maxValue:          REAL;

                            {Bit string length}
                            resolution:        INTEGER;

                            {Crossing Technique}
                            crossSites:        TCrossSites;
                            END;
```

There is no reason for different genes to be of the same length, or share any other characteristics in common. Below is an example of a solution set, the corresponding chromosome and the mapping between the two:

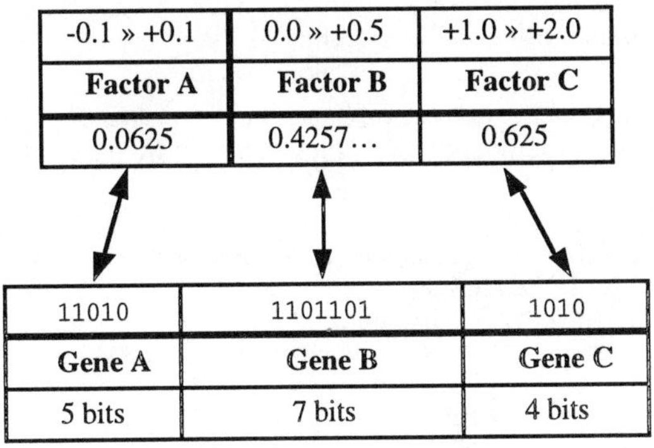

-0.1 » +0.1	0.0 » +0.5	+1.0 » +2.0
Factor A	**Factor B**	**Factor C**
0.0625	0.4257...	0.625

11010	1101101	1010
Gene A	**Gene B**	**Gene C**
5 bits	7 bits	4 bits

The bit string in the chromosome record would only need to be of 16 bits in length, and would be: 1101011011011010, which is the concatenation of the genes.

Running the genes together like this saves space but increases overhead in decoding. In the Genie implementation it was found to be easier to declare the type of TBitString to be:

```
TBitString = ARRAY[1..cMaxGenes] OF LONGINT;
```

Where `LONGINT` is a 32 bit integer and `cMaxGenes` is the maximum number of genes that can form a chromosome. 32 bits provides 4.29...E9 steps, and for the models TIP deals with, the range of a factor would never be more than -1.0 to +1.0. `cMaxGenes` was set to 10, but this is a largely artificial constraint. With the amount of memory available now it could be much larger; the real constraint however is the efficacy of the GA procedure in searching such a complex space.

1.8 Chromosomes

In the GA the individual is represented by the data structure:

```
TChromosome =                       RECORD
{A unique identifier for this chromosome}
chromID:                            INTEGER;

{The chromosome ID's of the parents}
parent1, parent2:                   INTEGER;

{The evaluated fitness this chromosome}
fitness:                            REAL;

{The bit representation of the chromosome}
bits:                               TBitStr;
END;
```

The first three fields are used mainly for tracing genealogies. The fitness is a function of the target factors in the model. The bit string is the 'genetic' representation of the solution. Each gene is encoded into a part of the bit string.

Below is the code for chromosome decoding.

This procedure converts the information in `theChromosome` to a vector of real numbers, the phenotype, which can then be passed to the simulation model.

```
PROCEDURE DecodeChromosome(theChromosome: TChromosome;
nGenes      : INTEGER;
phenoParams : TPhenoTypeParams;
VAR valueSet: ARRAY [1..cMaxGenes] OF REAL);
VAR
  geneNo :      INTEGER;
  nIntvls:      INTEGER;
  intvl  :      REAL;
```

```
BEGIN
FOR geneNo := 1 TO nGenes DO
  BEGIN
  {Calculate the interval size}
  nIntvls := 2**phenoParams[geneNo].resolution - 1;
  intvl := (phenoParams[geneNo].maxValue - phenoParams
                              [geneNo].minValue) / nSteps;
   {Calculate the value for this gene. Remember that
bitString is actually an array of INTEGER values}
  valueSet[geneNo] := phenoParams[geneNo].minValue + intvl
                              * theChromosome.bits[geneNo];
END;
```

1.9 Fitness

The fitness of an individual in a GA is the value of the objective function for its phenotype. In the case of Genie the objective function is the sum of the difference between the actual end values of the target factors and the desired values when the control factors have been initialized to the individuals solution.

To calculate fitness, the chromosome must be first decoded, and then sent to the simulation model. The model will return a value indicating fitness. Below is an outline of the fitness evaluation part of Genie.

> This function passes the values contained in `valueSet` to the model, the model runs the simulation and returns the fitness with respect to the target factors. For the purposes of the example it is assumed that the model knows how many values are in a solution set and what to do with them.

```
FUNCTION EvaluateSolution(valueSet: ARRAY[1..cMaxGenes] OF
                              REAL) : INTEGER;
```

The procedure `CalcFitness` is called every time a new individual is created. `phenoParams` supplies the phenotyping instructions.

```
PROCEDURE CalcFitness(VAR theChromosome: TChromosome;
phenoParams: TPhenoTypeParams);
VAR
valueSet: ARRAY[1..cMaxGenes] OF REAL;
BEGIN
{Get the solution set for this chromosome}
DecodeChromosome(theChromosome, phenoParams, valueSet);

{Get the fitness}
theChromosome.fitness := EvaluateSolution(valueSet);
END;
```

1.10 Populations

A population consists of a number of individuals being tested, the phenotyping parameters defining the individuals and some information about search strategies.

The Genie GA uses a fixed size population, which is initialized with randomly valued individuals. This is not the only technique for managing population size, but at the time it simplified the coding, and given the efficacy of the end result, it is doubtful if a more complex population schema would have been of much benefit. Other applications may require more sophisticated schemes. The population list is left unordered, but an index sorted by fitness value is maintained.

The search strategy is a combination of population size, breeding schema and termination criteria. These three components are discussed in detail below

1.11 Data Structures

The selection method determines how the parents are selected for crossing. The selection techniques are explained later.

```
TSelectMethod  =        (roulette, random, fitFit, fitWeak);
```

The replacement method determines how new individuals will be put into the population, and how individuals are to be eliminated. The replacement techniques are explained later.

```
TReplaceMethod =        (weakParent, bothParents,
                        weakestChromosome, random);
```

The convergence method determines when the search is to be terminated. The crossing techniques are explained later.

```
TConvergeMethod =       (fitnessSum, averageFitness,
                        bestChromosome, worstChromosome);
```

This is the population record.

```
TPopulation =       RECORD
{The maximum number of individuals to keep}
popMax:             INTEGER;

{To keep track of the ID's that have been allocated}
nextChromID:        INTEGER;

{The number of genes in a chromosome}
nGenes:             INTEGER;

{The structure of the genes}
phenoParams:        ARRAY [1..cMaxGenes] OF TPhenoTypeParams;

{The population}
chromosomes:        ARRAY [1..cMaxPopulation] OF TChromosome;

{Fitness index}
fitIndex:           ARRAY [1..cMaxPopulation] OF INTEGER;

{-- These fields set the search limits --}
{A generation count}
thisGen:            INTEGER;

{The maximum number of generations allowed}
maxGen:             INTEGER;

{The convergence schema to use}
```

```
cnvrgMethod:        TConverMethod;
cnvrgValue:         REAL;

{-- These fields set crossing criteria --}
{The selecion technique to use}
selectMethod:       TSelectMethod;

{The index of the last parent selected}
lastParentNo:       INTEGER;

{The bit mutation probability}
pMut:               REAL;

{The replacement technique to use}
replaceMethod:      TReplaceMethod;

{-- Population Statistics --}
maxFitness:         REAL;
minFitness:         REAL;
sumFitness:         REAL;
END;
```

1.11.1 Initialization

Initialization requires the creation of sufficient chromosomes to fill the population. The genes of these chromosomes are set to a random value and the fitness of the chromosome is evaluated.

This procedure creates the `fitIndex` list from the population. The body of the procedure is an index sort. The implementation doesn't have a bearing on the search and is not supplied.

```
PROCEDURE CreateFitIndex(VAR population : TPopulation);
```

It is assumed that the rest of the population record has been initialized, especially the `phenoParams`. The `Random` function produces a number in the range 0 to 1.

```
PROCEDURE InitPopulation(VAR population: TPopulation);
VAR
chromNo: INTEGER;
geneNo: INTEGER;
theChromosome:                  TChromosome;

BEGIN
{Initialise the chromosome ID's}
nextChromID := 0;

FOR chromNo := 1 TO population.popMax DO
  BEGIN
  {This is a founding member of the population,
   so no parents}
  theCromosome.parent1 := 0;
  theCromosome.parent0 := 0;

  {Set the chromosome ID}
  population.nextChromID := population.nextChromID + 1;
```

```
    theChromosome.chromID := population.nextChromID;

    {Set the genes to random values}
    FOR geneNo := 1 TO population.nGenes DO
      theChromosome.bits[geneNo] := TRUNC(16384 * Random);

    {Evaluate the fitness}
  CalcFitness(theChromosome,population.phenoParams);

    {Write the chromosome to the population}
    population.chromosomes[chromNo] := theChromosome;
    END;
  END;
```

1.11.2 Population Maintenance

Some routines are required for managing the population during the search. Replacing old individuals with new, and updating the population statistics and fitness index is the main problem. This area is dependent upon the type of housekeeping being performed, so only the procedure outline is provided.

This procedure replaces the old individual with the new individual in the `chromosome` list in `population`. The `fitList` is updated as well. It is not necessary to re-sort `fitList`, it suffices to insert the new individual in the appropriate position in the list.

```
PROCEDURE DoReplace(oldIndividual,
newIndividual : TChromosome;
VAR population : TPopulation);
```

This procedure replaces an individual in the population with a new individual. DoReplace does the work of actually replacing the old individual. This routine merely updates the population statistics.

```
PROCEDURE ReplaceIndividual(oldIndividual,
newIndividual : TChromosome;
VAR population : TPopulation);

BEGIN
DoReplace(oldIndividual,newIndividual,population);

{Update min and max fitnesses from fitList}
population.minFitness := population.fitList[1];
population.maxFitness := population.fitList
                    [population.popMax];
{Update fitness sum}
population.sumFitness := population.sumFitness -
                        oldIndividual.fitness;
population.sumFitness := population.sumFitness +
                        newIndividual.fitness;
END;
```

1.12 Search Strategies

The search or optimization process consists of initializing the population and then breeding new individuals until the termination condition is met. There can be several, somewhat opposed, goals for the search process, one of which is to find the global optima. With the types of models that GA's work well with, this can never be assured. There is always the chance that the next iteration in the search will produce a better solution. Alternatively, the search could run for years and not produce any better solution than it did in the first five iterations.

Another goal is quick convergence. When the objective function is expensive to run, quick convergence is desirable, however, the chance of converging on a local, and possibly quite substandard optima, is increased.

Yet another goal is to produce a range of diverse, but still good, solutions. When the solution space contains several distinct optima, which are similar in fitness, it is useful to be able to select between them, since some combinations of factor values in the model may be more feasible than others. In addition, some solutions may be more robust than others.

Which of these goals the GA will tend to fulfil is a function of population size, selection and replacement schema, and termination condition.

1.13 Population Size and Convergence

As stated previously, Genie uses a fixed size population. The population size for the search is set by the user to suit the requirements of the particular model.

Running the TIP simulation with large models can be expensive in terms of time, so a smaller population may be desirable, but if the population is too small then the loss of genetic diversity may compromise the search. Genetic diversity in GA's is important when the solution space is topographically rugged or convoluted[3]. A small population would be more likely to quickly converge on what may be a local optima; a comparatively fit individual in a small starting population will be selected for, and it and it's descendants will quickly come to dominate, further limiting genetic diversity. Also with the sparse spread of initial points, better optima may never be visited before the search converges on a poor local optima, or even if they are visited, the point may be comparatively unfit, and will not be given adequate opportunity to reproduce and start hill climbing.

At the other extreme, there are problems concomitant with excessively large populations. Firstly, the initialization becomes expensive. In some cases, the initialization of a large population is equivalent to a random search, and produces optimal or near optimal solutions, which makes the use of GA's somewhat moot.

Another potential problem with excessively large populations is that the search may flounder in the over abundance of genetic diversity. In a large population there may be many fit individuals; a problem arises when there are several local optima. The Genie implementation had no way of assessing genetic distance

---

[3]If it weren't then I suppose one wouldn't be using G.A.'s.

when selecting parents. If there are several groups of fit individuals clustering around different optima, the union of individuals from two distinct optima may not preserve the genetically 'good' part of either of the parents[4]. At worst, this type of miscegenation could depopulate the areas around the various optima. More likely is that the hill climbing around the optima will be retarded, which further increases the time cost. If the objective function is cheap, or the cost of failing to find the global optima is high, it may be worth bearing the extra time to increase the chance of finding the best possible solution.

1.14 Breeding

The breeding cycle is the heart of the GA It is here that the search process creates new and hopefully fitter individuals.

The breeding process consists of three steps: selecting parents; crossing the parents to create new individuals (children); and replacing old individuals in the population with the new ones.

1.14.1 Selection

Selection is the process of choosing two parents from the population for crossing. The reasons for selecting only two parents, and not more, are quite arbitrary. Firstly, Genie is only a simple implementation of GA's, and secondly, two parents seems to work adequately in the biological[5] model, on which GA's are based, in as much as two parents fulfils the requirement of structured interchange of genetic information.

---

[4]The problem of distinguishing between distinct distant optima is a serious one, which will be discussed (without resolution) later in the chapter.

[5]There is a danger in trying to adhere too faithfully to the biological, or natural model. GA's are a search technique based upon structured sharing of information. Nature deserves acknowledgement for showing the way, however, the optimisation problem confronting nature is vastly different, and more complex, than anything likely to be developed on a computer. Another marked departure from nature is the amount of time one is willing to wait for a solution (although in the case of nature, there is no end solution since the problem - environment - is always changing). It bears remembering that G.A.'s are merely a tool, and their aesthetic is their efficacy. The problems G.A.'s deal with may require techniques that would not be countenanced in 'nature', and vice versa, not all natural selection and reproduction techniques are applicable or useful to G.A.'s.

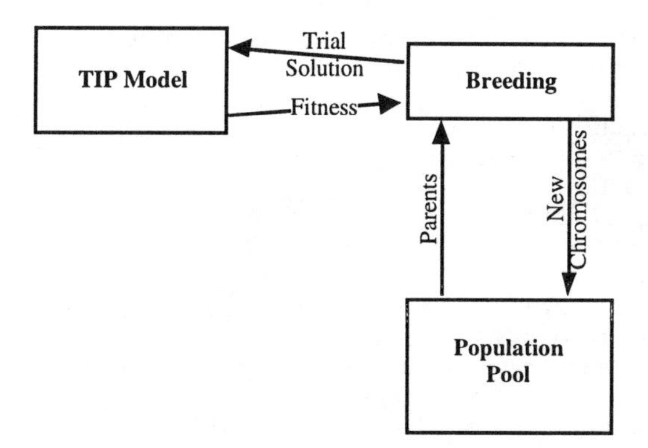

Determining which individuals should be allowed to interchange their genetic information has great bearing on the rate of convergence specifically and the success of the search in general. It is useful at this point to reflect upon the meaning of the genetic code carried in the chromosomes, particularly in the context of Genie. Each gene represents a scalar value. The chromosome represents a point in the solution space. The high bits of each gene will position the point coarsely in space, while the lower order bits will have more of a fine tuning effect. The result of a union between two fit, and genetically proximate, individuals will in all probability be as fit as the parents, but may not be different enough to be significantly better. This is more of a function of the crossing process, and will be dealt with in detail below, but it serves to illustrate the importance of selection.

The four selection techniques used by Genie span the range from genetically conservative to genetically disruptive. The selection technique to be used is stored in the `selectMethod` field of the population record.

1.14.1.1 Roulette

Roulette selection is one of the traditional GA selection techniques, and the version shown here is adapted from Goldberg. The principle of roulette selection is a linear search through a roulette wheel with the slots in the wheel weighted in proportion to the individual's fitness value. A target value is set, which is a random proportion of the sum of the fitnesses in the population. The population is stepped through until the target value is reached. This is only a moderately strong selection technique, since fit individuals are not guaranteed to be selected for, but have a somewhat greater chance. A fit individual will contribute more to the target value, but if it does not exceed it, the next chromosome in line has a chance, and it may be weak. It is essential that the population not be sorted by fitness, since this would dramatically bias the selection.

Genie added a couple of features to the roulette selection. The roulette wheel always starts from the position of the last selection, and the last parent chosen is excluded from selection so as not to cross an individual with itself.

> The starting point for the selection is taken from the field `lastParent` in the `population` record.

```
PROCEDURE RouletteSelect(VAR parent: TChromosome;
population : TPopulation);
VAR
  thisChrom: TChromosome;
  nextParent: INTEGER;
  targetValue: REAL;
  fitSum: REAL;

BEGIN
{Multiply the sum of the fitness by a value between 0 and 1
to get targetValue}
targetValue := population.sumFitness * Random;

fitSum:= 0; {Reset the running sum}

nextParent := population.lastParent;

{Step through the population until the sum exceeds the
target value}
REPEAT
  nextParent := nextParent + 1;
  IF nextParent > population.popMax THEN
    nextParent := 1;

  thisChrom := population.chromosmes[nextParent];

  {In the event that random returns a value close to 1 it
is necessary to step over the lastParent}
  IF nextParent <> population.lastParent THEN
    fitSum:= fitSum+ thisChrom.fitness;
UNTIL fitSum> targetValue;

{Update the lastParent field}
population.lastParent := nextParent;

{Return the selected chromosome}
parent := thisChrom;
END;
```

1.14.1.2 Random

This technique randomly selects a parent from the population. In terms of disruption of genetic codes, random selection is a little more disruptive, on average, than roulette selection.

```
PROCEDURE RandomSelect(VAR parent: TChromosome;
population : TPopulation);
VAR
  chromNo: INTEGER;
  thisChrom: TChromosome;

BEGIN
REPEAT
  {Get a random chromosome from the population, making
   sure that it isn't zero}
  chromNo := TRUNC(Random * (population.popMax - 1)) + 1);
```

```
    thisChrom := population.chromosomes[chromNo];
  UNTIL thisChrom.chromID <> population.lastParent;

  {Update the lastParent field}
  population.lastParent := lastID;

  {Return the selected chromosome}
  parent := thisChrom;
  END;
```

1.14.1.3 Fit-Fit

Fit-fit selection pairs an individual with the next fittest individual in the population by simply stepping through the ordered list of individuals. The population does not remain static for an entire cycle through the list, so not every individual will get an opportunity to breed, and some will breed twice, since individuals and their positions in the ordered list are being constantly replaced.

Fit-fit selection is highly conservative of genetic information and tends to converge rapidly to one solution. It does not foster great genetic diversity, however.

It should be noted that fit-fit selection still selects weak individuals for breeding, so this technique could not strictly be said to favour the strong in selection. However, the results of unions between strong individuals are more likely to be fit than unions between weak individuals. The weak individuals and their children will probably be quickly weeded out of the population.

If the population size is odd then once every cycle through the population, the weakest individual will be paired with the strongest.

```
PROCEDURE FitFitSelect(VAR parent: TChromosome;
population : TPopulation);
BEGIN
population.lastParent := population.lastParent + 1;
parent := population.chromosomes
[population.fitList[population.lastParent]];
END;
```

1.14.1.4 Fit-Weak

Fit-weak selection is maximally disruptive of genetic information. The fittest individual is paired with the least fit. The next fittest is paired with the next least fit, and so on. If the `lastParent` counter is in the lower half of the ordered list, it is assumed that the next parent should be from the same position in the upper half, and the opposite applies when the `lastParent` counter is in the upper half. In this way, alternately fit and unfit parents are chosen, until the middle of the list is reached, in which case the counter is reset. In an odd list the middle individual will be missed, but the GA search is robust enough to withstand this.

It is hard to quantify the effect of this upon convergence and the quality of the end solution. All that can be said is that this selection technique did not get used much in Genie.

```
PROCEDURE FitWeakSelect(VAR parent: TChromosome;
population : TPopulation);
VAR
  midPt : INTEGER;
  lastParent : INTEGER;

BEGIN
midPt := population.popMax DIV 2;
lastParent := population.lastParent;

IF population.lastParent < midPt THEN
  population.lastParent := population.popMax - lastParent +
1
ELSE IF lastParent > midPt THEN
  lastParent := lastparent - population.popmax + 1
ELSE
  lastParent := 1;  {Start at the beginning again}

parent := population.chromosome[population.fitList
                                [lastParent]];
population.lastParent := lastParent;
END;
```

1.14.2 Crossing

The crossing paradigm, while essentially the heart of the GA exhibits a less marked effect on convergence than either selection or replacement. Genie employs two gene splicing techniques combined with mutation.

1.14.2.1 One Site Splice

Splicing operates at the gene level. For each gene in each pair of parents a splice point is selected randomly. The first portion of the parents genes are then exchanged. In the diagram, bits belonging to parent 2 are greyed. The children are thus a combination of the parents. This process is repeated for each gene in the parents chromosomes.

It is possible for the splice point to be zero, or equal to the number of bits in the gene, in which case a straight copy is effected, where child A is equivalent to parent 1 and child B is equivalent to parent 2.

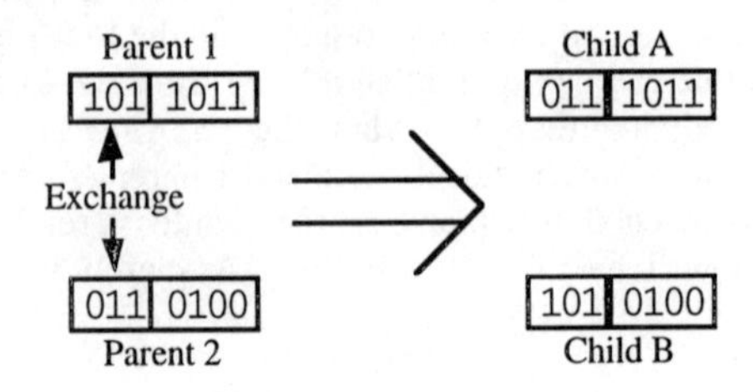

The effect of the splice depends a lot on where the splice occurs. To illustrate, let the range of the gene be from 0 to 100. The seven bit length provides 127 intervals of size 0.78740…. So if the splice were to be as shown:

Parent 1	= 71.6535
Parent 2	= 40.9449
Child A	= 46.4567
Child B	= 66.1417

The three high bits of each parent have been preserved in the children, so child B is reasonably close to parent A, and child A is reasonably close to parent B. If the splice were to occur immediately after the first bit from the left, then the following would be the result:

Child A	(0 101011)	= 21.2598
Child B	(1 110100)	= 91.3386

These children are markedly different from their parents. If the splice were to be two bits away from the right end of the gene the result would be very similar to the parents:

Child A	(0110 111)	= 43.3071
Child B	(1011 000)	= 69.2913

1.14.2.2 Two Site Splice

The two site splice is a variation on the one site splice, except that two splice points are chosen, and the bits between the two points exchanged.

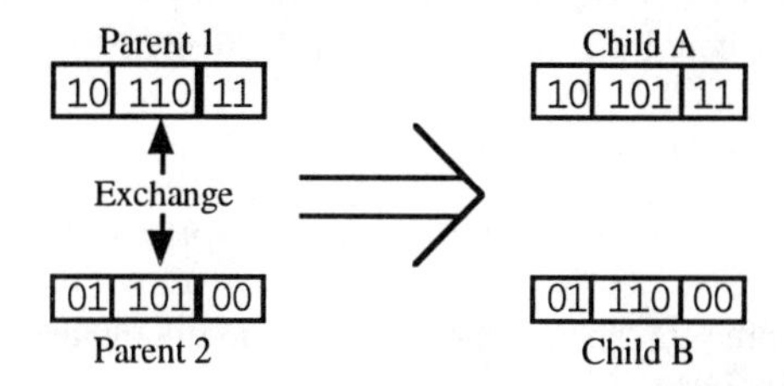

Using the same parents as in the one site splice examples, the crossing shown above has the result:

Parent 1	= 71.6535
Parent 2	= 40.9449
Child A	= 68.5039
Child B	= 44.0945

Once again the effect of the splice depends very much on where he splice points fall. In this example, the effect was minimal, since the middle order bits are quite similar. It is difficult to make a firm comparison of the effects of two site splices against one site splices. The experience with Genie suggests that there is not much to choose between them, but perhaps the two site splice is more disruptive on longer genes, since the segments being swapped are longer, and more likely to be substantially different.

One feature of the two site splice is that it can be used to make a one site splice by forcing one of the splice points to zero, which makes implementation a little easier.

1.14.2.3 Mutation

This is the process of randomly disturbing genetic information. Mutation operates at the bit copy level; when the bits are being copied from the parent to the child, there is a probability that each bit may become mutated. This probability is set by the user, usually to a quite small value, and stored in the `pMut` field of the population record. A coin toss mechanism is employed; if a random number between 0 and 1 is less than `pMut` the bit is inverted, so zeros become ones and ones become zeros. This helps lend a bit of diversity to the population by scattering the occasional point. This random scattering just might find a better optima, or even modify a part of a genetic code that will be beneficial in a later crossing. On the other hand it might produce a weak individual that will never be selected for crossing.

The mutation operation is integrated into the crossing mechanism.

1.14.2.4 Crossing Code

The procedure for crossing genes performs both one site and two site splice, with mutation occurring on the fly. Any procedure that accesses individual bits in a data type is unavoidably system specific. A description of the bit operators has been included. For the purposes of the example, it is assumed that bit zero is the right most bit. This takes care of problems with the gene being encoded as a 32 bit integer, where bit zero is the least significant bit.

This function returns true if the specified bit in the long integer is set, or false if it is not set.

```
FUNCTION bitsTst(l : LONGINT; bit : INTEGER) : BOOLEAN;
```

This procedure sets the specified bit in the long integer to 1 if `bool`is true or 0 if `bool`is false.

```
PROCEDUE bitSet(VAR l : LONGINT; bit : INTEGER; bool :
BOOLEAN);
```

This function simulates a coin toss. If a random number between 0 and 1 is less than the probability the function inverts the state of `bool`, otherwise it returns `bool`.

```
FUNCTION Mutate(bool : BOOLEAN; p : REAL) : BOOLEAN;
```

This function returns a random integer between `low` and `high`, inclusive.

```
FUNCTION randI(low, high : INTEGER) : INTEGER;
```

This is the crossing procedure. It takes as parameters two parent genes the population record and which gene this is, and returns two child genes.

```
PROCEDURE crossGenes(parent1, parent2              : LONGINT;
population          : TPopulation;
geneNo              : INTEGER;
VAR child1, child2 : LONGINT);
VAR
  geneLen : INTEGER;
  tmp     : INTEGER;
  bitNo   : INTEGER;

BEGIN
geneLen
population.phenoParams[genoNo].resolutions

{Select the first crossing site}
site1 := randI(0, geneLen- 1);

{Select the second crossing site}
IF population.phenoParams.crossSites = oneSite THEN
  site2 := geneLen  {One site only}
ELSE
  site2 := RandI(0,geneLen - 1);  {Two sites}

{Make sure the cross sites are in order}
IF site1 > site2 THEN
  BEGIN
  tmp   := site1;
  site1 := site2;
  site2 := tmp;
  END   ;

{Copy first section from parent to child}
FOR bitNo := 0 TO (site1 - 1) DO
  BEGIN
  {Copy parent1 bits to child1}
  BitSet(child1, bitNo,
           Mutate(BitTst(parent1,bitNo),population.pMut);

  {Copy parent2 bits to child2}
  BitSet(child2, bitNo,
           Mutate(BitTst(parent2,bitNo),population.pMut);
  END;

{Copy middle section from parent to alternate child. If
site1 were to be equal to site2 this wouldn't execute}
FOR bitNo := site1 TO (site2 - 1) DO
  BEGIN
  {Copy parent1 bits to child2}
  BitSet(child2, bitNo,
           Mutate(BitTst(parent1,bitNo),population.pMut);

  {Copy parent2 bits to child1}
  BitSet(child1, bitNo,
           Mutate(BitTst(parent2,bitNo),population.pMut);
  END;
```

```
{Copy last section from parent to child. This will not
execute
 when site2 = geneLen. ie. one site cross}
FOR bitNo := site2 TO (geneLen - 1) DO
  BEGIN
  {Copy parent1 bits to child1}
  BitSet(child1, bitNo,
            Mutate(BitTst(parent1,bitNo),population.pMut);

  {Copy parent2 bits to child2}
  BitSet(child2, bitNo,
            Mutate(BitTst(parent2,bitNo),population.pMut);
  END;
END;
```

1.14.3 Replacement

Replacement is the last stage of any breeding cycle. Two parents are drawn from a fixed size population, they breed two children, but not all four can return to the population, so two must be replaced. The technique used to decide which individuals stay in a population and which are replaced is on a par with the selection in influencing convergence.

1.14.3.1 Weak Parent

With weak parent replacement, a weaker parent is replaced by a stronger child. In effect, with the four individuals only the fittest two, parent or child, return to the population. This keeps improving the overall fitness of the population when paired with a selection technique that selects both fit and weak parents for crossing, but if weak individuals are discriminated against in selection the opportunity will never arise to replace them.

1.14.3.2 Both Parents

Both parents replacement is simple, the child replaces the parent. Under this schema each individual only gets to breed once. This keeps the population and genetic material moving around, but can lead to a problem when combined with a selection technique that strongly favours fit parents: the fit breed and then are disposed of. If their offspring are not as fit the population will be degraded.

1.14.3.3 Weakest Individual

This replacement schema replaces the two weakest individuals in the population with the children, as long as the children are fitter. The parents are included in the search for the weakest individuals. This technique rapidly improves the overall fitness of the population, and works well with large population where very unfit individuals would otherwise be left in for a long time.

1.14.3.4 Random

The children replace two randomly chosen individuals in the population. The parents are also candidates for selection. This can be useful for prolonging the search in small populations, since weak individuals can be introduced into the population.

1.14.3.5 Code

```
PROCEDURE Replacement(parent1, parent2 : TChromosome;
child1, child2 : TChromosome;
VAR population : TPopulation);
VAR
  chromNo : INTEGER;

BEGIN
CASE population.replaceMethod OF
  WeakParent:
    BEGIN
    IF parent1.fitness < parent2.fitness THEN
      BEGIN
           IF  (child1.fitness  <  child2.fitness)  AND
(child1.fitness > parent1.fitness) THEN
        BEGIN
        ReplaceIndividual(parent1,child1,population);

        {Check child2 against parent2}
        IF child2.fitness > parent2.fitness THEN
          ReplaceIndividual(parent2,child2,population);

        END
      ELSE IF child2.fitness > parent1.fitness THEN
        ReplaceIndividual(parent2,child2,population);
      END
    ELSE
      BEGIN
      IF (child1.fitness < child2.fitness) AND (child1.
                                fitness > parent2.fitness) THEN
        BEGIN
        ReplaceIndividual(parent2,child1,population);

        {Check child2 against parent1
        IF child2.fitness > parent1.fitness THEN
          ReplaceIndividual(parent1,child2,population);

        END
      ELSE IF child2.fitness > parent1.fitness THEN
        ReplaceIndividual(parent1,child2,population);
      END;
    END;

  bothParents:
    BEGIN
    ReplaceIndividual(parent1,child1,population);
    ReplaceIndividual(parent2,child2,population);
    END;

  weakestChromosome:
    BEGIN
    chromNo:= population.fitList[population.popMax]
    ReplaceIndividual(population.chromosomes[chromNo],
child1,population);
```

```
      {fitList will have been updated by the replace}
      chromNo:= population.fitList[population.popMax]
      ReplaceIndividual(population.chromosomes[chromNo],
child2,population);
      END;

    random:
      BEGIN
      chromNo := RandI(1,population.popMax);
      ReplaceIndividual(population.chromosomes[chromNo],
child1,population);

      chromNo := RandI(1,population.popMax);
      ReplaceIndividual(population.chromosomes[chromNo],
child2,population);
      END;
    END; {CASE}

END;
```

1.15 Search Termination

The termination, or convergence criteria finally brings the search to a halt. There is always a limit to the maximum number of generations in a Genie search since Genie keeps records of the search process, and these would overflow. Reaching this limit would indicate that the termination criteria were too stringent or that there is no satisfactory solution.

The termination techniques presented here are quite basic. Many more ways of ending the search could be devised, but these few worked adequately in Genie.

1.15.1 Fitness Sum

This termination scheme considers the search to have satisfactorily converged when the sum of the fitnesses in the entire population is less than or equal to the convergence value in the population record. This assures that virtually all individuals in the population will be within a particular fitness range, although it is a good idea to pair this termination technique with weakest gene replacement, otherwise a few unfit individuals in the population will blow out the fitness sum. When setting the convergence value, it is important to consider the population size.

1.15.2 Median Fitness

This is easy to implement since a sorted list is being maintained. At least half of the individuals will be better than or equal to the convergence value, which should give a good range of solutions to choose from.

1.15.3 Best Individual

Best individual termination stops the search once the minimum fitness in the population drops below the convergence value. This can bring the search to a speedy conclusion while guaranteeing at least one good solution.

1.15.4 Worst Individual

Worst individual terminates the search when the least fit individual in the population has a fitness less than the convergence criteria. This guarantees that the entire population will be of a minimum standard, although the best individual may not be significantly better than the worst. An overly stringent convergence value may never be met, in which case the search will terminate after the maximum generations has been exceeded.

1.15.5 Code

Termination is implemented as a function, returning true if the termination criteria have been met.

```
FUNCTION Terminate(population : TPopulation) : BOOLEAN;
VAR
  endSearch : BOOLEAN;

BEGIN
IF population.thisGen >= population.maxGen THEN
  endSearch := TRUE
ELSE
  CASE population.cnvrgMethod OF
    fitnessSum:
    endSearch := population.sumFitness <= population.
                            cnvrgValue;

    midPoint:
    endSearch := population.chromosomes[population.
                            fitIndex[population.popMax DIV
                            2]].fitness <= population.
                            cnvrgValue;

    bestIndividual:
    endSearch := population.minFit <= population.
                            cnvrgValue;

    worstIndividual:
    endSearch := population.maxFit <= population.
                            cnvrgValue;
  END;

Terminate := endSearch;
END;
```

1.16 Search Histories

Genie kept a record of the best, worst and average fitness for each generation. It was possible to display these graphically during the search, which gave the user visual feedback on some aspects of the convergence. In addition a record of the ten best ever individuals was maintained. This is worthwhile when the replacement schema might throw out fit genes.

1.17 Solution Evaluation

At the end of the search Genie displays the final population with their factor values and fitnesses, from which it is possible to select a solution and write it back to the model for detailed investigation. In large models it is not always practical to declare every factor of interest a target factor, or perhaps some factors were simply overlooked. By testing a potentially good solution in the model it is possible to check for side effects. Some solutions which fulfil the explicit requirements of the search may do so in a less than wholly satisfactory manner. For example, if one had as a target low unemployment in an econometric model, and was a little careless in declaring target factors, the GA may very well converge on a solution which increased the mortality rate dramatically; less people - less unemployed. A “good”, albeit politically untenable (one hopes!), solution.

1.18 After Genie

Genie/TIP II is under development while this paper is being written. As the title suggests, Genie I was a first attempt at writing a GA, and as such it performed tolerably well. Most of the effort in Genie II is going into refining the model, so that the GA can manipulate a wider range of characteristics of the model. The techniques described above are adequate for a simple GA, and provide a base for further development, but some areas require attention if the GA is to be used in a more sophisticated environment.

1.19 Dynamic Populations

While fixed size populations were fine for the first Genie, they are definitely on the way out in Genie II. A seed population of some size will always be necessary to start the search off, but by modifying the replacement technique to admit children to the population without necessarily replacing an existing individual it is relatively easy to implement a dynamic population.

1.20 Parallel Fitness Evaluation

This GA itself uses negligible processor time, the objective function, however, is much slower. The time to run even a small model on a mid-range 1993 desktop is measured in seconds. It is planned to distribute the fitness evaluation across several machines on a network, thereby improving the brute force (more is better) power of the algorithm. To achieve this, it is best if the fitness evaluation is processed asynchronously. The GA sends an individual to an evaluation pool, the pool finds a free machine to evaluate it, and then sends it back to the population when it is ready. This way the GA doesn’t have to sit idle while the model is running. A dynamic population size is essential to this scheme.

Asynchronous processing of individuals will inevitably introduce a lag between the GA’s selection of parents and its knowledge of how that selection turned out, however, this is no different to one of the traditional GA’s in which each generation the entire population's paired up and crossed, producing a new population of the children.

1.21 Niching

Niching describes a process of identifying when individuals are converging on distinct optima. In Genie I there is no way of checking for this multiple convergence which can result in acceptable solutions being discarded. Imagine a solution space with two peaks, one a little better than the other. The lesser peak may be well represented in the population, with several individuals clustered around the top. The better peak may have a few individuals on its lower slopes, but because they are less fit than those on the other peak, they stand a good chance of being passed over for crossing, or discarded. Alternatively, individuals from the different peaks may be selected for crossing together, with the result being individuals not on either peak, but in poor region of space.

By identifying when individuals are not climbing toward the same optima, it may be possible to take action to see that both optima are adequately developed. One option is to tag individuals with a family name, to prevent miscegenation. Another option is to split the different families into separate populations to allow them to thoroughly explore their region of the solution space. The families would be the seeds for a new population, which can then expand.

1.22 Search Refinement

Search parameter settings (selection, crossing and replacement) that are effective in the early stages of a search may not necessarily be the best toward the end of the search. Early in the search it is desirable to get good spread of points through the solution space in order to find at least the beginnings of the various optima. Once the population has started to converge on an optima it might be better to exercise more stringent selection and replacement to thoroughly cover that region of space.

Another refinement is the domain and resolution of the individual genes. A large range and a coarse resolution early in the search will help scatter the points. After a while it may become apparent that certain parts of space yield very poor results. It would then be appropriate to limit the gene ranges and increase the resolution to finely search the better regions.

It would be possible for the GA to monitor its performance and make alterations to the search parameters when the rate of convergence of fitness values has slowed or after a preset number of generations. Care would have to be exercised to avoid overly directing the search to early good results. A poor looking region of space may yet contain an undiscovered optima.

1.23 Robustness

The evaluation of solutions is not strictly a problem of the GA, but it is worth looking at. As mentioned before, there may be several comparable but more or less distinct optima. Differentiating these is important to the end use of the solutions. The evaluation of solutions in Genie I required that the user manually check each one in the TIP model and assess their viability. By niching the solutions the amount of checking required is reduced since a representative of each niche need only be checked.

Sensitivity of solutions is also important where it may not be possible to implement a solution accurately. Two distinct solutions may have comparable fitnesses and no undesirable ancillary effects, however, one may reside on a very steep optima while the other may lie on a broad mound. Obviously the solution on the broad mound will be less sensitive to errors in implementation than the one that is steep sided, where any small deviation will result in a plummeting fitness.

What is required is some sort of index of robustness. Keeping a record of discarded individuals helps by filling in the otherwise "black box" of space. The storage space required for this is probably not a problem, since only the phenotype, or factor values and fitness need be kept. The points that surround an optima should provide some idea of the stability of that optima. Exactly how to automate this is yet to be decided[6].

---

6 TIP models are designed to incorporate qualitative factors, which have traditionally been left out of models due to the difficulty in measuring them. It is possible to describe a general (and simple) relationship such as: if unemployement rises then the general level of contentment in the community will decline, and if the level of contentment in the community declines then the number of illnesses will tend to increase. This relationship can be expressed in a TIP model, with admittedly arbitary magnitudes for the strength of the cross effects.
The outcome of a search on a model of this type will make suggestions about how to influence, in more or less sophisticated ways, the control factors. Given such a general model and three control factors (A,B,C) all with a range of -0.1 to +0.1 how would one compare result of (0.05, 0.9, -0.8) against (0.1, -0.5, 0.04)? All that can (or should) be said is factor A stays about the same, factor B is going in a different direction and factor C may be changing direction, but maybe not. Now say the first set of values has a relatively good fitness value while a third set (0.3, 0.7, -0.3) has a relatively poor fitness value. What can be said about the two solutions, and their implications? The spatial proximity of two very different solutions probably renders them useless.

Chapter 2

Conor Ryan
Computer Science Dept.,
University College Cork, Ireland.

conor@csvax1.ucc.ie

Niche And Species Formation in Genetic Algorithms.

0-8493-2519-6/95/$0.00 + $.50

2.1 Introduction

Genetic Algorithms are usually described in terms of operating on a population of individuals and of using a global knowledge of every individual in that population to guide evolution. Typically, when selecting a partner for mating, an individual chooses a mate solely on the basis of the relative performance of that individual to all others in the population, a method known as Panmictic mating. In nature, however, an individual need not neccessarily be influenced only by the fitness of a potential mate, using instead different criteria to decide.

For instance, an individual may seek out individuals similar to itself, or it may actively solicit the attentions of those that differ from it. However, traditional Genetic Algorithms pay no heed to such personal opinions of individuals, instead using a central control to select mates; neither do traditional Genetic Algorithms pay attention to the geographic location of individuals, while in nature, it is more common for individuals to choose mates within a close proximity than to roam the landscape in search of their ideal mate. Moreover, individuals tend to settle for the best mate they can find at any particular time.

Recently, variants of the Genetic Algorithm have appeared which mimic nature in the above details. This chapter examines some of the methods used and describes the motivation behind and the rewards of these methods.

2.2 Motivation

The perennial problem with Genetic Algorithms is that of premature convergence, that is, a non-optimal genotype taking over a population resulting in every individual being either identical or extremely alike, the consequences of which is a population which does not contain sufficient genetic diversity to evolve further.

Simply increasing the population size may not be enough to avoid the problem, while any increase in population size will incur the twofold cost of both extra computation time and more generations to converge on an optimal solution.

Genetic Algorithms then, face a difficult problem. How can a population be encouraged to converge on a solution while still maintaining diversity? Clearly, those operators which cause convergence, i.e. crossover and reproduction, must be altered somehow.

Another problem often used as a criticism against Genetic Algorithms is the time involved in deriving a solution. Unlike more deterministic methods such as Neural Networks, hill-climbing, rules-based methods etc, Genetic Algorithms contain a large degree of randomness and no guarantee to converge on a solution within a fixed time. It is not unusual for a large proportion of runs not to find an optimal solution.

Fortunately, due to their very nature, Genetic Algorithms are inherently parallel, i.e. individuals can be evaluated in parallel as their performance rarely, if ever, affects that of other individuals. The reproduction phase, however, which commonly involves a sexual free-for-all, during which the individuals of a population dart about in a crazed frenzy vying with one another in attempts to mate as often as possible, represents a serious bottleneck in traditional Genetic Algorithms. The fitness of every individual must be known, and, despite the overwhelming and no doubt impatient ardour present in the population, only one reproduction/crossover may take place at a time. A method that could avoid this sort of bottleneck would lend itself very well to implementation on parallel machines, and hence speed up the whole process.

It is shown that there are two quite different approaches involving niches to the problems described above, isolation by distance and restricted mating. Isolation by distance puts individuals in explicit geographical locations with knowledge only of nearby individuals, while restricted mating is more like traditional Genetic Algorithms, in that a central control is still kept, but with some effort being made in either the selection of parents or those to be replaced to prevent premature convergence.

Generally speaking, both methods permit the evolution of individuals which fill differing environmental niches, with similar individuals congregating together. The correct biological name for such groups is an *ecotype*, but they tend to be referred to as different species in Genetic Algorithms. The use of the word "species" is not strictly correct, as individuals of each species may freely mate with each other. However, for the sake of consistency, the word species will be used with the meaning usually attributed to it in Genetic Algorithms. In this field, a *niche* usually refers to that which makes a particular group unique, e.g. having a common fitness rate, genotype etc, while *species* refers to the individuals in that group.

2.3. Isolation by Distance

Isolation by distance takes a direct parallel from nature by assigning each individual a geographic location. Each individual can either have a specific location in a single population, or individuals can be in distinct populations, each of which has a specific location relative to all other populations. So, just as an individual in say, Ireland, is more likely to choose an Irish person than one from say, Chad, to mate with, isolation by distance ensures that only individuals relatively close to one another may mate.

There are two main effects of this isolation, the first being that groups of similar or identical individuals congregate together, [Collins92] and second that they need only be aware of their immediate neighbours, and thus require only local knowledge, as opposed to the global knowledge required in traditional, Panmictic Genetic Algorithms.

Above, individuals in an isolation by distance model were likened to individuals as far apart from each other as Ireland and Chad. However, distance is not the only factor that prevents greater communication and integration between the Irish and the Chadiens, for there are many naturally occurring physical barriers separating the countries.

Some implementations of isolation by distance models, known as Islands models, take such barriers into consideration where individuals must explicitly travel from one country (island) to another if they are to find a foreign mate. The other main implementation, known as Spatial Mating, does not take natural barriers into consideration, and therefore the only barrier in force is that of distance.

The mating used in both these models is generally referred to as Spatial Mating, in comparison to the Panmictic mating of the traditional Genetic Algorithm.

2.3.1 The Islands Models

In the Islands models, individuals live on explicit islands. Each island can be as distinct as desired, each may have its own mating scheme or even fitness function, and each may have varying emigration rates. The islands are sometimes referred to as *demes*, with each deme considered as a separate subpopulation. Rather confusingly, however, this term is also used with a slightly different meaning in a later section, so, to avoid ambiguity this term will not be used in relation to the Island model.

Due to the isolation of the islands, each island may evolve at its own rate, to the extent that some islands may evolve in different directions, i.e. toward different solutions. It is because of this isolation that different environmental niches and thus different species may evolve in such a set up. Each island may fulfil its own niche, whereby individuals that inhabit that island are all quite similar. While the appearance of a different environmental niche on each island is possible, there is no reason why several islands shouldn't evolve to fill similar or even identical niches.

An island which has evolved in a suboptimal direction, or that has perhaps prematurely converged so that all evolution has effectively stopped, may be helped by the arrival of emigrants from other, better performing islands.

Each island can be looked upon as a separate implementation of a Genetic Algorithm to the problem at hand, with occasional "suggested" individuals being inserted through emigration which permits the appearance of individuals in the population who might not otherwise appear due to gene loss. It is because each island tends to evolve individuals which fill an environmental niche that stores of radically different individuals can be held.

2.3.1.1 Emigration in the Islands Model

To promote genetic diversity, and in essence, to allow the sharing of possible solutions, every so often some individuals emigrate from island to island. The question now arises of which individuals should be chosen to emigrate. There are two possible methods for choosing those who should emigrate. One can select randomly from the current population[Cohoon91] which, in effect, is giving the choice to the individuals, allowing those with a certain "wanderlust" or overriding desire to mate with a foreigner. However, because the selection is done randomly, it is impossible for a nomadic species to evolve, individuals of which would move about often, roaming the islands and various niches, spreading their genes as they do so. The advantage of randomly selecting is, of course, the greater mix of genes that will result.

A second method used by [Muhlenbein89] was to select the highest performing individual from each island to be copied to other islands. This would result in more directed evolution than the first case as the emigrant individuals would not be tainted by genes of lower performing individuals. This is not to say that the former method is worse, for the less directed a population is, the greater diversity it will contain. There is also the possibility that an individual may leave one island where it was considered just moderately successful only to find its fortune on another with its relative fitness being higher, allowing a smoother increase in the average fitness level of its new home than if a much fitter individual arrived.

A third possible method, so far implemented only in Spatial Mating [Todd91] and Panmictic [Ryan94a] models, is to include in each individual's genes a "wanderlust" gene. This gene could then be used to calculate how likely an individual is to leave its home to search for a mate. A possible result of this would be, as suggested above, the evolution of nomadic tribes of individuals.

As in selection, one has the choice of whether or not to replicate emigrating individuals, i.e. should individuals move to their new home or should a copy of them be sent there? The same risks are incurred as in selection, if one does not copy individuals it is possible that an island could be set back several generations in evolutionary terms by the mass emigration of its best performers. On the other hand, simply copying individuals across could lead to highly fit individuals dominating several populations.

In general, the decision of whether or not to copy individuals depends on how individuals are selected to emigrate. If the best individuals are to emigrate then removing them from a population could well harm its evolution, while in the case of individuals being chosen randomly, it is unlikely that all of the best individuals will be removed.

Due to the isolated nature of each island, this model may be very conveniently mapped onto a parallel architecture, in particular a coarse grained, i.e. relatively few processors (4-32) architecture, generally the more affordable parallel setup, may be used.

2.3.1.2 A Simple Implementation of the Islands model

What follows is a simple implementation of an islands model. This model can easily be mapped onto a coarse grained parallel architecture, but does not require one. This algorithm was designed with distributed evolution in mind, to aid the evolution of differing environmental niches, not to speed up the genetic algorithm through parallel programming.

```
REPEAT
FOR EACH POPULATION DO
FOR EACH INDIVIDUAL DO
TEST INDIVIDUAL
IF (TIME-TO-EMIGRATE) THEN
SELECT INDIVIDUALS TO EMIGRATE
EMIGRATE INDIVIDUALS
CULL POPULATION
PRODUCE NEW POPULATION
```

```
UNTIL END-CRITERION REACHED
```

The `TIME-TO-EMIGRATE` test above implies that individuals need not emigrate every generation. Too much emigration would overshadow any advantage of isolating a population as well as causing extra communication on parallel machines.

The `CULL POPULATION` step may be needed in models where individuals are copied as islands could well be overcrowded after the `EMIGRATE INDIVIDUALS` step, and, unless islands can accommodate a continually growing population, an appropriate number of individuals must be removed.

2.3.1.3 The Advantages of Islands

The Islands model has several advantages, the main one being that several solutions can actively be pursued by different islands in parallel in the form of different environmental niches, allowing the juggling of several solutions or the application of Genetic Algorithms to multi-modal functions which have several solutions. Secondly, highly directed selection methods can be used on each island, and one can then use emigration and the injection of new genes to counteract premature convergence. If individuals are copied from island to island rather than explicitly moved, then unviable matings between individuals from radically different niches cause no damage, as there is no loss of genetic material.

Finally, the islands are inherently parallel, the mapping of islands onto processors is a trivial task, the result being a massive speedup [Collins92] over populations executed on a serial machine. However, one should note that parallelism is an advantage of demes, not a prerequisite.

2.3.2 Spatial Mating

A somewhat more popular method for isolation by distance is the Spatial Mating model. Spatial Mating does not take into account any barriers other than physical distance, typically having individuals occupying a grid environment.

Various implementations of Spatial Mating differ in their approaches to selection and mating, some allow a limited form of movement for an individual seeking out a mate, while others adopt a strategy similar to that of the Island model. In the latter, individuals live in overlapping demes, and individuals can occupy a position in several demes. These demes are treated identically to the islands of the previous section, but with no explicit emigration - the overlapping nature of the demes automatically providing this.

Each of the overlapping demes in spatial mating are created around one particular individual, typically the individual to be replaced by a new one.

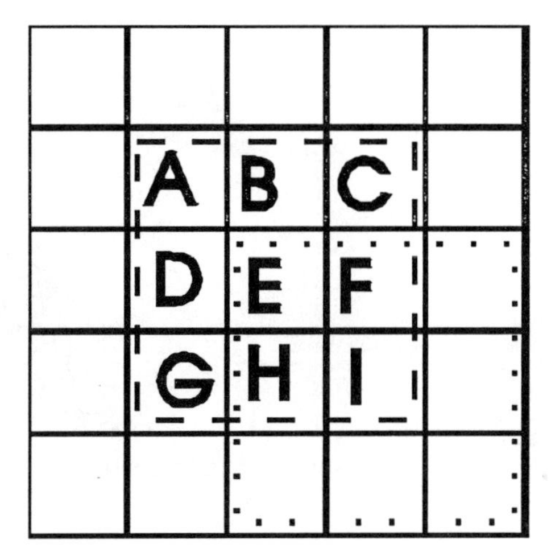

Fig 2.1: Individuals arranged in a 2 dimensional grid with overlapping demes.

Because a deme is created around each individual, there are as many demes in the population as individuals. Obviously, this creates much overlap with individuals appearing in several demes. The number of demes of which an individual is a resident depends on the size of a deme. In general, an individual appears in as many demes as there are individuals in a deme. A result of overlapping demes is a reduction of isolation, as neighbouring demes have a far greater influence over each other than those in the islands model described above. Generally, demes with a higher average fitness have a greater influence and the genes of their individuals propagate through the population, forming larger "islands of influence", each of which contains either identical or very similar individuals.

Like the Islands model, Spatial Mating is capable of maintaining several possible solutions in the population. Individuals containing the genes of each solution tend to congregate together, usually looked upon as a particular species or niche, with individuals on the edges of these "islands of influence" providing a smooth change of genotype, belonging completely to neither one nor the other.

These islands of influence are analogous to the islands of the previous section, and are commonly referred to as demes. The major difference between these demes and the islands is that, because they refer to the individuals within them and not to any physical land, the demes can grow and contract depending on their success. Relatively unfit demes will eventually disappear and the population will be taken over by one or more very fit demes. Note that the implementor does not need to specify either the starting number of islands or the final amount, they arise and dissipate automatically.

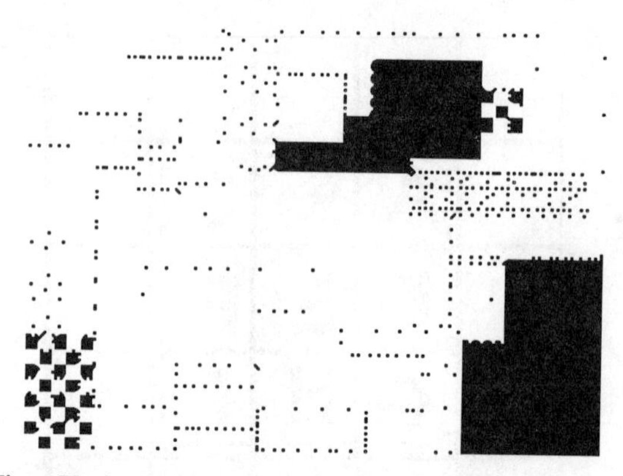

First: Various species in a population early on in a run.

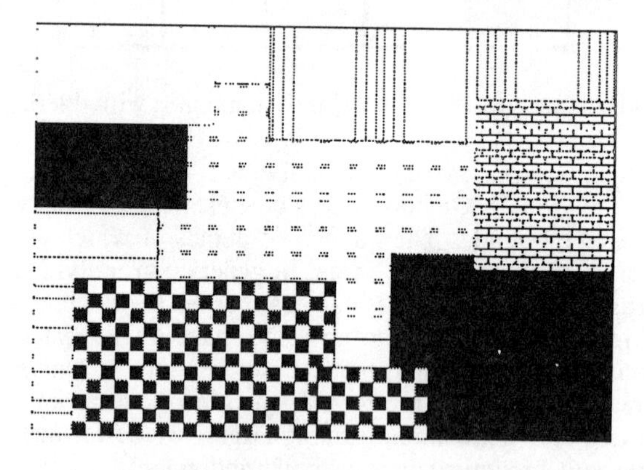

Fig 2.2. Second: Toward the end of a run, the population is dominated by a few demes.

2.3.2.1 The Geography of Spatial Mating.

So far, all examples have been done in terms of grids, where each individual has eight neighbours. Not all implementations of Spatial Mating rely on such a structure, however. Various geographies have been suggested, ranging from circular worlds to one dimensional worlds [Collins92].

The more practical of these are the circular or Wheel model and one dimensional world, both illustrated below in figure 2.3.

Both of these somewhat similar models have the advantage of being able to increase the deme (neighbourhood) size in smaller increments than the grid type. However, the grid model has neighbours in all directions, rather than just left and right of an individual in the one dimensional model, and has been shown to outperform the latter [Collins92]. No comparison has been done between the grid and Wheel models, but the wheel model suffers to some degree from the variation problem that one dimensional geographies do.

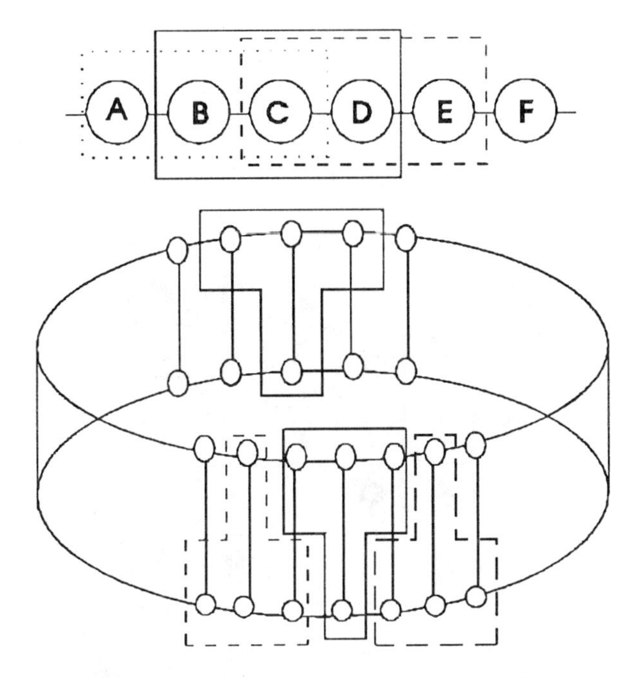

Fig 2.3: Two approaches to the geography of spatial mating.

2.3.2.2 Ireland and Chad revisited

Section 2.3 related the tale of an amorous Irishman travelling to Chad in search of a bride, a journey like which is made possible in the Islands model. Spatial Mating makes any such travelling difficult, individuals instead staying within their allotted demes, and with every individual travelling the same distance to procreate.

A technique which permits varying amounts of travel for individuals seeking mates is the Random Walk method [Collins92]. Using random walk, individuals move from their home location in random directions taking a set number of steps, when an individual has exhausted its steps it returns to the best prospective mate it encountered.

Random Walk, as well as being somewhat more biologically plausible than methods which resort to using roulette wheel selection in the demes, requires less computation than roulette wheel selection which involves sorting all other individuals, instead needing only to select the highest performing individual visited.

```
CURRENT-BEST = MY-SCORE
CURRENT-BEST-IND = ME
REPEAT
MOVE ONE STEP IN RANDOM DIRECTION
IF CURRENT-SCORE >= CURRENT-BEST THEN
CURRENT-BEST = MY-SCORE
CURRENT-BEST-IND = ME
```

```
UNTIL NUMBER-OF-STEPS EXHAUSTED
```

Before setting out on their search for a mate, an individual first examines themselves. If, on their travels, they do not encounter an individual who is a better performer than they are, then the individual will mate with themselves. In the case where two individuals of equal fitness are encountered, the *last* individual encountered is the one chosen as a mate.

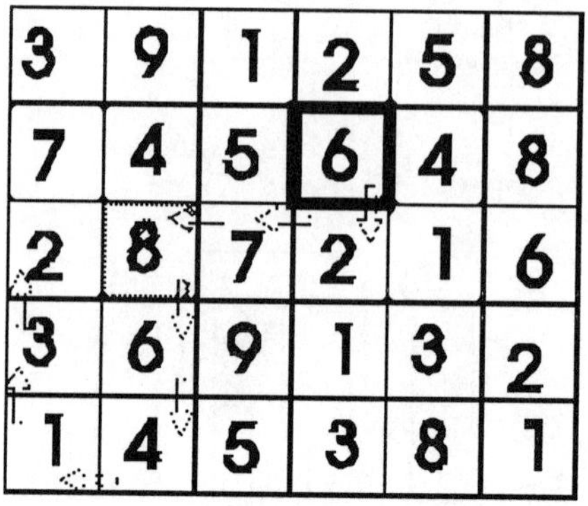

3	9	1	2	5	8
7	4	5	6	4	8
2	8	7	2	1	6
3	6	9	1	3	2
1	4	5	3	8	1

Fig 2.4: An example of a path chosen randomly by an individual seeking a mate.

2.3.2.3 A Simple Implementation of Spatial Mating

Below is a simple implementation of a Spatial Mating model. Like the Islands model above, this is an extremely paralellisable algorithm, except this algorithm can be mapped onto a very fine grained set of processors. Again, the algorithm was designed with distributed evolution in mind, and not parallel programming.

```
REPEAT
FOR EACH LOCATION DO
TEST INDIVIDUAL AT LOCATION
FOR EACH LOCATION DO
SELECT PARENTS
CREATE NEW INDIVIDUAL
UNTIL END-CRITERION REACHED
```

The `SELECT PARENTS` step is the only step that varies between implementations of Spatial Mating, and, depending on the implementation, varies from selection within demes to random walk.

2.3.2.4 The Advantages of Spatial Mating

Spatial Mating has practically the same advantages as the Islands model, as it too can maintain several solutions in the population. If overlapping demes are being used, favourable selection schemes can be implemented, relying again on emigration to prevent premature convergence. Species develop very easily in models like these, as a species' numbers will expand and contract as various species compete with each other to get control of their environment.

Parallelising Spatial Mating is slightly different to the Demes model as Spatial Mating has a very fine grained parallel structure, and every individual can be executed in parallel, both in evaluation and reproduction.

2.3.3 The Differences between Spatial Mating and the Island Model

The principal difference between Spatial Mating and the Island model is the amount of isolation for each sub-population. It was shown above how most Spatial Mating implementations can be looked upon as demes with overlapping populations, i.e. individuals can be in more than one deme, while Islands have no overlapping demes, using instead explicit emigration between subpopulations. It is important to note that the rate of emigration affects a run quite strongly so must be chosen with care while the Spatial Mating model does not require any setting of such a rate.

The importance of emigration is such that it has been described as being equivalent to mutation in Island models. If emigration is looked upon as being a replacement for mutation in these models, one must take the same care with it. Excessive emigration defeats the purpose of separate islands which degenerate into one large, panmictic population, while miserly use of the operation results in the search being equivalent to multiple runs of small populations.

The two, of course, are not mutually exclusive. Each island in the demes model can have any selection method, including spatial mating. Such an implementation would lead to isolation by distance not only on each island, but also between the islands themselves, permitting subspecies to evolve within each isolated species, a very natural state.

2.3.4 Summary

Genetic Algorithm implementations which employ isolation by distance are capable of juggling possible solutions to a problem through the use of different environmental niches, as well as coping with multimodal problems containing several solutions.

The absence of a central control in possession of total knowledge about every individual in the population means that a major bottleneck in traditional genetic algorithms can be avoided, leading to a speed up in computation time.

Finally, because of the elimination of this central control, isolation by distance is decidedly parallel, and presents a relatively simple task to be implemented on parallel architetures.

2.4 Panmictic Mating

Excessive centralised control becomes a problem in traditional genetic algorithms as the population increases. As the performance of every individual in the population must be rated relative to that of every other individual, large populations incur severe costs, typically countered by the strategies outlined above.

Much work has been done, particularly in the area of niche and speciation within single, panmictic populations, mainly in multimodal functions. These methods do not strive to reduce the control demanded in traditional genetic algorithms, in fact, they all rely on some control to balance genetic diversity in a population. Strategies relying upon this sort of operation might at first appear to be in conflict with the isolation by distance model described above, especially considering they both have a similar goal in mind - the balancing of genetic diversity in a population.

This, however, is not the case. For all its faults, panmictic mating is perfectly suitable for relatively small populations, and if evolution is to be highly directed, i.e. very elitist, the very structure of isolation by distance models acts against this.

2.4.1 Multimodal Functions

Genetic algorithms using panmictic applied to multimodal functions face two main problems, the first being that of distributing individuals evenly across all peaks in the solution as in figure 2.5.

Each peak in the solution landscape can be viewed as a separate environmental niche. A successful application of a genetic algorithm to a problem like this would have to result in several individuals spread across each of the environmental niches. What better way to do this than to permit the evolution of several species within the environment, each specialising in its own particular niche?

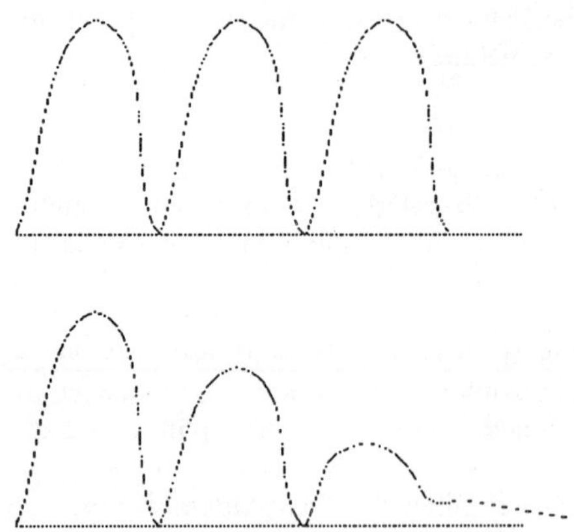

Fig 2.5: A multimodal function solution landscape.

Using niches and species in applications such as these can lend a more biological meaning to the word species. A second problem now arises as parents of differing species, i.e. from different environmental niches, tend to produce unviable children [Goldberg89] and so, parents occupying different niches must be discouraged from mating. Figure 2.6 below illustrates the problem.

2.4.1.1 Crowding

The first problem was addressed by [DeJong75] to prevent a single genotype from dominating a population. By ensuring that a newborn individual will replace one that it is genotypically similar to it, Crowding tries to maintain a balanced population. The mechanism for this replacement is quite simple: one randomly selects *CF* individuals and, through calculation of hamming distances, decides on a suitable victim.

An advantage of Crowding is how few individuals must be examined when choosing a victim, for *CF* is usually 2 or 3. This replacement of similar individuals acts to prevent one genotype from taking over a population completely, allowing other, possibly less fit, niches to form within the main population.

Crowding does not explicitly create niches, nor does it make any concentrated effort to encourage them, rather it allows them to form.

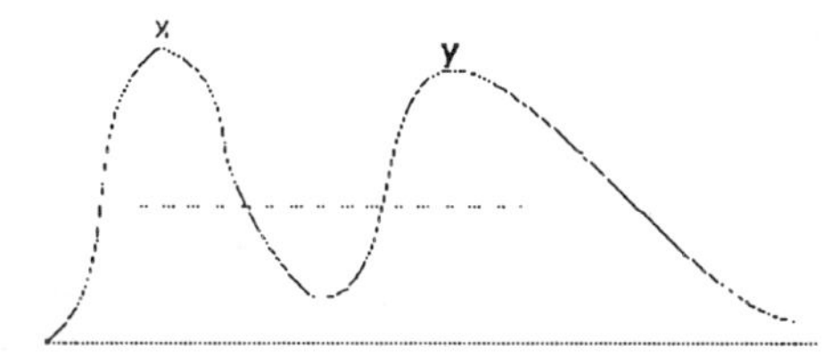

Fig 2.6: A possible result of mating parents from different peaks is this case, where the resulting offspring appears in a trough between its parents.

2.4.1.2 Sharing

A somewhat different approach was adopted by [Goldberg87] in the Sharing scheme, in that individuals in a population which uses Sharing face limited resources as they strive for fitness. To make life more difficult for them, individuals of the same environmental niche, in this case genotypically similar, are more inclined to search the same places for resources, so have a more difficult time than unique individuals.

In a similar manner to Crowding, domination of the population by a single genotype is discouraged by the punishing of individuals who are too similar to a large portion of the population.

Sharing, however, is not as simple to calculate as Crowding, and is very problem-specific as one must know in advance how many peaks there are in the solution landscape. Sharing does encourage the formation of niches and, to prevent the unsavoury prospect of individuals from different niches mating as in Figure 2.6 above, uses a form of restricted mating.

Although Crowding is far simpler than Sharing, both in its calculation and execution, the latter has been shown to be far more effective in the area of multimodal functions [Goldberg89], as the rather gentle powers of persuasion used by Crowding cannot prevent most individuals from ending up on only one or two peaks due to the few individuals that are examined each time. Sharing, on the other hand, aggressively encourages the development of new niches and consequently distributes individuals across all peaks in the landscape. The payoff is a simple one, Crowding is cheap and simple, while Sharing is relatively expensive yet successful.

2.4.2 Niche and Speciation in Unimodal Problems

The use of niches in multimodal problems is a very simple mapping, with the evolution of a different species for each peak in the solution landscape. The mapping is not quite so easy in unimodal problems which usually contain only one peak, or at the very least contain one peak higher than the others. Nevertheless, as seen in section 3, niches can maintain several possible solutions to a problem too.

All approaches to unimodal problems, involving niching, attempt to maintain a balanced population, either through restricted mating to prevent inappropriate parents mating, or through replacement methods which hinder the taking over of a population by a single genotype.

Several methods, which do not strictly use niches, but which do imitate their operation to some degree, exist. There are replacement methods, which ensure that newly born individuals are sufficiently different from the rest of the population before allowing them entry, e.g. Clone Prevention, Steady State Genetic Algorithms (SSGA). There are also special selection schemes which operate in a similar manner to the restricted mating described above, in that prospective parents are allowed to fulfil their conjugal rites only if they fulfil certain criteria. Restricted mating does permit the evolution of different niches, but typically forces parents to come from differing niches, thus allowing each niche to exert some influence on evolution.

Clone prevention and SSGA operate in a similar manner to Crowding in that before an individual is permitted entry to the population, it is compared to others to verify its uniqueness. While Crowding examines only a few individuals each time, these other methods guarantee uniqueness by comparing a new individual to every other individual in the population. Like Crowding, they do not attempt to explicitly create niches or species, but attempt to prevent the domination of the population by a single species.

Unfortunately, both methods incur high overheads, as the comparing of individuals is a costly affair. It is also fair to say that clones, the presence of which can retard evolution, do not always cause disaster, and in fact can sometimes even help direct evolution.

2.4.2.1 Incest Prevention

A similar view was taken by [Eshelman91] when he suggested the use of *Incest Prevention* which only discouraged clones, still permitting them to enter the population. Incest prevention attempts to "matchmake" parents with the intention of their offspring taking the best genes from their parents. It is by mating differing parents that diversity is kept in the population and thus further evolution permitted.

As a population evolves, its individuals become more and more similar, thus it becomes more difficult to find suitable parents. To avoid a situation where there are no such parents in a population, there is a difference threshold set, which can be relaxed if there is some difficulty in selecting parents. It is assumed that difficulty will arise if there is no change in the parent population, and, as incest prevention is used with elitism, i.e. a list of parents is maintained which individuals can only enter if their fitness is sufficiently high, it is a trivial matter to track any changes.

Again, differing species are not explicitly created, nor are guaranteed to appear, but if they do, Incest prevention encourages inter-species mating, as the fitness landscape in unimodal functions tends to be like that of figure 2.5 above.

```
SET THRESHOLD
REPEAT
FOR EACH INDIVIDUAL DO
TEST INDIVIDUAL
ENTER PARENT POPULATION
IF NO-NEW-PARENTS THEN LOWER THRESHOLD
FOR EACH INDIVIDUAL DO
REPEAT
SELECT PARENTS
UNTIL DIFFERENT()
UNTIL END-CRITERION REACHED OR THRESHOLD=0
```

As soon as an individual is tested it attempts to enter the parent population, as described above, this step is only successful if the individual is fitter than the least fit member of the parent population. After all the new individuals have been tested, one checks to see if the parent population has been changed. An unaltered population will lead to the difference threshold being reduced.

The `DIFFERENT()` test simply calculates the hamming distance between parents and ensures that, if they are to breed, the difference will be above the threshold. As can be seen from the last step in the algorithm, it is possible for a run to end for reasons other than the reaching of some end criterion. In this case it is common to "reinitialise" the population by mutation of the best performing individuals encountered so far.

2.4.2.2 The Pygmy Algorithm

Although incest prevention avoids the cost of clone prevention, there is still the cost of finding a satisfactory couple each time mating is to be performed. To reduce the cost as much as possible another method, the Pygmy Algorithm [Ryan94b] has been suggested, which does not explicitly measure differences between parents, but merely suggests that the parents it selects are different.

The Pygmy Algorithm is typically used on problems with two or more requirements, e.g. the evolution of solutions which need to be both efficient and short. Niches are used by having two separate fitness functions, thus creating two species. Individuals from each species are then looked upon as being of distinct genders, and when parents are being chosen for the creation of a new individual, one is drawn from each species with the intention of each parent exerting pressure from its own fitness function.

Typically, there is one main fitness function, say efficiency, and a secondary requirement such as shortness. Highly efficient individuals would then enter the first niche, while individuals who are not suited to this niche undergo a second fitness test, which is simply their original fitness function modified to include the secondary requirement. These individuals then attempt to join the second niche, and failure to accomplish this results in a premature death for the individual concerned.

The use of two niches maintains a balanced population and ensures that individuals who are fit in both requirements are produced. Below is the pseudo code for the Pygmy Algorithm.

```
REPEAT
FOR EACH INDIVIDUAL DO
TEST INDIVIDUAL WITH MAIN FITNESS FUNCTION
ENTER PARENT POPULATION #1
IF UNSUCCESSFUL
TEST INDIVIDUAL WITH SECONDARY FITNESS FUNCTION
ENTER PARENT POPULATION #2
FOR EACH INDIVIDUAL DO
SELECT PARENT FROM POPULATION #1
SELECT PARENT FROM POPULATION #2
CREATE NEW INDIVIDUAL
UNTIL END-CRITERION REACHED
```

Each niche is implemented as a separate, elitist group, because of the elitist nature of each niche, which maintains individuals on a solution landscape similar to Figure 2.5 above, there is much pressure on newly born individuals to appear between its parents, and thus outperform them. It is also possible, of course, that a child may be endowed with the worst characteristics of its parent. A child like this will be cast aside by the Pygmy Algorithm but its parents, because they have the potential to produce good children are maintained, outliving their luckless offspring.

2.4.3 Summary

Spatially located populations help, but are not a prerequisite for, the maintanence of diversity in a population. The problems caused by the bottleneck of a central control in large populations are not so exaggerated in smaller populations, so much so that the existence of a central control can be exploited so its global knowledge helps to prevent premature convergence.

Niches and species can be evolved to have similar effects as those in the isolation by distance models, i.e. maintain diversity, track multimodal functions etc, and outperform traditional Genetic Algorithms which have difficulty maintaining a balance between directed evolution and diversity.

2.5 Conclusion

As Genetic Algorithms stem directly from natural methods, it is perhaps unsurprising that there are so many benefits to be derived from copying nature once more. Differing niches and species can be evolved and maintained in a number of ways, ranging from decentralised models as close as possible to nature, to highly controlled methods.

Most importantly, once subpopulations have established their environmental niches, they can be put to many uses. Several solutions can be maintained in the population at a time, a diverse array of individuals and, indeed species, can easily be persuaded to coexist with one another, thus easing the pressure toward premature convergence.

These models, in particular the isolation by distance variety, allow convenient mapping on to parallel architectures, which results in a massive speed up in processing time.

Bibliography

Cohoon, J.P. (1991) A Multi-population Genetic Algorithm for Solving the K-Partition Problem on Hyper-cubes. *Proceedings of the 4th International Conference on Genetic Algorithms.* R.K. Belew, L.B. Booker, Eds. San Mateo, CA : Morgan Kauffmann.

Collins, R. (1992) *Studies in Artificial Life.* Doctoral Dissertation, Computer Science Department, University of Alabama.

DeJong, K. (1975) *An analysis of the behaviour of a class of genetic adaptive systems.* Doctoral Dissertation. University of Michigan.

Eshelman (1991) Preventing premature convergence in Genetic Algorithms by preventing incest, in *Proceedings of the 4th International Conference on Genetic Algortihms*, R. K Belew, L.B. Booker, Eds. San Mateo, CA : Morgan Kaufmann.

Goldberg, D. (1987) Genetic Algorithms with sharing for multimodal function optimization. *Proceedings of the 2nd International Conference on Genetic Algorithms.* San Mateo, CA: Morgan Kauffmann.

Goldberg, D. (1989) Niche and Species formation in Genetic Function Optimization. *Proceedings of the 3rd International Conference on Genetic Algorithms.* J.D. Schaffer, Ed. San Mateo, CA: Morgan Kauffmann.

Muhlenbein, H. (1989) The Parallel Genetic Algorithm as a function optimiser. *Proceedings of the 3rd International Conference on Genetic Algorithms.* J.D. Schaffer, Ed. San Mateo, CA: Morgan Kauffmann.

Ryan, C. (1994a) Racial Harmony in Genetic Algorithms. *Proceedings of Genetic Algorithms within the framework of Evolutionary Computing, KI'94 Workshop.* Germany: Springer. *To appear.*

Ryan, C. (1994b) Pygmies and Civil Servants. *Advances in Genetic Programming.* K. Kinnear Ed. Cambridge MA: MIT Press.

Todd, P. (1991) On the Sympatric Origin of Species: Mercurial Mating in the Quicksilver Model. *Proceedings of the 4th International Conference on Genetic Algorithms.* R.K. Belew, L.B. Booker, Eds. San Mateo, CA: Morgan Kauffmann.

Chapter 3

Steve G. Romaniuk
Department of Information Systems and Computer Science
National University of Singapore
10 Kent Ridge Crescent
Singapore 0511

stever@iscs.nus.sg

Construction of Neural Networks

0-8493-2519-6/95/$0.00 + $.50

Abstract
The ability to automatically construct neural networks is of importance, since it supports reduction in development time and can lead to simpler designs than, for example, traditionally hand-crafted networks. Automation is further required to take the step toward a more autonomous learning system. In this chapter, we outline one possible approach for building an automatic network construction algorithm, which utilizes evolutionary processes to locally train network features using the simple perceptron rule. In the first part of this chapter, emphasis is placed on determining the effectiveness of several types of crossover operators in conjunction with varying the population size and the number of epochs in which individual perceptrons are trained. Under investigation are simple, weighted and blocked crossover operators. The second part outlines how co-evolution of multiple populations can have a beneficial impact on speeding-up the learning process by increasing selective pressure. Both case studies are accompanied by experiments which involve learning several instances of the even-parity function.

3.1 Introduction

3.1.1 What are Neural Networks?

Neural networks are in general biologically inspired information processing systems (Rummelhart and McClelland 1986; Beale 1991; Wasserman 1989; Freeman 1991), which are composed of neurons and connections. Since the human brain is considered to be comprised of millions of neurons and billions of connections, it comes as no surprise that brain researchers believe intelligent behavior emerges from interactions involving huge numbers of neurons and their interconnectivity. An individual neuron is a simple element (unit) capable of some limited operation. Neurons interact with one another via weighted connections, where the magnitude of the weight indicates the strength of a connection. Positive and negative weights act as reinforcing and inhibiting forces, respectively. In various neural models the *activation* of a neuron is calculated as follows:

$$A_i = \sum_{\forall j} W_{i,j} A_j \qquad (3.1)$$

Here, Ai refers to the activation of the ith neuron, and *Wi,j* is the connection weight between the ith and jth neuron. Naturally, due to this connectivity of neurons, we can think of a network being layered and in a feedforward neural network, the activation impulse travels from the input layer, through the hidden layer, until it reaches the final output layer. Figure 3.2 displays a simple 3-layer neural network. The units (neurons) labeled I1,..., I5 represent the inputs, whereas elements labeled H1 and H2 are considered hidden units (on different layers) and element H3 represents an output unit. The output of a unit is calculated, by passing the units' activation value through what is known as a transfer function:

$$O_i = T(A_i) \qquad (3.2)$$

Transfer functions can come in all forms and shapes: from the simple linear function, the widely used sigmoid (logistic) function or the non-linear threshold function.

One of the most important benefits of utilizing neural models is their ability to learn, which has made them an attractive alternative to more classical artificial intelligence learning strategies. Today there is a wide variety of neural algorithms. One of the most popular learning algorithms is backpropagation (22). In order to utilize a learning algorithm such as backpropagation, several choices must be made by the user. These include, but are not limited to the number of hidden layers, the number of hidden units, the choice of transfer function and an assortment of learning parameters such as momentum term and learning rate.

In general, determining an effective architecture for much-layered feedforward networks can be both time-consuming and frustrating. Striving for optimality among potential network architectures in terms of hidden units and connections, is in general of high priority, since deriving minimal configurations can curb total training time as well as the time required to test new patterns (during classification).

3.1.2 Automatic Network Construction and its Problems

To this date several methods have been proposed to automatically construct neural networks and drive for reduction in network complexity, that is determine the appropriate number of hidden units, layers, etc. (1,2,7,8,12,13,17,19,21,20, 23). Many of these approaches utilize rather complicated algorithms and network structures. The simplest of these approaches is the Extentron (2) algorithm which is based on the perceptron rule (18). Unfortunately, the algorithm is ineffective in determining suitable network structures for higher dimensional problems. For example, learning 2- and 3-bit parity requires a single hidden unit. Learning 5-bit parity already requires 5 hidden units, whereas for 6-bit parity the number shoots up to 11 hidden units. Other algorithms such as Upstart (8) or Cascade Correlation (7) require about N hidden units to learn N-bit parity, but these algorithms are also increasingly more complicated.

By closely observing the perceptron learning rule for more difficult parity problems, it was noticed that by presenting the complete training set far fewer examples could be correctly recognized as opposed to removing a few prior to training. In other words, some of the examples interfered with the remaining ones, causing a sort of *confusion* within the perceptron to maximize recognition of all training patterns. This observation gave rise to further investigate the importance of providing the perceptron - at every stage of network construction - with the *right* set of training examples.

3.2 Merging Neural Networks and Genetic Algorithms

3.2.1 Constructing Networks with the Help of Evolutionary Processes

Genetic algorithms are in general simple adaptive search algorithms, which create new solutions to a given problem by exploiting past performance of older solutions in a manner similar to evolutionary processes as found in nature. The solution space from which the individual solutions are drawn is represented in the form of a finite length string of chromosomes. Through use of genetic operators such as crossover and mutation, modifications to the individual chromosomes of a population are introduced in a systematic fashion based on optimizing some criteria. Binary encoding schemes of chromosomes have been popular in past GA use and are based on a fundamental theorem of GAs introduced by Holland (14). According to Holland's schema theory binary coding schemes provide for low-cardinality in obtaining a maximum number of similarities in what is encoded. Due to their simplicity and their past success in finding sub-optimal solutions to NP-hard problems (such as Traveling Salesman, (10,11)), GAs seem ideal candidates for determining the *right* partitions of the training set during the various stages of network construction. Next, we point out how GAs can be utilized to develop effective partitions of the original train set.

Representing individual chromosomes of a population is straightforward. Assume we are given a set of training examples T. The size of each chromosome is then given by |T| and presence/absence of a training example in the current partition is encoded by a binary 1/0, respectively.

The function to optimize is simply the number of examples of T that are correctly recognized by the current perceptron. We can express this more formally as,

$$\Im_{opt} = \max_{\forall C_i \in P}(O_{P_i}(T)) \quad (3.3)$$

where C_i is a chromosome (one particular partition of the original train set) and an element of the total population P. O_{P_i} is the ith perceptron trained on the partition of training examples represented by C_i.

The basic genetic operations made use of by the EGP algorithm can be summarized as follows:

Reproduction: Half of the population is selected for further reproduction. The half includes those chromosomes that have the highest fitness according to the function which is optimized (as stated earlier).

Crossover: This function creates new offspring from the original parents. In this particular implementation the parent chromosomes are grouped into pairs and from each of these pairs two new child chromosomes are created by a process of combination. Consider the following two chromosomes which have been selected for reproduction:

$$\begin{array}{lcllllll} C_i & : & x_1 & x_2 & x_3 & \ldots & x_{n-1} & x_n \\ C_{i+1} & : & y_1 & y_2 & y_3 & \ldots & y_{n-1} & y_n \end{array} \quad (3.4)$$

Here, the x_i, y_i represent individual genes. Now assume we select the crossover point for chromosome C_i as $P_i = x_{n-1}$ and for C_{i+1} as $P_{i+1} = y_2$. Hence, we obtain the succeeding two offspring:

$$\begin{array}{lcllllll} C'_i & : & x_1 & x_2 & x_3 & \ldots & x_{n-1} & x_n \\ C'_{i+1} & : & y_1 & y_2 & y_3 & \ldots & y_{n-1} & y_n \end{array} \quad (3.5)$$

In its most simplest form, crossover occurs at a single point, as opposed to having different crossover points for each chromosome as demonstrated above. In a later section we introduce two crossover operators which utilize 2 point crossover.

Mutation: A mutation factor determines the rate at which a chromosomes' gene value is altered (toggled). Mutations add new information in a random way to the genetic search process, and ultimately help avoid GAs from getting stuck at local optimums. In the herein represented GA model two parameters determine the rate at which chromosome's genomes are modified, that is they determine how many chromosomes and how many genomes are changed during each generation.

3.2.2 Traditional Approach

Before we more formally outline the EGP algorithm, it is worthwhile to briefly look at a more standard technique aimed at optimizing neural networks. Generally, the designer provides an initial network structure, that is the connectivity of the network is predefined as well as the range of values a connection weight can take on. The later restriction can be somewhat relaxed, if a non-binary representation of the connection weights is chosen.

Once the encoding scheme has been determined, the weights of a specific network are encoded and the resulting binary string exemplifies a single member of the population. New offspring are created in the usual manner and their performance is measured by decoding the genotype back into a collection of network weights, and measuring the performance of the resulting network on the train set. The measure takes on the role of the fitness function, which decides how well a single population component is performing. For some work in this area the interested reader is referred to (16,25,4,5,6,15).

One of the problems with this methodology is, it is still necessary for the user to select the network structure and its connectivity. On the other hand, savings are enjoyed by having eliminated some of the learning parameters which are required by backpropagation. Also, the more complicated backpropagation algorithm is replaced by a simpler weight adaptation mechanism. Unfortunately, for larger problems serial implementation of this genetic approach can become prohibitive. These high costs stem mainly from the encoding and decoding phases required after every generation to determine the fitness of a chromosome. Note, that for EGP no such costs are incurred. Finally, exploiting this method of merging fixed sized neural networks trained by a genetic algorithm, stands in contrast to applying genetic algorithms to decide how to appropriately train (select the right training set) a perceptron, without predetermining the network's connectivity (EGP).

3.3 Evolutionary Growth Perceptrons

The network constructed by EGP is similar to the ones generated by Cascade Correlation (7) and Extentron (2) algorithms. Only one hidden unit is created for each layer. Once a unit has been created it receives inputs from all previously generated hidden units, inputs and outputs. The latter is in accordance with Divide & Conquer networks (19) and represents a departure from the previous two approaches. Since outputs are treated just like any other hidden feature, this requires training them one at a time. Once a hidden element has been created, it is trained for a fixed number of epochs on a given partition of the train set. The unit's performance at the end of training is used to measure the chromosome's fitness. For each chromosome within the population a perceptron is trained. After the fitness has been determined the evolutionary operators of crossover and mutation are executed. The above outlined process is continued until all outputs have been trained. Figure 3.1 displays high level pseudo-code describing EGP.

Let us first take a closer look at the `TRAIN_OUTPUT` routine, which is called whenever a new output needs to be trained for a given problem. To start training a new hidden unit is recruited and receives its input connections from all previous input, hidden, and output units in the network (Step 1.1). Next, evolutionary processes are invoked by a call to routine `EVOLUTION` (Step 1.2). Upon return from this routine the performance of the recently installed hidden unit on the training set is measured (Step 1.3). If the unit recognizes all patterns correctly, it is designated as an output unit and the training process for the current process comes to a successful close. If one or more patterns are not correctly recognized by feature CF, a new hidden unit is recruited and the above process repeated (Step 1).

During `EVOLUTION` the succeeding steps are executed: First, the fitness of the current hidden unit (feature) is determined by calling routine `DETERMINE_FITNESS` (Step 1). Next, the members of the population are sorted by decreasing fitness (Step 2). In the last step (Step 4), crossover and mutation are initiated, and the fitness of the population is measured (Step 4.1 and 4.2). Again, we test, if the chromosome with the highest fitness (representing the current feature) recognizes all patterns in the training set (Step 4.5). If so `EVOLUTION` is completed and control returns to routine `TRAIN_OUTPUT`. The same transition is made, if after `MAX_RECESSION` generations no improvement in error reduction is observed. Finally, the last step (4) is repeated for `MAX_GENERATIONS`. The third and final routine that requires discussion, is

`DETERMINE_FITNESS`. Its purpose is to calculate the performance of each of the population members. This is accomplished as follows:

The current feature unit is trained on all patterns encoded by each chromosome of the population for at most `GEN_CYCLE_TIME` epochs using the perceptron learning rule (Step 1). After feature *CF* has been trained for one epoch on the patterns encoded in chromosome *CHi* (Step 1.1), its performance is measured and compared to its earlier performance (previous epoch, Step 1.1.3). If it has improved, the connection weights of feature unit *CF* are saved (Step 1.1.3.1). Furthermore, if all the training patterns encoded in chromosome *CHi* have been learned, further training is interrupted (Step 1.1.3.2). Otherwise, a new iteration is initiated. After either `GEN_CYCLE_TIME` epochs have passed, or all the training patterns encoded by chromosome *CHi* have been learned, the best feature weights are restored (Step 1.2) and assigned to element *CF*. Finally, the unit is tested on the training patterns and its performance is recorded as its fitness (Step 1.3).

Nomenclature for EGP:
T: set of training examples.
P: set of population.
CF: feature cell.
CH_i: ith chromosome of population P.
$A(C_F, T)$: number of examples in T correctly classified by C_F.
MAX_GENERATIONS: maximum number of generations allowed.
MAX_RECESSION: number of generations for which there is no improvement, end evolution.
GEN_CYCLE_TIME: maximum number of epochs perceptron is trained.

TRAIN_OUTPUT():
(1) While *T RUE* Do
(1.1) Create Feature Cell *CF*
(1.2) *EVOLUTION(CF)*
(1.3) If *A(CF, T)*= $|T|$ Then
(1.3.1) Exit
(2) Halt

EVOLUTION (CF)
(1) *DETERMINE_FITNESS(CF)*
(2) Sort Population P by decreasing fitness (3) *RecessionCnt = 0*
(4) For *MAX_GENERATIONS* Do
(4.1) Perform Cross-Over on Population P
(4.2) If *TimeToMutate* Then
(4.2.1) Perform Mutation of Population P
(4.3) *DETERMINE_FITNESS(CF)*
(4.4) Sort Population P by decreasing fitness
(4.5) If *A(CF, T)* = |*T*| *V RecessionCnt* = *MAX_RECESSION* Then
(4.5.1) Exit
(5) Return

DETERMINE_FITNESS (CF):
(1) For all Chromosomes *CHi* Do
(1.1) For *GEN_CYCLE_TIME* Do
(1.1.1) Train Feature Cell *CF* using Perceptron Rule for one epoch
(1.1.2) Measure Performance of *CF*, Find *A(CF, CHi)*
(1.1.3) If *A(CF, CHi)* = *maxvk,k<i(A(CF, CHk))* Then
(1.1.3.1) Save Feature *CF* for later use
(1.1.3.2) If *A(CF, CH,)*= |*CHi*| Then
(1.1.3.2.1) Exit
(1.2) Restore saved Feature Cell *CF*
(1.3) Measure Performance of *CF*, Find *A(CF, T)*
(1.4) *A(CF, CH,)= A(CF, T)*
(2) Return

Figure 3.1: High Level Description of EGP Algorithm

Figure 3.2 displays a representative network constructed by EGP for the 5-input even-parity function. Initially, the hidden unit H_1 is recruited by EGP and connections are made from all input units. After training (perform EVOLUTION) there are still some examples which are not correctly recognized by feature H_1, hence a new hidden unit H_2 is installed. This element receives its input connections from the previous hidden unit H_1 and all 5 inputs I_1,... I_5. Again the element is trained using the perceptron rule. After training there is still some residual error left and a final hidden element H_3 is added, receiving its inputs from the previous 2 hidden units and the 5 input units. This time training succeeds, that is, all training patterns are correctly recognized by feature H_3 and it is installed as an output unit. The parity-5 problem has been learned. Table 3.1 contains the training patterns for parity-3.

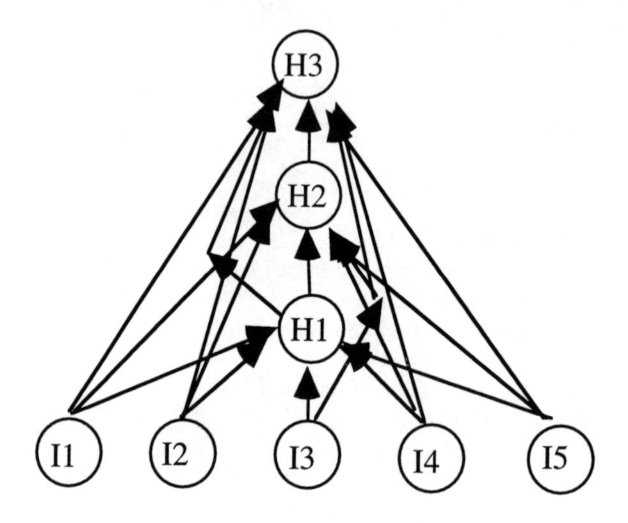

Figure 3.2: Parity-5 Function as learned by EGP

I_1	I_2	I_3	Output
0	0	0	0
0	0	1	1
0	1	0	1
0	1	1	0
1	0	0	1
1	0	1	0
1	1	0	0
1	1	1	1

Table 3.1. Parity-3 function training patterns

3.4 Types of Crossover Operators

The purpose of this section is to outline 3 approaches to constructing crossover operators which can be invoked by EGP.

3.4.1 Simple Operator

This widely applied operator makes use of a single crossover point which is selected by a random process. In other words, a pseudo-random number generator determines a sole crossover point at which the exchange of genetic information occurs between two parent chromosomes.

3.4.2 Weighted Operator

Here the crossover point for each of the parents is chosen as a function of their relative fitness. The more fit of the parent chromosomes is, the more of its original genetic code is passed on to its children. Crossover always generates two new offspring. The ordering of genetic information is preserved during the crossover process. Crossover points between two parent chromosomes C_i and C_{i+1} are determined as follows:

$$CP_{Ci} = \frac{|T|*(|T|-F(C_i))}{2*|T|-(F(C_i)+F(C_{i+1}))} \quad (3.7)$$

Here, CP_{Ci} represents the crossover point for chromosome C_i (calculated analogously for C_{i+1}). $F(C_i)$ indicates the fitness of chromosome C_i. The following may serve as an example of weighted crossover.

Assume the subsequent quantities are provided: $|T| = 4$, $F(C_i) = 1$ and $F(C_{i+1}) = 3$. We can then compute,

$$\begin{aligned} CP_{Ci} &= \frac{4(4-1)}{2*4-(1+3)} = 3 \\ CP_{Ci+1} &= \frac{4(4-3)}{2*4-(1+3)} = 1 \end{aligned} \quad (3.8)$$

Hence, given the succeeding two parent chromosomes,

$$\begin{array}{llllll} C_i & : & x1 & x2 & x3 & x4 \\ C_{i+1} & : & y1 & y2 & y3 & y4 \end{array} \quad (3.9)$$

and the earlier derived crossover points, the offspring is given as:

$$\begin{array}{llllll} C'_i & : & x1 & x2 & x3 & x4 \\ C'_{i+1} & : & y1 & x2 & x3 & x4 \end{array} \quad (3.10)$$

3.4.3 Blocked Operator

This approach attempts to create barriers between chunks of genes within a chromosome. These barriers act to prevent accidentally breaking up potential co-occurrences of genes that may have formed during the evolutionary process. By assigning a blocking factor to every location of a gene within a chromosome, a simple decision matrix (outlined in Table 3.1) is used to decide whether a randomly selected crossover point for each of the parent chromosomes is a suitable choice. In Table 3.2 the column labeled P_i represents the randomly chosen crossover point for chromosome C_i. The < and ≥ entries indicate whether the blocking factor associated with crossover point P_i is below a blocking boundary or not. The final column contains a decision indicating whether a split point P_i can serve as a crossover point for chromosome C_i or not. For example, if the left and right neighbors of P_i (P_{i-1} and P_{i+1}) and P_i itself have blocking factors which are equal to or above the blocking boundary, a split does not occur (Last entry of Table 3.2). In this case a new random split point is selected and the test repeated. If the test succeeds, P_i is chosen as a crossover point for chromosome C_i. Blocking factors for individual genes of a chromosome are maintained as simple counters. Whenever 2 parents mate, the blocking factor of a gene is incremented, if its current value (with respect to a genes position in the chromosome) is preserved in the offspring after crossover has occurred, it is reset to 0.

P_{i-1}	P_i	P_{i+1}	Split
<	<	<	T
<	<	≥	T
<	≥	<	T
<	≥	≥	F
≥	<	<	T
≥	<	≥	T
≥	≥	<	F
≥	≥	≥	F

Table 3.2. Decision Matrix for crossover point P_i's selection

3.5 Empirical Results

In the succeeding experiments we consider 2 values for the blocking boundary of the blocking method: blocking bounds 4 (B-4) and 10 (B-I0). EGPs performance is evaluated on the well known N-parity function (requires output to be set such that the total number of I's for inputs and outputs is the same; even parity), based on 3 different methods to construct crossover operators. The parity functions considered are 2- through 7-parity.

In the experiments detailed in this section, 10 trial runs were executed for learning each of the 6 parity functions. Figures 3.3 through 3.6 display the performance behavior for the different crossover operators. Results are provided that correlate the number of generations versus the number of hidden units generated. Here, the label S refers to the simple operator, W to the weighted operator and B-n to the blocked operator with n being the blocking factor. The first four graphs are for the case MR=20, were MR (Max Recession) indicates after how many unsuccessful generations (no improvement in error reduction) the evolutionary process is aborted.

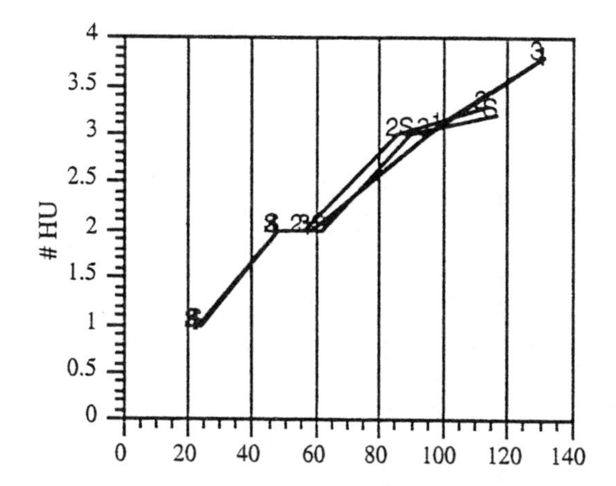

#HU for MR=20, PS=8, E=30 (Figure 3.3a)

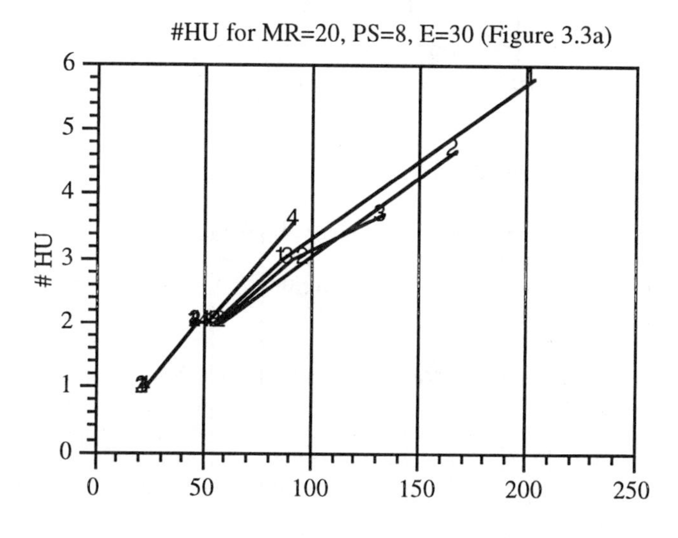

#HU for MR=20, PS=8, E=130 (Figure 3.3b)

For a population size PS of 8 and a maximum number of epochs E of 30 (Figure 3.3a), the weighted and blocked method (with a low blocking factor of 4) provide the best results in terms of reducing the number of hidden units and the number of generations required to derive an acceptable solution. As the number of epochs is increased to E=130 (Figure 3.3b), the weighted method again displays superior results, but the total number of generations required for the cases E=30 and E=130 are about equal. This, even though the training time for every individual perceptron is increased by more than a factor of 4! Incidentally, the same result holds for the other crossover operators. In fact, the B-4 operator yields a higher generation count for E=130 than for E=30. Also, the number of hidden units remains about the same, or is even slightly higher for E=130. This indicates that for small population size, increasing the number of epochs may not translate into a decrease in the final network structure or the time required to create that structure.

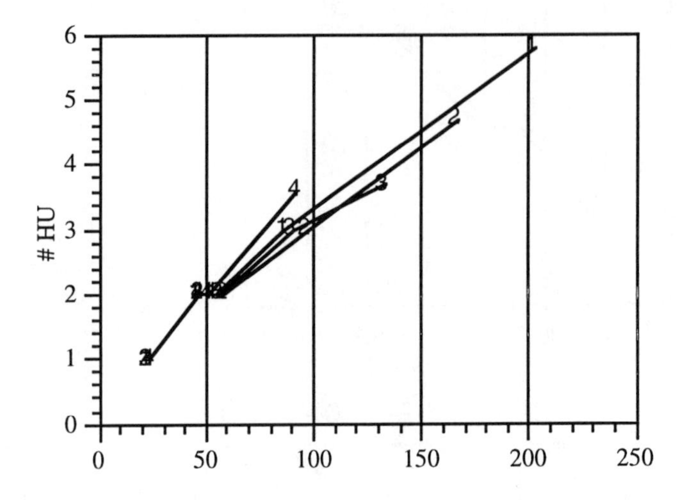

#HU for MR=20, PS=64, E=30 (Figure 3.4a)

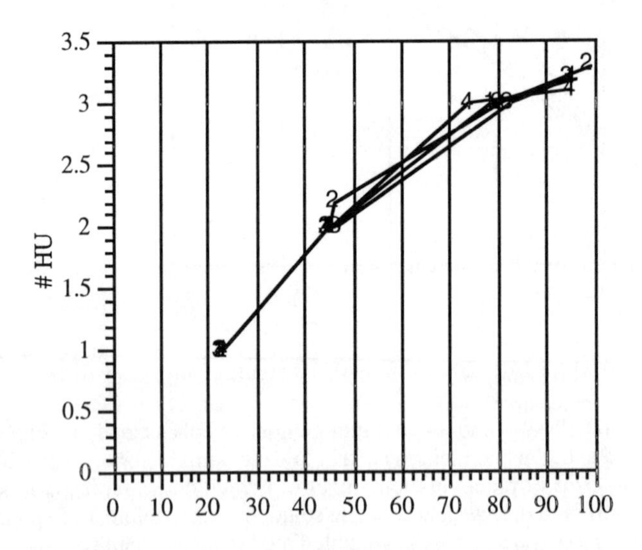

#HU for MR=20, PS=64, E=130 (Figure 3.4b)

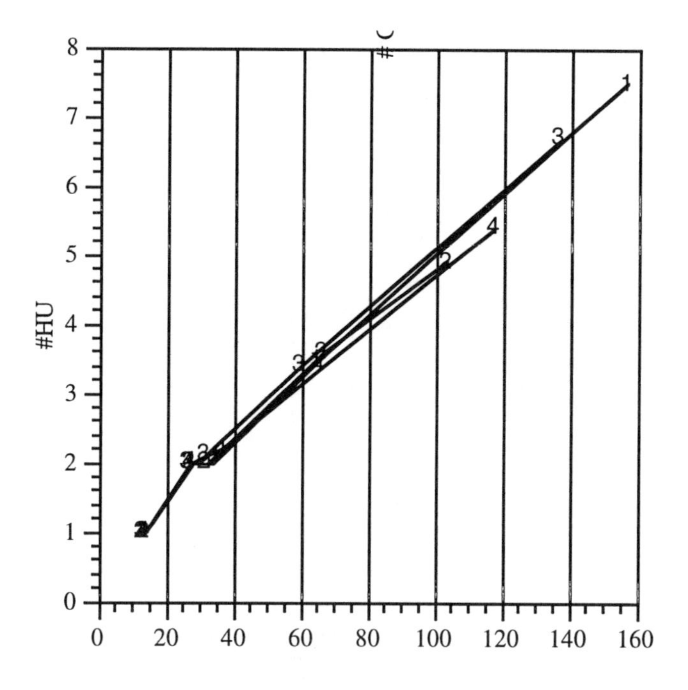

#HU for MR=10, PS=8, E=30 (Figure 3.5a)

As we increase population size to PS=64 (Figure 3.4) previous findings are completely reversed. First, for all crossover operators, except the simple operator, the total number of generations is significantly reduced from about 120 generations (E=30) to less than 100 generations. Again, similar to the previous case, increasing the parameter E has neither an effect on the final network structure nor the number of generations. Additionally, the difference in results between blocked and weighted crossover has diminished even further. This emphasizes that population size, if increased sufficiently, can offset differences in crossover operators (with the exception of the simple operator). On the other hand, for small population size the choice of crossover operator plays a more dominant role than, for example, the number of epochs employed to train a perceptron. Finally, increasing PS also yields simpler network structures. For 6-parity and 7-parity the average number of hidden units drops from about 3.5 to 3.

For the second set of experiments we let MR=10. We now observe for PS=8 (Figure 3.5) a substantial change in the performance of various crossover operators. This time the difference between blocked B-4 and weighted operators is even more evident. Even though the number of generations is about the same as for the case MR=20, we notice an apparent increase in the number of hidden units, regardless of the operator used. The hidden unit count jumps to about 7 for 7-parity (B-10 and simple operator). Now, as the number of epochs is increased, we detect that the performance of the various crossover operators is almost identical. As a matter of fact, the B-4 and weighted operators actually experience a deterioration in their performance. From the combined results of the previous experiments we can conclude, that increasing the number of epochs has little or even detrimental effects on EGPs performance. Here performance entails both the number of generations and number of hidden units required to learn N-parity.

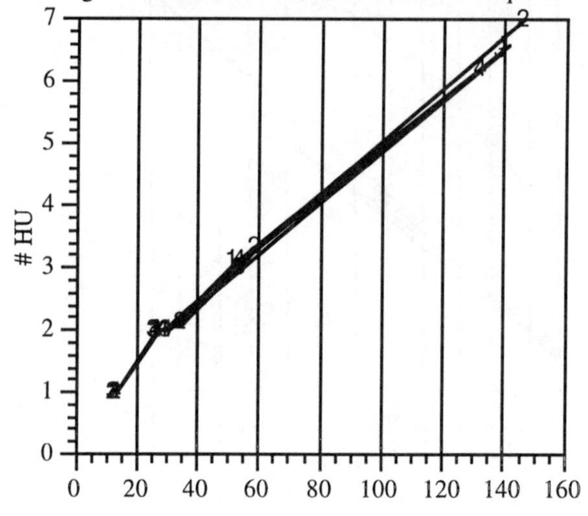

#HU for MR=10, PS=8, E=130 (Figure 3.5b)

By increasing the population size to 64 (Figure 3.6) we again note a substantial decrease in the number of generations (over 50%) for all operators compared to the case PS=8. Also, similar to MR=20 there is no apparent difference in performance of crossover operators. In every instance the final network architecture for 7-parity is around 3 hidden units. No apparent change in results is observed as E is increased.

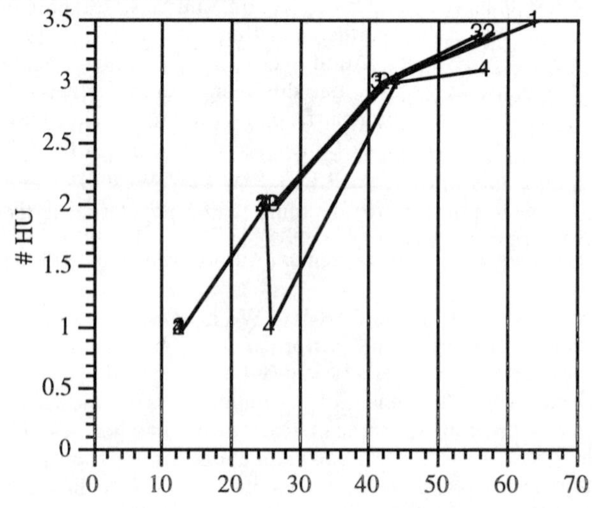

#HU for MR=10, PS=64, E=30 (Figure 3.6a)

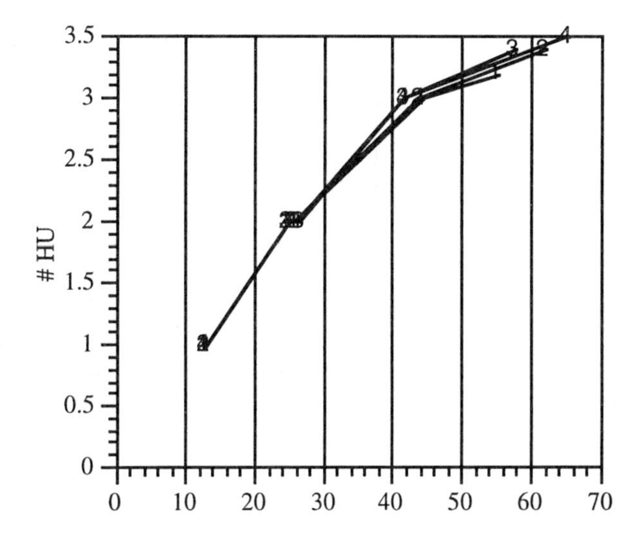

#HU for MR=10, PS=64, E=130 (Figure 3.6b)

We summarize our findings as follows: Clearly, EGP strongly benefits from the size of its population. The choice of crossover operator only plays a role, if population size is relatively small. On the other hand, increasing the number of epochs for training a perceptron has no positive effects.

3.6 Co-Evolution of Populations

3.6.1 Overview

So far we have looked at applying a single population of chromosomes to the task of optimizing the neural network construction process. By allowing only the best population elements (those with highest fitness measure) to reproduce, selective pressure is thus asserted. Besides allowing rank based allocation for reproduction, we can inject additional pressure by introducing competition between 2 or more populations. Co-evolution involves several populations competing for resources in a constrained environment. This competition for resources not only helps to improve the quality of individual population elements, but may also be responsible for deciding which population is more suitable whenever the environment undergoes a change.

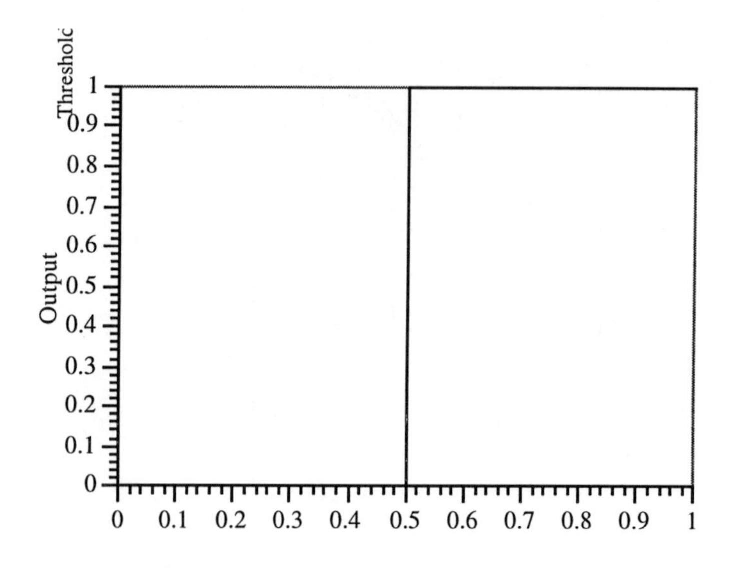

Figure 3.7a

During the individual phases of network construction the actual *environment* changes whenever a new hidden unit is added to the current network structure. In other words, the original problem is recast by the change of patterns which are recognized by the so far constructed network. It is quite easy to imagine that to solve these recasted problems can be more readily achieved with one method than another. For example, the choice of local learning rule can be of importance depending on the type of problem under consideration. Another candidate is the type of transfer function utilized by hidden units.

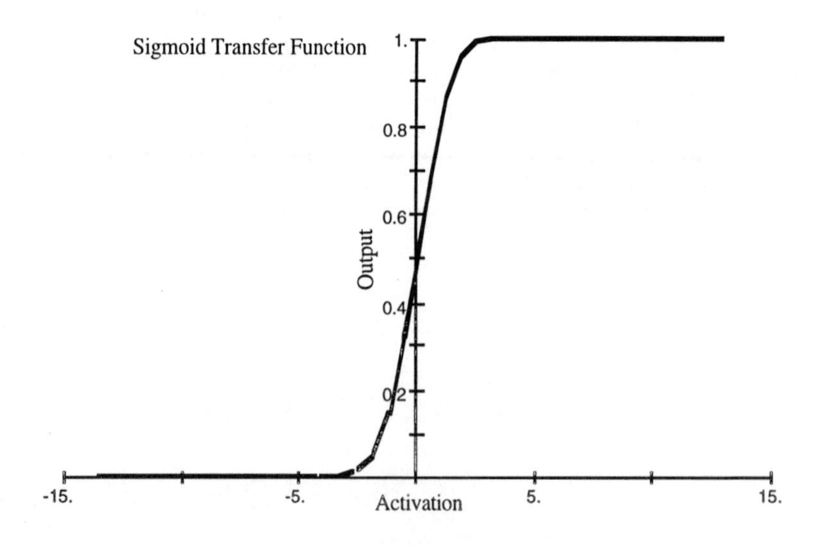

Figure 3.7b

In this section we attempt to determine the overall affect different transfer functions can have on training time reduction. To accomplish this task, we investigate co-evolution between 2 populations of chromosomes. Of course, the results can be generalized by incorporating more than just 2 populations. For this particular study we chose to apply both threshold as well as sigmoid units. Figure 3.7 displays the activation behavior for the 2 transfer functions.

Co-evolution simply requires maintaining 2 populations. One population consists of perceptrons that have been trained using threshold transfer functions, whereas the other incorporates sigmoid functions. Both populations are of fixed size and are allowed to compete separately from one another for a pre-determined amount of time. Co-evolution occurs at a fixed frequency, which we denote by `CO_EVOLVE`. After every `CO_EVOLVE` generations has passed, both populations undergo competition, that is their members fight for the right to stay alive and proceed to the subsequent generation.

The decision which members survive from the pool of 2 populations is made simply by comparing the member's fitness in terms of overall error reduction. In case of a threshold unit, error is measured as being either 1 or 0, since this function only returns these 2 possible values. The sigmoid function on the other hand, can provide a graded response from the interval [0, 1]. In this case error is measured as the square difference between actual and desired output of a network unit.

Initially, every population has *PS*/2 elements. After co-evolution occurs, exactly *PS* members are allowed to survive. Hence, one population may consist of more than *PS*/2 elements, if its members are fitter than those of the other population. In order to avoid one population from becoming extinct, the 2 best (highest fitness) population members are always allowed to survive and reproduce during future generations, until both populations again engage in a new round of competition. Finally, when the evolutionary process comes to a halt, the best member is picked from the pool provided by both populations.

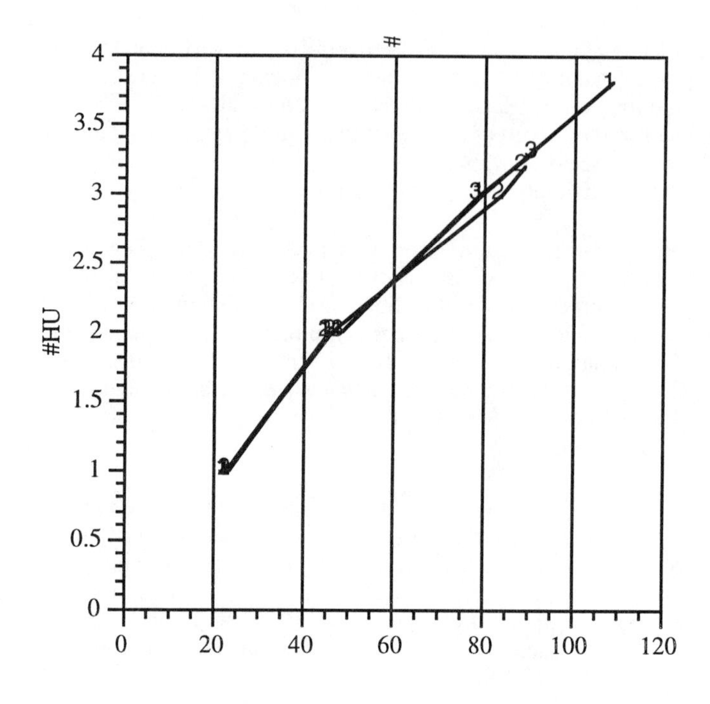

#HU for MR=20, PS=8, E=30 (Figure 3.8a)

3.6.2 Experiments

For the following set of experiments we chose to fix the maximum number of epochs an individual perceptron is trained at E = 30 epochs and utilize the weighted crossover operator only. In the following four figures we report 3 graphs, each obtained for a different value of CO_EVOLVE *(CE).* Figure 3.8a displays the performance results for the case *MR* = 20. Performance for cases *CE* = 10 and *CE* = 20 are about the same, whereas for *CE* = 5 both an increase in number of generations as well as hidden units is observed.

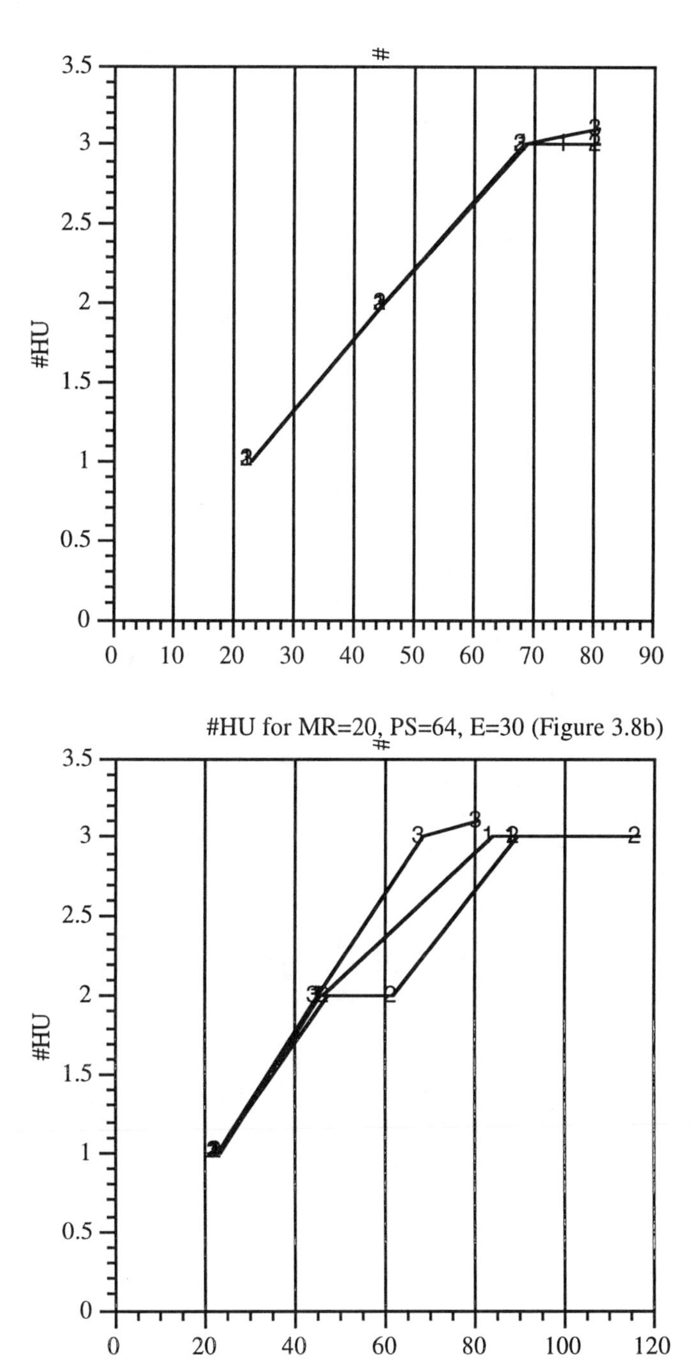

#HU for MR=20, PS=64, E=30 (Figure 3.8b)

#HU for MR=20, PS=8, E=30 (Figure 3.9a)

As the population size is increased to 64 (Figure 3.8b), performance is very much the same, regardless of the frequency at which competition between the 2 populations occurs. For a low value of *CE* = 5 the total number of generations is slightly reduced when compared with the other 2 cases. Similar to earlier findings, increasing population size can help to smooth out differences due to various selection criterias. In Figure 3.9 we directly compare the weighted crossover operator for *PS* = 8 with and without co-evolution. This time we observe a clear drop in the number of generations from about 120 to less than 90 generations (*CE* = 10) after applying co-evolution. The results are mirrored even as the population size is increased to 64. Figure 3.9b points out this difference. This time we observe a drop from about 95 to 75 generations (*CE* = 5).

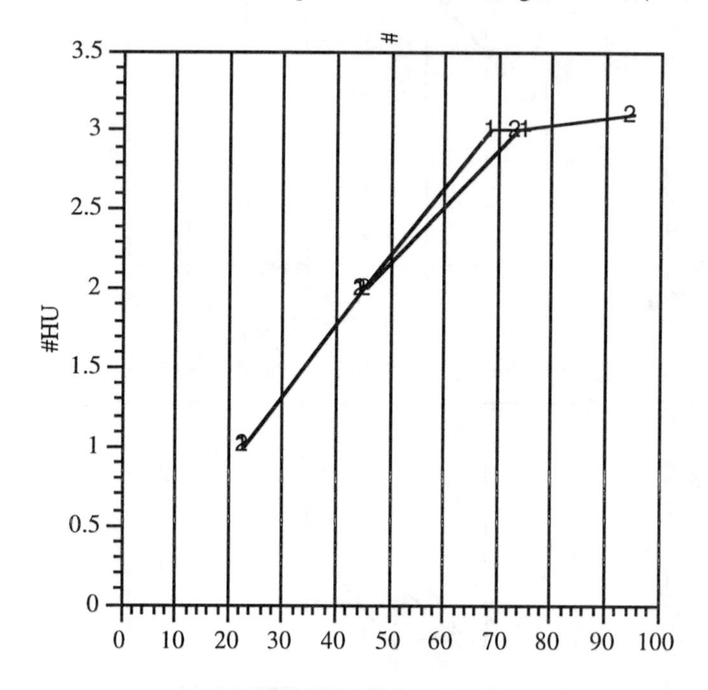

#HU for MR=20, PS=64, E=30 (Figure 3.9b)

In the final experiment we set MR = 10. Figure 3.10a displays the results for *PS* = 8. We notice clear differences in performance for various frequencies *CE*. Superior results are obtained for *CE* = 20, followed by *CE* = 10 and trailed by *CE* = 5. This seems to indicate that less competition is more beneficial in reducing overall training time, hence high values of *CE* provide best results. As we increase the population size to *PS* = 64 (Figure 3.10b) we again observe very similar behavior. For all 3 frequencies *CE* the total number of generations is less than 50 generations. This represents a clear improvement for the case *CE* = 5. From Figure 3.11a we can glean that for the case *PS* = 8 the weighted crossover operator improves from around 120 generations to just above 50 generations *(CE*

= 20). This significant drop in training time is also observed for the case *PS* = 64 (Figure 3.11b).

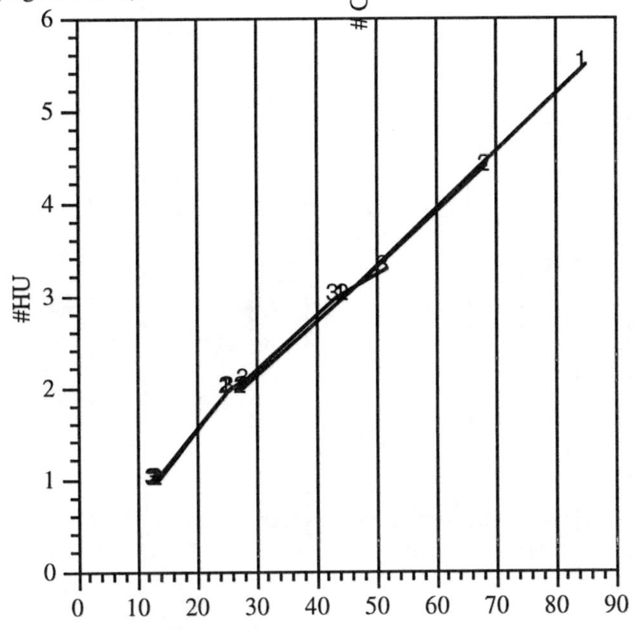

#HU for MR=10, PS=8, E=30 (Figure 3.10a)

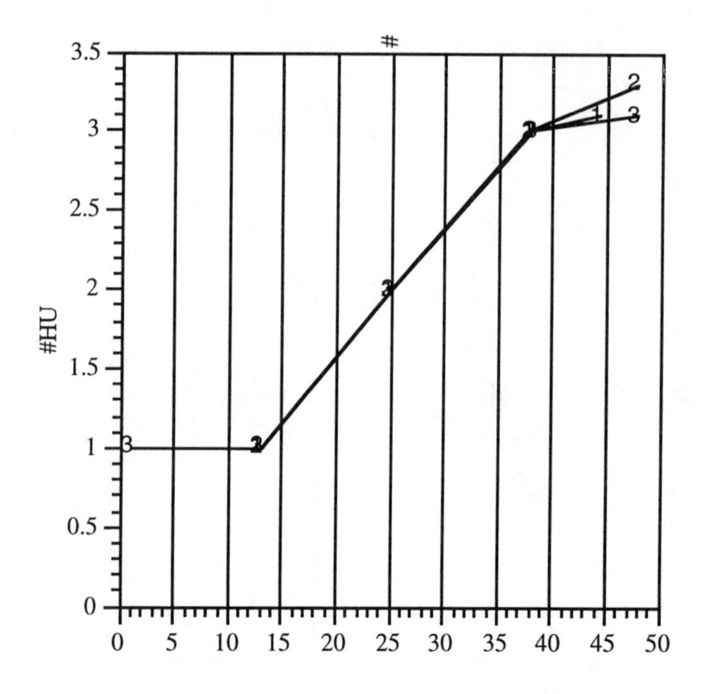

#HU for MR=10, PS=64, E=30 (Figure 3.10b)

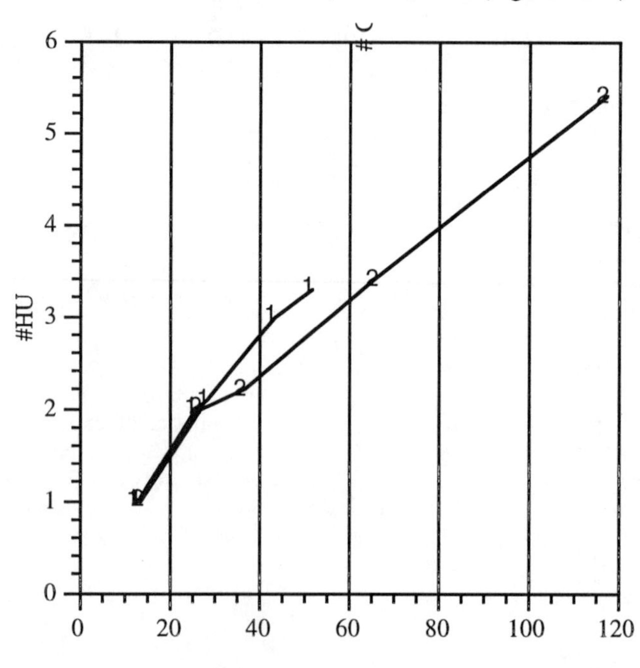

#HU for MR=10, PS=8, E=30 (Figure 3.11a)

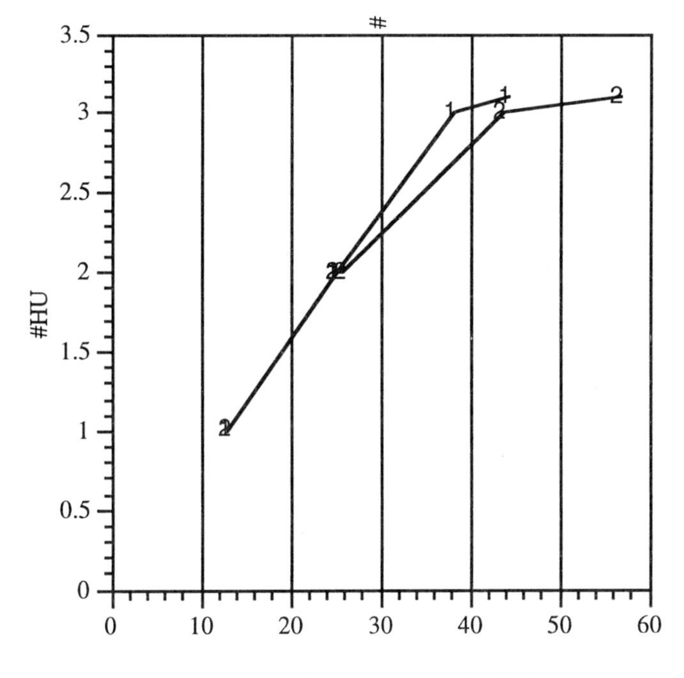

#HU for MR=10, PS=64, E=30 (Figure 3.11b)

As the results clearly point out, involving co-evolution can result in substantial improvement in reduction of total training time. This performance enhancement is more apparent for small populations than larger ones, again indicating that insufficiencies resulting from overly large training times and networks can be overcome by simply increasing population size.

3.7 Summary

Viewing units within a network solely as features and learning as a process of bottom-up feature construction were deemed necessary notions to develop a feasible implementation of an automatic network construction algorithm. Basing local feature training on the simple perceptron rule, combining evolutionary methods to effectively create training partitions, and pointing out the relative advantages of various crossover operators, have substantially contributed toward the development and understanding of EGP. One of the main observations made is that differences in crossover operators and number of epochs fade in the presence of larger population sizes with the exception of the simple operator. Additionally, increasing the number of epochs has hardly any positive effect in reducing training time or simplifying the final network structure. Finally, it was pointed out how co-evolution can help decrease training time by exerting additional selective pressure and utilizing different transfer functions for hidden and output units. These improvements are predominantly observed for small population sizes, but can also be observed to a lesser degree for larger population sizes. Empirical results further underline that applying genetic algorithms as implemented by EGP can result in simpler network architectures and help reduce training time, compared to traditional algorithms and other hybrid systems that integrate genetic algorithms and neural networks.

References

[1] Ash, T. (1989). Dynamic Node Creation in Backpropagation Networks. (Tech. Report ICS Report 8901), Inst. for Cognitive Science, University of California, San Diego, La Jolla, Ca.

[2] Baffes, P.T. and Zelle, J.M (1992). Growing Layers of Perceptrons: Introducing the Extentron Algorithm, Proceedings of the 1992 International Joint Conference on Neural Networks (pp. II-392- II-397), Baltimore, MD, June.

[3] Beale, R. and Jackson, T. (1991). Neural Computing: An Introduction, IOP Publishing Ltd., N.Y., N.Y.

[4] Belew, R.K., McInerney, J., and Schraudolph, N.N. (1991). Evolving Network: using the genetic algorithm with connectionist learning, In C.G. Langton, C. Taylor, J.D. Farmer and S. Rasmussen (Eds.) Artificial Life 11, Redwood City, CA:Addison-Wesley.

[5] Bellgard, M.I. and Tsang, C.P. (1991). Some Experiments on the Use of Genetic Algorithms in a Boltzmann Machine. International Joint Conference on Neural Networks, pp. 2645-2652, Singapore.

[6] de Garis, H. (1991). GenNETS: Genetically Programmed Neural Nets. International Joint Conference on Neural Networks, pp. 1391-1396, Singapore.

[7] Fahlman, S.E. and Lebiere, C. (1990). The Cascade-Correlation Learning Architecture, In D. Touretzky (Ed.), Advances in Neural Information Processing Systems 2 (pp. 524-532). San Mateo, CA.: Morgan Kaufmann.

[8] Frean, M. (1991). The Upstart Algorithm: A Method for Constructing and Training FeedForward Neural Networks, Neural Computation, 2, 198-209.

[9] Freeman, J.A. and Skapura, D.M. (1991). Neural Networks, Algorithms, Applications and Programming Techniques. Reading, MA.: Addison-Wesley.

[10] Grefenstette, J.J., Gopal, R., Rosamira, B., Gucht, D.V. (1985). Genetic Algorithms for the traveling salesman problem. Proceedings of International Conference on Genetic Algorithms and their Application, pp. 160-165, Lawrence Erlbaum & Associates, New Jersey.

[11] Grefenstette, J.J. (1987). Incorporating problem specific knowledge into genetic algorithms. In Genetic Algorithms and Simulated Annealing, Eds. L. Davis, Morgan Kaufmann, Los Altos, CA.

[12] Hall, L.O. and Romaniuk, S.G. (1990). A Hybrid Connectionist, Symbolic Learning System, AAAI-90, Boston, Ma.

[13] Hirose, Y., Koichi, Y., Hijiya, S. (1991). Back-Propagation Algorithm Which Varies the Number of Hidden Units. Neural Networks, 4, pp. 61-66.

[14] Holland, J.D. (1975). Adaption in Natural and Artificial Systems. University of Michigan Press, Ann Arbor, MI.

[15] Marti, L. (1992). Genetically Generated Neural Networks II: Search for an Optimal Representation, IJCNN'92, Vol. II, Baltimore, June.

[16] Miller, G.F., Todd, P.M., and Hedge, S.U. (1989). Designing neural networks using genetic algorithms, In J.D. Schaffer (Ed.) 3rd ICGA, San Mateo, CA:Morgan Kaufmann.

[17] Martinez, T.R. and Campbell, D.M. (1991). A self-adjusting Dynamic Logic Module, Journal of Parallel and Distributed Processing, 11, 303-313.

[18] Minsky, M.L., and Paperr, S.A. (1988). Perceptrons, Expanded Edition, MIT Press.

[19] Romaniuk, S.G., and Hall, L.O. (1993). Divide and Conquer Networks. To appear Neural Networks.

[20] Romaniuk, S.G., and Hall, L.O. (1993). SC-net: A Hybrid Connectionist, Symbolic System. Information Sciences, 71, 223-268.

[21] Romaniuk, S.G. (1993). Evolutionary Growth Perceptrons. In S. Forrest (Ed.) Genetic Algorithms: Proceedings of the 5th International Conference, Morgan Kaufmann.

[22] Rummelhart, D.E., McClelland, B. L., (Eds.) (1986). Parallel Disributed Processing: Exploration in the Microstructure of Cognition, Vol I, Cambridge, Ma. MIT Press.

[23] Sanger, T.D. (1991). A Tree-structured Adaptive Network for Function Approximation in High Dimensional Spaces. IEEE Transactions on Neural Networks,V. 2, No. 2, pp. 285-293.

[24] Smotroff, I.G., Friedman, D.H., and Connolly, D. (1991). Self Organizing Modular Neural Networks. *International Joint Conference on Neural Networks,* (II-187-II-192), Seattle, Wa.

[25] Srinivas, M., and Patnaik, L.M. (1991). Learning Neural Network Weights using Genetic Algorithms - Improving Performance by Search-Space Reduction *International Joint Conference on Neural Networks,* pp. 2331-2336, Singapore.

[26] Wasserman, P.D. (1989). Neural Computing, Theory and Practice, N.Y., N.Y.: Van Nostrand Rheinhold.

Chapter 4

Marc Andrew Pawlowsky
c/o 22/307/844/TOR
IBM Canada Limited
844 Don Mills Rd.
North York, Ontario
Canada M3C 1V7

marcap@vnet.IBM.COM

Crossover Operators

0-8493-2519-6/95/$0.00 + $.50

The crossover mechanism is the primary means of search in Genetic Algorithms (GA's). In this chapter we will be exploring with working code several crossover operators.

GA's can be viewed as trying to maximize a function, called the evaluation function, by evaluating several solution vectors the crossover's purpose is to create new solution vectors by combining other solution vectors that have shown to be good temporary solutions. In GA's we call the solution vectors chromosomes. The vectors that are used as the source for the vectors are called parents, and the new vectors are called children. The components of the solution vectors are called genes, and are referred to as genetic material.

Holland [2] has shown that GA's perform best when the solution vectors are binary. Therefore this chapter will give code that works with binary vectors, though all the code presented here with the exception of variable crossover will work with higher-order alphabets. Crossover may produce one or two children. The techniques for producing one child is conceptually the same as producing two children and discarding one. The source code given here produces two children.

4.1 Source Code

We will be examing C++ source code for performing crossover in a bottom up manner. All the code given has been successfully compiled and executed on a RS/6000 using the C SET ++ compiler version 2.1. Rather than just present the code and have the user use it as a library, the code will be explained in great detail, with the ideas behind it also explained. The explanations are long on C++ mechanisms, and short on GA theory. It is hoped that those that need to examine the theory will use the references mentioned, and implementors of the code will have a good enough understanding on why things were implemented the way they were, and are able to modify the code to suit there own needs.

Generous use is made of the **alloca** operator, to create arrays on the stack. On some platforms such as real mode MS-DOS, which have a limited sized stack, the **alloca** operator should be replaced with new operations, and delete statements should be added to the end of the functions.

While C SET ++ supports exception handling and nested classes, other popular C++ compilers do not, therefore the code included uses neither of these language features. Experienced C++ programmers may want to change the assertion statements to statements that use exception handling, for a more user friendly product.

The code as given here has many assertion statements that are used to verify that the parameters are correct. The assertion statements have two purposes: to reduce debugging time by catching errors, and to provide documentation on the expected values. When code development has been completed and speed becomes an issue the user of the code should compile the code with NDEBUG as documented in the proposed ANSI C++ standard.

4.2 Random Numbers

There are many arguments in the GA community about how important the properties of the psuedo-random number generator used are. I leave it to the reader to experiment if their application is affected by different generators, and use the standard system supplied generators.

Two common operations in GA's are: to select **m** members of a set of length **n**, and to find a random permutation. We will be discussing two algorithms that can be used to solve both these problems.

Functions **random_n_fast** and **random_n_small** both have the same parameters, and can be used interchangeably. The functions return in the argument **chosen**, **nb** integers chosen from the range [0, **max**-1]. For finding a permutation the user should set **nb** to **max**.

Of the two functions **random_n_fast** is the more interesting of the two. The idea behind the algorithm is to set up an array of all the eligible numbers to be chosen (lines 170-176 See appendix C). Then for each number that is to be chosen a random element of the array is chosen, and remembered. The element is then removed from the array with the array being compacted. This is equivalent to spreading a deck of cards out on a table and choosing one card at a time.

As the experienced programmer will quickly realize, the step of compacting the array would be incredibly expensive. So what is done instead is to move the last unchosen element in the array to the position of the number that has been chosen (lines 181-182), and consider only the reduced array size (line 180).

As the name implies, this algorithm tends to be very fast, especially when the number of elements to be chosen is close to the range of the elements to choose from. The disadvantage of this algorithm is the array of the numbers to choose from must be created and initialized. Therefore when **nb** is much less than **max** the overhead might overtake the savings in time. Also for some platforms the amount of space taken for the array might be too large.

random_choose_n_small is a more straightforward implementation of choosing members from a set. The algorithm simply consists of repeatedly choosing a random number in the range (line 221), and checking to see if it has been previously chosen (line 223). If it has been chosen, then another number is picked. The obvious deficency of this algorithm is that the odds of picking an unchosen number decreases as the numbers previously chosen increases.

The source code uses the inline function **random_choose_n** which chooses between **random_choose_n_fast** and **random_choose_n_small** depending on the number of elements to be chosen, and the size of the set to choose from. The function uses a simple heuristic and should be evaluated more carefully depending on the host platform, and the random number generator used.

4.3 Array

class Array is an implementation of a delayed copied array. Its purpose is that instead of keeping multiple copies of equivalent data, only one copy is kept, and all references refer to it.[7] When the array is modified, then if needed, a copy is made before the modification is made.

Readers who do not use C++, may want to skip the remainder of this section, and think of an **Array** as simply an array, whose contents are copied when passed as a parameter, or assigned to a variable. Access to the array is through the functions **get** and **put** which retrieve and set elements of the array respectively.

The array is implemented in two parts. First, **class Array_Storage** which contains a pointer to the data being stored (**contents**), the size of the array (**size**) and the number of references to the array (**nb_references**).

Array_Storage is not meant to be used by any other class other than by the member functions of **class Array**, and therefore has all its member functions protected, and has **class Array** as a friend.

What is interesting to note is that **Array_Storage** does not have a copy constructor, or a valid assignment operator, since all references to the data are supposed to be through an instance of **class Array**.

class Array provides the public Interface to the data stored, and consists simply of a pointer to the storage (**store**, defined on line 421). When an array is created, the space for it is allocated through **Array_Storage**, and the number of references is set to 1. The copy operator (lines 330-337) simply sets the pointer to the same data, and increments the reference counter. When a reference to an array is no longer used the destructor array (lines 340-350) is called. The destructor function, reduces the number of references and determines if there are any more references to the array. If there are none, then the storage is freed.

The assignment operator (lines 353-374) is a combination of destruction and copying. Since the old reference will no longer be used, the reference count to the data is decremented and freed (lines 359-364). The new reference is then set, and the reference count incremented (lines 357-371).

Access of the elements of the array is controlled through the member functions **put** and **get**, which check to see that the index in the array to ensure that it is valid (lines 377-406).

get simply returns a constant copy of the element, to ensure that the element can not be modified, without notifying the array through the use of the function **put**.

put is responsible for ensuring that if there are multiple references to the same data, the modification of the data only affects one reference. This is done by checking how many references there are to the data (line 397).

---

7 A reference is either a pointer to the array or a variable containing the array.

If the reference count is more than one, meaning there is a reference other than the current one, the contents of the entire array are copied to new storage which is used by the current reference (lines 400-401), and the reference count for the old storage is decremented (line 402). The element to be stored is then entered into the array (Figure 4.1).

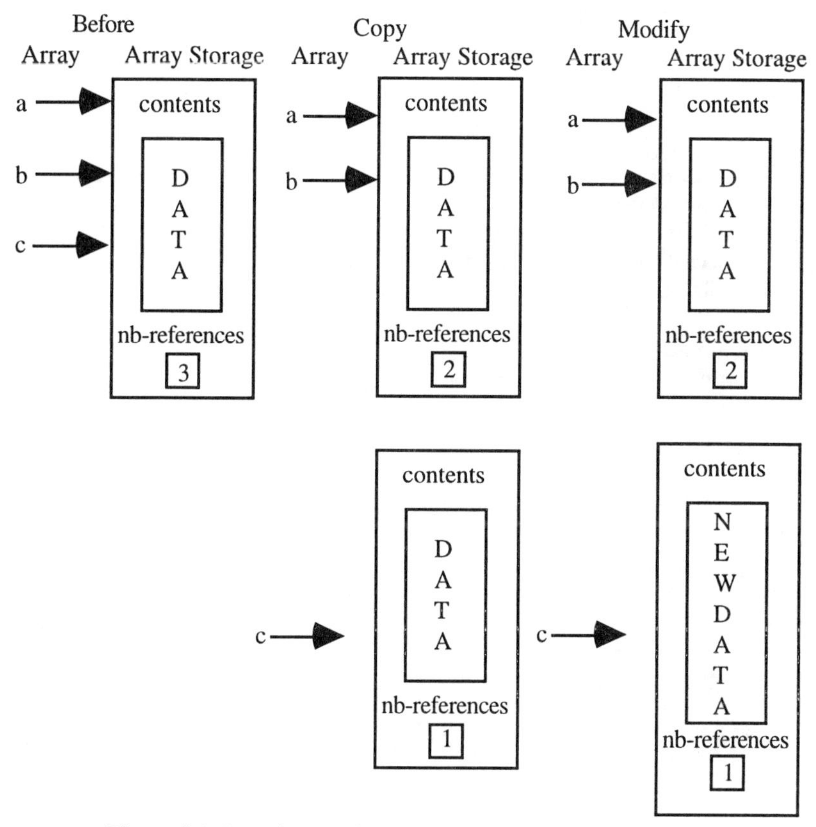

Figure 4.1: Inserting an element into a delayed copied array.

4.4 Chromosome

Before we can discuss the code implementing crossover operators, we need to define the data structure that represents chromosomes.

class Chromosome is an abstract class that defines a chromosome based on any alphabet (template parameter **Genome**). A chromosome simply is an ordered set of genes (**gene_data** on line 525).

As given **class Chromosome** is insufficient for solving any maximization problems. In particular there is no way for the fitness function to determine the merits of each chromosome. **class Chromosome_Ordered** augments **Chromosome** with the method **compare** inherited from **class Ordered**, and is to be used for solving problems where the result of the evaluation function is fully ordered.[8] To solve partially ordered problems see Schaffer [6] . Member function **compare** takes as its parameter another chromosome and returns a negative number if the current chromosome is less than the parameter, 0 if they are equal, and a positive number if' the parameter is less than the current chromosome.

This definition by itself is enough for GA's that use ranking (Greffenstette and Baker [1]). Methods that do not use ranking may also need the results of the evaluation function.

4.5 Crossover

4.5.1 n-point crossover

The first, crossover operator that we shall examine is n-point crossover. The idea behind n-point crossover is to align two chromosomes together (called the parents), cut the chromosomes at identical places into n+1 fragments, and create two new children by mixing the fragments.

A little more formally, let p^0 and p^1 represent the two parents, and c^0 and c^1 represent two children. The first step is to align the two parents. For examples purposes let the length of the chromosomes be of size 10, with indexes 0 to 9.

$$p^0 = p_0^0 \quad p_1^0 \quad p_2^0 \quad p_3^0 \quad p_4^0 \quad p_5^0 \quad p_6^0 \quad p_7^0 \quad p_8^0 \quad p_9^0$$
$$p^1 = p_0^1 \quad p_1^1 \quad p_2^1 \quad p_3^1 \quad p_4^1 \quad p_5^1 \quad p_6^1 \quad p_7^1 \quad p_8^1 \quad p_9^1$$

The second step is to find n points to cut the chromosome into n+1 fragments. In this example let n = 3, and the cuts come after the second, sixth and seventh gene.

$$p^0 = \left| p_0^0 \quad p_1^0 \right\| p_2^0 \quad p_3^0 \quad p_4^0 \quad p_5^0 \left\| p_6^0 \right\| p_7^0 \quad p_8^0 \quad p_9^0 \right|$$
$$p^1 = \left| p_0^1 \quad p_1^1 \right\| p_2^1 \quad p_3^1 \quad p_4^1 \quad p_5^1 \left\| p_6^1 \right\| p_7^1 \quad p_8^1 \quad p_9^1 \right|$$

The final step is to create the children, by copying the fragments into the children. The first child is created by copying the first parent from the first, parent, the second fragment from the second parent, the third fragment from the first parent, and the fourth fragment from the second parent. If there were more fragments, then the copying process would continue in the same manner, with the parents alternating between each other. The second child is created in a likewise manner, with the second parent becoming the source for the first and third fragments, and the first parent being the source for the second and fourth fragments.

---

[8] A set is fully ordered if every two elements in the set are comparable, e.g. a < b or a = b or a > b, for all elements a and b.

$$
\begin{array}{lcll|llll|l|l|ll}
p^0 & = & p_0^0 & p_1^0 & p_2^0 & p_3^0 & p_4^0 & p_5^0 & p_6^0 & p_7^0 & p_8^0 & p_9^0 \\
p^1 & = & p_0^1 & p_1^1 & p_2^1 & p_3^1 & p_4^1 & p_5^1 & p_6^1 & p_7^1 & p_8^1 & p_9^1 \\
\hline
c^0 & = & p_0^0 & p_1^0 & p_2^1 & p_3^1 & p_4^1 & p_5^1 & p_6^0 & p_7^1 & p_8^1 & p_9^1 \\
c^1 & = & p_0^1 & p_1^1 & p_2^0 & p_3^0 & p_4^0 & p_5^0 & p_6^1 & p_7^0 & p_8^0 & p_9^0
\end{array}
$$

Notice that the source for each fragment (and therefore each gene), is opposite from each other. The number of crossover points that should be used, is heuristically determined, but best results have been reported with 3 or less points. The location of the points where the chromosomes should be cut, called crossover points, should be chosen at random.

One of the advantages/disadvantages of n-point crossover is that the distance between genes affects the probability that they will be passed together. For example, consider a chromosome of length 10, with two "good" genes at the first and second locations. For l-point crossover, the two genes would be passed together to a child if the crossover point is anywhere other than after the first gene. Giving a 7/8 probability of the genes being copied together. Now consider if the good genes were at the extremes, the first and last gene position. Then all 8 crossover points will seperate the two genes between the children.

n-point crossover is implemented by the function **crossover_n_point (const Chromosome, const Chromosome, Chromosome*, Chromosome*, const int)** (lines 744-784), along with the helper functions **crossover_n_point(const Chromosome, const Chromosome, Chromosome*, Chromosome*, const int, const int** [])**)** (lines 698-741), and **copy_genes** (lines 668-694).

Lines 755 to 759 ensure that the size of the chromosomes are positive, and that they all have the same size; a necessary requirement for the algorithm to work.

The first step of aligning the chromosomes is not necessary, since the chromosomes have the same index numbers for the same genes, and is therefore skipped.

Lines 771 to 780, determine the location of the crossover points. A crossover point of i means that the cut for a fragment is made after the i-th gene and before the $i+l$-th gene. If the length of a chromosome is l, then it, stands to reason that no cut can be made before the first gene (index 0), and after the last gene (index $l-1$). Therefore, the problem is to choose n random locations from the range 0 to $l-2$ inclusive.

To have clean code that does not worry about whether a fragment is at an end-point or not, we add two extra points to the array that stores the crossover points. Lines 773-780 sort the crossover points and add crossover points before the first gene (index -1), and after the last gene (index $l-1$).

Lines 782-783 call function **crossover_n_point(const Chromosome, const Chromosome, Chromosome*, Chromosome*, const int, const int crossover_points[])** which takes the locations of the fragments. Fragment *i-1* starts at **crossover_points[i]+1** and ends at **crossover_points[i+1]**.[9] The fragments are alternately copied between the parents to the children (lines 728-740).

4.5.2 n-point shuffle crossover

n-point shuffle crossover (sh-n) is similar to n-point crossover, as the name indicates. The procedure is identical, except that the genes are shuffled before crossover, and unshuffled after crossover.

Using the same symbols as before let p^0 and p^1 represent the two parents, and c^0 and c^1 represent the children. The first step is to align the two parents.

$$\begin{array}{lcllllllllll} p^0 & = & p_0^0 & p_1^0 & p_2^0 & p_3^0 & p_4^0 & p_5^0 & p_6^0 & p_7^0 & p_8^0 & p_9^0 \\ p^1 & = & p_0^1 & p_1^1 & p_2^1 & p_3^1 & p_4^1 & p_5^1 & p_6^1 & p_7^1 & p_8^1 & p_9^1 \end{array}$$

The second step is to shuffle the parents such that if the *i*-th gene is in position *j* in the first parent, the *i*-th gene is also in the *j*-th position in the second parent.

$$\begin{array}{lcllllllllll} p^0 & = & p_0^0 & p_5^0 & p_6^0 & p_9^0 & p_0^0 & p_7^0 & p_8^0 & p_4^0 & p_3^0 & p_1^0 \\ p^1 & = & p_2^1 & p_5^1 & p_6^1 & p_9^1 & p_0^1 & p_7^1 & p_8^1 & p_4^1 & p_3^1 & p_1^1 \end{array}$$

The third step is to find n points to cut the chromosome into $n+1$ fragments.

$$\begin{array}{lcll|llll|l|lll} p^0 & = & p_0^0 & p_5^0 & p_6^0 & p_9^0 & p_0^0 & p_7^0 & p_8^0 & p_4^0 & p_3^0 & p_1^0 \\ p^1 & = & p_2^1 & p_5^1 & p_6^1 & p_9^1 & p_0^1 & p_7^1 & p_8^1 & p_4^1 & p_3^1 & p_1^1 \end{array}$$

The fourth step is to create the children, by copying the fragments into the children. Like n-point, crossover, the parents alternate copying the fragments to the children,

$$\begin{array}{lcll|llll|l|lll} p^0 & = & p_2^0 & p_5^0 & p_6^0 & p_9^0 & p_0^0 & p_7^0 & p_8^0 & p_4^0 & p_3^0 & p_1^0 \\ p^1 & = & p_2^1 & p_5^1 & p_6^1 & p_9^1 & p_0^1 & p_7^1 & p_8^1 & p_4^1 & p_3^1 & p_1^1 \\ \hline c^0 & = & p_2^0 & p_5^0 & p_6^1 & p_9^1 & p_0^1 & p_7^1 & p_8^0 & p_4^1 & p_3^1 & p_1^1 \\ c^1 & = & p_2^1 & p_5^1 & p_6^0 & p_9^0 & p_0^0 & p_7^0 & p_8^1 & p_4^0 & p_3^0 & p_1^0 \end{array}$$

The final step is to unshuffle the genes so that they have the original order.

---

[9] Remember in C++, array indices start at 0.

$$
\begin{array}{lclllllllll}
p^0 & = & p_0^0 & p_1^0 & p_2^0 & p_3^0 & p_4^0 & p_5^0 & p_6^0 & p_7^0 & p_8^0 & p_9^0 \\
p^1 & = & p_0^1 & p_1^1 & p_2^1 & p_3^1 & p_4^1 & p_5^1 & p_6^1 & p_7^1 & p_8^1 & p_9^1 \\
\hline
c^0 & = & p_0^1 & p_1^1 & p_2^0 & p_3^1 & p_4^1 & p_5^0 & p_6^1 & p_7^1 & p_8^0 & p_9^1 \\
c^1 & = & p_0^0 & p_1^0 & p_2^1 & p_3^0 & p_4^0 & p_5^1 & p_6^0 & p_7^0 & p_8^1 & p_9^0
\end{array}
$$

The location of the crossover points like with n-point crossover are also chosen at random.

Notice that like n-point crossover the source for each fragment (and therefore each gene), is opposite from each other. Also that while the genes will be different, the number of genes copied from each parent to each child would be same as n-point crossover when the crossover points are at the same locations. The major advantage of shuffle crossover over n-point crossover is that the distance between genes does not affect the probability of being passed together, since the genes are reordered randomly.

Shuffle crossover is implemented by two different sub-routines and algorithms. The first, **crossover-n_point_shuffle_slow** is a rather straightforward implementation of the algorithm given above. The second **crossover-n_point_shuffle** uses mathematical properties discussed below, that trade off picking random numbers, for some off-line calculations.

crossover-n_point_shuffle_slow starts off by calculating the order of the shuffled genes, by permuting the numbers 0 to l-1. Like with n-point crossover, the gene fragments are calculated by picking n points in the range 0 to l-2, and adding values for the end-points (lines 853-862).

The chromosome fragments are copied in lines 865-877, which uses the function **copy_shuffled_genes**. Instead of actually shuffling and unshuffling the genes, the genes that are in each fragment are mapped from the permutation, and placed directly into the children.

The alternative algorithm **crossover_n_point_shuffle** uses the function **table_crossover_probability**.

In Pawlowsky [3], the formula for calculating the probability of copying c genes, for n-point shuffle crossover, of a chromosome of length l is given as:

$$
\frac{\binom{c-1}{c-k}\binom{l-c-1}{n-k}+\binom{l-c-1}{k-1}\binom{c-1}{n-k}}{2\binom{l-1}{l-n-1}}
$$

where k = (n+1)/2 when n is odd, k = (n+2)/2 when n is even (Figure 4.2).

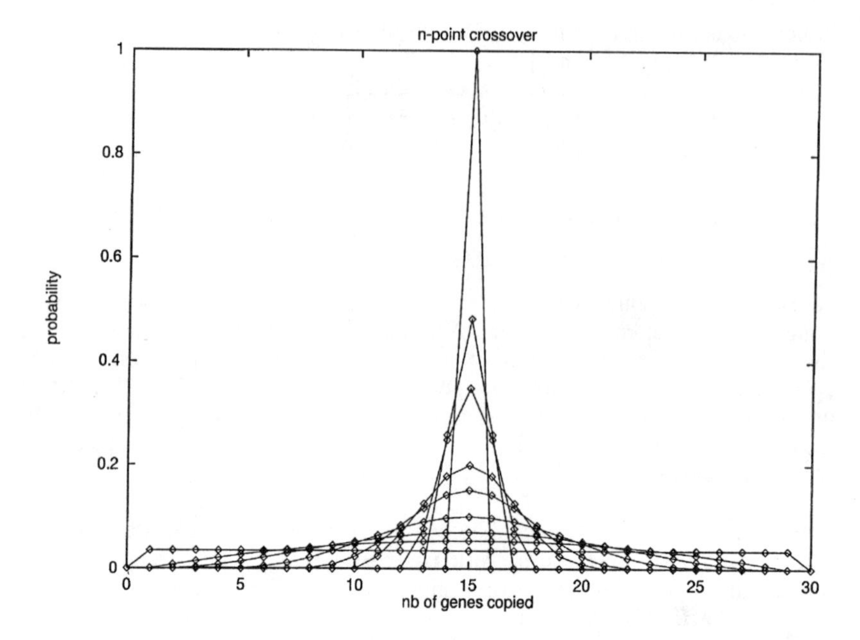

Figure 4.2: Probabilities of copying c genes for a chromosome of length 30.

table_crossover_probability takes as inputs, the length of the chromosome and the number of crossover points, and returns an array, which contains the probability of copying c or less genes, where c is the index in the table. Rather than actually calculate the table every time, or read it in from disk, the table is calculated when first called and saved in memory so it can be reused as necessary. The data structure used is an array of pointers indexed by chromosome length, to an array of pointers indexed by the number of crossover points, that points to the table. The data structure is sparse, with the pointers being NULL (0), if a chromosome length is not used (Figure 4.3).

crossover_n_point_shuffle starts by retrieving the probability table (line 945). A random real number is then chosen from [0,1), and a binary search is made of the table to find the element *mid,* where the random number is greater than the *mid*-l-th entry in the table and less than or equal to the *mid-th* entry (lines 949-970). The function **crossover_shuffle_copy_c_genes** is then called with *mid* as a parameter, indicating the number of genes to copy from the first parent to the second child.

crossover_shuffle_copy_c_genes starts by copying the genes of the first parent to the first child, and the genes from the second parent to the second child. *mid* indices are chosen at random, and the corresponding genes from the second parent are copied to the first child, and the first parent genes are copied to the second child.

Since, the permutation of the genes in the shuffling operation is totally random, copying c genes chosen independently from one parent and l -c genes from the other is functionally equivalent to performing the shuffling, the crossover, and unshuffling. The advantage is that only $c+1$, or l -c+1 random numbers (whichever is lower) have to be generated instead calculating the permutation (l -1 random numbers) and n random numbers.

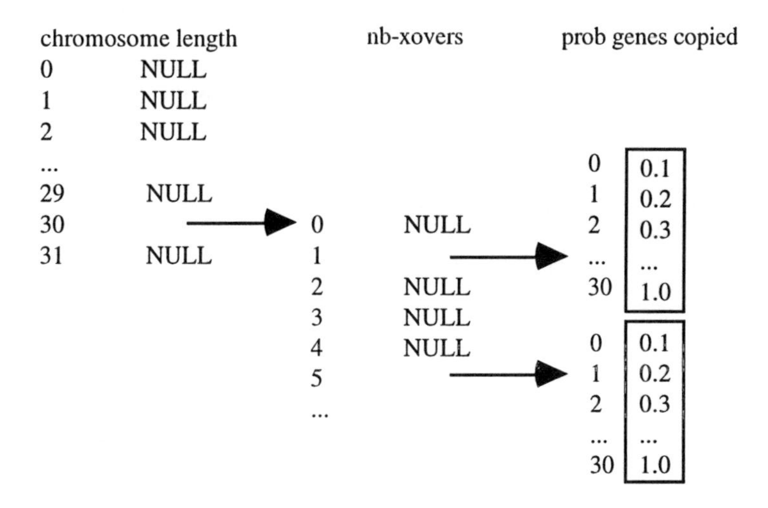

Figure 4.3: Example of the data structure used for storing the probabilities of copying c genes.

One interesting aspect to the algorithm as given here, is that any probability distribution for determining the number of crossover points could be used. One interesting aspect for future research may be a distribution with high likelihood of extrema occurring, and low likelihood for middle values.

4.5.3 uniform crossover

Uniform crossover is fully described by Syswerda [7]. The definition is straightforward. Given two parent chromosomes of length 1, each parent copies *1/2* genes to each child, with the selection of the genes being chosen independently. Again like n-point crossover, the source for each gene for a child is opposite its sibling.

Uniform crossover is implemented by the function **crossover_uniform** which simply calls **crossover_shuffle_copy_c_genes** with 1/2 as a parameter.

It is interesting to note that uniform-crossover is equivalent to n-point shuffle crossover, when n is l-1. The explanation is based on the fact that l fragments are generated for l-1 crossover points, and each fragment can only have one gene in it. Since the copying of fragments alternates between the parents, each parent would copy *7l/2* fragments to one child and *l/2* fragments to the other child (Pawlowsky [3]).

4.5.4 modified uniform crossover

Modified uniform crossover was introduced by Pawlowsky [3]. The idea behind it is to copy all the genes that are the same in both parents to the children first, then copy half of the remaining genes from each parent to each child (Figure 4.4). The purpose behind modified uniform crossover is to reduce the amount of variation between crossovers by concentrating the selection of genes that differ. Once again, the source of a gene in a child is opposite the source of its sibling.

The operator is implemented in the function **crossover_uniform_modified**. The first step is to look for genes that are the same and copy them to the children (lines 1036-1040). If the genes are different then their index is stored in the array **to_do**.

In a manner similar to **random_choose_n_fast**, *l/2* entries from **to_do** are chosen, with the gene for the first child being copied from the second parent, and the gene for the second child being copied from the first parent (lines 1047-1055). The remaining genes are copied with the first child getting the genes from the first parent, and the second child getting the genes from the second parent.

4.5.5 variable crossover

Given the above crossover operators, how does a programmer choose which one to use? Empirical results seem to show that the choice of which crossover operator is dependent on the problem being solved, as well as the parameters being used. Pawlowsky [3] and Rissmann [4] described a technique called variable crossover. We will be discussing a variant of his method.

In variable crossover, each chromosome is prefixed with extra bits, with the extra bits determining the crossover method that is chosen. The operator is defined on lines 1114-1180.

Initial

p^0	=	0	0	0	0	0	0	0	0	0	0
	=	1	1	0	1	1	1	1	1	0	1

Copy duplicate genes to children

p^0	=	0	0	0	0	0	0	0	0	0	0
p^1	=	1	1	0	1	1	1	1	1	0	1
c^0	=			0						0	
c^1	=			0						0	

Copy half of first parent genes to each child

p^0	=	0	0	0	0	0	0	0	0	0	0
p^1	=	1	1	0	1	1	1	1	1	0	1
c^0	=		0	0	0	0		0		0	
c^1	=	0		0			0		0	0	0

Copy half of second parents genes to each child

p^0	=	0	0	0	0	0	0	0	0	0	0
p^1	=	1	1	0	1	1	1	1	1	0	1
c^0	=	1	0	0	0	0	1	0	1	0	1
c^1	=	0	1	0	0	1	0	1	0	0	0

Figure 4.4: Modified uniform crossover.

In this variation, crossover is limited to n-point shuffle crossover and n-point crossover. The function takes, as usual, the parents and the addresses of the children that are to perform in crossover. Also included as parameters is the location offset and size of the extra bits that indicate the preferred method of crossover.

The idea is that along with finding the optimal solution vector, the GA can also find the optimal method of crossover.[10]

Function **crossover_variable** implements the operator and uses **decode_crossover**. Lines 1151-1152 call the helper routine **decode_crossover** (lines 1070-1111), which examines the first bit and if it is set indicates that shuffled crossover is to be used (line 1084). The remaining encoded genes are treated as a binary number, and converted to a fraction by dividing by $2^{(size-1)}$. The fraction is then multiplied by the maximum number of crossover points (size of chromosome -1). (Lines 1090-1098).

The last step in decoding the bits is to apply a heuristic to the number of crossover points (lines 1104-1107). The heuristic is based on the idea that n-point crossover works well for low values of n, yet uniform crossover (which is l-1-point shuffle crossover, where l is the chromosome size) also works well. To not have the "goodness" of the operator depend on one bit, we reverse the meaning of the decoded number of crossover points if shuffle crossover is to be used.

4.6 Which operator to use?

Unfortunately there is no one best operator in all circumstances. The performance of a crossover operator is problem specific, as well as dependent on the parameters that are used for mutation, population size, etc. The best thing a practitioner in GA's can do is to try and evaluate the various operators for the specific problem at hand. Luckily all the operators given in this chapter do perform well, and the main difference between using one operator or another will be the number of evaluations made before a "good" answer is found.

---

[10] For a similar idea see punctuated crossover. [6]

References

[1] Grefenstette, John J. and James E. Baker. (1989) How genetic algorithms work: A critical look at implicit parallelism. Proceedings of Third International Conference On Genetic Algorithms.

[2] Holland, John Henry. (1975) Adaption in Natural and Artificial Systems. University of Michigan Press.

[3] Pawlowsky, Marc Andrew. (1992) Modified Uniform Crossover and Desegregation in Genetic Algorithms. Masters Thesis. Concordia University, Montreal, Quebec, Canada.

[4] Rissmann, T. (1993) A Continous Evolving Portfolio of Crossover Operators for Function Optimizing Genetic Algorithms. Honors Project. Carleton University, Ottawa, Ontario, Canada.

[5] Schaffer, J. David (1985) Multiple objective optimization with vector evaluated genetic Algorithms. Proceedings of the First International Conference on Genetic Algorithms and Their Applications.

[6] Schaffer, J. David, and Morishima A. (1987) An adaptive crossover mechanism for genetic algorithms. Proceedings of the Second International Conference on Genetic Algorithms.

[7] Syswerda, Gilber (1989) Uniform crossover in genetic algorithms. Proceedings of the Third International Conference on Genetic Algorithms.

Chapter 5

Joseph L. Breeden
Prediction Company
320 Aztec St., Suite B
Santa Fe, NM 87501

breeden @predict.com

Optimal State Space Representations via Evolutionary Algorithms: Supporting Expensive Fitness Functions

0-8493-2519-6/95/$0.00 + $.50

Abstract
We have developed a procedure for finding optimal representations of experimental data. Criteria for optimality vary according to context; an *optimal* state space representation will be one that best suits one's stated goal for reconstruction. We consider an ∞-dimensional set of possible reconstruction coordinate systems that include time delays, derivatives, and many other possible coordinates; and any optimality criterion is specified as a real valued functional on this space. The optimization problem is solved using a customized evolutionary algorithm. To effectively optimize an expensive fitness function, parameters are grouped into data structures and a context-specific grammar is incorporated into the genetic operators. Diversity enforcement through a fitness penalty and a *graveyard* of past genomes are also incorporated.

5.1 Introduction to the Problem

In this chapter we will see how aspects of genetic algorithms (GA) [1, 2] and evolutionary programming (EP) [3, 4] can be applied to a fundamental problem in time series analysis: What is the optimal state space representation of the data? In most engineering and physics applications, a state space is viewed as a set of variables which can be used at time t to uniquely determine the state of the system at time $t^{t'} > t$. For example, if we know the position and velocity of an idealized pendulum, we know what the position and velocity will be one second later.[11] If we have a multivariate dynamical system from which we observe one variable, we can reconstruct the state space from transformations of the observed time series, e.g. for the pendulum, we could measure the velocity, compute it from the position time series by approximating the derivative, or use a delayed value of the position as the second variable. Thus for an unknown system, we must ask what dimension our state space should be and what variables we should use to construct it.

In general, there is no single best state space representation of a system for all problems. One must first ascertain the purpose of the analysis. The *optimal* representation must be the one which meets the practitioner's needs. The lack of a unique solution for all purposes is due in part to the presence of noise[12] and the finite length of the data set. Even different algorithms for achieving the same goal often have different optimal representations. No single representation will be optimal for all possible objectives.

What is less known is that we can consider goals which are not topological invariants. For some optimality criteria, different representations are not equivalent even for *perfect* data, i.e. an infinitely long, noiseless data set. This

---

11 A state space composed of x(t) and derivatives thereof is known as a phase space.

12 Some researchers take the view that noise does not exist – only high dimensional dynamical systems which are not properly resolved. All of the present arguments concerning noise are still valid if the noise is considered to be a contamination of our primary system by a signal from an unresolved high-dimensional system.

is true for limited-access, open loop control from a reconstructed state space [5] where the stability of the control is optimized in *the reconstructed state space.*

As an example, say we were concerned about disruptions in communications due to solar activity. We can start with a time series of solar flare activity which we wish to predict and then choose from transformations on this variable and other related time series (included based upon our physical intuition) to reconstruct our state space. We can make this selection based upon minimizing the total forecast error given a specified forecasting algorithm. However, since for this thought experiment we are concerned about communications disruptions, then perhaps we just need to minimize our forecast error for large events. This is an optimization of only a few regions of state space, and could definitely alter the coordinates chosen--even the dimension of the representation. Figure 5.1 shows how coordinate choice can alter one's view of the underlying system.

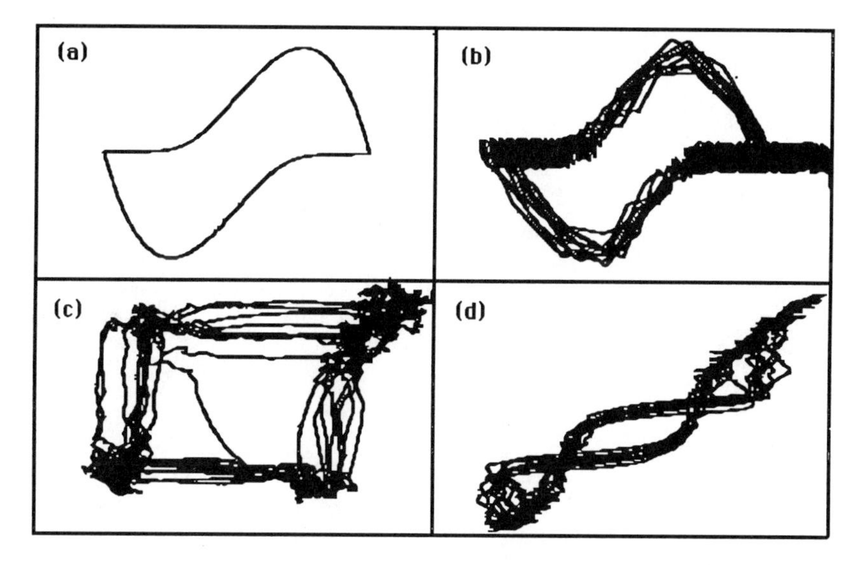

Figure 5.1: Four representations of the van der Pol oscillator, $\dot{x} = y$, $\dot{y} = 10(1 - x^2) y - x$ are shown. In (a) the state space is plotted from x vs. y. In (b)-(d), a uniform dynamical noise of $\epsilon = 0.1$ was added. These reconstructions are (b) x vs. $\dot{x}$, (c) x(t) vs. x(t - 2.0), and (d) x(t) vs. $\int_{-2.0}^{2.0} x(t)dt$.

Our problem can be summarized as an attempt to determine the number and type of transformations on our observed time series to construct a state space which is optimally suited to the desired analysis [6]. This is not a new objective, but with an evolutionary algorithm we can greatly expand the range of possibilities beyond those available in *projection pursuit* [7] and nonlinear dynamics techniques [8, 9, 10, 11]

5.2 Introduction to the Method

5.2.1 Suitability of Evolutionary Search

The problem described above is obviously non-trivial. The stated goal is to optimize some unknown function over possible state space representations. The available transformations are parameterized sometimes with real-valued parameters. Thus the search landscape is of unknown dimension and complexity, and contains an infinite number of possible solutions. This problem is not amenable to most standard search techniques. Although it may be possible to approach this via another algorithm, it is easily implemented within an evolutionary search framework.

One feature common to all current evolutionary search procedures is the use of populations of possible solutions. While this has definite advantages within the search, it also allows us to generate multiple solutions for our stated objective. Often in real applications, the single best answer from the search is not the best solution in practice. For instance, if we were doing a control system optimization, the best solution from the search may be costly to implement, whereas the second or third is almost as good and easily implemented. Sometimes it is also advisable to use multiple solutions in concert to solve a problem [12].

Evolutionary algorithms are a kind of directed random search. Often researchers are suspicious of the efficiency of such searches, but they perform surprisingly well. Consider a purely random search, i.e. random points are chosen in the search landscape. If there are a finite number of possible states to select, N, how many random selections, m, must we make to be guaranteed with probability p of finding one in the top q fraction of the possible solutions?

If $q = 1 - \in$

$$p(N,m,q) \cong 1 - \prod_{i=1}^{m} \frac{(1-q)N}{(N-i+1)} = 1-(1-q)^m N^m \frac{(N-m+1)!}{N!} \qquad (5.2.1)$$

If m << N,

$$p \cong 1-(1-q)^m \qquad (5.2.2)$$

or

$$m = \frac{\ln(1-p)}{\ln(1-q)} \qquad (5.2.3)$$

Thus if we wanted to be 99% certain of finding a solution in the top 1% of possible solutions, we would need to test only 458 possibilities.

5.2.2 Non-standard Aspects

Some aspects of this problem do not map directly onto either the GA or EP paradigms.[13] The variable dimension of the search is not a feature of most standard algorithms, but is similar to a direction suggested by Goldberg, Deb, and Korb with *messy GA* [12, 13].

A greater difficulty is the expense of the fitness function. A quick analysis of the problem we propose to solve shows that each evaluation of the fitness function could be very cpu intensive. There are examples where cpu-minutes are required for a single evaluation on a modern workstation. Most of the literature on GA and EP implicitly assumes that fitness evaluation time is negligible. In this problem, the probable expense of the fitness evaluation will drive most of the design choices. Avoiding the evaluation of redundant, hopeless, or insufficiently dissimilar trial solutions is of utmost importance. According to the literature on GA, this may lead to a non-optimal search, but it will save us enough cpu time to make the problem tractable. A possibly less optimal solution is better than no solution.

5.2.3 Relation to GA/EP

Genetic Algorithms use a bit-string representation of the parameters. New members to the population are generated by crossing over the genes of two parents at a random point within the bit-string, and a small amount of mutation. All the members of the population are evaluated and the best survive to the next generation.

Evolutionary Programming uses real-valued parameters. For new member generation, crossover is not allowed, and mutation shifts a parameter by a random amount according to a distribution, usually Gaussian. Survivors to the next generation are chosen randomly with the probability determined by their relative fitness.

These two procedures are different in their details and philosophical motivations, but are in large part functionally equivalent. For example, if we represent our parameters as positive integers with N bits, the crossover of GA can be viewed partially as a mutation on a single parameter where the perturbation, Δ, has a magnitude of roughly 2^n.

$$p(n) = \frac{1}{N-1}, n \in [1, N-1] \qquad (5.2.4)$$

so

$$p(\Delta | n) = \frac{1}{2^n}\Theta(2^n - 1 - \Delta) \qquad (5.2.5)$$

---

[13] A tremendous number of modifications to the original GA framework have been proposed in the last several years. Whenever a comparison is made to a GA in this chapter, it is always to the simple GA algorithm.

Thus,

$$p(\Delta) = \sum_{n=1}^{N-1} p(\Delta|n)p(n) \qquad (5.2.6)$$

$$= \frac{1}{N-1}\sum_{n=1}^{N-1}\frac{1}{2^n}\Theta(2^n - 1 - \Delta) \qquad (5.2.7)$$

Let k be the minimum number of bits needed to represent Δ, i.e.

$$k = [\ln_2(1+\Delta)] + 1 - \delta(\ln_2(1+\Delta)(\text{mod}\,1) \qquad (5.2.8)$$

Then

$$p(\Delta) = \frac{1}{N-1}\left\{\sum_{i=1}^{N-1} 2^{-i} - \sum_{i=1}^{k-1} 2^{-i}\right\} \qquad (5.2.9)$$

$$= \frac{1}{N-1}\left(\frac{1}{2^{k-1}} - \frac{1}{2^{N-1}}\right) \qquad (5.2.10)$$

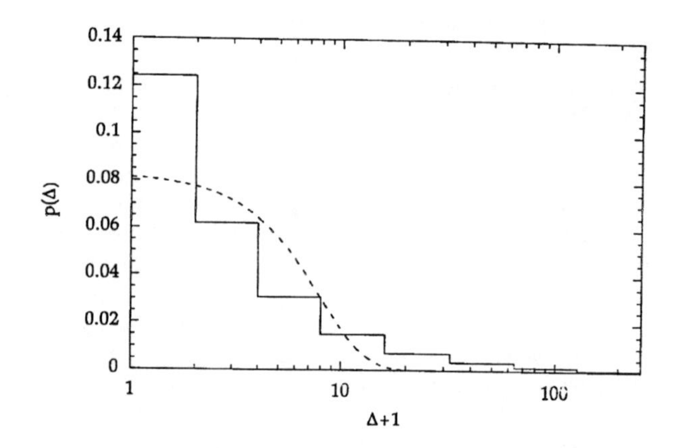

Figure 5.2: GA perturbations compared to Gaussian perturbations.

This is to be compared to the Gaussian perturbations in EP, figure 5.2. The preceding analysis assumes that all the bit-strings needed to cause $\Delta \in [1, 2^{n-1}]$ are represented in the population. This is typically not the case if N > 4 because population sizes tend to be much smaller than 2^{n-1}. Since only a small amount of bit-flip mutation is usually employed in GA, there is a danger of premature convergence of the search [14, 15, 16]. This form of stagnation is not a danger in EP. The other effect of crossover is to swap groups of parameters on either side of the one which is mutated in the previously described manner. This functionality is not present in EP. From the above discussion, it seems clear that a single algorithm could be written which would include all the functionality of both GA and EP, if we simply use Gaussian perturbation (or some other distribution function) and allow crossover on parameter boundaries.

The key to optimizing the speed of our search will be to represent our parameters in data structures so that we can integrate a grammar into the operators [17]. This is similar to previous methods of incorporating heuristic information into a GA [18]. Some work in constrained optimization has taken the direction of incorporating penalties into the fitness function to account for forbidden parameter values [19, 20]. This is a reasonable approach, but it was thought to be too cpu intensive for the present application.

A single transformation may require several parameters, so these are grouped into a data structure. The combination of data structures and context-sensitive operators is very much like the work of Michalewicz [21]. Mutation will be context specific (parameters may be boolean, integer, or real-valued), and hierarchical (changing one parameter may invalidate another). Because of the logical grouping of parameters, crossover makes intuitive sense, e.g. a transformation or set of transformations may be a recurrent component of good solutions. This represents a context specific implementation of the discussion above.

Although the stochastic survival of EP may more accurately reflect biological evolution, we cannot afford to discard good solutions since we may not be able to run the search long enough to randomly regenerate them.

5.2.4 Other Specializations

Several other specializations are added to enhance the efficiency of the search. A simple improvement is not to allow any new members which are identical to existing members of the population. Also, existing members are not re-evaluated from one generation to the next. Taking this a little further, we find that a memory of previously rejected members, *the graveyard,* can be very useful. We disallow the creation of any new members which are identical to ones in the graveyard. This could lead to problems if *deceased* members contain information which is useful at some latter stage of the search. Therefore, we can maintain a fixed size for the graveyard. When new members are added to a full graveyard, the oldest deceased members are discarded and thus can be reincarnated.

Another important enhancement is the use of similarity penalties. The merits of this procedure are still being debated, but with expensive fitness functions, the argument in favor is more compelling. With as few evaluations as possible, we want to sample as large an area in the search landscape as possible. The idea is to penalize a member for being too phenotypically similar to a more fit member of the population.

5.3 Algorithm Overview

Consider the situation of an experimenter who has made a sequence of measurements of a physical system providing a single time series of fixed length. The algorithm developed is a general procedure for searching the space of possible dimensions and coordinates with which to reconstruct the state space. This search attempts to optimize a quality function, relating to the experimenter's objectives, which is designed to place a numerical value upon the usefulness of a given trial reconstruction. Simpler search schemes are often ineffective, because situations arise in which the best D-dimensional representation is not an exact subspace of the best (D + 1)-dimensional representation [22], but is nearby in the search landscape. In such cases, building a representation by sequentially adding new coordinates, ie. a greedy algorithm, will be ineffective.

Once we have decided upon the space of coordinates over which to search and a quality function with which to compare representations, we implement our search algorithm. For this, a representation will be described by a genome with each gene identifying a coordinate. For example, $\{\mathbf{G_1}, \mathbf{G_2}, \mathbf{G_3}\}$ is a 3-dimensional representation where $\mathbf{G_i}$ is a vector of parameters indicating how the ith coordinate is generated from x(1): for derivatives, G^0 is the order of the derivative; for delays, G^0 is the time delay; for smoothed coordinates, G^0 is the delay and G^2 the size of the averaging window; etc.

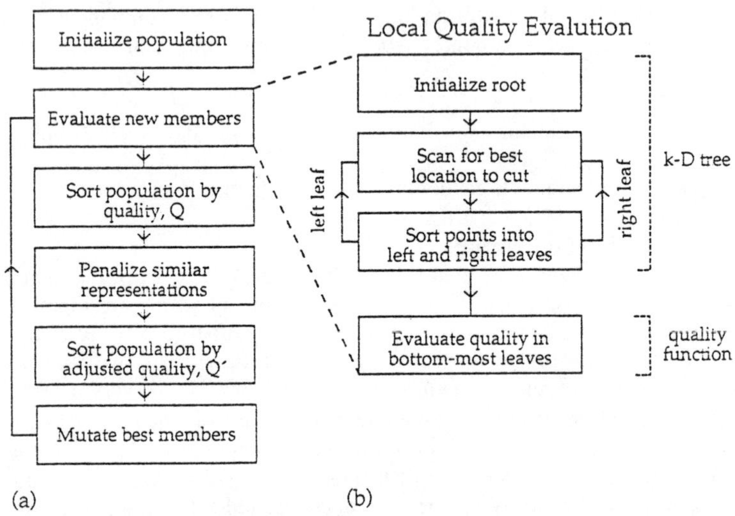

Figure 5.3: A diagram showing the major components of the learning algorithm, (a). Part (b) outlines the quality function used for local optimization.

The search can be seeded with representations of physical interest or an initial random population of genomes. These representations are evaluated and ranked according to the quality function (Figure 5.3). A penalty based upon phenotypic similarity is incorporated to produce the adjusted fitness scores. The highest quality members of the population according to the adjusted fitness are mutated using the operators defined in figure 5.4. These new members are then evaluated, ranked, and so forth. Eventually the most fit members of the population should converge to an optimal representation.

The iterative procedure of mutation, evaluation, and ranking continues until the population converges. The highest quality member of the population is used to obtain the necessary coordinates for an optimal representation of the data. Note that at no point have we specified a minimum number of data points or maximum noise level permitted for this procedure to succeed, because this technique finds the optimum representation for the data at hand.

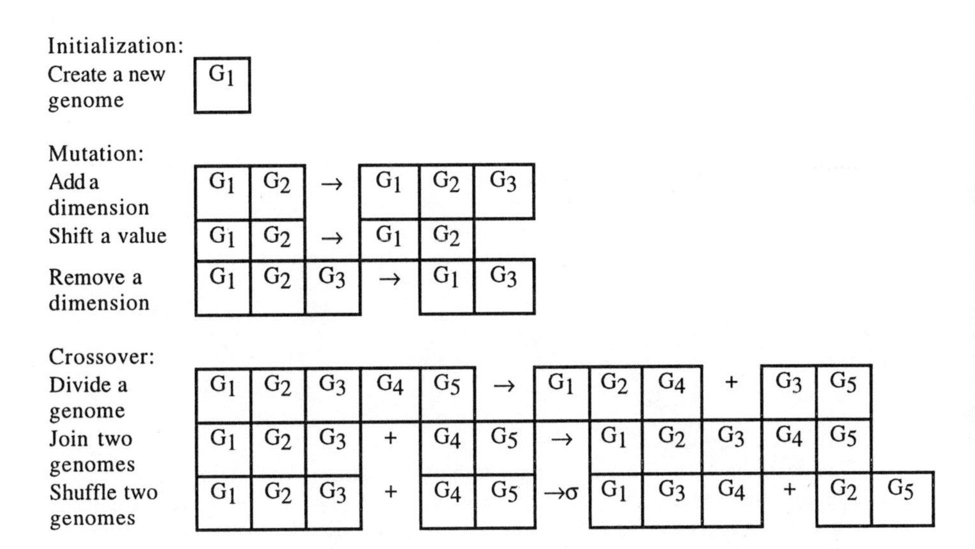

Figure 5.4: The operators used for our genetic algorithm-based search are shown. A particular representation is encoded by a genome, where each gene specifies a choice for one coordinate. The *mutation* operators move a genome a small distance in the quality landscape, while the *crossover* operators can generate large moves in the landscape. The divide and join operators help in searching quality landscapes with high symmetry. The shuffle operator is equivalent to crossover operators in standard genetic algorithms.

5.4 The Code Framework

The various stages of the search algorithm will be discussed in the following sections with C++ code examples. Headers are given for the basic classes and source code for a few non-trivial functions. Understanding the code is not essential to an understanding of the concepts to be discussed.

The concept of a population is central to studying evolutionary algorithms. A population contains members (trial state space representations) which contain a genetic description of the transformations used and their fitness scores.

```
class Representation
{
    public:
        Representation(const Genome& genome);
        double distance(const Representation& rep); //Used in
        diversity penalty
    private:
        Genome _genome; //Contains the parameters describing the
        transformations
        double fitness;
        double _discounted_fitness;
}
```

The following is an example of a container class for representations.

```
class Population
{
    public:
        Population(int dim);

        void add_to_pop(Representation* rep);
        int in_population(const Representation& rep);

    private:
        Representation **_reps;
        int _dim;
}
```

5.5 The Genome

The genome is the part of the representation under control of the search. It contains an array of objects, each of which completely specifies a transformation. Also, it has member functions to perform the genetic operations required by our search.

```
class Genome
{
    public:
      Genome(Gene **genes, int dim);
      Genome(int dim); //Construct a random genome with dim
      genes.

      int dim(void) { return(_dim); }
      int in_genome(const Gene& gene); //Test to see if a gene is
      already in the genome

      //Mutation-like operators:
      void add_dim(void);
      void remove_dim(void);
      void mutate(void);

      // Crossover-like operators:
      Genome *split ();
      void join(const Genome& genome);
      Genome *shuffle(const Genome& genome);

    private:
        Gene **_genes;
        int _dim;
}
```

Each gene contains the parameters necessary to specify a transformation of the input data.

```
class Gene
{
    public: Gene( );
    virtual void mutate(void);
```

```
    virtual Gene* spawn() {return new Gene(*this); )

    private:
         int _column; //Column to use from the input data.
         int _power;//Coordinate will be raised to this power;
}
```

We want to allow for many different types of transformations. Some of the possibilities are

- *Delays (Lags)* - A value from an input time series a specific amount into the past.
- *Derivatives (Differences)* - Finite differences on an input time series, or an approximation of the derivative.
- *Fuzzy Delays* - A delay value chosen with a tolerance window about the requested delay [23].
- *MA Coordinates* - Used in linear regression. A delay coordinate on the forecast error.
- *Smoothed Coordinates* - A value from a smoothed version of an input time series.
- *Delayed Coordinates* - Any coordinate may be delayed by some amount, such as a delayed derivative.
- *Polynomials* - Coordinates can be raised to some power
- *Other Functions* - Any function on an input time series which can be parameterized is a possibility.

The transformations can be encoded as derived classes from the Gene base class. These are examples for delay, derivative, and smoothing transformations.

```
class Delay: public Gene
{
    public:
        Delay();
        void mutate(void);
        virtual Gene* spawn() {return new Delay(*this); }

    private:
   int _delay; //Delayvalue.
}

class Derivative: public Gene
{
    public:
```

```
        Derivative( );
        void mutate(void);
        virtual Gene* spawn() {return new Derivative(*this); }

    private:
      int _order; //Order of the derivative.
}

class Average: public Gene
{
    public:
        Average();
        void mutate(void);
        virtual Gene* spawn() {return new Average(*this); }

    private:
      int _delay; //Delay value.
   int _window;/ /Width of the averaging window.
}
```

For a transformation like derivative, we could parameterize the algorithm for computing the derivative (differencing, polynomial fitting, etc.) and place this under genetic control.

5.6 New Member Generation

In our search, we have three basic types of operators: random new genome, mutate (add a dimension, remove a dimension, and mutate a parameter), and crossover (split a genome, join two genomes, and shuffle two genomes).

Random new genomes are most useful when diversity enforcement is being used, section 5.7. Otherwise, they are rarely fit enough to survive in a population which is several generations old. In our code, the default constructor for a gene is assumed to generate a random new transformation.

The mutation operators represent relatively small moves in the fitness landscape. They are assumed to be context sensitive. The following is example code for the Average class.

```
int Genome::mutate(void)

        genes[Rnd(0,- dim- 1)]Æmutate();
}

void Average::mutate(void)
{
   double x = Rnd(0,1); //Choose a uniform random number in
   [0,1].

   if (x< 0.2) //Change column.
   {
       _column = Rnd(0,numcols-1 );

       _delay = Rnd(0,maxdelay-1);
```

```
        _window = Rnd(0,-delay-1);
        _power = Rnd(1,maxpower);
    }
    else if (x< 0.5) //Change delay.
    {
        do
        }
            _delay += GaussianRnd(0,sigma);
        } while (_delay < 0.0 || _delay > maxdelay);
    }
    else if (x< 0.8) //Changewindow
    }
       do
    {
            _window += GaussianRnd(0,- delay- 1);
        } while (_window < 0.0 || _window > _delay);
    }
    else // Change polynomial order.
    {
            _power = Rnd(1 ,maxpower);
    }
}
```

The fraction of the time the different parameters are mutated can be fixed according to numerical experiments or made to change over generations. In this way we can control the average perturbation amount as is done in simulated annealing. There would appear to be definite advantages to combining the two approaches.

Note that the perturbations for the **_delay** and **_window** parameters are Gaussian, but uniform for **_power**. This is another context-specific decision. Also, if the **_column** is changed, then we have no reason to believe that the previous values for **_delay**, **_window**, or **_power** have any information. Thus, we choose new ones. Also, the allowed values for **_window** are bounded by **_delay** (to prevent future information from entering a historical transformation). This is an example of how the grammar of a specific problem may be encoded into the transformations to prevent generation of genomes which we know *a priori* to be faulty. If one wishes to maintain the biological analogy, there are precedents in nature for this kind of rejection of genetically flawed individuals, e.g. miscarriage.

Because of the natural hierarchy in the parameters, we could try more involved approaches to mutation [24, 25]. We could even pursue hybrid approaches such as hill-climbing for lower level parameters and GA/EP for higher level parameter sets. These possibilities are intriguing, but have not been investigated here.

We implement crossover-like operators to effect large moves in the search landscape while still preserving the grouping of parameters within a gene. For state space representation problems, the ordering of the transformations is irrelevant, so our crossover operators will be order independent. They also allow for varying length genomes to be manipulated. For join and shuffle where a mate is required, one is randomly chosen from the surviving population. The shuffle operator is similar to uniform crossover in GA for ordered genomes [26, 27]. There are several possible alternatives to this procedure which will not be debated here.

The following is example code for a shuffle operator.

```
int Genome::shuffle(const Genome& genome)
    Gene **long_genome;
    Gene **short_genome;
    int maxdim, mindim;

    if (_dim > genome._dim)
        long_genome = _genes;
        short_genome = genome_genes;
        maxdim = _dim;
        mindim = genome._dim;
    }
    else
    {
        long_genome = _genes;
        short_genome = genome_genes;
        maxdim = _dim;
        mindim = genome_dim;
    }

    Gene **genes1 = new Gene* [maxdim];
    Gene **genes2 = new Gene* [mindim];

    for (int i = 0, dim1 = dim2 = 0; i < mindim; i++)
    {
        if (Rnd(0,1) < 0.5)
        {
                genes1 [dim1 ++] = short_genome[i]Æspawn();
                genes2[dim2++] = long_genome[i]Æspawn( );
        }
        else
        }
            genes1[dim1++] = short_genome[i]Æspawn();
            genes2[dim2++] = long_genome[i]Æspawn( );
        }
    }

    for (;i < maxdim; i++)
    {
      if (Rnd(0,1) < 0.5) genes1[dim1++] = long-genome[i]
              Æspawn();
       else
    genes2[dim2++] = long-genome[i]Æspawn();
    }

    free (_genes);
    _dim = dim1;
    _genes = new Gene* [_dim];
    for (int i = 0;i < _dim; i++) _genes[i] = genes1[i];
```

```
    free (genes1);

    return (new Genome(genes2,dim2));
}
```

Whenever a new genome is generated, it is first checked to see if it is already present in the graveyard. The graveyard is another instantiation of the Population class. If it is found in the graveyard, it is immediately rejected and another genome generated.

5.7 Diversity Enforcement

The motivations for diversity enforcement have been described previously. Implementation is somewhat less obvious. In this algorithm, we take the direct approach of adding a penalty term to the fitness function as opposed to more implicit approaches [28]. The first step is to define a distance measure between the two representations. This is necessarily problem specific and need not be a true metric, i.e. $d_{ij} \neq d_{ji}$ is acceptable and often advantageous when dealing with variable length genomes. For convenience, we will assume that all distances are between 0 and 1. The distance function can measure similarity in genotypic or phenotypic space. Phenotypic distance measures are preferable in state space representation problems because different transformations could contain the same underlying information.

To compute the discounted fitness, we first sort the population according to the fitness. For the best member, $f' = f$. For all the other members, we want the distance between that member and the best one. We then compute

$$f_k' = f_k(1-\alpha_n(1-d)),\ \alpha_n = \left(\frac{N_G - n}{N_G}\right)^{\beta} \qquad (5.7.1)$$

α_n is a tuning parameter to give an effect similar to simulated annealing (the "annealing schedule" above is but one possibility); n is the generation number; and N_G is the total number of generations the search will be run.

The population is now sorted by f', the top two members fixed, and we recompute the distance based upon the top two members. Thus, after i iterations,

$$d_k = \min_{l \in [1,k-1]} d_{k,l}. \qquad (5.7.2)$$

When i equals the number of members in the population, we are finished.

The following code shows a faster implementation of the diversity enforcement algorithm. This code assumes we are maximizing the fitness and all fitness scores are greater than or equal to 0.

```
void diversify(Population& pop)
{
    int m = pop.size();
```

```
    Matrix dfit(m,m);  // Any vector & matrix classes will do

    for (int i = 0;i < m; i++)
         for (int j = 0; j < m; j++)
         {
             if (i == j)dfit(i,j)= pop[i].fitness();
             else dfit(i,j) = pop[i].distance(&pop[j]);
         }

    Vector cols(m);

    for (j= 0; j < m; j++) //Doesn't assume population is sorted
         if (dfit(j,i) > dfit(cols(0),cols(0)) cols(0) = j;

    pop[cols(0)].discounted-fitness(pop[cols(0)].fitness());

    // This is a cheap way to remove a model from consideration
    // once it has been selected.
    for (j= 0; j < m; j++) dfit(cols(0),j) = -BIGNUM;

    for (int ncol = 1; ncol < m; ncol++)
    {
        Vector bestdfit(m);
        int site = 0;

        for (j= 0; j < m; j++)
            bestdfit(j) = min(dfit.row(j).pick(cols));
        site = max-index(bestdfit); //Index of the max value in
the vector.

        pop [site]. dfit (bestdfit (site));
        cols(ncol) = site;
        for (j= 0; j < m; j++) dfit(site,j) = -BIGNUM;
    }
}
```

5.8 Relation to Simulated Annealing

There are many places within evolutionary algorithms to incorporate aspects of simulated annealing (SA) [29, 30]. One option which has received attention is to adapt the genetic operator probabilities over search-time [31, 32, 33, 34, 35]. A transition from predominantly crossover-like operators to predominantly mutation-like operators gives some of the flavor of SA and has proven to be useful in many situations. Adapting the width of the Gaussian distribution used to generate perturbations of the parameters is very similar to SA. Also, the penalty factor described above incorporates a tuning factor which removes the penalty over time, thereby allowing more of the search effort to be focused upon refining the few best members toward the end of the search.

Although many aspects of the search can be made to resemble SA, there are important differences between them. In typical simulated annealing implementations, the search is conducted on a single trial solution. By introducing noise in the landscape, the simulated annealing algorithm aims to smooth the local minima, causing the search to find the global minimum. Then reducing the noise allows the search to settle onto the best solution.

In contrast, the learning algorithm described above maintains a population of possible solutions. The best available solution is always defined according to the unaltered quality landscape, so that if it is near the global optimum the best solution can be found rapidly. Solutions of lesser quality which are near higher quality representations are penalized. This creates an effective quality landscape for the less fit representations in which the optima found by higher quality representations are removed. Thus, each member of our population tries to find a separate optimum in the landscape. As the parameter a_n goes to zero, more of the effort of the algorithm focuses upon searching near the best optima which have been found.

5.9 Stopping Conditions

Usually with search algorithms we are interested in finding the global optimum. As we use an evolutionary search, we observe asymptotic convergence in the fitness of the best member, figure 5.5. How long the search is conducted is a function of the researcher's patience and hardware resources.

When we are learning optimal state space representations, the fitness function is based upon a noisy, finite-length data set. Therefore, we must be concerned about statistical fluctuations which create optima in the fitness landscape but do not correspond to structure which can be exploited out of sample. This is the generalization problem. Sub-optimal solutions may generalize better out of sample than the global optimum. Thus, we need a stopping condition for the search. This is not a new problem, and has found most recent application in training neural networks [36, 37]. Most of those methods are also applicable to evolutionary searches.

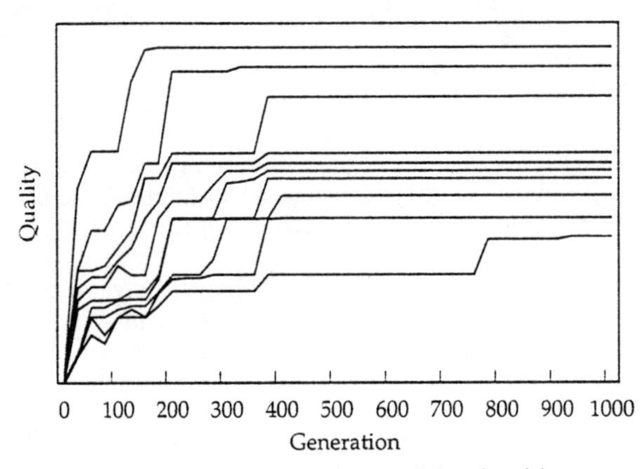

Figure 5.5: A typical plot for the convergence of the algorithm over generations. The lines show the qualities of the ten best members of a population. Diversity is enforced throughout the search. There is an asymptotic approach to the "optimal" representation.

5.10 Examples

5.10.1 State Resolution, Determinism, and Forecasting

We begin the examples by comparing the results of optimizing state resolution (mutual information), forecasting, and determinism.

Some calculations, such as fractal dimension, work best when the state of the system is well resolved. Typically, this corresponds to the data being well spread in the state space. This can be optimized for nonlinear systems by optimizing the mutual information [38],

$$M(\tau) = \sum_{leaves} p(x(t), x(t+\tau)) \log \frac{p(x(t), x(t+\tau)}{p(x(t)) p(x(t+\tau))} \qquad (5.10.1)$$

This was used to find the optimal two-dimension representation of the Mackey-Glass equation

$$\dot{x} = -\gamma x + \beta x_{\tau} \frac{\theta^n}{\theta^n + x_{\tau}^n} \qquad (5.10.2)$$

which is a model for the white blood cell population in humans [39, 40]. For this study, we have used $\gamma = 0.1$, $\beta = 0.2$, $\theta = 1$, $n = 10$, and $\tau = 30$. The optimal representation is shown in figure 5.6a.

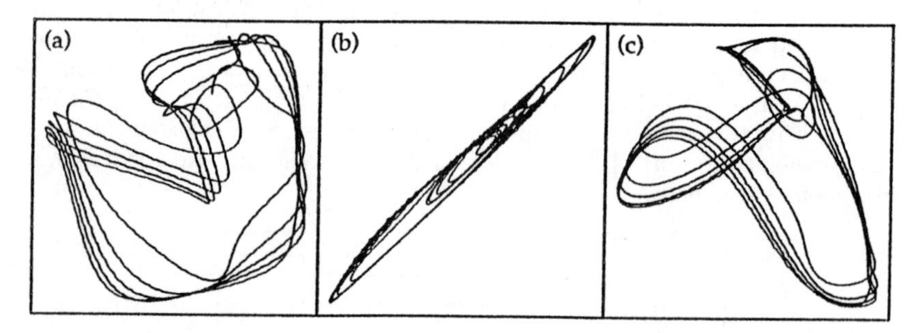

Figure 5.6: The delay coordinate reconstructions shown were selected as optimal using (a) mutual information, (b) complete local linear maps, and (c) forecasting only x(t).

An extensive literature exists on forecasting. Methods have been developed in recent years involving state space modeling for predicting nonlinear and even chaotic systems over short times. These methods have shown promise, but can benefit from a reconstructed state space optimized for predictability.

We can devise many modeling schemes, but for illustration a local mapping with k-D tree partitioning [41] is considered. The standard partitioning schemes are based upon maintaining an equal probability of a reference point being in any of the leaves. This implementation places the partition so as to minimize the range variance in the two new leaves [42]. The point lookup efficiency of the tree is reduced, but its predictive ability is maximized. In particular, when significant noise is present, we can isolate regions of poor predictability from more predictable ones. For example, the corners in figure 1b are strongly distorted by dynamical noise and thus are only slightly predictable. By adjusting the partitions to isolate those regions, the predictability of the remaining partitions will not be degraded. This is also effective in the high noise limit for determining if any predictable regions are present.

This is implemented by fitting a map to the points within each partition,

$$x(t') = M_i(\xi(t);\mathbf{p}), \qquad (5.10.3)$$

where x is the observable variable and ξ is the reconstructed state space vector including x. The mean square error may be measured by

$$\eta^2 \equiv \frac{1}{N}\sum_{i=1}^{L}\sum_{j=1}^{N_i} x_j(t') - M_i(\xi(t);\mathbf{p}))^2 \qquad (5.10.4)$$

N is the total number of points, L is the number of leaves, Ni is the number of points in the ith leaf, and **pi** are the best fit parameters to the points in the ith leaf. The function M can have many forms, but is commonly taken to be affine,

$$x(t') = \mathbf{m} \bullet \xi(t) + \mathrm{b} \qquad (5.10.5)$$

with **m** and b as free parameters. The optimal representation is the one which minimizes η. For the Mackey-Glass data, the optimal two-dimensional representation is shown in figure 5.6c.

Finally, we compare the optimization of determinism. The quest for deterministic state space representations is usually an attempt to determine the number of degrees of freedom in the original system--an important first step in developing a theory for the dynamics. Also, the calculation of some dynamical properties, such as Lyapunov exponents, are best estimated with a deterministic representation.

The difficulty with real data lies in defining determinism. For the present, we will simply use the forecasting method previously described to predict all the state space coordinates rather than just $x(t)$. Thus

$$\eta'^2 = \sum_{k=1}^{D} \eta^2 k \qquad (5.10.6)$$

For ideal data, this should select a deterministic representation. Otherwise, it will generate the best possible approximation. For the Mackey-Glass data, if we again restrict the search to two dimensions, the optimal representation is shown in figure 5.6b.

These three examples illustrate the fundamental motivation for this algorithm: no single representation is optimal for all purposes. A general tool is needed which can generate a state space reconstruction satisfying any specified goal.

One other interesting (even if overused) example is the yearly average sunspot numbers [43, 44] taken from [45, 46, 47]. By optimizing mutual information, we find little improvement with representations above 4 dimensions. For illustration, we show the best three dimensional delay coordinate reconstruction obtained, figure 5.7.

5.10.2 Equations of Motion

Constructing equations of motion for a time series refers to building a single set of globally defined equations to describe the dynamics. The goal here is typically to gain physical insight into the system being observed. Therefore, the coordinate transformations to be considered must be *physically meaningful* to the scientist.

Both global and local approaches are possible. The global method generally employs derivatives to form a set of ordinary differential equations [48]. The coefficients to the model equations are fit parametrically to the data. The learning algorithm chooses the number of equations and derivatives to minimize the modeling error.

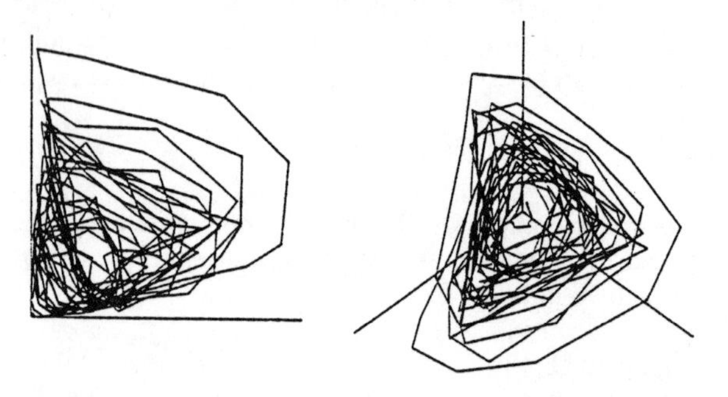

Figure 5.7: Two views of the yearly sunspot numbers in the reconstructed state space {x(t), x(t - 3), x(t - 7)}

The global method works in principle, but it can be numerically difficult and computationally costly. An alternate local approach within the learning algorithm framework may be more effective. We begin by constructing the best possible local linear model by searching for the best derivatives, integrals, redundant coordinates, or whatever are deemed physically relevant transformations. This results in a model

$$\xi(t') = \mathbf{a} \bullet \xi(t) + \mathbf{b}$$

where ξ is the state space vector and **a** and **b** are linearly state dependent coefficients. We can use the fit locally in the state space.

As an example of fitting state dependent parameters, we tried to predict x^{n+1} from the standard map,

$$x^{n+1} = x^n - \frac{\alpha}{2\pi}\sin(2\pi y^n)$$
$$y^{n+1} = y^n + x^{n+1}$$

where $\alpha = 2.0$. The best representations for minimizing rms error in predicting x^{n+1} were of the form $\{x^n, y^n, (y^n)^2, (y^n)^3, ...\}$. If we prune the coordinates which have small weights in the modeling, the even terms start to drop out. The same affect could probably have been achieved by implementing a quality function with a penalty term on the number of parameters. The best representations are consistent with a Taylor series expansion of the original equation.

5.10.3 Control

Recent work has shown that it is possible to control a low dimensional dynamical system without feedback by exploiting convergent regions of the state space [49, 50, 51]; sometimes called entrainment control. Further, it is sometimes possible to control a system when only one variable is observable and controllable using a reconstructed state space [5]. We refer the full description of the technique to previous work, but show here that we can allow the learning algorithm to search for a set of delay coordinates to best solve the problem.

The key to entrainment control is to locate convergent regions in the state space. If the dynamics can be modeled as $\mathbf{x}^{n+1} = M(\mathbf{x}^n)$, we can compute the local characteristic exponents, μ_j. If $[\mu_j] < 1 \ \forall j \in [1,D]$, then it is a convergent region. Figure 5.8 shows the convergent regions for the Henon map,

$$x_{n+1} = y_n - 1.4x_n^2 + 1$$
$$y_{n+1} = 0.3x_n$$

To find a stable goal dynamics to which we can entrain the system, it is sufficient but not necessary to choose a goal within the convergent regions.

If we phrase our optimality criterion as, 'find the most stable period-p goal dynamics to which to entrain x^n,' we can write the optimality function as

$$z^i = x^i + F(x^n, x^{n-\tau_1}, x^{n-\tau_2}, ...)$$
$$\eta^2 = \frac{1}{N}\sum_{i=1}^{N}(z^i - z^{i-p})^2$$

where F is the driving force as described by Jackson [50] and z is the driven system. In minimizing V, we will be finding both the best state space representation and the specific goal dynamics within that representation. Note that this is the first optimality criterion which distinguishes between reconstructed state spaces even for perfect data and models.

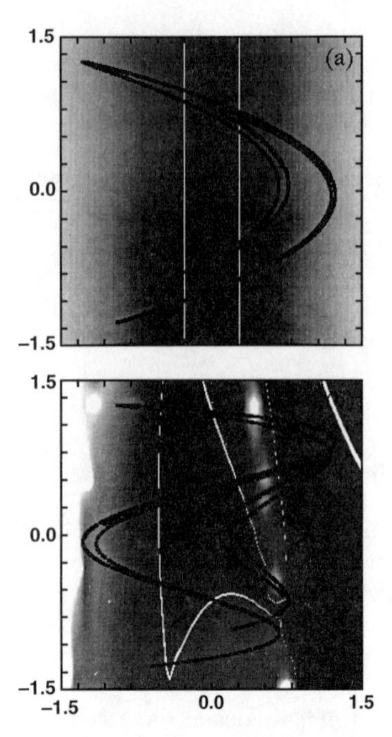

Figure 5.8: Convergent regions in reconstructed state spaces of the Henon map with the reconstructed Henon map superimposed. We show two state space reconstructions: a) $x(t^n)$ vs. $x(t^{n-1})$ and b) $x(t^n)$ vs. $x(t^{n-2})$. Dark-grey indicates convergent regions; light-grey indicates non-convergence. The white line indicates the boundary between the two regions. The dark color indicates stronger convergence. For the top figure, the convergent region is simply $|x| < 0.25$. For the bottom figure the multiple convergent regions are nontrivial.

We applied this procedure to the Henon map where we tried to drive x^n to a fixed point dynamics (when observed at the driving interval) by applying a driving force only to x^n. We show those solutions selected for the $\{x^n, x^{n-2}\}$ state space reconstruction, figure 5.9, for comparison with the convergent regions in figure 5.8. The solutions shown which do not lie within the convergent regions were lesser quality solutions which we kept by employing a large population size in the learning algorithm. This is a useful feature of our learning algorithm, since the best solutions are not always physically feasible.

5.11 Conclusions

We have presented a method of generating optimal representations of experimental data from a generalized set of reconstructed coordinates. This method is based upon the genetic algorithm and is sufficiently general to accommodate a variety of optimality criteria.

The goals discussed here encompass a wide range of possibilities. Little or no physical insight is gained when optimizing forecasting, control, or noise reduction. However, determinism, state resolution, and equations of motion modeling all have some explanatory value. These distinctions are important, because the required representations are very different. As the field of nonlinear dynamics influences more areas of science, new goals and accompanying coordinate transformations will surely be added.

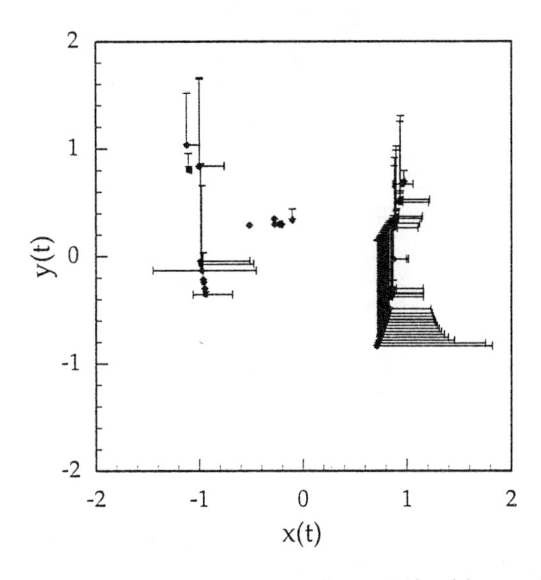

Figure 5.9: Stable goals found by searching figure 5.8b with a model made only from the data on the attractor represented in the original state space. The error bars give a rough approximation of the basin of entrainment of these goals. Note that some of these goals do not lie within the convergent regions. The convergent regions are a sufficient but not necessary condition when accurate models are used.

The algorithmic framework for this problem was a synthesis of Genetic Algorithms, Evolutionary Programs, and aspects of Simulated Annealing. The design choices were driven by the extreme computational expense of the fitness function. Most of the speed enhancements were gained by incorporating a context-sensitive grammar into the genetic operators and organizing the parameters into data structures. While this approach may not be theoretically appealing to some in the GA community, it is practical and effective.

References

[1] J. Holland, *Adaptation in Natural and Artificial Systems.* University of Michigan Press, 1975.

[2] D.E. Goldberg, *Genetic Algorithms in Search, Optimization, and Machine Learning.* Addison-Wesley, 1989.

[3] L.J. Fogel, "Autonomous Automata", *Industrial Research,* 4, 14-19, 1992.

[4] D.B. Fogel, "An analysis of evolutionary programming", in: D.B. Fogel and W. Atmar, eds, *Proceedings of the First Annual Conference on Evolutionary Programming,* San Diego, CA, 1992. Evolutionary Programming Society.

[5] J.L. Breeden and N.H. Packard, "Model-based control of nonlinear systems", preprint, 1991.

[6] J.L. Breeden and N.H. Packard, "A learning algorithm for optimal representation of experimental data", *Bifurcations and Chaos*, **4**, April, 1994

[7] J. Friedman and J. Tukey, "A projection persuit algorithm for exploratory data analysis", *IEEE Transactions on Computers,* c-23(9), 881-890, 1974.

[8] A.M. Fraser and H.L. Swinney, "independent coordinates for strange attractors from mutual information", *Phys. Rev. A,* 33, 1134-1140, 1986.

[9] D.S. Broomhead and G.P. King, "Extracting qualitative dynamics from experimental data", *Physica D,* 20, 217-236, 1986.

[10] J. Benedetto and M. Frazier, *Wavelets: Mathematics and Applications.* CRC Press, 1993.

[11] T. Buzug and G. Pfister, "Optimal delay time and embedding dimension for delay-time coordinates by analysis of the global static and local dynamical behavior of strange attractors", *Physical Review A,* 45(10), 7073-7084, 1992.

[12] R.T. Clemen, "Combining forecasts: A review and annotated bibliography", *International Journal of Forecasting,* 5, 559-583, 1989.

[13] D. Goldberg, B. Korb, and K. Deb, "Messy genetic algorithms: Motivation, analysis, and first results", *Complex Systems,* 3, 493-530, 1989.

[14] D. Goldberg, K. Deb, and B. Korb, "Don't worry, be messy", in: R. K. Belew and L.B. Booker, eds, *Proceedings of the Fourth International Conference on Genetic Algorithms,* San Mateo, CA, 1991. Morgan Kaufmann.

[15] L. Eshelman and J. Schaffer, "Preventing premature convergence in genetic algorithms by preventing incest", in: R.K. Belew and L.B. Booker, eds, *Proceedings of the Fourth International Conference on Genetic Algorithms,* pages 115-122, San Mateo, CA, 1991. Morgan Kaufmann.

[16] K.A. De Jong, "The analysis of the behavior of a class of genetic adaptive systems", Doctoral dissertation, Univ. of Michigan, Ann Arbor, MI.

[17] D.B. Fogel, "A brief history of simulated evolution", in: D.B. Fogel and W. Armar, eds, *Proceedings of the First Annual Conference on Evolutionary Programming,* San Diego, CA, 1992. Evolutionary Programming Society.

[18] S. Bagchi, S. Uckun, Y. Miyabe, and K. Kawamura, "Exploring problem specific recombination operators for job shop scheduling", in: R.K. Belew and L.B. Booker, eds, *Proceedings of the Fourth International Conference on Genetic Algorithms,* San Mateo, CA, 1991. Morgan Kaufmann.

[19] J.Y. Suh and K.V. Gucht, "Incorporating heuristic information into genetic search", in: J.J. Greffenstette, ed, *Genetic Algorithms and Their Applications: Proceedings of the Second International Conference on Genetic Algorithms,* Hillsdale, N J, 1987. Lawrence Erlbaum Associates.

[20] J. Richardson, M. Palmer, G. Liepins, and M. Hilliard, "Some guidelines for genetic algorithms with penalty functions", in: J.D. Schaffer, ed, *Proceedings of the Third International Conference on Genetic Algorithms,* Los Altos, CA, 1989. Morgan Kaufmann.

[21] Z. Michalewicz and C. Janikow, "Handling constraints in genetic algorithms", in: R.K. Belew and L.B. Booker, eds, *Proceedings of the Fourth International Conference on Genetic Algorithms,* San Mateo, CA, 1991. Morgan Kaufmann.

[22] Z. Michalewicz, *Genetic Algorithms + Data Structure = Evolution Programs.* Springer-Verlag, 1992.

[23] F.C. Richards, T.P. Meyer, and N.H. Packard, "Extracting cellular automaton rules directly from experimental data", *Physica,* 45D, 189-202, 1990.

[24] J.L. Breeden and N. Packard, "Nonlinear Analysis of Data Sampled Nonuniformaly in Time", *Physica D,* 58, 273-283, 1992.

[25] R. Reynolds and W. Sverdlik, "Solving problems in hierarchically structured systems using cultural algorithms", in: D.B. Fogel and W. Atmar, eds, *Proceedings of the Second Annual Conference on Evolutionary Programming,* San Diego, CA, 1993. Evolutionary Programming Society.

[26] L. Shu and J. Schaeffer, "Hcs: Adding hierarchies to classifier systems", in: R.K. Belew and L.B. Booker, eds, *Proceedings of the Fourth International Conference on Genetic Algorithms,* Sam Mateo, CA, 1991. Morgan Kaufmann.

[27] G. Syswerda, "Uniform crossover in genetic algorithms", in: J.D. Schaffer, ed, *Proceedings of the Third International Conference on Genetic Algorithms,* Los Altos, CA, 1989. Morgan Kaufmann.

[28] W. Spears and K.D. Jong, "On the virtues of parameterized uniform crossover", in: R.K. Belew and L.B. Booker, eds, *Proceedings of the Fourth International Conference on Genetic Algorithms,* San Mateo, CA, 1991. Morgan Kaufmann.

[29] Y. Ichikawa and Y. Ishii, "Retaining diversity of genetic algorithms for multivariable optimization and neural network learning", in: *1993 IEEE International Conference on Neural Networks,* pages 1110-1116, Piscataway, NJ, 1993. IEEE.

[30] L. Davis, ed, *Genetic Algorithms and Simulated Annealing.* Morgan Kaufmann Publishers, Inc., 1987.

[31] D. Adler, "Genetic algorithms and simulated annealing: A marriage proposal", in: *1993 IEEE International Conference on Neural Networks,* pages 1104-1109, Piscataway, N J, 1993. IEEE.

[32] L. Davis, "Adapting operator probabilities in genetic algorithms", in: J.D. Schaffer, ed, *Proceedings of the Third International Conference on Genetic Algorithms,* Los Altos, CA, 1989. Morgan Kaufmann.

[33] C. Shaefer and A. Morishima, "The ARGOT strategy: Adaptive representation genetic optimizer technique", in: J.J. Greffenstette, ed, *Genetic Algorithms and Their Applications: Proceedings of the Second International Conference on Genetic Algorithms,* Hillsdale, N J, 1987. Lawrence Erlbaum Associates.

[34] D. Whitley, "Using reproductive evaluation to improve genetic search and heuristic discovery", in: J.J. Greffenstette, ed, *Genetic Algorithms and Their Applications: Proceedings of the Second International Conference on Genetic Algorithms,* Hillsdale, N J, 1987. Lawrence Erlbaum Associates.

[35] T. Fogarty, "Varying the probability of mutation in the genetic algorithm", in: J.D. Schaeffer, ed, *Proceedings of the Third International Conference on Genetic Algorithms,* Los Altos, CA, 1989. Morgan Kaufmann.

[36] D.B. Fogel, L.J. Fogel, W. Atmar, and G. B. Fogel, "Hierarchic methods of evolutionary programming", in: D.B. Fogel and W. Atmax, eds, *Proceedings of the First Annual Conference on Evolutionary Programming,* San Diego, CA, 1992. Evolutionary Programming Society.

[37] P. Baldi and Y. Chanvin, "Temporal evolution of generalization during learning in linear networks", *Neural Computation,* 3, 589-603, 1991.

[38] J. Moody, "The effective number of parameters: an analysis of generalization and regularization in nonlinear learning systems", in: J. Moody, S. Hanson, and R. Lippmann, eds, *Advances in Neural Information Processing Systems* 4, pages 847-854, San Mateo, CA, 1992. Mogan Kaufmann.

[39] A.M. Fraser, "Information and entropy in strange attractors", *IEEE Trans. Information Theory,* 5(2), 245-262, March 1989.

[40] M.C. Mackey and L. Glass, "Oscillation and chaos in physiological control systems", *Science,* 197, 287, 1977.

[41] L. Glass and M.C. Mackey, *From Clocks to Chaos.* Princeton University Press, 1988.

[42] S.M. Omohundro, "Efficient algorithms with neural network behavior", J. *Complex Sys.,* 1,273-347, April 1987.

[43] J.L. Breeden, "Optimal Representation of Experimental Data", Doctoral dissertation, Univ. of Illinois, Urbana-Champaign, IL.

[44] R.N. Bracewell, "The sunspot number series", *Nature,* 171(4354), 649-50, April 11, 1953.

[45] R.N. Bracewell, "Sunspot number series envelope and phase", *Aust. J. Phys.,* **38,** 1009-25, 1985.

[46] H.T. Stetson, *Sunspots in Action.* Ronald Press Co., New York, 1947.

[47] M. Waldmeier, *International Astronomical Union Quarterly Bulletin on Solar Activity.* EidgenSssische Sternwarte Zurich, 1961.

[48] D.M.A. Chinnery, "Solar-geophysical data: Part I (prompt reports)", U.S. Dept. of Commerce, September 1986, No. 505.

[49] J.L. Breeden and A. Hübler, "Reconstructing equations of motion using unobserved variables", *Phys. Rev. A,* 42, 5817-5826, 1990.

[50] A. Hübler and E. Lüscher, "Resonant stimulation and control of nonlinear oscillators", *Naturwissenschaften,* 76, 67, 1989.

[51] E. Jackson and A. Hubler, "Periodic entrainment of chaotic logistic map dynamics", *Physica D,* 44, 407-420, 1990.

[52] E.A. Jackson, "The entrainment and migration controls of multiple-attractor systems", *Phys. Lett. A,* 151(9), 478-484, 1990.

Chapter 6

Arthur L. Corcoran
Roger L. Wainwright

Department of Mathematical and Computer Sciences
The University of Tulsa
600 South College Avenue
Tulsa, OK 74104-3189

corcoran|rogerw@penguin.mcs.utulsa.edu

Using LibGA to Develop Genetic Algorithms for Solving Combinatorial Optimization Problems*

* Research partially supported by OCAST Grant AR2-004 and Sun Microsystems, Inc.

0-8493-2519-6/95/$0.00 + $.50

Abstract

In this paper we provide an introduction to genetic algorithms and how they are used to solve combinatorial optimization problems. We describe LibGA, a genetic algorithm development library. LibGA is used to solve three simple combinatorial optimization problems: bin packing, the traveling salesman problem, and multiprocessor scheduling. Sufficient details of LibGA are provided to enable the reader to easily use LibGA as a tool for solving additional combinatorial optimization problems.

6.1 Introduction

Combinatorial optimization problems are among the most difficult problems faced by Computer Scientists. These problems collectively belong to the class of NP-complete problems. The genetic algorithm is a search technique which borrows ideas from natural evolution to effectively find good solutions for combinatorial optimization problems.

This chapter is organized as follows: Section 6.2 provides a brief introduction to genetic algorithms. Section 6.3 describes how combinatorial optimization and genetic algorithms are related. Section 6.4 introduces LibGA, a genetic algorithm development library developed for solving combinatorial problems. Section 6.5 describes three different combinatorial optimization problems: bin packing, the traveling salesman problem, and multiprocessor scheduling. Code and sample output are provided for a simple implementation of these problems using LibGA. Conclusions are presented in Section 6.6.

6.2 Genetic Algorithms

A genetic algorithm (GA) is an adaptive search technique based on the principles and mechanisms of natural selection and 'survival of the fittest' from natural evolution. GAs grew out of Holland's [21] study of adaptation in artificial and natural systems. By simulating natural evolution in this way, a GA can effectively search the problem domain and easily solve complex problems. Furthermore, by emulating biological selection and reproduction techniques, a GA can perform the search in a general, representation-independent manner.

The genetic algorithm operates as an iterative procedure on a fixed size population or pool of candidate solutions. The candidate solutions represent an encoding of the problem into a form that is analogous to the chromosomes of biological systems. Each chromosome represents a possible solution for a given objective function. Associated with each chromosome is a fitness value, which is found by evaluating the chromosome with the objective function. It is the fitness of a chromosome which determines its ability to survive and produce offspring. Each chromosome is made up of a string of genes (whose values are called alleles). The chromosome is typically represented in the GA as a string of bits. However, integers and floating point numbers can easily be used.

Figure 6.1 illustrates a 'canonical' genetic algorithm. The GA begins by generating an initial population, $P(t = 0)$, and evaluating each of its members with the objective function. While the termination condition is not satisfied, a portion of the population is selected, somehow altered, evaluated, and placed back into the population. At each step in the iteration, chromosomes are probabilistically selected from the population for reproduction according to the principle of the 'survival of the fittest'. Offspring are generated through a process called crossover, which can be augmented by mutation. The offspring are then placed back in the pool, perhaps replacing other members of the pool. This process can be modeled using either a 'generational' [20, 21] or a 'steady-state' [39] genetic algorithm. The generational GA saves offspring in a temporary location until the end of a generation. At that time the offspring replace the entire current population. Conversely, the steady-state GA immediately places offspring back into the current population.

The genetic algorithm relies on genetic operators for selection, crossover, mutation, and replacement. The selection operators use the fitness values to select a portion of the population to be parents for the next generation. Parents are combined using the crossover and mutation operators to produce offspring. This process combines the fittest chromosomes and passes superior genes to the next generation, thus providing new points in the solution space. The replacement operators ensure that the 'least fit' or weakest chromosomes of the population are displaced by more fit chromosomes.

```
procedure GA
begin
 t = 0;
 initialize P(t)
 evaluate structures in P(t);
 while termination condition not satisfied do
 begin
  t = t+1;
  P(t) = select from P(t-1)
  alter structures in P(t);
  evaluate stuctures in P(t);
 end
end.
```

Figure 6.1: Genetic algorithm

While the fundamental concepts of genetic algorithms are fairly simple and straightforward, there are numerous implementation variations and options to incorporate into a genetic algorithm. For example, there are numerous ways to parameterize a model and encode it into a finite length chromosome. There are numerous selection techniques for determining chromosomes for crossover. There are literally dozens of possible crossover operators that have been developed in recent years depending on the problem type and chromosome encoding scheme. There are also several techniques for introducing some random changes to a chromosome (i.e., mutation).

6.3 Combinatorial Optimization

Methods to solve difficult combinatorial problems can be divided into two types. The first type includes those methods which try to find optimal solutions through an 'intelligent' exhaustive search. This includes techniques such as backtracking, branch and bound, implicit enumeration, and dynamic programming. Such techniques are only useful for solving combinatorial problems with small sizes. The other type of method for solving combinatorial problems relies on optimization. That is, rather than finding the absolute optimal solution, a 'good' solution is desired within an acceptable time period. These are known as combinatorial optimization techniques. These methods usually employ heuristic algorithms which are problem specific. Since the true optimal is often unknown and impossible to determine, the 'optimal' solution is usually considered the best one obtainable.

Combinatorial optimization methods have as their goal the minimization or maximization of a problem. They are composed of three parts. First, there is a set of problem *instances.* Second, for each problem instance, there is a finite set of *candidate solutions.* Finally, there is a function which assigns to each instance and candidate solution a positive rational number called the *solution value* for the candidate solution. Notice how these elements correspond to those found in genetic algorithms. For a particular problem instance, the GA maintains a set of candidate solutions which are evaluated by the problem specific evaluation function. The solution value returned by the function is used by the GA to measure the relative fitness of that candidate solution. This information is used with the idea of 'survival of the fittest' to conduct the genetic search. As a result GAs are very successful in finding good near-optimal solutions for combinatorial optimization problems.

6.4 LibGA

The LibGA software package [11] was developed primarily because of the noticeable deficiencies of existing GA packages at the time. LibGA is a collection of routines written in the C programming language. It can run on a variety of workstations and PC's, however, since everything in LibGA is in double precision and since genetic algorithms are inherently CPU bound, we elected to implement LibGA in a workstation environment using Unix. We found that executing LibGA on a PC was extremely slow. LibGA provides a user-friendly workbench for genetic algorithm research. It is especially useful for working with combinatorial problems which are often *order-based.* That is, problems for which the representation depends on a certain order being preserved, as with a permutation representation, for example. LibGA includes a rich set of genetic operators for selection, crossover and mutation. An important feature of LibGA is the ability to implement both generational and steady-state genetic algorithms using the genetic operators. This allows researchers the ability to compare between the two approaches. Other features of LibGA include a generation gap, elitism, and the ability to implement a dynamic generation gap. Other routines are provided for initialization, reading a configuration file and generating various statistical reports. LibGA has been requested and sent to locations all over the world. In addition, many have obtained LibGA via anonymous ftp from the 'GA Digest' archive. Information for obtaining LibGA can be found at the end of this paper.

The operators in LibGA include selection, replacement, crossover, and mutation. Selection and replacement can be augmented with elitism. This ensures that the best member of a population survives into the next generation. The selection operators included in LibGA are: *uniform-random, roulette,* and *rank-biased.* Uniform-random selection picks a member of the pool at random, completely ignoring fitness or other factors. Thus each chromosome in the pool is equally likely to be selected. Roulette is the classic selection method used in generational GAs and rank-biased is the classic selection method used in steady-state GAs.

The replacement operators in LibGA are: *append, by-rank, first-weaker,* and *weakest.* The append replacement operator appends new chromosomes to an existing pool. This operator is used in the classical generational GA to place offspring in the new pool. In the by-rank operator the pool is ranked by sorting the fitness values. If the chromosome has a 'high' fitness, it will be placed in the pool, displacing 'weaker' chromosomes. If its fitness is worse than the weakest member of the pool, it dies and is not placed in the pool. Note the weakest and first-weaker operators are somewhere between the append and by-rank operators.

LibGA's crossover operators include *simple, uniform, order1, order2, position, cycle, PMX,* and *asexual.* Simple crossover and uniform crossover are used for traditional bit string encodings of the chromosome. In simple crossover, a random crossover point is selected which divides each parent chromosome into two parts. Alternate parts are contributed by each parent to generate two offspring. This is also known as single point crossover. Uniform crossover selects genes uniformly from either parent to create offspring. The choice of parent is determined randomly for each gene and each parent is equally likely to be selected. Note, however, these crossover operators do not work for order-based problems, since order is not preserved. The other crossover operators preserve order information. Order1, order2, position, cycle, and PMX operators are described in Starkweather *et al.* [35]. The asexual operator is a simple swap of two randomly selected genes, which also is suitable for order-based problems.

LibGA currently offers the following mutation operators: *simple-invert, simple-random,* and *swap.* Simple-invert and simple-random are both used with bit string representations. They both randomly select a gene for mutation. The difference is that simple-invert inverts the bit while simple-random selects a random bit value for the gene, which may be the same as the original bit value. Thus, simple-random has an effective mutation rate of half of the mutation rate for simple-invert. Swap mutation can be used for any representation. It simply swaps two randomly selected genes, which is similar to asexual crossover.

We developed LibGA to be straightforward and easy to use. Figure 6.2 and Figure 6.3 show **ga-test.c**, one of the files provided with LibGA. This file provides the best illustration of how a GA is developed with LibGA. The goal of this test program is to find a sequence of genes which is in sorted order. That is, it is a GA for sorting.

Figure 6.2 shows the main function. The file **ga.h** must be included to obtain all of the necessary definitions for the LibGA library routines. A forward definition for the objective function is found here as well. The main program begins with a call to **GA_config()**, which reads the configuration file **ga_test.cfg** and registers the objective function **obj_fun()**. The complete text of **ga_test.cfg** can be found in the Appendix. Memory is allocated for all of the global information required by LibGA and a pointer to this information is returned by **GA_config()** and is assigned to **ga_info**. If a command line argument is specified, it is assumed to be the name of a crossover function to use instead of the one set by **GA_config()**. In this case, the crossover is changed by using the LibGA internal library routine **X_select()**. When the desired configuration is established, **GA_run()** is then called to run the genetic algorithm.

The remainder of Figure 6.2 has been commented out through the use of the preprocessor directive. This code illustrates the possibility of rerunning the GA with LibGA. Calling **GA_run()** a second time restarts the GA with the same pool and configuration which was present at the end of the previous run. The GA can be reset and a new configuration established by first calling **GA_reset()**, setting any additional parameters, and then calling **GA_run()** again. In Figure 6.2, the GA is reset and run with a different chromosome length.

```
# include "ga.h"

int obj_fun(); /*---  Forward declaration ---*/

/*--- main () ---*/

main(argc, argv)
   int  argc;
   char *argv[];
{
   GA_Info_Ptr ga_info;

    /*--- Initialize the genetic algorithm ---*/
    ga_info = GA_config("ga-test.cfg", obj_fun);

    /*--- Select crossover ---*/
    if(argc > 1) {
        X_select(ga_info, argv[1]);
    };

    /*--- Run the GA ---*/
    GA_run(ga_info);

#if 0
    /*--- Rerun the GA ---*/
    GA_run(ga_info);

    /*--- Reset and rerun the GA ---*/
    GA_reset(ga_info, "ga-test.cfg");
    ga_info->chrom_len = 15;
    GA_run(ga_info);
#endif
}
```

Figure 6.2: ga-test. c (part 1)

Figure 6.3 shows the objective function. This function gives the lowest fitness value when the values in the chromosome are ordered by nondescending allele value. A penalty is used when genes are encountered which are not in the proper position. At the beginning of the objective function, the penalty is set to 1. A 'fudge factor' is computed as l/(10L) where L is the length of the chromosome. The fudge factor is a small fractional value between 0 and 0.1 which is used to indicate how far off a gene is from the desired position. This helps the GA by giving it better 'resolution' when comparing two chromosomes. The *for* loop in Figure 6.3 performs a comparison between the allele value and the one it expects to be in each gene position. If there is no match, it computes a positive distance for how far off the allele is from the desired value. This is multiplied by the fudge factor to get a fractional value which is combined with the penalty and added to the fitness. The final fitness of the chromosome then is a real number where the integer part is the number of alleles out of place, and the fractional part is a measure of how far off each allele is from the desired position. Higher fitness values indicate more undesirable chromosomes, thus, the GA's goal should be to minimize the objective function.

```
/* obj_fun() - user specified objective function */

int obj_fun(chrom)

      Chrom_Ptr chrom;
{
      int i, how_far_off;
      double val = 0.0, penalty, fudge_factor;

     /*--- Penalty for not being in correct position ---*/
     penalty = 1.0;

     /*---  Fudge factor for variance from optimal
         Ensure this is never more than penalty ---*/
     fudge_factor = (1.0 / (double)(chrom->length * 10.0));

     /*--- Fitness is number of genes out of place for sorted
     order ---*/ for(i = 0; i < chrom->length; i++) {
         if(chrom->gene[i] != i+1) {
             how_far_off = chrom->gene[i] - (i+1);
             if(how_far_off < 0) how_far_off = -how_far_off;
             val += penalty + how_far_off * fudge_factor;
         }
      }
     chrom->fitness = val;
}
```

Figure 6.3: ga-test.c (part 2)

Figure 6.4 shows the output of the test program, **ga-test**. LibGA first prints the configuration used in this run. A value of 1 was used to seed the random number generator. An integer permutation representation was used. The initial population was generated randomly with a pool size of 100 and a chromosome length of 10. The GA's objective was to minimize the objective function and elitism was used. A generational GA was run using roulette selection, position based crossover, no mutation, and append replacement. A short report was generated with output interval set at one generation. Next LibGA prints out statistics during the run. Iteration zero indicates the statistics from the initial pool, and statistics for successive iterations follow. When the GA terminates, a report is made indicating the best chromosome ever encountered as well as its fitness. Note that in generation 6 in Figure 6.4 the best fitness in the pool becomes zero. This value corresponds to a chromosome which is ordered by nondescending allele value, that is, a 'most desirable' chromosome. Unfortunately, a fitness value of zero spells disaster for a proportional selection scheme like roulette. However, LibGA automatically scales all of the fitnesses by one when computing the percentage of total fitness, so roulette can still be used. The fitness values printed by LibGA are the unscaled values. As seen in Figure 6.4, the GA converged after 10 generations with a final fitness of 0, which for this problem is optimal. This is apparent when examining the final resulting chromosome, which is sorted.

```
GA    Configuration Information:
Basic Info
      Random Seed                  : 1
      Data Type                    : Integer Permutation
      Init Pool Entered            : Randomly
      Chromosome Length            : 10
      Pool Size                    : 100
      Number of Trials             : Run  until convergence
      Minimize                     : Yes
      Elitism                      : Yes
      Scale Factor                 : 0
Functions
      GA                           : generational (Gap = 0)
      Selection                    : roulette
      Crossover                    : position (Rate = 1)
      Replacement                  : append
Reports
      Type                         : Short
      Interval                     : 1
```

Gener	Min	Max	Ave	Variance	Std Dev	Tot Fit	Best
0	4.16	10.46	9.29	1.44	1.2	928.74	4.16
1	4.08	10.44	8.21	2.66	1.63	821.04	4.08
2	4.08	10.44	6.13	3.51	1.87	713.28	4.08
3	3.04	9.32	6.11	3.16	1.78	611.14	3.04
4	2.02	9.3	5.05	2.1	1.45	505.38	2.02
5	2.02	6.26	4.16	1.08	1.04	415.78	2.02
6	0	5.18	3.47	1.09	1.04	346.86	0
New	scale	factor	= 1				
7	0	5.14	2.63	1.41	1.19	263.18	0
8	0	5.14	1.27	2.12	1.46	126.66	0
9	0	3.14	0.205	0.512	0.715	20.52	0
10	0	0	0	0	0	0	0

The GA has converged after 10 iterations.

Best: 1 2 3 4 5 6 7 8 9 10 (0)

Figure 6.4: ga-test output

6.5 Examples

In this section, we present three simple, related combinatorial optimization problems: bin packing, the traveling salesman problem, and multiprocessor scheduling.

6.5.1 Bin Packing

The Bin Packing Problem is one of the classic NP-complete problems. Except for trivial cases, it is impossible to optimally solve any of these problems. As a result, researchers have focused on approximation techniques which provide efficient, near optimal solutions. Some of these techniques, which are applicable to bin packing and related problems, include heuristic techniques, simulated annealing, neural networks, tabu search and genetic algorithms. Papadimitriou and Steiglitz [30] and Parker and Ratdin [31] present several classical techniques for solving the bin packing problem.

In the classic bin packing problem, a finite collection of packages is packed into a set of bins. The packages and bins are characterized by their weights and capacities, respectively. The problem can be stated either as a decision problem or as an optimization problem. In the decision problem it is necessary to determine whether or not there is a disjoint partitioning of the set of packages such that each partition fits into a bin. Thus, given an integer number of bins, determine if all of the packages fit into the bins. On the other hand, the optimization problem attempts to minimize the number of bins required, or equivalently, to minimize the amount of wasted bin capacity in the packing.

The bin packing problem is a generalization of the Partition Problem. The partition problem is stated as follows: Let $P_1, P_2, ..., P_n$ be a set of real numbers each between 0 and 1; the goal is to partition the numbers into as few subsets as possible such that the sum of numbers in each subset is at most 1.

In the bin packing problem, the bin represents a partition size and the packages must be optimaly placed in these partitions. In most cases, the standard approximation algorithms (first fit, next fit, best fit, etc.) are nearly optimal. However, there are worst case examples which are far from optimal. Some algorithms like the Modified First Fit Decreasing algorithm have tried to improve absolute bounds by special treatment of these worst cases. See Hu [23] for more details on bin packing.

The bin packing problem is applicable in a variety of situations. In computer systems it is used in allocation problems, such as allocating core memory to programs, or space on a disk or tape. The two dimensional bin packing is equivalent to solving the problem of multiprocessor scheduling with time and memory constraints. The packages represent the time and memory requirements of tasks, and the bins represent processors. The knapsack problem is another closely related problem. In other disciplines, bin packing can be used in such problems as packing trucks, allocating commercials to station-breaks on television, and cutting pipe from standardized lengths. The two dimensional problem can be used for stock cutting, where the packages are patterns which must be cut from a fixed width roll of material (the bin). Other classical problems related to bin packing include Job Scheduling, Network Routing, various other layout problems, and the vehicle routing problem (VRP) [9].

The vehicle routing problem (VRP) involves determining minimum cost vehicle routes for a fleet of vehicles originating and terminating from a central location. The fleet of vehicles services a set of customers with a known set of constraints. All customers must be assigned to vehicles such that each customer is serviced exactly once and each vehicle cannot exceed its capacity. The vehicle routing problem has been studied extensively. Bodin *et al.* [5] provides a comprehensive survey of VRP and several variations.

The vehicle routing problem with time windows (VRPTW) adds the additional constraint to the VRP where each customer provides a time window for servicing. Hence, in the presence of time windows, optimization of routing and scheduling of vehicles involves not only total distance traveled, but time costs for waiting when a vehicle has arrived too early to service a customer. Time windows arise frequently in business applications. Examples include bank deliveries, postal deliveries, school bus routing, industrial refuse collection, over night delivery services, passenger and freight operations such as airline, railway and bus routing and scheduling. Time windows also arise in most retail distribution systems. This is an extremely practical problem; efficient routing and scheduling can save industry and government millions of dollars each year. Solomon [33, 34] provides an excellent survey of the vehicle routing problem with time windows. The VRPTW is a variation of the bin packing problem, where each vehicle corresponds to a 'bin' and each customer to be serviced corresponds to a 'package'. The need for multiple vehicles to satisfy the requirements corresponds to the need for multiple 'bins'. The goal is the same: to minimize the number of vehicles (bins) required, while meeting all of the constraints. Blanton and Wainwright have developed special GA crossover operators called MX1 and MX2 specifically for the VRPTW [4].

The classic bin packing problem is expressed using one dimensional packages. This approach blindly generates partitions so that the sum of the one dimensional package parameters in each partition does not exceed the bin capacity. This parameter is typically stated as the package weight. For every package attribute there is a corresponding capacity or maximum value for the bin. Generally, there can be no single package with an attribute exceeding the maximum value set by the bin. Otherwise, the package could never be packed. Interesting cases occur when the parameters differ in relation to one another. For example, when the bin is rectangular and the packages are all square.

In the classic problem, the bin has a fixed capacity. For multiple dimensions, the analogous bin would have fixed capacity in every dimension. Thus, a two dimensional bin would define a rectangle, and a three dimensional bin would define a 'closed box'. A common variation of the classic problem is to use a single, open-ended bin. Used primarily for two or more dimensions, the problem is to minimize the value of the open dimension subject to all other constraints.

When using the level technique, this method can be transformed to the classical problem. This is done by packing each level of the single bin into the multiple closed bins, as if each level were a single package. A less common variation uses multiple, dissimilar bins. The distribution of the bins could be like the distribution of package sizes. They could be random, uniform, skewed, etc. It is analogous to packing a fleet of trucks of different sizes and capacities.

Many near-optimal heuristic techniques have been developed for the one-dimensional bin packing problem. None of these techniques guarantee an optimal packing. However, many techniques approximate the optimal packing within a constant bound. The three best-known approximation algorithms for bin packing are Next Fit, First Fit, and Best Fit. There are many other algorithms, however, most are variations or refinements of these basic methods and only offer modest improvements in packing efficiency. For more details, see Floyd and Karp [17], Garey and Johnson [19] and Johnson *et al.* [25].

The Next Fit heuristic is described as follows: beginning with a single bin, the packages are taken from the list in order, and placed in the next available position. If there is not enough room to pack the current package, the bin is considered 'closed' and a new bin is started. Packages are considered in the order they appear in the list, and once a bin is 'closed' no additional packages can be placed into it. Waste can obviously occur since a new bin is started even when a package may fit into a 'closed' bin. This is a linear algorithm in both time and space. Furthermore, it has been shown that the Next Fit algorithm generates packings no worse than twice optimal. The First Fit algorithm places each package in the first bin in which it will fit. A new bin is added only when all of the previous bins have been examined and no space can be found for a package. In this algorithm there are no 'closed' bins. This algorithm can use quadratic time in the worst case and is O(n log n) on average for n packages. However, it has been shown the packings produced are no worse than 1.7 times optimal.

The Best Fit algorithm places each package in the 'best' bin in which it will fit. The best bin is the one with the least amount of space left over when the package is added. Surprisingly, this algorithm's asymptotic performance and packing efficiency is identical to that for First Fit. Minor differences in packing efficiency are related to package distribution. That is, for package sizes larger than 1/6 of the bin size (distributed uniformly in the range [1/6..1]), Best Fit is more efficient than First Fit. For package sizes larger than 1/5 of the bin size (distributed uniformly in the range [1/5..1]) the packing efficiencies are identical. Sorting the packages before applying these methods can lead to improved results. For example, sorting by decreasing package size before applying First Fit results in packings which are no worse than 11/9 times optimal, a 28% improvement. This variation is called First Fit Decreasing. Similar improvements can be found in Next Fit Decreasing and Best Fit Decreasing [9].

Expanding the problem to two dimensions (rectangle packing) demands different techniques. The first technique uses a 'bottom up - left justified' packing rule, or simply 'bottom-left'. Each package is packed as close to the bottom of the bin and as far to the left of the bin as it can go. This differs from the one dimensional cases where there exists a permutation of the packages for which the methods generate an optimal packing. There are instances where the best bottom-left method produces packings which are 5/4 times optimal. That is, no matter how the packages are ordered, the optimal packing cannot be found. The best absolute packing bounds are obtained by sorting the packages by decreasing width. Classical two dimensional packing generally requires the packing to be orthogonal and disallows rotation of the packages. However, some applications may allow rotation or translation of the packages. Packing efficiency may be improved if each package is rotated so that its width exceeds its height, or vice versa. Several theoretical and practical results are presented in Baker *et al.* [1], Carpenter and Dowsland [7], Coffman *et al.* [8], Dowsland [14] and Leung *et al.* [27].

When extending the problem to three dimensions, it is desirable to apply the results of two dimensional research to obtain similar efficiency. Ideally, the packages would be presorted, then placed level by level, using a two dimensional method to pack each level. Unfortunately, it is difficult to extend the packing efficiency in this way. For example, sorting by decreasing height does not guarantee decreasing width or length. Thus, the two dimensional packing may be inefficient. On the other hand, ordering the packages to make the two dimensional packing efficient may cause wasted space to appear in the height. Unlike the purely two dimensional problem, the two dimensional packing stage must deal with a boundary on the second dimension (the length). Clearly, three dimensional packing is a very practical problem, yet proves to be a very difficult problem to solve [9].

In this chapter, we will concentrate on the simple example of solving the one-dimensional bin packing problem using LibGA. To apply genetic algorithms to the one dimensional packing problem, one must define the encoding of the chromosome, the evaluation function, and the recombination operator. The most natural encoding is to use a string of integers which form an index into the set of packages. The first package is denoted '1', the second '2', and so on. A random reordering of the string represents a random permutation of the packages. The evaluation function returns the number of bins obtained by applying a one dimensional next fit packing algorithm.

The recombination operator must produce a permutation of the packages using partial orderings contained in the two parents. The resulting chromosome must include all of the packages with no duplicates. Fortunately, there are several general purpose crossover functions which meet the requirement. The possible crossover functions that can be used for order based problems include Order1, Order2, Cycle, Position, and Partially Mapped Crossover (PMX). These are described by Whitley and Starkweather [40]. In addition. we have developed an asexual crossover operator which simply exchanges two packages in the list. This is precisely what is done by swap mutation.

Figure 6.5, Figure 6.6, and Figure 6.7 show **gabp.c**, a genetic algorithm for one dimensional bin packing. This program is adapted from **ga-test.c**. Figure 6.5 shows the main program. The GA is configured with **GA_config()**, using the configuration file **gabp.cfg** and registering the objective function **next_fit()**. The configuration file is nearly identical to **ga-test.cfg**. After the

GA has been configured, the packages are read from the data file specified by the **user_data** directive in **gabp.cfg**. The chromosome length is set to the number of packages read. Finally, **GA_run()** is called to run the GA.

```
/* Genetic Algorithm For One Dimensional Bin Packing */

#include "ga.h"

int next_fit(); /* Objective function */
#define MAXPKGS 100 /* Maximum number of packages */
float  Pkgs[MAXPKGS], /* Packages */
       Sum_Pkgs;     /* Sum of all package weights */
int  Num_Pkgs;    /* Actual number of packages */

/* Entry point */
main(argc, argv)
   int argc;
   char *argv[];
{
   GA_Info_Ptr ga_info;

    /*--- Initialize the genetic algorithm  ---*/
    ga_info = GA_config("gabp.cfg", next_fit);

    /*--- Read packages from data file ---*/
    read_packages(ga_info->user_data);

    /*--- Set chromosome length to number of packages ---*/
    ga_info->chrom_len = Num_Pkgs;

    /*--- Run  the GA ---*/
    GA_run(ga_info);

    /*--- This gives us some idea of optimal ---*/
    printf("Sum of package weights = %f\n\n", Sum_Pkgs);
}
```

Figure 6.5: gabp. c (part 1)

Figure 6.6 shows the objective function, **next_fit()**. The chromosome in this problem is a permutation of the list of packages read. The objective function packs these packages using a simple next fit heuristic and returns the total number of bins required in the packing for the chromosome's fitness. In the trivial case of no packages the objective function returns a fitness of zero. Otherwise, it initializes a single empty bin. As seen in the *for* loop in Figure 6.6, the weight of the next package to add to the bin is determined by the index provided by each gene in the chromosome. The weight is a fractional value between 0 and 1 which represents a value normalized with respect to the bin capacity. If the package fits into the bin it is placed there. Otherwise, a new bin is initialized and the package is placed there. The final fitness is the number of bins and the GA's goal is to minimize this value.

```
* Use simple next fit heuristic for objective function */
```

```
int next_fit(chrom)
   Chrom_Ptr chrom;
{
   int i, num_bins;
   float pkg_weight, weight;

/*---  Trivial case: no packages ---*/
    if(chrom->length < i) {
        chrom->fitness = 0;
        return;
    }
    /*--- Initialize ---*/
    num_bins = i;  /* First bin */
    weight  = 0;  /* Its empty */
    /*---  Place each package using next fit ---*/
    for(i  = 0; i < chrom->length; i++) {
        pkg_weight = Pkgs [(int) chrom->gene [i] -1];
        if(weight + pkg_weight > 1.0) {     /* Oops, too big */
                weight = pkg_weight;
                num_bins++;
        }
    else
        {       /* Ahh, it fits */
                weight += pkg_weight;
        }
    }
    /*--- Goal is to minimize the number of bins ---*/
    chrom->fitness = num_bins;
}
```

Figure 6.6: gabp. c (part 2)

Figure 6.7 shows the code used to read the data file. The data file consists of a single line indicating the number of packages, n, followed by n lines which are the normalized package weights. As the packages are read, their weights are summed to give an indication of what the optimal might be.

```
/* Read packages from data file */

read_packages(filename)
    char *filename;

    FILE *fid;
    int i;
    /*--- Open data file ---*/
    if((fid  =  fopen(filename,"r"))  ==  NULL)  {
       printf("Error opening package data file <%s>\n",
       filename);
       exit(1);
    }
    /*--- Get number of packages ---*/
    fscanf(fid,"%d"',  &Num_Pkgs);
    if(Num_Pkgs < 1 || Num_Pkgs > MAXPKGS) {
       printf("Number of packages, %d, out of bounds [1..%d]\n",
           Num_Pkgs, MAXPKGS);
       exit(1);
```

```
    }

    /*--- Get package weights and sum them ---*/
    Sum_Pkgs = 0;
    for(i=0; i < Num_Pkgs; i++) {
        fscanf(fid,"%f", &Pkgs[i] );
        Sum_Pkgs += Pkgs[i];
    }

    /*--- Close data file ---*/
    fclose(fid);
}
```

Figure 6.7: gabp.c (part 3)

Figure 6.8 shows the output from running **gabp** on a 50 package data set. The configuration is the same as for **ga-test** with the exception of user data **r50.bp**, asexual crossover, and report interval 10. The data file used, r50.bp, was a randomly generated set of 50 packages. From Figure 6.8 we see the GA begins with a pool of solutions with fitnesses in the range from 31 bins to 38 bins. The average is 34.5 bins. After 62 generations, the GA has converged to a solution with fitness value of 28 bins. The sum of package weights is nearly 26, indicating an optimal packing can use no less than 26 bins. Thus, the GA has found a solution that is within 2 bins of optimal in this case. Since the true optimal is unknown and impossible to calculate in a reasonable time, and the random packages may not necessarily fit in 26 bins, it is quite possible that the GA has found the optimal. In any case, the GA has found a very good solution. Additional details about genetic algorithms for bin packing can be found in Smith [32], and in our previous work [9, 10, 11].

GA Configuration Information:

Basic Info

User Data	: r50.bp
Random Seed	: 1
Data Type	: Integer Permutation
Init Pool Entered	: Randomly
Chromosome Length	: 50
Pool Size	: 500
Number of Trials	: Run until convergence
Minimize	: Yes
Elitism	: Yes
Scale Factor	: 0

Functions

GA	: generational (Gap = 0)
Selection	: roulette
Crossover	: asexual (Rate = 1)
Replacement	: append

Reports

Type	: Short

Interval : 10

Gener	Min	Max	Ave	Variance	Std Dev	Tot Fit	Best
0	31	38	34.5	1.46	1.21	17256	31
1	31	37	33.9	0.938	0.968	16968	31
10	30	32	30.9	0.226	0.476	15455	30
20	29	31	29.9	0.1	0.317	14956	29
30	28	30	29.1	0.193	0.439	14573	28
40	28	29	28.7	0.214	0.462	14346	28
50	28	29	28.1	0.104	0.323	14059	28
60	28	29	28.0	0.002	0.0447	14001	28
62	28	28	28.0	0	0	14000	28

The GA has converged after 62 iterations.

Best: 36 48 41 10 49 43 37 39 25 26 1 11 22 33 5 2 40 24 6 29
21 47 23 35 14 50 19 15 45 12 4 20 8 31 42 32 44 7 30 3
34 46 27 13 28 16 18 9 17 38 (28)

Sum of package weights = 25.928408

Figure 6.8: gabp.c output

6.5.2 Traveling Salesman Problem

Another well known combinatorial optimization problem is the Traveling Salesman Problem (TSP). Given a set of n points in a plane corresponding to the location of n cities, find the minimum distance closed path that visits each city exactly once. This is called the traveling salesman problem. The traveling salesman problem belongs to a class of minimization problems for which the objective function has many local minima. The objective function is simply the total length of the tour. The traveling salesman route can be thought of as a circular arrangement of n cities, or as a permutation of a list of n cities. Solving this problem requires *O(n!)* computation time since the number of possible tours for n cities is (n- 1)!. In this chapter it is assumed each city is directly connected to every other city by Euclidean distance. That is, distances satisfy the triangle inequality, which means that the direct route between any two cities is never more than an indirect route between two cities. This assumption helps in the design of approximation algorithms for this problem.

We chose the traveling salesman problem to demonstrate LibGA because it is a representative problem for a wide variety of combinatorial optimization problems where the solution space is all permutations of n objects. Other combinatorial optimization problems that fall into this category include the bin packing problem, job scheduling problems, stock cutting, vehicle routing and transportation scheduling problems, etc. Developing efficient genetic algorithms to solve this problem will have direct applications for solving a host of other practical combinatorial optimization problems [3, 37, 38, 41].

Figure 6.9, Figure 6.10, and Figure 6.11 show **gatsp.c**, a genetic algorithm for the traveling salesman problem. This program is adapted from **gabp.c** and **ga_test.c**.

The main program is shown in Figure 6.9. As in **gabp**, the GA is configured with **Gl_config()**, the city information is read from the data file specified in **gatsp.cfg**, and the chromosome length is set to the number of cities read. Finally, the GA is run by calling **GA_run().**

```
/* Genetic Algorithm For The Traveling Salesman Problem */

#include "ga.h"

int eval_tour();                        /* Objective function */
#define MAXCITIES 100           /* Maximum number of cities */
struct {
       float x, y;

} City[MAXCITIES];     /*      Cities */
int  Num_Cities;       /*      Actual number of cities */

/* Entry point */
main(argc, argv)
   int  argc;
   char *argv[];
{
       GA_Info_Ptr ga_info;

       /*--- Initialize the genetic algorithm---*/
       ga_info = GA_config("gatsp.cfg", eval_tour);

       /*--- Read cities from data file ---*/
       read_cities(ga_info->user_data);

       /*--- Set chromosome length to number of cities ---*/
       ga_info->chrom_len = Num_Cities;

       /*--- Run the GA ---*/
       GA_run(ga_info);
}
```

Figure 6.9: gatsp.c (part 1)

The objective function, **eval_tour()**, is shown in Figure 6.10. In the trivial case of no cities, the fitness returned is zero. Otherwise, the Euclidean distance between each successive city in the list is added to the fitness. In the *for* loop in Figure 6.10, each gene is an index into the original list of cities. After the loop, the distance from the last city indexed by the chromosome to the first city indexed by the chromosome is added to the fitness. Thus, the fitness is the total cost of the tour. The GA must minimize this total cost.

```
/* Objective function evaluates tour cost based on Euclidean
distance */

int eval_tour(chrom)
     Chrom_Ptr chrom;
{
     int i, idx1, idx2;
        float dx, dy, cost;
    /*--- Trivial case: no cities ---*/
    if(chrom->length < 1) {
    chrom->fitness = 0;
    return;
    }
    /*--- Initialize ---*/
    cost = 0;

    /*---  Add Euclidean distance from each city to next city ---
         */
    for(i = 0; i < chrom->length - 1; i++) {
        idx1 = (int)chrom->gene[i] - 1;
        idx2 = (int)chrom->gene[i+1] - 1;
        dx   = City[idx1].x - City[idx2].x;
        dy   = City[idx1].y - City[idx2].y;
        cost += sqrt(dx * dx + dy * dy);
    }
    /*---  Add cost from last city to first ---*/
        idx1 = (int)chrom->gene[chrom->length-1] - 1;
        idx2 = (int)chrom->gene[0] - 1;
        dx   = City[idx1].x - City[idx2].x;
        dy   = City[idx1].y - City[idx2].y;
        cost += sqrt(dx * dx + dy * dy);

    /*--- Goal is to minimize the tour cost ---*/
    chrom->fitness = cost;
}
```

Figure 6.10: gatsp.c (part 2)

Figure 6.11 shows the routine which reads the data file. This file is much like the one used in the bin packing problem. The first line indicates the number of cities, and the remaining lines give the x and y coordinates for each city. These coordinates are assumed to be normalized so they fall in the range from 0 to 1.

```
/*--- Read cities from data file ---*/

read_cities(filename)
   char *filename;
{
   FILE *fid;
   int i;

    /*--- Open data file ---*/
    if((fid = fopen(filename,"r")) == NULL) {
        printf("Error opening city data file <%s>\n", filename);
        exit(1);
```

```
    }

    /*--- Get number of cities ---*/
    fscanf(fid."%d", &Num_Cities);
    if(Num_Cities < 1 || Num_Cities > MAXCITIES) {
       printf("Number of cities, %d, out of bounds [1..%d]\n",
          Num_Cities, MAXCITIES);
       exit(1);
    }

    /*--- Get city coordinates ---*/
    for(i=0; i < Num_Cities; i++) {
       fscanf(fid,"%f  %f",  &City[i].x,  &City[i].y);
    }

    /*--- Close data file ---*/
    fclose(fid);
}
```

Figure 6.11: gatsp.c (part 3)

Figure 6.12 shows the output of **gatsp**. The parameters are the same as before in Figure 6.4 and Figure 6.8 except for user data **r50.tsp** and report interval 100. The data file **r50.tsp** is a randomly generated set of 50 cities. As Figure 6.12 shows, the initial pool contained solutions whose fitness ranged from about 21 to about 30. After 1309 generations, the GA converged to a fitness of about 6.

```
GA Configuration Information

Basic Info
    User Data                : r50.tsp
    Random Seed              : 1
    Data Type                : Integer Permutation
    Init Pool Entered        : Randomly
    Chromosome Length        : 50
    Pool Size                : 500
    Number of Trials         : Run  until convergence
    Minimize                 : Yes
    Elitism                  : Yes
    Scale Factor             : 0

Functions
    GA                       : generational (Gap = 0)
    Selection                : roulette
    Crossover                : asexual (Rate = 1)
    Replacement              : append

Reports
    Type                     : Short
    Interval                 : 100

Gener   Min      Max      Ave    Variance  Std Dev  Tot Fit   Best
  0   21.0887  30.2088  26.7     2.24      1.5    13374.2  21.0887
```

1	21.0887	29.5356	25.7	1.77	1.33	12873.9	21.0887
100	10.9397	12.3754	11.6	0.0843	0.29	5819.41	10.9397
200	9.4759	10.1167	9.82	0.0231	0.152	4906.61	9.4759
300	8.89632	9.5162	9.27	0.018	0.134	4633.49	8.89632
400	8.38697	8.93954	8.65	0.00876	0.0936	4324.46	8.38697
500	8.15736	8.48956	8.3	0.00555	0.0745	4149.75	8.15736
600	8.05966	8.25668	8.12	0.0031	0.0557	4061.96	8.05966
700	6.96856	8.15455	8.05	0.00178	0.0422	4025.26	6.96856
800	6.4698	6.9195	6.6	0.00759	0.0871	3799.99	6.4698
900	6.40255	6.66445	6.51	0.00219	0.0468	3753.72	6.40255
1000	6.34219	6.48982	6.42	0.00236	0.0486	3706.87	6.34219
1100	6.34219	6.42519	6.35	0.000118	0.0109	3674.05	6.34219
1200	6.34219	6.3614	6.35	4.85E-05	0.00696	3672.66	6.34219
1300	6.34219	6.3614	6.34	3.66E-06	0.00191	3671.19	6.34219
1309	6.34219	6.34219	6.34	0	0	3671.1	6.34219

The GA has converged after 1309 iterations.

Best: 7 30 41 33 45 15 27 42 16 24 38 9 22 37 5 13 17 12 35 49
28 14 47 40 43 31 8 4 25 46 26 29 10 34 23 19 36 48 20 1
6 32 50 18 39 21 11 44 2 3 (6.34219)

Figure 6.12: gatsp output

Figure 6.13 shows an example random tour for the data. Figure 6.14 shows the final tour found by the GA. We see again that the GA has been able to find a very good solution.

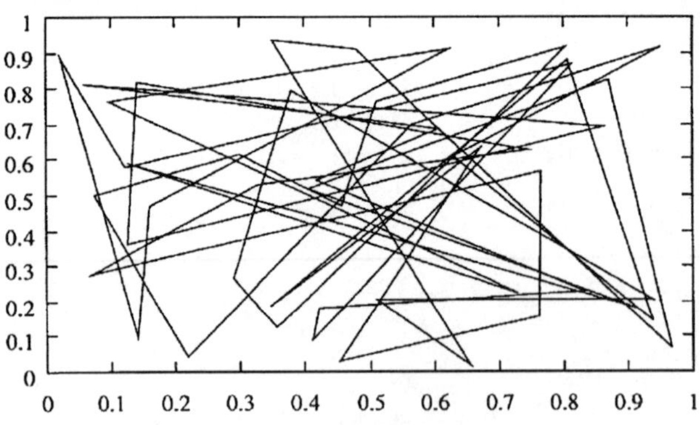

Figure 6.13: Random Tour.

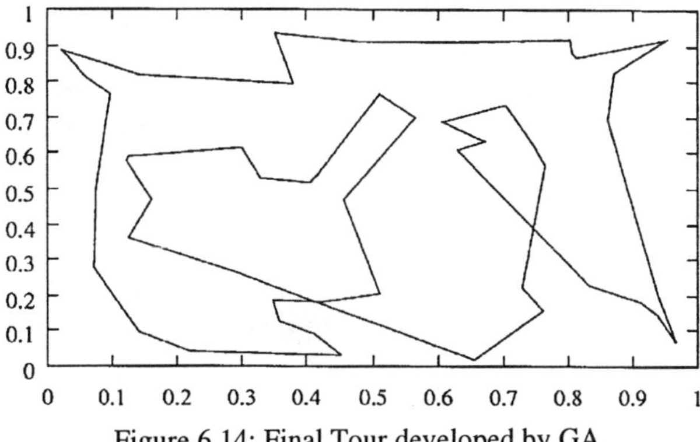

Figure 6.14: Final Tour developed by GA.

6.5.3 Multiprocessor Scheduling

The Multiprocessor Scheduling Problem is defined as follows: a set of n jobs is to be scheduled on a set of m identical processors. Each job J is specified as J = (t, c), where c is the capacity (memory) requirement of the job and t is its running time. Note that only nonpreemptive job scheduling is considered. That is, once a job is started it remains in the processor until it is finished. The objective is to determine a schedule of jobs on the machine so as to minimize the total processing time. For example, consider the case where m = 3, each processor has memory capacity = 5, and the 14 jobs to be scheduled are:

Job	1	2	3	4	5	6	7
	(2,2)	(1,3)	(1,3)	(3,2)	(4,1)	(3,1)	(3,4)

Job	8	9	10	11	12	13	14
	(2,3)	(1,4)	(2,2)	(1,2)	(2,2)	(2,l)	(3,1)

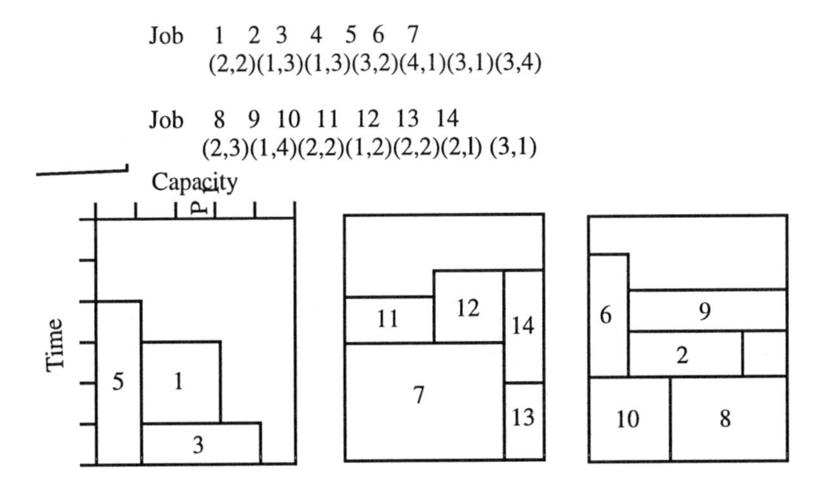

Figure 6.15: An Example Schedule.

An example schedule is shown in Figure 6.15 requiring 5 time units. An optimal schedule requiring 4 time units is shown in Figure 6.16. Note the Job Shop Scheduling problem is a variation of the multiprocessor scheduling problem. In the job shop scheduling problem each job, J_i requires the completion of several tasks, $T_{j,1}$, $T_{j,2}$, ..., $T_{j,n}$. The tasks for any job J_i are to be carried out in the order 1, 2, 3, ..., etc., where each task j cannot begin until task j - 1 (j > 1) has been completed [22]. In the job shop scheduling problem, the processor capacity is not considered. It is assumed every task in every job uses the entire capacity of a given processor.

The multiprocessor scheduling problem is also similar to the two dimensional bin packing problem, where each processor is a bin, and each job is a package. The multiprocessor scheduling problem and its variations of other scheduling problems are economically very important problems, especially in industrial applications.

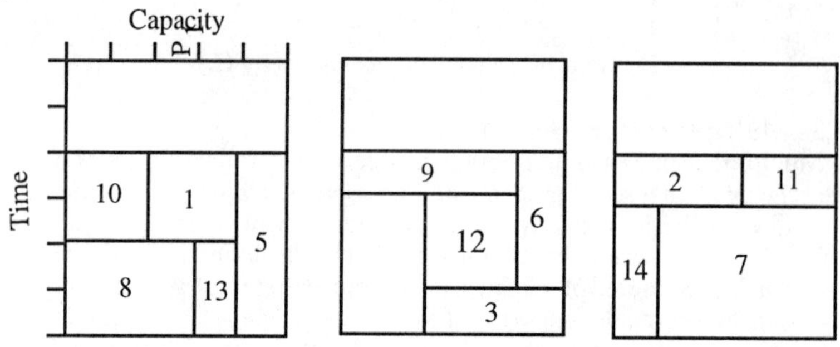

Figure 6.16: An Optimal Schedule.

```
#include "ga.h"

int eval_tasks();                 /*       Objective function */

#define MAXTASKS 100  /*       Maximum number of tasks */

struct  {
     float time,      /*   Time requirement */
     mem;             /*   Memory requirement */
} Task[MAXTASKS];     /*   Tasks */
int  Num_Tasks;       /*   Actual number of tasks */

double Sum_Area;      /*   Sum of task "area" (time * memory) */

/* Entry point */
main(argo, argv)
   int  argc;
   char *argv[];
{
   GA_Info_Ptr ga_info;

   /*--- Initialize the genetic algorithm---*/
   ga_info = GA_config("gams.cfg", eval_tasks);

   /*--- Read tasks from data file ---*/
```

```
   read_tasks(ga_info->user_data);

   /*--- Set chromosome length to number of tasks ---*/
   ga_info->chrom_len = Num_Tasks;

   /*--- Run the GA ---*/
   GA_run(ga_info);

   /*--- This gives us some idea of optimal ---*/
   printf("Total task area = %G\n", Sum_Area);
}
```

Figure 6.17: gams.c (part 1)

```
/* Objective function evaluates task list using next fit
   placement */

int eval_tasks(chrom)
   Chrom_Ptr chrom;
{
      int i;
   float tot_time, max_time, tot_mem, task_time, task_mem;

/*--- Trivial case: no tasks ---*/
   if(chrom->length < 1) {
      chrom->fitness = 0;
      return;
   }

   /*--- Initialize ---*/
   tot_time = 0;
   max_time = 0;
   tot_mem  = 0;
   /*---      Place each task using next fit ---*/
   for(i      = 0; i < chrom->length; i++){

      task_time = Task [ ( int ) chrom->gene [i] -1]. time;
      task_mem = Task[(int)chrom->gene[i]-1] .mem;

      /*--- Place task on a "level" ---*/
      if(tot_mem + task_mem > 1.0) { /* Oops too much memory */
            tot_mem  = task_mem;
            tot_time += max_time;
            max_time  = 0;
      } else {                        /* Ahh, it fits */
            tot_mem += task_mem;
}
/*--- Find longest task time on a "level" ---*/
if(task_time > max_time) max_time = task_time;
}
/*--- Goal is to minimize the total time ---*/
chrom->fitness = tot_time;
}
```

Figure 6.18: gams.c (part 2)

```
/* Read tasks from data file */

read_tasks(filename)
   char *filename;
{
   FILE *fid;
   int i;

    /*--- Open data file ---*/
    if((fid = fopen(filename,"r")) ==  NULL) {
        printf("Error opening task data file <%s>\n", filename);
        exit(1);
    }

    /*--- Get number of tasks ---*/
    fscanf(fid,"%d", &Num_Tasks);
     if(Num_Tasks < 1 || Num_Tasks > MAXTASKS){
        printf("Number of tasks, %d, out of bounds [1..%d]\n",
            Num_Tasks, MAXTASKS);
        exit(1);
    }

    /*--- Get task time and memory requirements ---*/
    Sum_Area = 0;
    for(i=0; i < Num_Tasks; i++){
        fscanf(fid,"%f %f", &Task[i].time, &Task[i].mem);
        Sum_Area +=Task[i].time * Task[i].mem;
    }

    /*--- Close data file ---*/
    fclose(fid);
}
```

Figure 6.19: gams.c (part 3)

For the interested reader, Yamada and Nakano [42] present a GA implementation for large-scale job shop problems. Davidor *et al.* [13] investigated GAs as a technique for solving the job shop scheduling problem. Kidwell [26] developed a GA to schedule distributed tasks on a bus-based system. Li and Cheng [28] developed a job shop scheduling algorithm to partition a mesh connected system where jobs require meshes and the system itself is a square mesh of size a power of two. For other related work see [6, 15, 16, 24, 36, 41]. The LibGA implementation for the multiprocessor scheduling problem is shown below. This is an extremely simplified version of the one the authors studied in more detail in [12].

Figure 6.17, Figure 6.18, and Figure 6.19 show the GA for multiprocessor scheduling, **gams.c**. As before, the GA is configured with **GA_config()**, the set of tasks is read from the data file, and the chromosome length is set to the number of tasks. The objective function in Figure 6.18 evaluates the total time required for the task list ordering indicated by the chromosome. In the trivial case of an empty task list, the fitness is zero. Since this problem is equivalent to a two dimensional bin packing problem, each task is otherwise placed using a two dimensional, level oriented next fit heuristic. Note, the 'width of the bin' in this case corresponds to the memory capacity of the processors. The goal of the GA is to minimize the total time. Figure 6.19 shows the routine to read the initial task list, which is straight forward.

Figure 6.20 shows the output of **gams**. The configuration is the same as before except the user data is **r50.ms** and the report interval is 30. The GA begins with an initial pool of chromosomes which represents solutions whose total time is in the range from about 16 to about 21 time units. After 297 generations, the GA has obtained a solution of about 14 time units, which is quite close to the optimal's lower bound of about 12 time units.

```
GA      Configuration Information:

Basic Info
      User Data                       : r50.ms
      Random Seed                     : 1
      Data Type                       : Integer Permutation
      Init Pool Entered               : Randomly
      Chromosome Length               : 50
      Pool Size                       : 500
      Number of Trials                : Run until convergence
      Minimize                        : Yes
      Elitism                         : Yes
      Scale Factor                    : 0

Functions
      GA                              : generational (Gap = 0)
      Selection                       : roulette
      Crossover                       : asexual (Rate = 1)
      Replacement                     : append

Reports
      Type                            : Short
      Interval                        : 30
```

Gener	Min	Max	Ave	Variance	Std Dev	Tot Fit	Best
0	16.1001	20.7399	18.6	0.566	0.752	9311.39	16.1001
1	15.7861	20.0496	18.1	0.344	0.587	9074.3	15.7861
30	14.2964	15.199	14.8	0.0236	0.154	7412.93	14.2964
60	13.9174	14.2645	14.1	0.0035	0.0592	7036.35	13.9174
90	13.8126	14.0378	13,9	0.00139	0.0372	6958.99	13.8126
120	13.7852	13.8808	13.8	0.000255	0.016	6903.08	13.7852
150	13.6964	13.7908	13.7	0.0015	0.0388	6862.18	13.6964
180	13.6655	13.6964	13.7	0.000219	0.0148	6841.08	13.6655
210	13.6655	13.6964	13.7	2.36E-05	0.00486	6833.32	13.6655
240	13.6655	13.6655	13.7	1.37E-12	1.17E-06	6832.73	13.6655
270	13.6655	13.6655	13.7	6.71E-13	8.19E-07	6832.73	13.6655
297	13.6655	13.6655	13.7	0	0	6832.73	13.6655

```
The GA has converged after 297 iterations.

Best:   28 43 42 12 24 31 14 41 33 40 49 13 30 29 36 38 10 46 45
        27 15  2  7 11 5 34 26 20  6 21 8 35 22 3 39  9 19 23
         4 50 25 48 32 47 44 17 16 37 18 1 (13.6655)

Total task area = 12.4629
```

Figure 6.20: gams output

6.6 Conclusions

We have given several examples to show that genetic algorithms are ideally suited for solving combinatorial optimization problems. We used LibGA to implement three such problems: bin packing, the traveling salesman problem, and multiprocessor scheduling. The use of LibGA allows the problems to be coded with a minimal knowledge of genetic algorithms. Parameters can be easily changed by simply editing a configuration file. One of the objectives of this paper is to show how easy LibGA is to use, and provide enough example code to allow the reader the ability to easily begin to use LibGA.

Acknowledgements

This research has been partially supported by OCAST Grant AR2-004. The authors also wish to acknowledge the support of Sun Microsystems, Inc. We also extend heartfelt thanks to those who have taken the time to use LibGA and have sent us comments, questions, and suggestions.

LibGA Availability

LibGA is available at no cost by sending an e-mail request to the authors. The authors respective e-mail addresses are *corcoran@penguin.mcs.utulsa.edu*, and *rogerw@penguin.mcs.utulsa.edu*. Conventional mail should be addressed to the authors at:

Department of Mathematical and Computer Sciences
University of Tulsa
600 South College Avenue
Tulsa, OK 74104-3189
USA

LibGA can also be obtained via anonymous ftp from *ftp.aic.nrl.navy.mil* as */pub/galist/src/ga/libga100.tar.Z.*

References

[1] B.S. Baker, E.G. Coffman, and R. L. Rivest. Orthogonal packings in two dimensions. *SIAM Journal of Computing,* 9(4):846-855, Nov. 1980.

[2] R. K. Belew and L.B. Booker, editors. *Proceedings of the Fourth International Conference on Genetic Algorithms,* San Diego, California, 1991. Morgan Kaufmann.

[3] J.L. Blanton and R.L. Wainwright. Vehicle routing with time windows using genetic algorithms. In W.A. Coberly, editor, *Proceedings of the Sixth Oklahoma Symposium on Artificial Intelligence,* pages 242-251, Tulsa, Oklahoma, Nov. 1992.

[4] J.L. Blanton and R.L. Wainwright. Multiple vehicle routing with time and capacity constraints using genetic algorithms. In Forrest [18], pages 452-459.

[5] L. Bodin, B. Golden, A. Assad, and M. Ball. Routing and scheduling of vehicles and crews: The state of the art. *Comput Opns. Res.,* 10:62-212, 1983.

[6] R. Bruns. Direct chromosome representation and advanced genetic operators for production scheduling. In Forrest [18].

[7] H. Carpenter and W.B. Dowsland. Practical considerations of the pallet-loading problem. *Journal of the Operational Research Society,* 36(6):489-497, 1985.

[8] E.G. Coffman, M.R. Garey, D.S. Johnson, and R.E. Tarjan. Performance bounds for level-oriented two dimensional packing algorithms. *SIAM Journal of Computing,* 9(4):808-826, Nov. 1980.

[9] A.L. Corcoran and R.L. Wainwright. A genetic algorithm for packing in three dimensions. In H. Berghel, E. Deaton, G. Hedrick, D. Roach, and R. Wainwright, editors, *Proceedings of the 1992 ACM/SIGAPP Symposium on Applied Computing,* pages 1021-1030, New York, 1992. ACM Press.

[10] A.L. Corcoran and R.L. Wainwright. A heuristic for improved genetic bin packing. *Information Processing Letters,* May 1993. Submitted.

[11] A.L. Corcoran and R.L. Wainwright. LibGA: A user-friendly workbench for order-based genetic algorithm research. In E. Deaton, K.M. George, H. Berghel, and G. Hedrick, editors, *Proceedings of the 1993 ACM/SIGAPP Symposium on Applied Computing,* pages 111-118, New York, 1993. ACM Press.

[12] A.L. Corcoran and R.L. Wainwright. A parallel island model genetic algorithm for the multiprocessor scheduling problem. In E. Deaton, K.M. George, H. Berghel, and G. Hedrick, editors, *Proceedings of the 1994 ACM/SIGAPP Symposium on Applied Computing,* New York, 1994. ACM Press.

[13] Y. Davidor, T. Yamado, and R. Nakano. The ECOlogical framework II:: Improving GA performance at virtually zero cost. In Forrest [18].

[14] K.A. Dowsland. An exact algorithm for the pallet loading problem. *European Journal of Operational Research,* 31:78-84, 1987.

[15] F.F. Easton and N. Mansour. A distributed genetic algorithm for employee staffing and scheduling problems. In Forrest [18].

[16] H.-L. Fang, P. Ross, and D. Corne. A promising genetic algorithm approach to job-shop scheduling, re-scheduling, and open-shop scheduling problems. In Forrest [18].

[17] S. Floyd and R.M. Karp. FFD bin packing for item sizes with uniform distributions on [0, 1/2]. *Algorithmica,* 6(2):222-239, 1991.

[18] S. Forrest, editor. *Proceedings of the Fifth International Conference on Genetic Algorithms,* Urbana-Champaign, Illinois, July 1993. Morgan Kaufmann.

[19] M.R. Garey and D.S. Johnson. Approximation algorithms for bin packing problems: A survey. In G. Ausiello and M. Lucertini, editors, *Analysis and Design of Algorithms in Combinatorial Optimization,* pages 147-172. Springer-Verlag, New York, 1981.

[20] D.E. Goldberg. *Genetic Algorithms in Search, Optimization, and Machine Learning.* Addison-Wesley, Reading, Massachusetts, 1989.

[21] J.H. Holland. *Adaptation in Natural and Artificial Systems.* The University of Michigan Press, Ann Arbor, Michigan, 1975.

[22] E. Horowitz and S. Sahni. *Fundamentals of Computer Algorithms.* Computer Science Press, 1984.

[23] T.C. Hu. *Combinatorial Algorithms.* Addision-Wesley, 1982.

[24] P. Husbands and F. Mill. Simulated co-evolution as the mechanism for emergent planning and scheduling. In Belew and Booker [2].

[25] D.S. Johnson, A. Demers, J.D. Ullman, M.R. Garey, and R. L. Graham. Worst-case performance bounds for simple one-dimensional packing algorithms. *SIAM Journal of Computing,* 3(4):299-325, Dec. 1974.

[26] M.D. Kidwell. Using genetic algorithms to schedule distributed tasks on bus-based system. In Forrest [18].

[27] J.Y. Leung, T.W. Tam, C.S. Wong, G.H. Young, and F.Y. Chin. Packing squares into a square. *Journal of Parallel and Distributed Computing,* 10:271-275, 1990.

[28] K. Li and K.-H. Cheng. Job scheduling in a partitionable mesh using a two-dimensional buddy system partitioning scheme. *IEEE Transactions on Parallel and Distributed Systems,* 2(4):413-422, Oct. 1991.

[29] R. Manner and B. Manderick, editors. *Parallel Problem Solving from Nature, 2.* North-Holland, Amsterdam, 1992.

[30] C.H. Papadimitriou and K. Steiglitz. *Combinatorial Optimization: Algorithms and Complexity.* Prentice-Hall, Englewood Cliffs, New Jersey, 1982.

[31] R.G. Parker and R.L. Ratdin. *Discrete Optimization.* Academic Press, New York, 1988.

[32] D. Smith. Bin packing with adaptive search. In J.J. Grefenstette, editor, *Proceedings of the First International Conference on Genetic Algorithms and Their Applications,* pages 202-207, Hillsdale, New Jersey, 1985. Lawrence Erlbaum.

[33] M.M. Solomon. Algorithms for the vehicle routing and scheduling problems with time window constraints. *Operations Research,* 35(2):254-265, 1987.

[34] M.M. Solomon and J. Desrosiers. Time window constrained routing and scheduling problems: A survey. *Transportation Science,* 22(1):1-11, 1988.

[35] T. Starkweather, S. McDaniel, K. Mathins, D. Whitley, and C. Whitley. A comparison of genetic sequencing operators. In Belew and Booker [2], pages 69-76.

[36] H. Tamaki and Y. Nishikawa. A paralleled genetic algorithm based on a neighborhood model and its application to the jobshop scheduling. In Manner and Mandrake [29].

[37] S.R. Thangiah.*GIDEON: A Genetic Algorithm System for Vehicle Routing with Time Windows.* PhD thesis, North Dakota State University, May 1991.

[38] S.R. Thangiah, K.E. Nygard, and P.L. Juell. Gideon: A genetic algorithm system for vehicle routing with time windows. In *Proceedings of the Seventh Conference on Artificial Intelligence Applications,* pages 322-325, Miami, Florida, 1991.

[39] D. Whitley and J. Kauth. GENITOR: A different genetic algorithm. In *Proceedings of the Rocky Mountain Conference on Artificial Intelligence,* pages 118-130, Denver, Colorado, 1988.

[40] D. Whitley and T. Starkweather. GENITOR II: A distributed genetic algorithm. *Journal of Experimental and Theoretical Artificial Intelligence,* 2:189-214, 1990.

[41] D. Whitley, T. Starkweather, and D. Fuquat. Scheduling problems and traveling salesman: The genetic edge recombination operator. In J.D. Schaffer, editor, *Proceedings of the Third International Conference on Genetic Algorithms,* Arlington, Virginia, 1989. Morgan Kaufmann.

[42] T. Yamado and R. Nakano. A genetic algorithm applicable to large-scale job-shop problems. In Manner and Manderick [29].

Chapter 7

Lance Chambers
140 Treasure Road
Queens Park
Western Australia 6107

Strategic Modelling using a Genetic Algorithm Approach

0-8493-2519-6/95/$0.00 + $.50

7.1 Introduction

In the modelling paradigm used (refer to Chapter 1 for a description of the modelling system employed) for the following examples the terms impact and growth are somewhat interchangeable, with impact tending toward the immediate sense of a single period, and growth being more of a cumulative value, e.g. over several periods of a simulation. Also growth doesn't necessarily imply increase, it can be negative. The term Cross Impact is usually used in the sense of the factor in question as being dependent upon another factor, and the cross impact being the effect of the latter upon the former. A model is a model, a graph is a graph, and a population is the set of genes, and their associated parameters being employed in a Genetic Search, which is what the whole exercise is for.

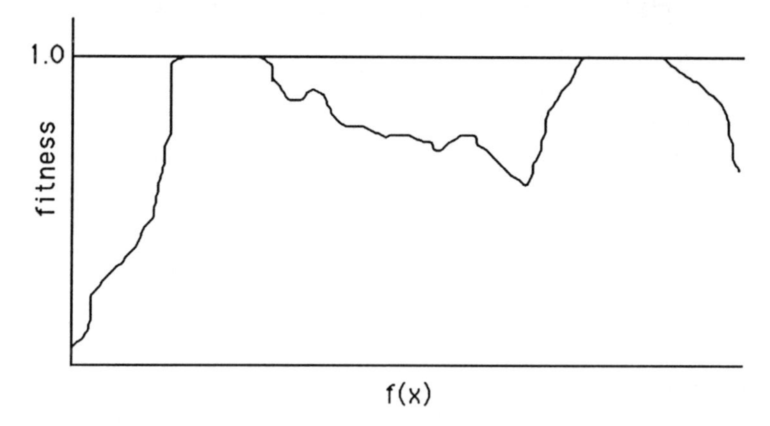

Figure 7.1 Example of multi-optima function.

7.2 Structure of a Model

What is a model? The foundation of the model is the factors represented by the Model Window (everything not evident here will be explained below). A factor is a name, some values, some connections and a set of attributes. The primary value of a factor is its self impact, the rate at which the factor will grow each simulation period, if unrestrained/encouraged. Another value is its start value. This is the initial value assigned in the first period of the simulation. A more important value, which only a few factors will use, is the target value. This is the desired result for the factor at the end of the simulation. This forms the objective function for the search procedure. A factor has the following set of attributes:

Controllable: Is the factor part of the solution for the search procedure?

Delayed Impact: If the factor is active only for a subset of the simulation time.

Bounds: Limits on the total growth of a factor.

Thresholds: The impact values which can influence a factor. Impacts less than the lower threshold have no effect, and impacts greater than the upper threshold are limited. Mainly used for cross impacts

System Shocks: Exceptional growths for specific periods.

Graphs: If the factors growth during the simulation is graphed.

Cross Impacts: The proportion of another factor's growth in a period which caused a growth in the specified factor.
Delayed Cross Impacts: When the growth of a factor for a period is carried over to the next period. This is mainly useful if a factor has a zero self impact, but benefits from cross impacts form other factors. This enables other dependent factors to in turn benefit from the cross growths in this factor... understand?

The model, apart from being the sum of the factors, contains a few bits of other information, mostly to do with default values for new factors, but also some global flags. Specifically, bounds, thresholds and system shocks can be turned off, and also cross impacts can be forced to delayed or immediate if desired.

7.3 A Simulation

A simulation is the self impacts and cross impacts being exercised over a specified number of periods. Given the nature of cross impacts, recursive cross impacts and the discrete nature of simulation time in particular, it wouldn't be a good idea to get too picky over whether a particular event occurs in, say, period 9 or period 10. Some things get delayed for a period, but over a number of periods, it all works out. The simulation saved the cumulative growths for each factor for each period, which can later be graphed or saved for further analysis. Old simulation results which have been saved cannot be used again by Genie.

7.4 Graphs

Factors which have the graph attribute set are graphed at the end of each simulation run. The graphs are displayed in a reduced form in the Graph Browser window, and can be viewed at full size if desired. The graphs are useful for setting target values for searching, and also for comparing growths.

7.5 Populations

A population is a collection of genes and information as to how to manipulate them. Some of this information is local to each gene, or even to the chromosomes within the genes, and some is universal to the population. The details are discussed below in ‘Population Windows’.

Populations are attached to and saved with a model. A population cannot migrate across models. Many populations can be attached to one model.

7.6 The Menus

7.6.1 File Menu

7.6.1.1 New Model/ Open Model:

These are only active when no model is currently open. They do exactly what they say they do.

7.6.1.2 Close:

This will close the topmost window. If the window is a Model or Population window and is in need of saving, a prompt will be issued. Closing the model window closes all other windows as well.

7.6.1.3 Save:
Saves the topmost window. This is only applicable to Model and Population windows.

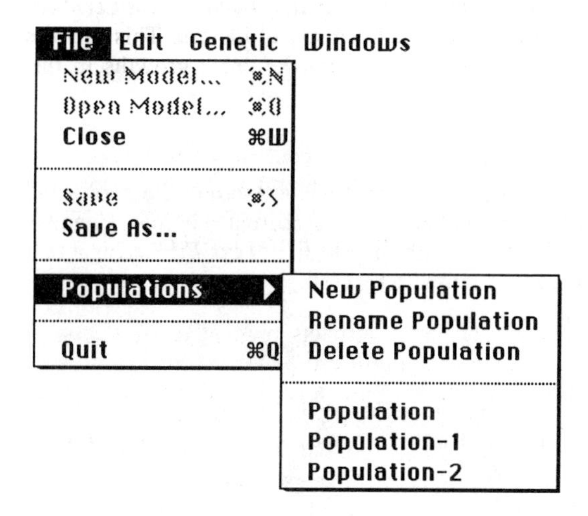

7.6.1.4 Save As:
This is the main avenue for save text representations of models and populations, it also performs the more mundane task of copying models and populations. See the detailed 'Save As', below.

7.6.1.5 Populations:
This is only active when a model is open. The top half of the submenu deals with creating and managing existing populations. The items are:

7.6.1.6 New Population:
Creates, but doesn't save, a new Population window, with default settings.

7.6.1.7 Rename Population:
See 'Population Windows', below.

7.6.1.8 Delete Population:
See 'Population Windows', below.

The bottom half of the menu lists existing populations; those saved with the model, and those created, and still existing, in the current session. Selecting one of these opens and/or brings the window to the front.

7.6.1.9 Quit:
Prompts the user to save unsaved windows, sounds taps, brushes its teeth and goes to bed. What do you think its supposed to do?

7.6.1.10 Save As Menu Item

As mentioned, this command does quite a lot. Its function is different for each window. For all windows excepting population windows, a modified Standard Put File dialog is presented. A pop-up menu with the different save options appears alongside the prompt. For a fuller explanation of the windows, see the respective sections below.

7.7 Model Window:

Genie File
Text File

Two options exist for model windows. **Genie File** copies the current window to a new file and this becomes the current window. Changes made since the last save are not saved with the old file. **Text File** saves a text description of the model. This is not reusable by **Genie.**

7.7.1 Graph Browser:

PICT
TEXT Data

PICT saves the complete window in, you guessed it, 'PICT' format. **Text Data** saves the results from the most recent simulation in a tab delimited text file.

7.7.2 Graph Window:

Same options as for Graph Browser, but only saves either the graph picture, or the data for the factor it represents.

7.7.3 Population Window

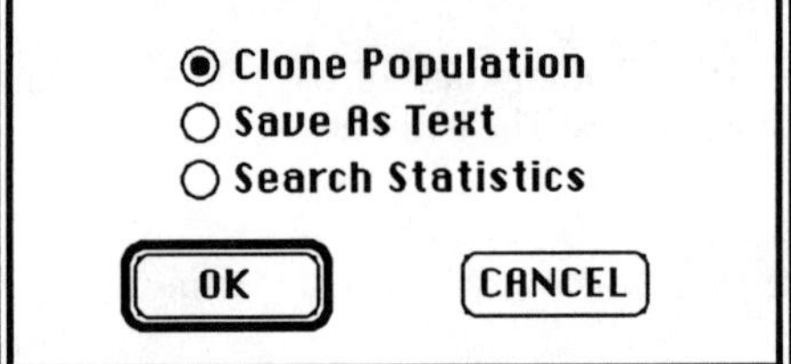

Clone Population creates a copy of the Population window. **Save As Text** writes all the population and search parameters followed by any existing chromosomes, as a text file. **Search Statistics** saves the best, worst and average fitnesess for each generation of the most recent search in a tab delimited text file.

7.8 Edit Menu

The edit menu doesn't really get used much. The Graph and Graph Browser windows can be copied to the clipboard with it, and, if the Graph Browser is the topmost window, 'Select All' will select all the cells, but beyond that, not much.

7.9 Window Menu

This menu lists all the current windows. Items beneath the dimmed 'Graphs' are, of course, graphs. The first item is the model and beneath that are any open population windows.

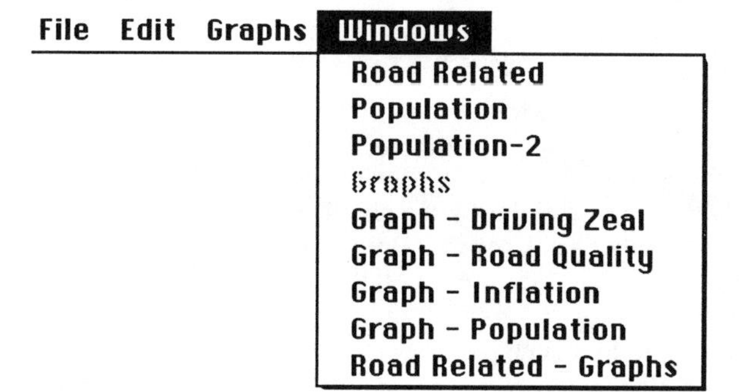

7.10 The Windows

7.10.1 Model Window

The foundation of any **Genie** session is the model window. There can be many windows open, but only one model window can be open at any time.

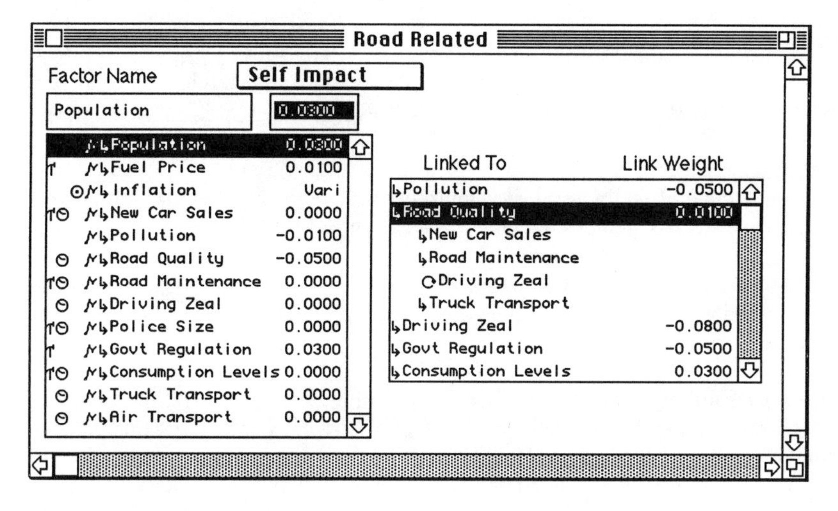

7.10.2 Factors List

There are two lists in the model window. The left list is the factors currently defined for the model. The factor name is in the centre. To the left of the factor name are attribute flags. The symbols are as follows:

- Hammer - Factor forms part of the solution during searching;
- Clock - Cross impacts carried over to next period during simulation;
- Target - Factor is a search target;
- Graph - Factor produces graph after simulation;
- ↳Link - Factor is affected by other factors.

The rightmost number displays the value indicated by the pop-up menu above the list.

Self Impact
Search Values
Start Value
Active Targets
All Targets

7.10.2.1 Self Impact:
The growth value applied each period during the simulation. 'Vari' means that the growth value is defined by the 'VariGrowth' factor attribute.

7.10.2.2 Search Values:
The self impacts derived from the search process. These values are set from the search window. If 'Search Values' is selected, then a simulation will use these values instead of defined self impacts during simulation. If the factor is not active in the search, no value is displayed.

7.10.2.3 Start Value:
An arbitrary starting value.

7.10.2.4 Active Targets:
This is the target value for the factors defined as current search targets. If the factor is not a current search target, no value is displayed.

7.10.2.5 All Targets:
The target value of the factor, if it were to be declared as a current search target. Changing this value includes the factor as a current search target.

The two text fields above the factors list are for changing the factor name and displayed value. The tab key changes between the fields, as will clicking on the field.

Clicking on a factor selects that factor for editing. When an existing factor is selected, the value field is made active. Only the factor name and value can be edited directly in the window. Double clicking on a factor brings up the 'Cross Impacts' dialog box, as explained below.

7.10.3 Cross Impacts List

This is a hierarchical list of the factors that influence the factor selected in the Factors List through cross impacts. Indented entries indicate the hierarchy. The format of the list is similar to the Factors List, with a few exceptions. There are only two symbols displayed, the ↳ symbol, if the factor is in turn subject to cross impacts, and the ↻ symbol, when the factors cross impacts are recursive. The value is the cross impact. Only the values for the top level factors are listed.

Double Clicking on a factor in this list causes the factors that influence it to be listed, indented. If a factor is free from cross impacts, nothing will occur (aside from a chastising beep, perhaps). Double Clicking again on the factor will cause the sub factors to disappear.

The length of the lists is pinned the the window size. As the window gets longer, so do the lists. If the window is reduced in size so as to partly cover one of the scroll bars, the scroll bar will disappear, meaning that if you want to scroll the list, you must drag with the mouse. There is a good reason for this.

7.11 Model Menu

File Edit **Model** Windows

New Factor	⌘K
Remove Factor	
Reset Factor	
Cross Impacts...	⌘I
Factor Attributes...	⌘T
Model Preferences...	⌘M
Go	⌘G

This is the menu associated with the model window.

7.11.1 New Factor:
Unsurprisingly, this creates a new factor, if the maximum number of factors is not exceeded, that is. The factor is given a default name and attributes. See 'Limits and Defaults' below.

7.11.2 Remove Factor:
The factor, and all references to it are deleted.

7.11.3 Reset Factor:
The factor is restored to its default values and attributes, and all cross impacts are removed.

7.11.4 Cross Impacts:
See 'Cross Impacts Dialog' below.

7.11.5 Factor Attributes:
See 'Factor Attributes Dialog' below.

7.11.6 Model Preferences:
See 'Model Preferences Dialog' below.

7.11.7 Go:
If all the prerequisites are in order (i.e. factors exist) then the simulation is run for the number of periods defined in the Model Preferences Dialog. The results of the simulation can be saved as a tab delimited text file from the 'Graph Browser'

window which saves all factors or individual 'Graph' windows, which saves only the data for the factor shown. See 'Save As' menu item above.

7.12 Cross Impacts Dialog

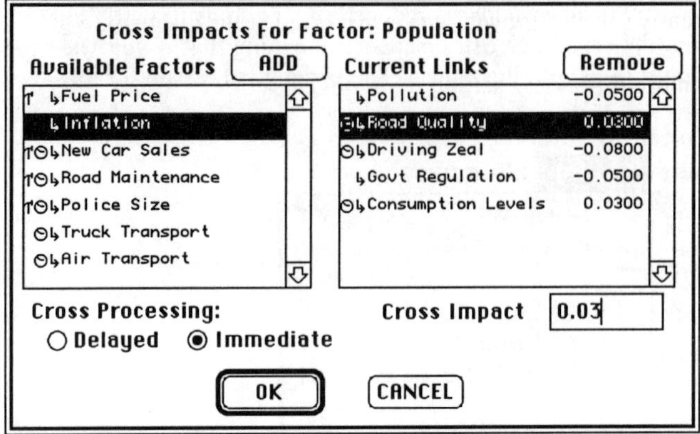

The cross impacts are set through this dialog box. In the left hand list are the available factors, and in the right hand list the factors already defined as cross impacts. A subset of attribute symbols appears in both lists. To add a cross impact, select a factor from the list of available factors, and click the 'Add' button, or alternatively, double click the item. To remove a factor, select an item in the 'Current Links' list, and click the 'Remove' button, or double click the item. Use the 'Cross Impact' text field to enter the value of the impact. Zero values imply no impact, and the factor will not be included as a cross impact when the dialog box is closed.

The 'Cross Processing' radio button determines whether the Cross Impact growths are carried across to the next period or not.

7.13 Factor Attributes Dialog

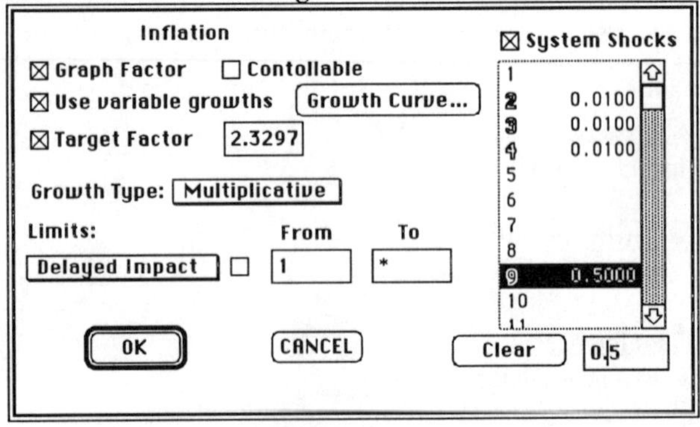

Most of a factor's attributes are set from here. The dialog is largely self explanatory. The pop-up menu for growth type looks like this:

The currently selected growth type is displayed in the dialog.

A little more difficult is setting the limits.

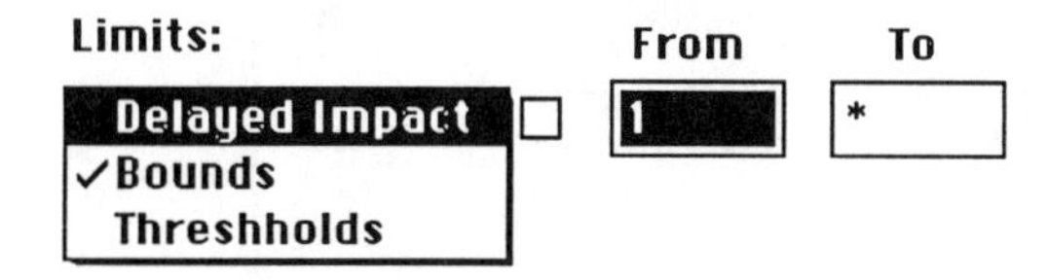

The check box turns the selected limit on or off, and the boxes show the bounds. Active limits are checked in the menu. An asterix '*' in a text field indicates that the default value is to be used, usually +/- infinity.

The system shocks list specifies a limited number of special shocks to apply during a simulation. Periods with shocks are displayed in outlined text, along with the shock value. To set a shock, select the desired periods, and enter a shock value in the text field below the list. To clear shocks, select the periods, and click the 'Clear' button.

If VariGrowths are set, then the button 'Growth Curve' becomes active. Clicking on this button allows the definition of variable self impacts. If the factor has not previously had VariGrowths defined, then this dialog will appear first:

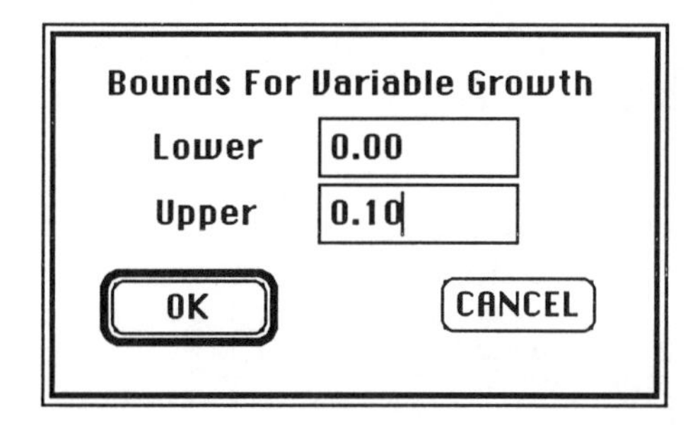

The lower and upper bounds are possible limits of the self impacts. The VariGrowths dialog looks like:

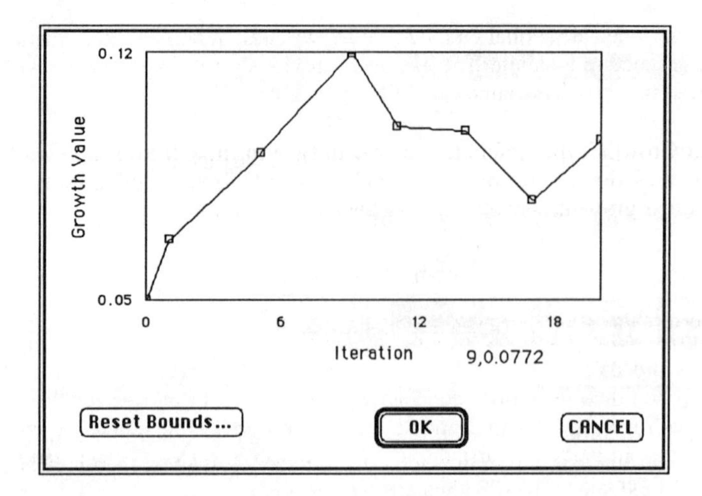

Points can be defined within the range by clicking. The first and last points are always tied to the first and last period and cannot be deleted. Clicking within a point removes it, and clicking at a different growth value on the same period as another point repositions it. This can get a bit tricky when the number of iterations is large, as one pixel may cover several periods, so don't be too exacting about where your points go.

The 'Reset Bounds' button brings up the bounds dialog box. If the bounds are changed, the points already defined are scaled to the new bounds. This is also true for the iterations. If the number of iterations is later changed, the points are scaled to fit.

7.14 Model Preferences Dialog

Model Parameters

Iterations 20 | ☒ Shocks On
Display Precision 4 | ☒ Bounds On
Default Start Val. 1.0000 | ☒ Threshholds On

Default Growth Type: Multiplicative
Force Cross Impacts: None

OK CANCEL

This dialog sets the model defaults and parameters. The 'Iterations' field is the number of periods in a simulation. The 'Display Precision' sets the number of decimal places used in the various outputs. 'Default Start Val.' is the value that new factors use for their starting values.

The check boxes are global enable/disable switches. When checked, factors can use the respective attribute if it has been previously set. When unchecked, the attribute is blocked. The names are self explanatory.

'Default Growth Type' refers to the growth type for new factors, and works the same way as the pop-up menu in the 'Factor Attributes' dialog (see above). Forced cross growths is another global switch. The options are:

'None' is the default option, and factors are allowed to carry over their cross impacts as set in the 'Cross Impacts' dialog above. If 'Immediate' is selected, then no cross impacts are carried over. If 'Delayed' is selected, all cross impacts are carried over to the next period.

7.15 Graph Browser Window

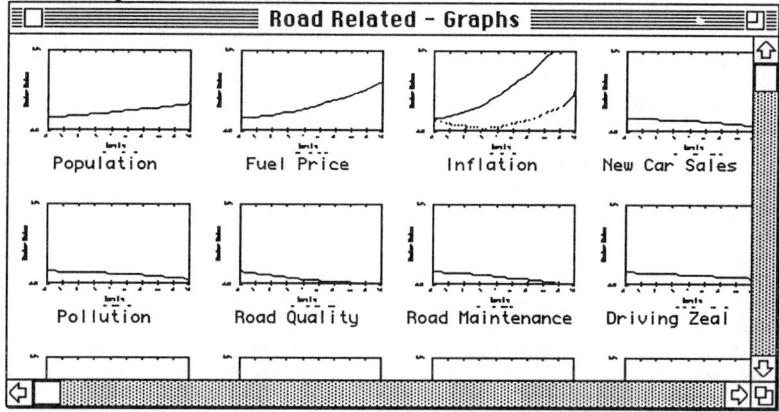

This window presents the graphs from the last simulation run in a reduced form. The window is actually a list and can be manipulated as such. To open the full size graph, double click it. It is possible to rescale the graphs from the Graph Browser window, using the 'Graphs' Menu.

7.15.1 Rescale Graph:

A dialog is brought up asking the user for new scale bounds. Only the growth scale can be changed. Data points outside of the new bounds are not shown. If more than one graph is selected then all are scaled to the new bounds.

7.15.2 Auto Scale:

The selected graphs are scaled to the maximum and minimum of the data. If more than one graph is selected, the user is asked whether to scale all graphs alike, in which case the maximum and minimum are derived from the data for the selected graphs (factors), or to scale the graphs independently.

The 'Graph Browser' window is updated after each simulation run. The window can be copied to the clipboard or saved to a file. It is possible to save the window in a 'PICT' file, or to save the data (all of it) in a tab delimited text file. See 'Save As' above.

7.16 Graph Window

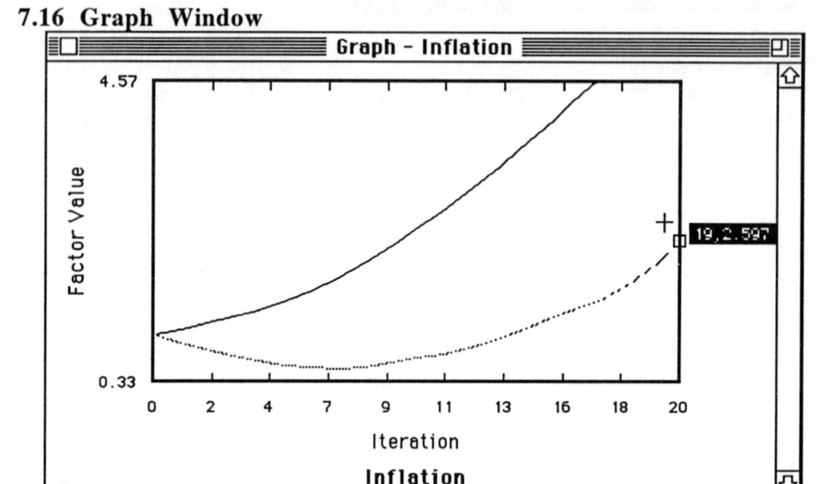

This is the full sized representation of a factor's behaviour during a simulation run. The same scaling options as in the 'Graph Browser' are applicable: Rescale and Auto Scale. This window can be copied to the clipboard. It is also possible to save the graph as a 'PICT' image, or to save the data represented by the graph. See 'Save As' above.

What can be done in this window, however, is the setting of target values. There is a roving coordinate to the right of the graph. When the cursor approaches the right boundary of the graph, this becomes highlighted. Clicking now will set the target value to the y value in the coordinate. An new grey curve will appear. The new curve has the same origin as the original curve, and terminates at the target. What happens in between is not particularly important and should be taken as a serious expectation of factor behaviour.

If the factor is not already an active target, the target value can be changed by selecting a new value. If the cursor enters the small target box, it becomes highlighted. Clicking in this box removes the target.

7.17 Population Window

7.17.1 Population Structure

Simply put: A population consists of chromosomes; chromosomes are made up of genes and genes are constituted of allele. Now to work back up the scale. Each allele is represented by a bit and a string of bits forms a gene. Currently the maximum length of a gene is 28 alleles, which is ridiculously accurate.This comes about more because its convenient to store these things as LongInts (32 bits, the Integer type is 16 bits, which may not be sufficient for all eventualities). What happened to the last 4 bits? They're being reserved for later fancy stuff like switching factors on and off, etc. The number of allele may change later. A chromosome is made up of these genes and also some information specific to the chromosome. Each chromosome is given a unique ID number. The ID's of its parent are stored in the chromosome, which could be useful for tracing ancestry. The chromosome also stores its own fitness value.

At the population level is the phenotype template. This is essential to the interpretation of the genes. For each gene is stored:

The associated factor from the model;
The range of values that the gene can represent;
The resolution of the gene (number of allele);
The coding method;
The mating method.

Currently there is only the binary range coding available. Mating methods available at the moment are one site or two site splices. Pretty ordinary, but then we evolved from the primeval sludge, so there's hope yet.

Also at the population level is the selection technique. Young hopefuls are selected by one of the following:

Roulette - Random selection weighted toward fitter chromosomes;
Random - Straight random selection
Fit/Fit - With the population sorted by fitness, successively fit chromosomes are selected. Fit goes with fit, weak goes with weak; 'and like shall go with like…'
Fit/Weak - The fittest chromosome mates with the weakest, the next fittest with the next weakest, etc.

<u>7.17.1.1 Available Replacement Schema:</u>
Weak Parent - If the child is fitter than the parent, the parent is replaced;
Both Parent - Both parents are replaced by the children;
Weakest Gene - The weakest chromosome in the population gets it;
Random - A randomly selected chromosome is replaced by the child.

<u>7.17.1.2 Convergence Schema:</u>
Fitness Sum - Converged if the sum of the finesses over the population falls below the convergence criteria;
Average Fitness - When the population average falls below the criteria;
Best Gene - When the chromosome of best fit is under the criteria;
Worst Gene - When the worst chromosome is under the criteria.

7.18 The Population Window

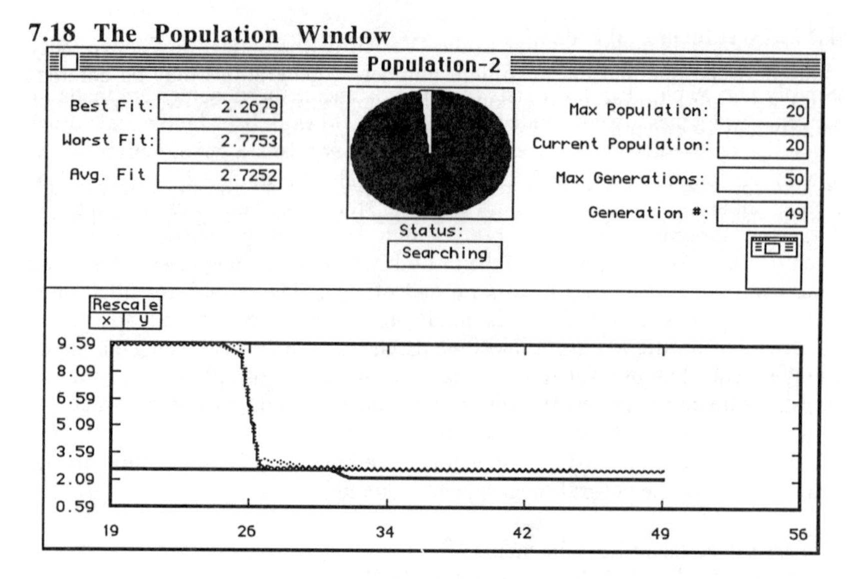

Above is the full sized population window. Two sizes are available, half size, which displays only the statistics for the current generation, and full size, which displays a running graph of best fit, worst fit and average fit up to the current generation. The top half of the window is self explanatory. The oval in the centre is a progress indicator, and the status box underneath the oval shows one of the following:

Stopped - No search in progress.

Seeding - The seed population is being generated. The indicator shows the progress of the seeding operation.

Searching - A search is in progress. The indicator displays the proportion of the maximum generations completed.

Paused - The search has been suspended, and is able to be restarted if desired. This gives an opportunity to modify some noncritical parameters (see 'Population Limits,' below).

Clicking on this icon switches back to the half window. It is replaced by another icon which when clicked upon, restores the graph.

The graph is a running graph of the current search. Since the domain of the solution space is unknown, and more to the point, some of the bad fits are really bad, it is necessary to be able to rescale the graph. Zooming in on a portion of the graph is accomplished by click-dragging a rectangle over the desired area. The boundaries of the rectangle will become the new graph. Going the other direction, or being more precise is achieved by this box: Rescale x y. Clicking on **Rescale** rescales the entire graph to the global minima and maxima. Clicking **x** brings up the graph bounds dialog, similar to the 'VariGrowth Bounds' box above. Any value between one and the maximum number of generations may be entered. Clicking **y** also brings up the bounds box, and any value between zero and infinity may be entered.

A text copy of the population parameters, and existing chromosomes or the search statistics can be obtained through the 'Save As' menu item in the file menu, discussed above.

7.19 The Genetic Menu

File Edit **Genetic** Windows

Reset Parameters	
Population Parameters...	⌘L
Meet The People...	⌘M
Go	⌘G

This menu is associated with population windows.

7.19.1 Reset Parameters:
Reinitialises the population and search parameters to the default settings.

7.19.2 Population Parameters:
See 'Population Parameters,' below.

7.19.3 Meet The People:
To view the population, kiss babies, open hospitals, etc. See 'Meet The People' below.

7.19.4 Go:
This starts a genetic search. If a previous search has been suspended, or, there exists a population, the the user is asked if they wish the population to be reinitialised. In the case of a suspended search, not reinitialising continues on from the last generation. With a population left over from a completed search, not reinitialising uses the old population as the seed population. Reinitialising clears the population and resets the statistics.

7.20 Population Limits Dialog

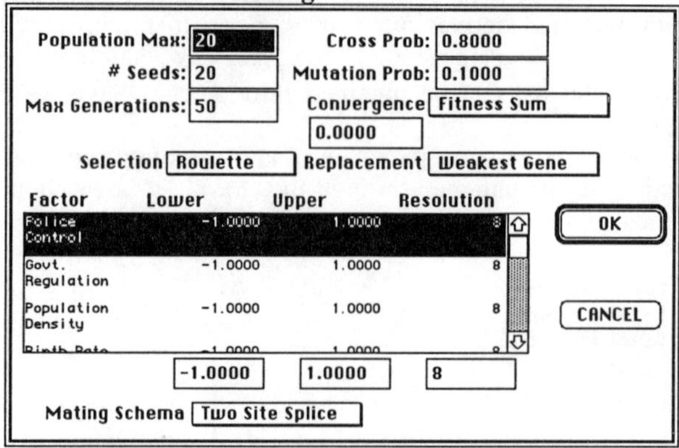

All the parameters for a population and search are contained here. This dialog may be accessed while a search is suspended, but changing some of the parameters will cause the population to be cleared. These are: Population Max; Selection Criteria; Phenotype Parameters.

The various option for the selection, replacement and convergence criteria have been discussed above in 'Structure of a Population'. The list box, and the fields below it are concerned with the phenotype parameters. The values in the edit fields are for the selected line in the list, and changing these effects a change in the list.

7.21 Meet The People Dialog

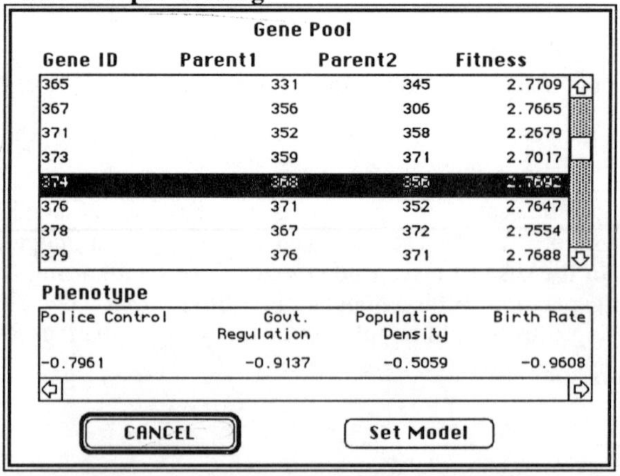

Gene Pool

Gene ID	Parent1	Parent2	Fitness
365	331	345	2.7709
367	356	306	2.7665
371	352	358	2.2679
373	359	371	2.7017
374	368	356	2.7692
376	371	352	2.7647
378	367	372	2.7554
379	376	371	2.7688

Phenotype

Police Control	Govt. Regulation	Population Density	Birth Rate
-0.7961	-0.9137	-0.5059	-0.9608

CANCEL Set Model

Here is where we get a more detailed look at the population behind the search. The upper box presents the surviving chromosomes, with their immediate parentage and fitness. Selecting a chromosome puts its gene values into the lower box. Clicking **Set Model** writes these values into the 'Search Values' field of the model. It is then possible to run the model with these values and analyse the effects of the changes more completely. If Fit/Fit or Fit/Weak chromosome selection was employed, the list will be sorted on fitness, otherwise, it is unsorted (sorry).

A text copy of the list can be obtained through the 'Save As' menu item, discussed above.

7.22 Defaults And Limits

Model
Maximum Factors:100
Factor Name Length:20
Default Start Value:1.0
Maximum System Shocks:12 (per factor)
Maximum VariGrowth Points:12 (incl. ends)
Maximum Simulation Periods:100
Default Periods20
Maximum Graphs:50

Populations
Maximum Population Size:100
Default Size:100
Minimum Population Size:10
Maximum Genes:10
Maximum Resolution:30
Default Resolution:8
Self Impact Range:-1.0 ... 1.0
Maximum Targets:10
Maximum Generations:500
Default Generations:50

7.23 Model Construction and Interpretation of Results

7.23.1 Global Warming

The first model is one that was constructed to test the theory of a coming Ice Age as proposed by Professor Hamaker of Purdue University (note that any misinterpretation of his theory is purely the responsibility of the author and in no way can be attributed to Professor Hamaker).

The table is read as follows:

The factor descriptor is followed by a list of factors that are affected by the factor described. For example Vege Growth Rates will cause a reduction in Atmos CO2 levels and an increase in food supplies. The values given for each affected factor is a measure of the extent these factors are affected. A value of 0.5 implies that the affected factor moves at a rate of 50% of the movement rate of the affecting factor, therefore the model assumes that for a 100% increase in Vege Growth Rates there will be a decrease of 2% in Atmos CO2 and there will be a 3% increase in food supplies.

Description of Model Ice Age
No. of Factors: 14

1 Atmos CO2 levels	
Atmos CO2 levels	0.0010
Vege Growth Rates	0.0300
World Temperature	0.0200
Extreme Weather	0.0200
Acid Rain	0.0100
2 Vege Growth Rates	
Atmos CO2 levels	-0.0200
Food Supplies	0.0300
3 Food Supplies	
Atmos CO2 levels	-0.0700
Population	0.0030
Birth Rate	0.0030
Rainfall	0.0300
Surface Water	-0.0030
4 Population	
Atmos CO2 levels	0.0500
Food Supplies	-0.0500
Population	0.0300

Birth Rate	-0.0030
Degredation of Soil	0.0700
Forest Fires	0.0200
Acid Rain	0.1000
Surface Water	-0.0030
5 Birth Rate	
Population	0.0050
6 Degredation of Soil	
Vege Growth Rates	-0.0600
Degredation of Soil	0.0100
Forest Fires	0.0600
7 Size of Arctic Icecap	
Degredation of Soil	-0.0010
Size of Arctic Icecap	0.0200
World Temperature	-0.0200
Volcanic Activity	0.0700
Extreme Weather	0.0400
Surface Water	-0.0500
8 Rainfall	
Vege Growth Rates	0.0200
Degredation of Soil	0.0050
Size of Arctic Icecap	0.0200
Surface Water	0.0300
9 World Temperature	
Vege Growth Rates	0.0020
Size of Arctic Icecap	0.0020
Rainfall	0.0300
Extreme Weather	0.0200
Forest Fires	0.0300
10 Volcanic Activity	
Atmos CO2 levels	0.0200
World Temperature	0.0010
Volcanic Activity	0.0200
Extreme Weather	0.0300
Forest Fires	0.0010
Acid Rain	0.0300
11 Extreme Weather	
Vege Growth Rates	-0.0030
Food Supplies	-0.0050
Size of Arctic Icecap	0.0200
Rainfall	0.0020
World Temperature	-0.0020
Extreme Weather	0.0150
12 Forest Fires	
Atmos CO2 levels	0.0200
Extreme Weather	0.0100
Forest Fires	0.0100
13 Acid Rain	
Vege Growth Rates	-0.0500
Degredation of Soil	0.0200
Surface Water	-0.0020
14 Surface Water	
Rainfall	0.0400

The following results show the changes that occur in the model factors over the period of the model run.

Period	Atmos CO2 levels	Vege Growth Rates	Food Supplies	Population	Birth Rate	Degredation of Soil	Size of Artic Icecap
1	1.000	1.000	1.000	1.030	1.000	1.012	1.020
5	1.017	0.997	0.993	1.159	1.000	1.063	1.106
10	1.035	0.993	0.984	1.344	0.999	1.130	1.223
20	1.078	0.983	0.960	1.806	0.998	1.283	1.494
30	1.132	0.971	0.929	2.427	0.996	1.464	1.826
40	1.200	0.957	0.886	3.261	0.993	1.679	2.232
50	1.287	0.939	0.830	4.383	0.990	1.939	2.728
60	1.397	0.918	0.754	5.890	0.985	2.255	3.333
70	1.541	0.892	0.653	7.915	0.979	2.643	4.072
80	1.727	0.860	0.517	10.637	0.971	3.123	4.975
90	1.972	0.820	0.336	14.294	0.959	3.724	6.078
100	2.292	0.769	0.092	19.209	0.944	4.482	7.425

Period	Rainfall	World Tempurature	Volcanic Activity	Extreme Weather	Forest Fires	Acid Rain	Surface Water
1	1.000	1.000	1.021	1.017	1.011	1.000	1.000
5	1.000	0.998	1.112	1.085	1.057	1.015	0.995
10	0.999	0.996	1.236	1.178	1.119	1.036	0.989
20	0.998	0.991	1.527	1.389	1.253	1.089	0.975
30	0.997	0.985	1.886	1.638	1.406	1.160	0.957
40	0.995	0.977	2.330	1.933	1.580	1.253	0.935
50	0.993	0.968	2.878	2.283	1.780	1.377	0.907
60	0.990	0.956	3.553	2.697	2.011	1.542	0.873
70	0.986	0.942	4.387	3.189	2.279	1.762	0.831
80	0.980	0.925	5.416	3.772	2.593	2.055	0.779
90	0.973	0.904	6.686	4.466	2.963	2.445	0.715
100	0.963	0.878	8.252	5.290	3.401	2.966	0.635

The following graphs show (not to scale) the general directions of movement for the various factors.

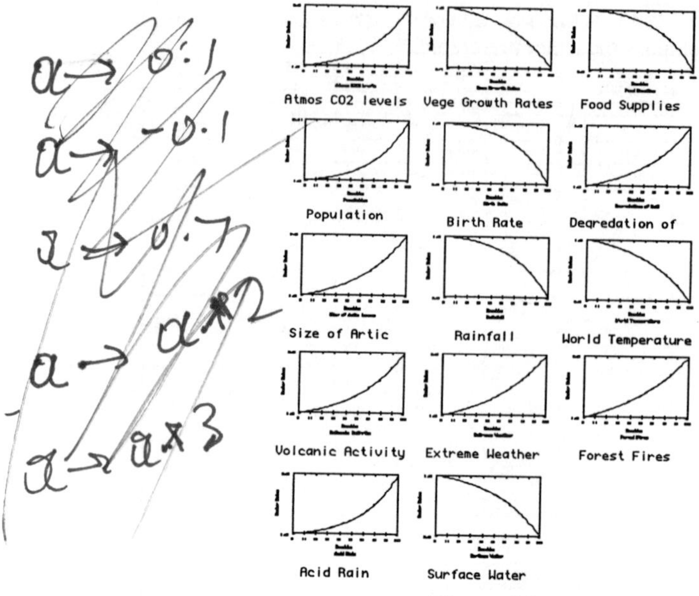

Figure 7.2

After model GA run.
The task given to the GA was to maintain world temperature, the "best" (closest result achievable) was selected for display. As you will note, the temperature was almost maintained, but at what cost?

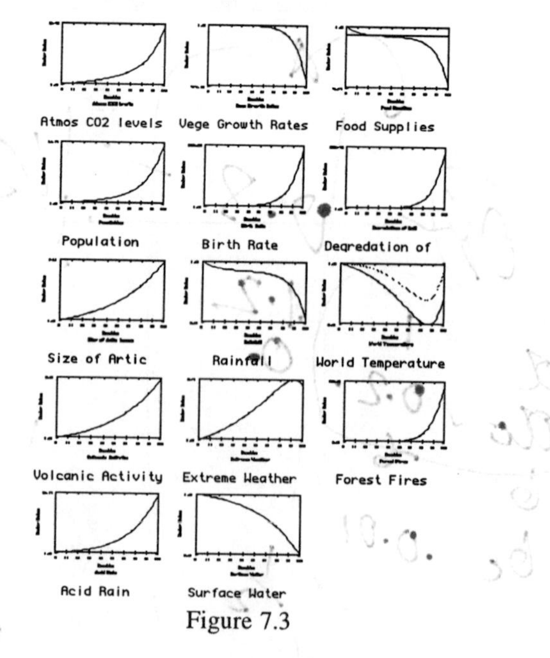

Figure 7.3

The table below supplies the actual raw data. As you will see the GA has selected a future that although possible is not preferable.

Period	Atmos CO2 levels	Vege Growth Rates	Food Supplies	Population	Birth Rate	Degredation of Soil	Size of Artic Icecap
1	1.023	1.000	0.905	1.030	1.083	1.096	1.020
5	1.118	0.968	0.602	1.161	1.487	1.577	1.105
10	1.237	0.911	0.358	1.348	2.211	2.481	1.222
20	1.498	0.690	0.112	1.826	4.892	6.101	1.494
30	1.810	0.174	0.010	2.488	10.829	14.929	1.826
40	2.197	-1.054	-0.050	3.417	23.977	36.436	2.231
50	2.693	-3.996	-0.117	4.755	53.090	88.792	2.727
60	3.360	-11.089	-0.239	6.753	117.557	216.196	3.332
70	4.322	-28.248	-0.506	9.877	260.312	526.154	4.070
80	5.850	-69.846	-1.123	15.049	576.426	1280.148	4.972
90	8.561	-170.821	-2.581	24.156	1276.427	3114.154	6.072
100	13.901	-416.124	-6.070	41.167	2826.509	7574.975	7.411

Period	Rainfall	World Tempurature	Volcanic Activity	Extreme Weather	Forest Fires	Acid Rain	Surface Water
1	1.000	1.000	1.021	1.016	0.950	1.033	1.000
5	0.990	1.000	1.112	1.084	0.783	1.195	0.996
10	0.982	0.999	1.236	1.176	0.638	1.431	0.990
20	0.973	0.998	1.527	1.386	0.541	2.050	0.975
30	0.970	0.997	1.886	1.635	0.742	2.931	0.955
40	0.967	0.995	2.330	1.929	1.478	4.183	0.931
50	0.965	0.993	2.877	2.271	3.403	5.961	0.900
60	0.961	0.990	3.553	2.661	8.157	8.488	0.861
70	0.952	0.988	4.387	3.083	19.759	12.083	0.812
80	0.934	0.986	5.416	3.483	47.991	17.211	0.749
90	0.891	0.986	6.686	3.711	116.658	24.557	0.669
100	0.793	0.993	8.251	3.381	283.657	35.165	0.563

The next set of graphs are from another solution as recommended by the GA. Remember that the graphs are not to scale and that the world temperature remained within 20% of what was desired. The graphs are presented to illustrate the significant different types of recommended alternatives that can be presented by the GA.

There has been a reduction in the recommended birth rate and in the setting of forest fires, these two recommendations have helped reduce the pressure on the weather systems and have helped maintain world temperatures.

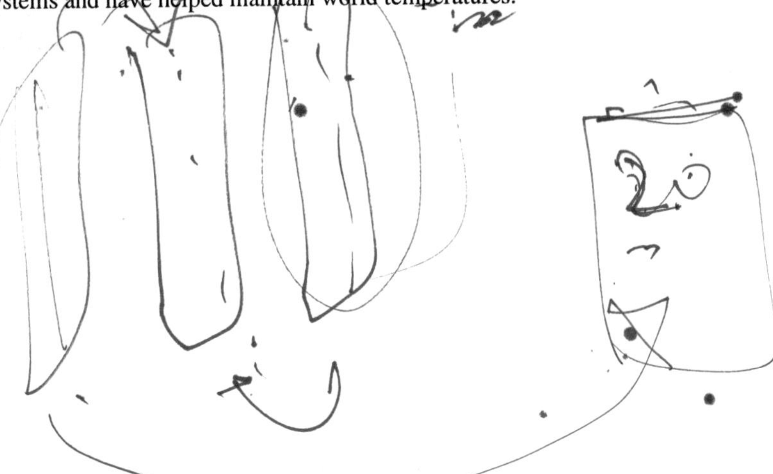

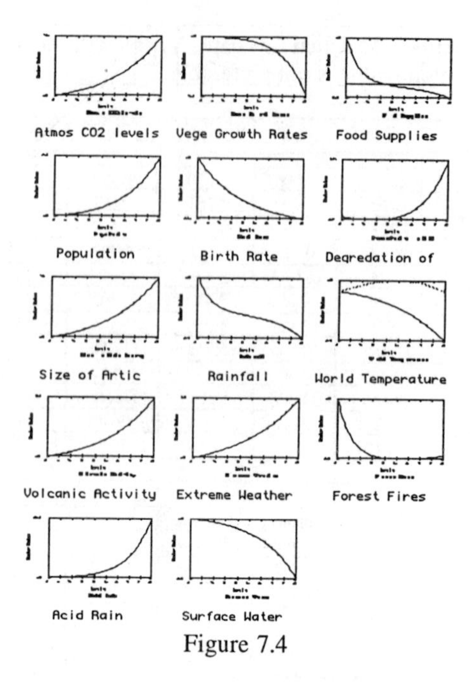

Figure 7.4

In the following examples of the test of the Hamaker theory we would like to present the following.

In these tests we, as in the earlier examples, allowed the GA to modify the levels of Atmospheric CO2, Food Production, the Birth Rate, Degredation of Soil, Forest Fires, and Acid Rain. However, in the earlier examples we allowed the GA to determine the level of these factors to achieve its objective of maintaining world temperature. In this example we set certain limits on the GA. These limits were that food production could not be reduced and that none of the other factors could be increased, in other words we are constraining the factors to moving in directions that we believe people would state were the preferred directions of movement.

We have selected the 'best' and the 'worst' that the GA had to offer after 20 generation runs.

The first gave a fitness of 0.057, maintaining world temperature rather well, the suggested changes to the controllable factors were:

CO2	-0.057 reduce acid rain
Food	0.052 increase food production
Birth	0.000 maintain the existing birth rate (remember we constrained the birth rate so that it would not go positive. It looks like it went as high as the constraint would allow and may have preferred to go positive)
Soil	-0.087 maintain soil fertility, expend significant effort to do so
Fires	-0.029 reduce burn-off, but not dramatically
Rain	-0.055 reduce acid rain

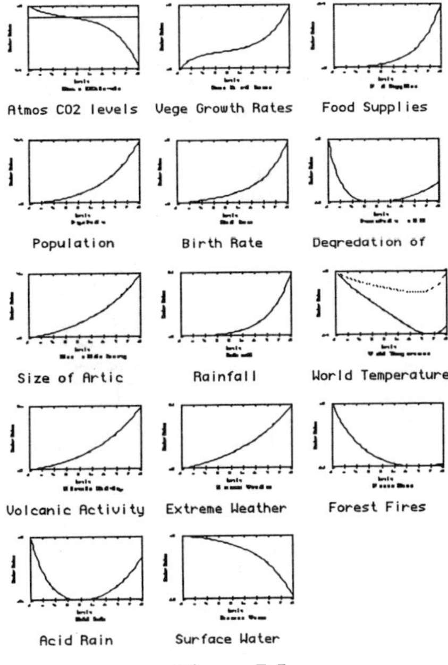

Figure 7.5

The second example achieved a fitness of 0.143, not nearly as good as the previous example, but still not an ice age. The GA suggested the following factor changes:

CO2	-0.1 reduce CO2 generation dramatically
Food	0.031 increase food production
Birth	-0.1 reduce the birth rate dramatically
Soil	-0.049 reduce soil degredation
Fires	-0.094 significantly reduce burn-off
Rain	-0.098 reduce acid rain dramatically

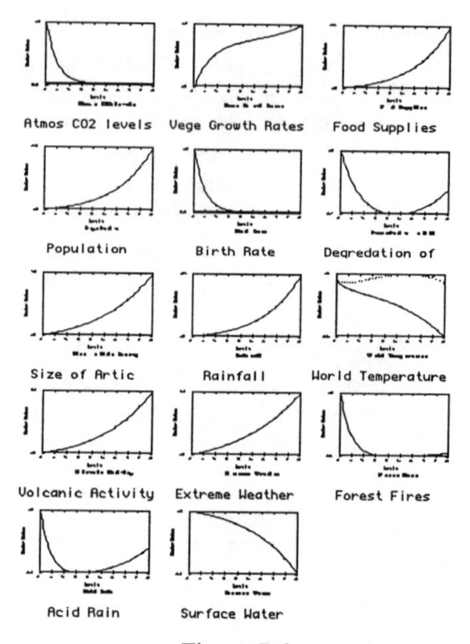

Figure 7.6

The degree of factor changes suggested by the GA are those that we feel sure would be generally believed are required to 'save the world,' however, the model suggests that if these extreme measures are taken then we would end up in a worse state than if we take less extreme measures.

The following graphs represent various data from the Ice Age runs. The first plot is of the volume representing the points, in 3-D space, that have accurate data in the measure of fitness.

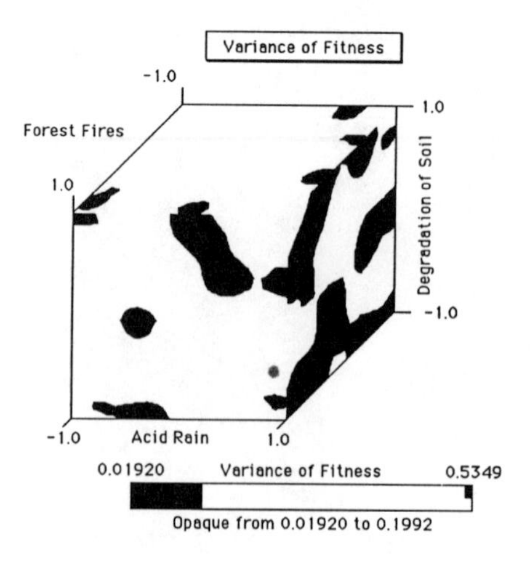

Figure 7.7

The dark colored volume (above) represents, in the following graphs, the volume that has highly accurate data for those factors labelled on the axis. This graph is presented to show that data, of all forms, is not consistently accurate over the solution space in a GA environment.

Although data points not within the dark volume (displayed above) have variances greater than 0.2, the maximum variance is only 0.5349. These data allow us to calculate the error associated with each section in the following graph.

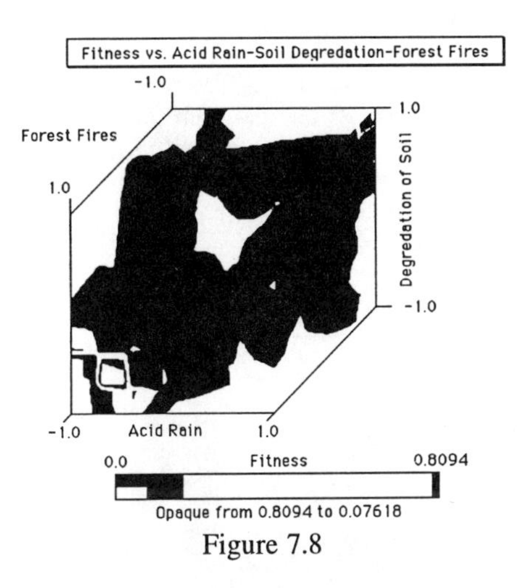

Figure 7.8

This graph, keeping in mind the variance of the data as displayed in the previous graph, shows the volume of bad fitness, this is the volume that the GA has attempted to breed out ('avoid' in this graphical context).

You will note that the volumes bred away from by the GA are complex in shape and would test any algorithm's abilities.

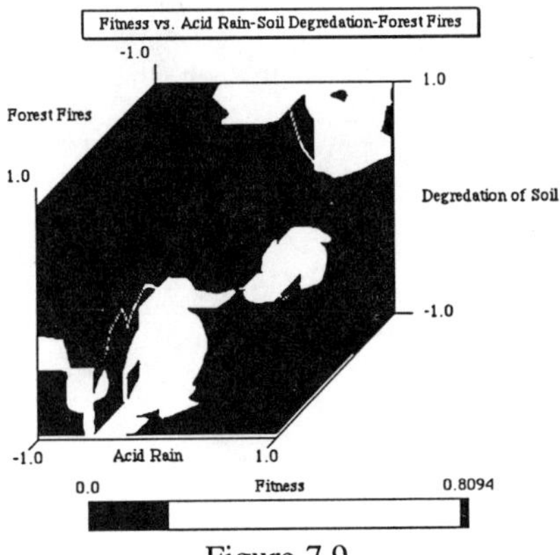

Figure 7.9

This, and all subsequent graphs, show the volume searched through for the three factors labelled on the axis.

As you progress through the following graphs you will come to have an appreciation of the great complexity of the problem space represented by even this relatively simple problem, the Ice Age model.

If we then consider models of far greater complexity, that GAs can still solve with relative ease, the tremendous value of GA in solving these types of problems can be really appreciated.

Sadly, given the limitations of our abilities to represent n-dimensional spaces, it is not possible to demonstrate on one diagram the true complexity of even this simple problem. What the GA is doing, in real terms, is isolating spaces, in n-dimensions, that are all of very high fitness. The GA does not search these one at a time, as some processes do, but rather solves in all n-dimensions at once. Here is where one of the great strengths of GAs resides.

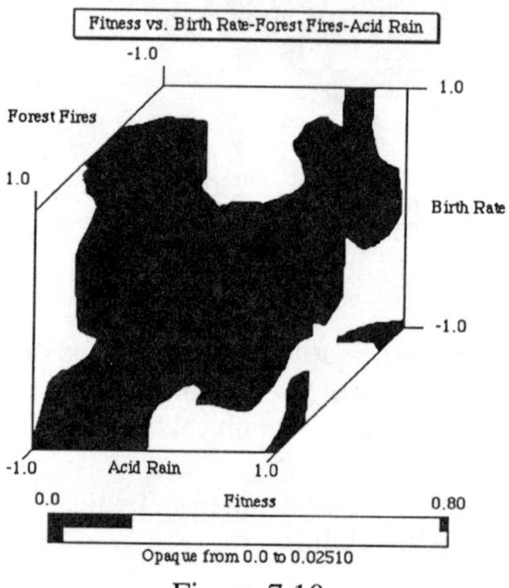

Figure 7.10

The volume shown in figure 7.10 represents the volume of good statistical fitness.

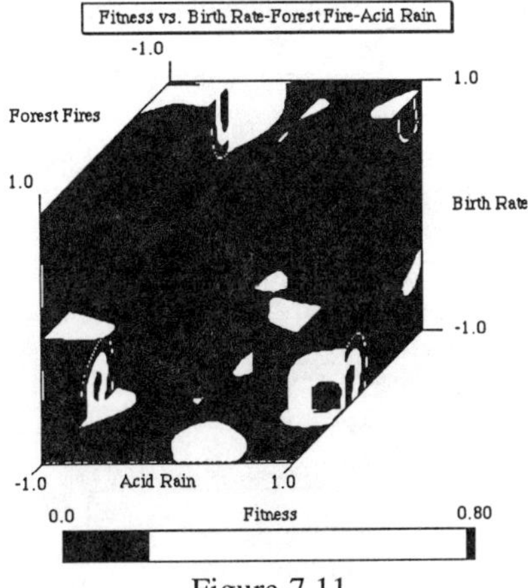

Figure 7.11

Graph in figure 7.11 represents the total problem space from the previous plot.

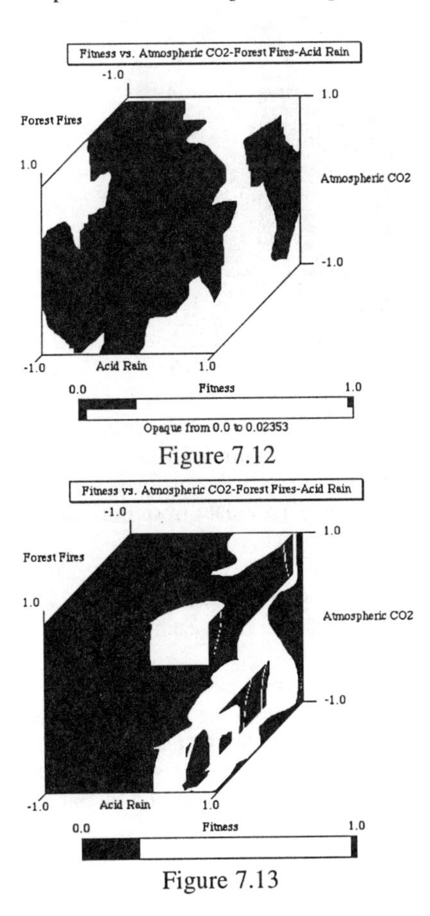

Figure 7.12

Figure 7.13

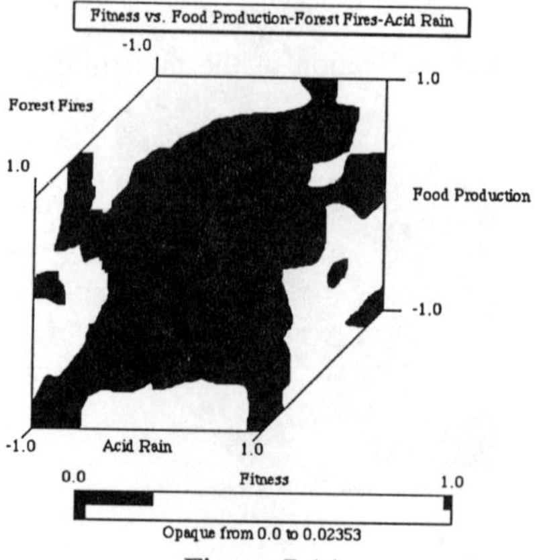

Figure 7.14

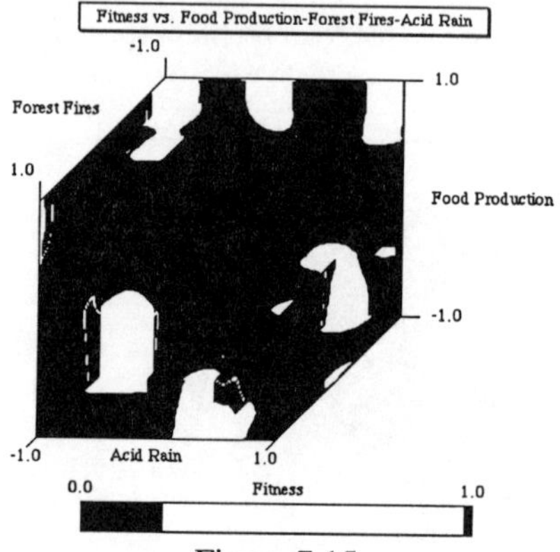

Figure 7.15

The collection of plots presented in the last few pages are presented to illustrate the differences between the volumes of high fitness (good fitness represented by the dark color) that are related to each set of factors that comprise a particular model. In this case we have displayed data from our 'Ice-Age' representation of Hamaker's theory of Global cooling.

You will note that each set, two graphs representing the same set of three factors plus fitness, are significantly different from each other.

This further represents the complex problem domains that can be developed by even the simplest of models. None of the models represented in this chapter represent real world levels of complexity and yet even these trivial models generate problem domains that would require significant computing power and very sophisticated algorithms to operate upon in a tractable manner, if GA were not the tool of choice.

7.23.2 The Gulf War

The next example of the application of the modelling system is to the recent Gulf War. The question that we wanted to answer was: was it necessary for the US to go to war to protect its oil supplies?

Description of Model Gulf War
Simulation Iterations: 20
System Shocks: On
Cross Impacts Processing: All Delayed

No. of Factors: 18

1 Escal Rhetoric

Escal Rhetoric	0.0500
US Oil Supplies	-0.1000
US Costs	0.2000
Anti Muslim Backlash	0.1000
Level of Int'l Terro	0.0700
Muslim Fundamental	0.1000
Civilian Deaths	0.2000
Combatant Losses	0.0200
Anti-War Feelings	-0.0500
Cost of Oil	0.0700
World Costs	0.0400
Saddam Elimination	0.1000

2 Escal Conflict

Escal Rhetoric	0.5000
US Oil Supplies	-0.0300
US Costs	0.0500
Anti Muslim Backlash	0.0300
World Oil Supply	-0.1000
Combatant Losses	0.1000
Surgical Strikes	-0.0800
Anti-War Feelings	-0.0200
World Costs	0.0600
Saddam Elimination	-0.3000
Destruc Iraq War Cap	-0.2000

3 US Oil Supplies

Escal Rhetoric	-0.0200
Escal Conflict	-0.0800
World Oil Supply	0.2000
Muslim Fundamental	-0.0500
Cost of Oil	-0.0200

4 US Costs

Escal Conflict	0.2000
Combatant Losses	0.1000
Cost of Oil	0.0700

5 Iraq Costs

Escal Conflict	0.1000
US Oil Supplies	-0.1500
Muslim Fundamental	-0.0700
Combatant Losses	0.2000
Surgical Strikes	0.4000
Cost of Oil	-0.0700
Destruc Iraq War Cap	0.2500

6 Anti Muslim Backlash

Escal Rhetoric	0.0800
Escal Conflict	0.0800
US Costs	0.0050
Anti Muslim Backlash	0.0200
Anti-War Feelings	-0.0020
World Costs	0.0300

7 Level of Int'l Terro

Escal Rhetoric	0.0050
Escal Conflict	0.0200
Iraq Costs	0.0500
Anti Muslim Backlash	0.1000
Anti-War Feelings	0.0020
Saddam Elimination	0.0300

8 World Oil Supply

Escal Conflict	-0.1000
World Costs	-0.0400
Destruc Iraq War Cap	0.1000

9 Muslim Fundamental

Escal Rhetoric	0.0400
Escal Conflict	0.2000
Iraq Costs	0.1000
Anti Muslim Backlash	0.1000
Level of Int'l Terro	0.0050
Muslim Fundamental	0.0500
Combatant Losses	0.0500
Saddam Elimination	-0.0200
Destruc Iraq War Cap	-0.1000

10 Weapons Technology

Escal Rhetoric	0.0500
Escal Conflict	0.1500
Weapons Technology	0.0100

11 Civilian Deaths	
Escal Conflict	0.0600
Level of Int'l Terro	0.0002
Weapons Technology	-0.7000
Surgical Strikes	-0.1500

12 Combatant Losses	
Escal Conflict	0.1500
Weapons Technology	0.1000
Surgical Strikes	0.2500
Destruc Iraq War Cap	0.1000

13 Surgical Strikes	
Escal Rhetoric	0.2500
Escal Conflict	0.8000
Weapons Technology	0.2000
Civilian Deaths	0.4000
Combatant Losses	0.1000
Surgical Strikes	0.0500
Anti-War Feelings	0.2000

14 Anti-War Feelings	
Escal Conflict	0.0300
US Costs	-0.0400
Level of Int'l Terro	-0.0030
Muslim Fundamental	-0.0300
Civilian Deaths	0.1500
Combatant Losses	0.1000
Surgical Strikes	-0.0300
Cost of Oil	-0.0060

15 Cost of Oil	
Escal Conflict	0.1000
Iraq Costs	0.1000
World Oil Supply	-0.0400
Cost of Oil	0.0100

16 World Costs	
Escal Conflict	0.1000
Level of Int'l Terro	0.0010
World Oil Supply	-0.0800
Cost of Oil	0.0750

17 Saddam Elimination	
Escal Conflict	0.0250
US Oil Supplies	-0.0800
US Costs	0.0030
Iraq Costs	0.0010
World Oil Supply	-0.0040
Muslim Fundamental	-0.0020
Weapons Technology	0.0300
Combatant Losses	0.0300
Surgical Strikes	0.4000
Destruc Iraq War Cap	0.0200

18 Destruc Iraq War Cap	
Escal Conflict	0.0300
US Oil Supplies	-0.2000
World Oil Supply	-0.0020
Weapons Technology	0.2000
Surgical Strikes	0.8500

The following graphs are those that fall out of the model.

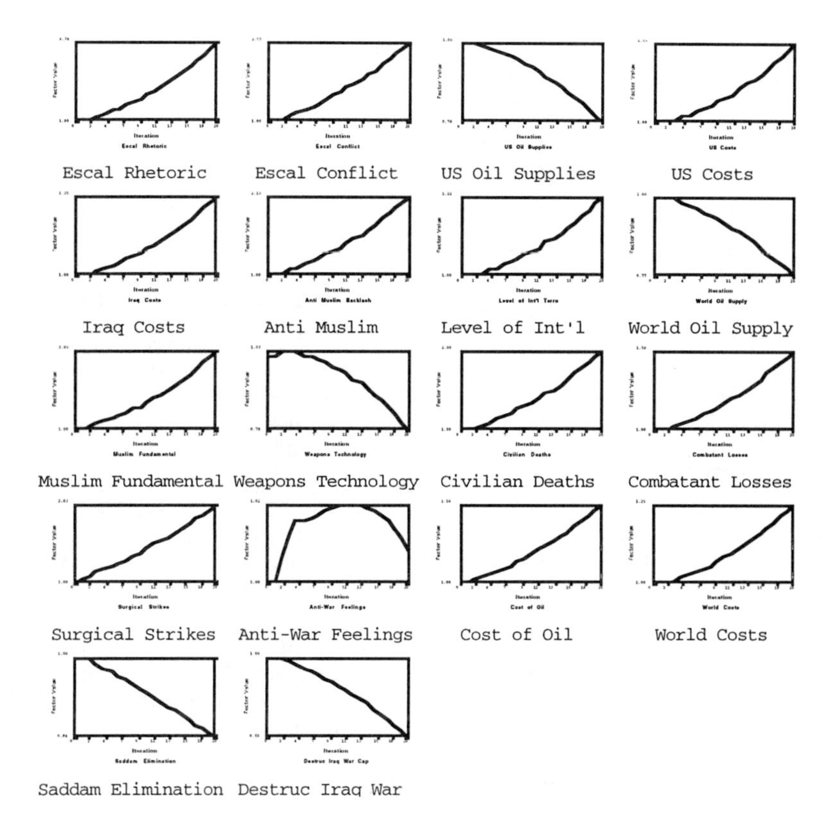

Figure 7.16

You will note that the model shows that there will be a continuing decrease in the US supplies of oil given the situation as it was just before the war erupted.

If we now run the GA with the intent purpose of maintaining US oil supplies and by having Escalation in Rhetoric, Escalation in Conflict, Weapons Technology, Surgical Strikes, Elimination of Saddam, and Destruction of the Iraqi War Capacity under our control what could we have been done?

Gene ID	Parent 1	Parent 2	Fitness	Escal Rhet	Escal Conf'ct	Weapons Techn'gy	Surgical Strikes	Saddam Elim'tion	Destruc Iraq Cap
Startup Population									
133	0	0	238255	0.9922	-0.5843	-0.5294	0.9843	0.9686	-0.8196
134	0	0	4967	0.1294	0.1843	0.2157	-0.7412	0.7882	0.2471
135	0	0	8239	0.6392	0.6549	0.8667	0.1922	-0.9529	-0.3961
136	0	0	63695	0.8353	0.8745	-0.8745	0.7961	-0.6784	0.4745
137	0	0	4811	-0.4196	0.2000	-0.7255	-0.5137	0.5765	0.7098
138	0	0	856	-0.7569	-0.9686	0.9843	-0.2235	0.0980	0.0275
139	0	0	27957	0.0039	-0.6078	0.4353	-0.0510	-0.3882	0.8667
140	0	0	11.016	-0.1373	-0.5294	-0.2235	0.2078	0.2784	0.0902
141	0	0	2600.3	-0.2392	0.6314	0.5059	0.9137	-0.0118	-0.1608
142	0	0	252.03	0.5373	-0.5294	-0.2863	-0.7490	-0.6471	-0.5373

Average			35164						

After 10 generations									
143	134	135	13.462	0.1294	0.4039	0.2157	-0.6706	0.0275	0.2471
148	145	146	115.66	-0.1765	0.3804	0.6627	0.3647	-0.1686	-0.4745
151	149	142	39.690	-0.1922	-0.6235	-0.1608	-0.4196	0.3569	0.2157
154	148	152	20.790	-0.1451	0.4431	-0.4980	0.3333	-0.5059	-0.1137
155	148	152	65.302	-0.3882	-0.2157	0.6784	-1.0000	-0.6784	-0.2549
156	143	148	12.085	0.1451	0.4039	0.0588	-0.6706	0.1216	0.2157
157	154	156	3.3564	-0.6471	0.3882	-0.2157	0.3333	-0.5059	-0.0588
158	154	143	13.271	-0.1529	0.4431	-0.4353	-0.6706	-0.5059	-0.0824
159	154	143	17.508	0.2000	0.4039	-0.9765	0.2706	0.2784	0.1373
Average			30.112						

After 20 generations									
140	0	0	11.016	-0.1373	-0.5294	-0.2235	0.2078	0.2784	0.0902
143	134	135	13.462	0.1294	0.4039	0.2157	-0.6706	0.0275	0.2471
156	143	148	12.085	0.1451	0.4039	0.0588	-0.6706	0.1216	0.2157
157	154	156	3.3564	-0.6471	0.3882	-0.2157	0.3333	-0.5059	-0.0588
158	154	143	13.271	-0.1529	0.4431	-0.4353	-0.6706	-0.5059	-0.0824
161	154	156	1.3770	-0.2078	0.3176	-0.6706	-0.5373	-0.6627	-0.0510
164	154	143	11.507	-0.3333	0.3804	-1.0000	0.2706	-0.5373	0.2314
166	164	161	0.4157	-0.3333	0.2549	-0.8667	0.2627	-0.7255	0.0431
167	164	166	0.0444	0.1294	-1.0000	-0.4980	-0.9843	-0.7490	-0.6235
168	167	166	2.4955	-0.9608	0.3176	-0.9922	0.0118	0.2784	-0.4667
Average			6.903						

As you will note the fitness of the genes approaches zero (100% fitness) very rapidly. This is the demonstrable power of the GA when used in searching complex problem domains, of which these problems are but one example.

To analyse the recommendations we proceed as follows:

Find the gene with the best fitness. In this example this is gene 167.

The recommendations are to Escalate the Rhetoric, eliminate any Escalation of the Conflict, reduce significantly the effects of modern Hi-Tech Weaponry, eliminate any of the benefits of Surgical Strikes, almost eliminate any actions that could be construed as a move to Eliminate Saddam Hussein, and almost eliminate any actions that could be construed as a move to Destroy Iraqi Capacity for War.

What these results tell us is that the UN went to war, in the Gulf, to do more than preserve US oil supplies. This simulation shows that there were alternatives to war to preserve those supplies. We assume that the US administration was as aware of the alternatives as we are.

Now have a look at the worse gene, 133. You will note that the suggestions for the worse possible answer to solving the oil supply problem was to escalate the Rhetoric dramatically, use Surgical Strikes to a very high level and attempt to Eliminate Saddam Hussien. We seem to remember that these were some of the recommended suggestions flying around at the time, thank God for the USA that the objective was not to maintain oil supplies but was rather a desire to punish what was perceived as an international hostile act.

The following table is a listing of the gene fitnesses calculated at the end of each new generation bred during the GA run. You will note that the GA fitness values converge very rapidly, as was commented upon earlier.

Generation	Best Fit	Worst Fit	Avg. Fit
1	11.01623558	27957.65668383	6089.14987396
2	11.01623558	8239.30649409	2569.00695988
3	11.01623558	4811.39796482	1557.06624375
4	11.01623558	2600.35901301	850.53836343
5	11.01623558	1547.92497163	417.65501364
6	11.01623558	856.93771551	280.64495406
7	11.01623558	573.97578307	141.39157534
8	11.01623558	252.03735164	85.20252544
9	3.35636835	177.82437580	60.33442711
10	3.35636835	115.66070433	31.21444350
11	3.35636835	65.30252198	22.40531725
12	3.35636835	65.30252198	22.40531725
13	1.37699122	35.02074353	15.54583091
14	1.37699122	35.02074353	15.54583091
15	1.37699122	27.56944182	13.59792724
16	1.37699122	27.56944182	13.59792724
17	1.37699122	20.79070334	11.99177016
18	1.37699122	17.50795444	11.65771167
19	0.04442229	15.54170683	7.20792003
20	0.04442229	13.46285493	6.90329596

We now present a gene genealogy. You will note that it is possible to trace 'good' bloodstock through the generations. Follow the trail from gene 143 and 156 to the best gene in the population, gene 167.

In fact nine of the ten best genes can be traced back to genes 143, 156, and 154 also gene 154 has gene 142 in its parentage which is the tenth gene in the list of ten best.

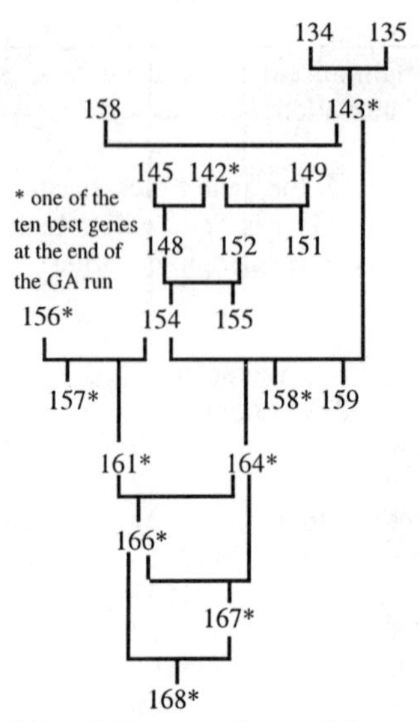

Figure 7.17 Plot of the geneology of the ten best genes

This is a common genealogical pattern. Better genes are bred from good genes, this is why we have hog, cattle, horse, corn, potato, melon, and so it goes with breeders, because we know we can breed stronger and better stock if we can control the 'bloodlines.'

The following sets of graphs represent the solution surfaces for all manipulable factors in the Gulf War model. The dark areas are unfit solutions, the clear areas are fit. As you go through these graphs be aware of the similarities and differences between these surfaces.

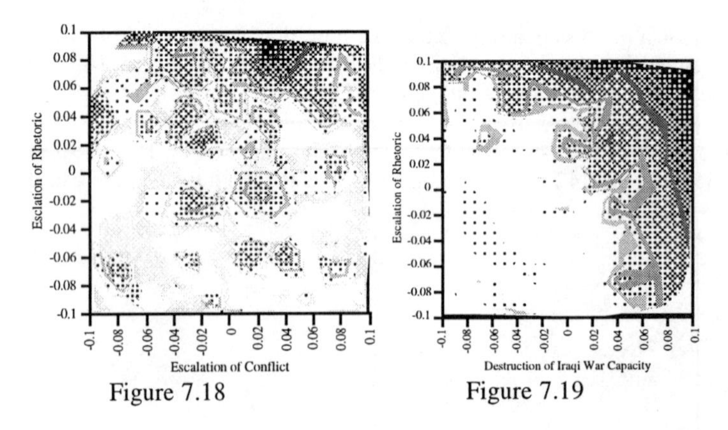

Figure 7.18

Figure 7.19

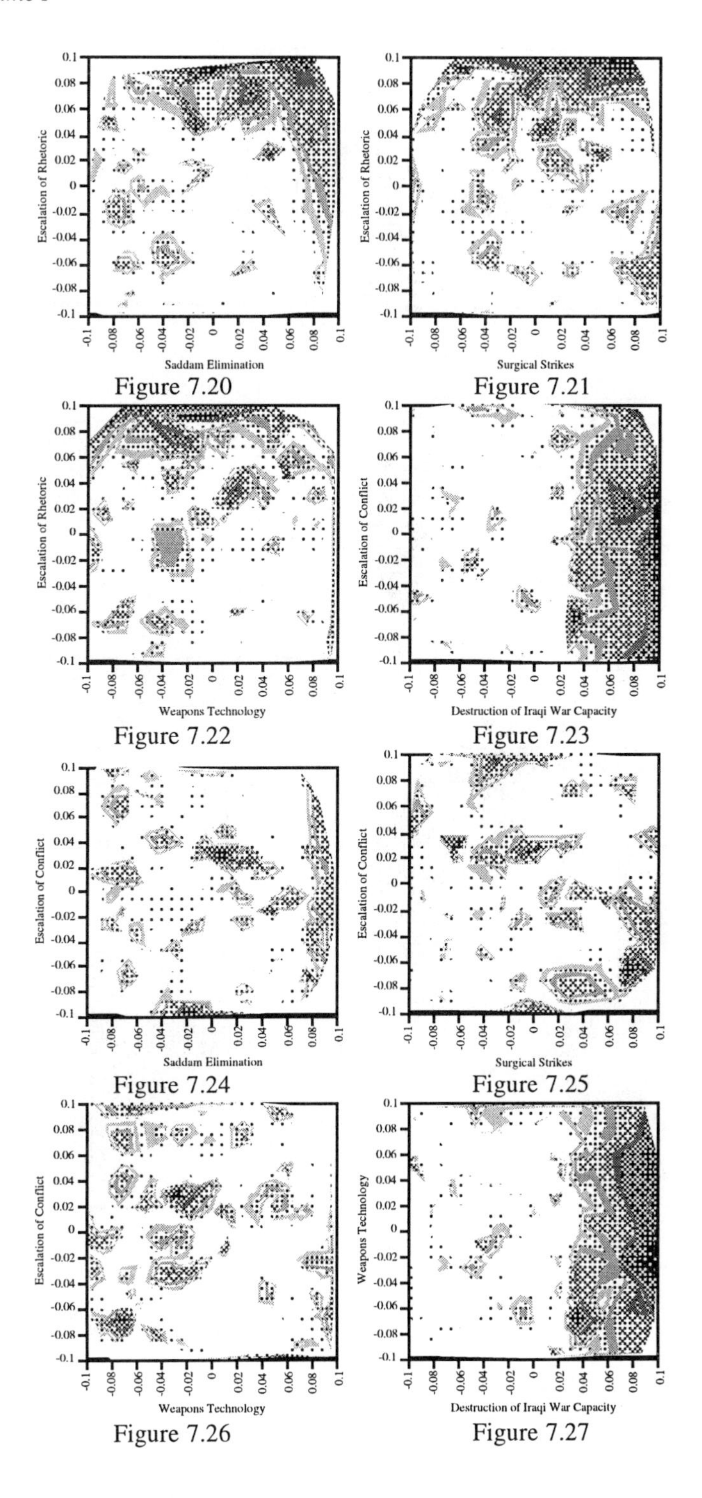

Figure 7.20

Figure 7.21

Figure 7.22

Figure 7.23

Figure 7.24

Figure 7.25

Figure 7.26

Figure 7.27

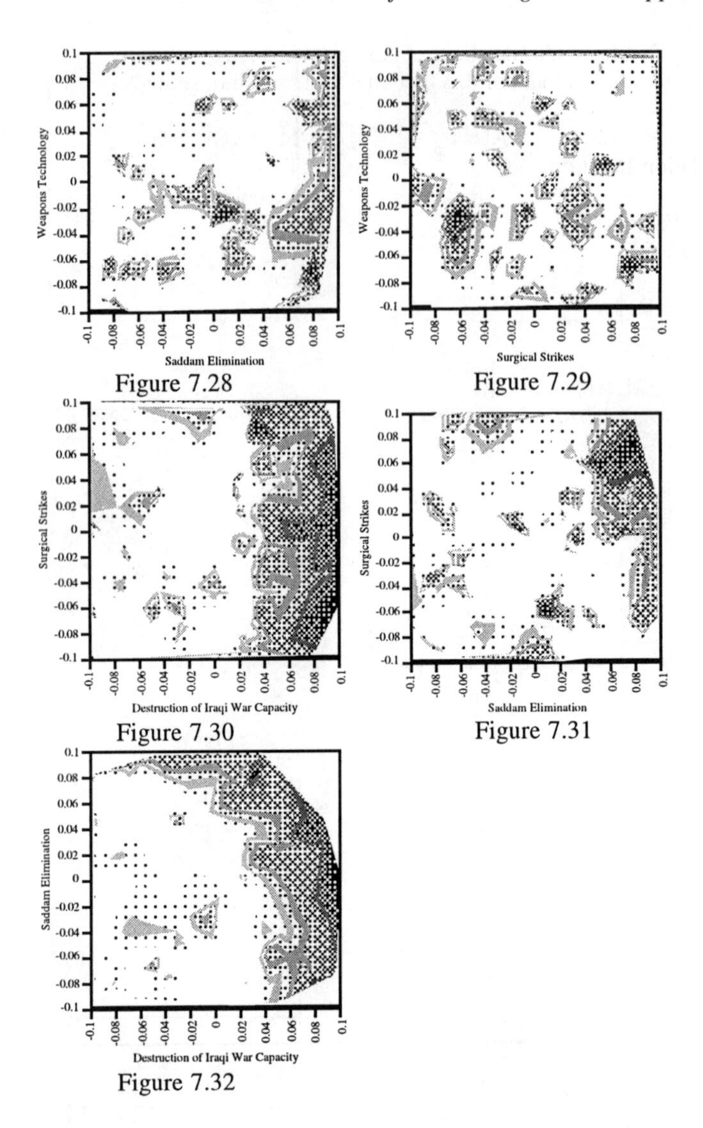

Figure 7.28

Figure 7.29

Figure 7.30

Figure 7.31

Figure 7.32

7.23.3 Western Australian Transport Model

And now the last model of this type that will be demonstrated.

Description of Model WADoT
Simulation Iterations: 20

No. of Factors: 35

1 Congestion	
Road Supply	-0.0020
Decen Wrk Place	-0.0060
Improved Techn	-0.0020
Use of P/T	-0.0050
Flexible Wrk Hours	-0.0300
Vehicle Usage	0.0400
Bicycle Usage	-0.0050
Walking	-0.0050

2 Fuel Costs	
Fuel Availablity	-0.0060
Alternative Fuels	-0.0080
Taxes	0.0400

3 Fuel Availablity	
Improved Techno	0.0020
Alternative Fuels	0.0080

4 Road Supply	
Congestion	-0.0030
Urban Sprawl	0.0080

5 Road Speeds	
Congestion	-0.0300
Fuel Costs	-0.0020
Road Supply	0.0040
Improved Techno	0.0100
Size Police Force	-0.0200

6 Population	
Population	0.0010
Family Size	0.0400
Migration	0.0300

7 Veh Purch Costs	
Improved Techno	-0.0050
Taxes	0.0200

8 Veh Running Costs	
Fuel Costs	0.0300
Urban Sprawl	0.0060
Decen Wrk Place	0.0020
Improved Techno	-0.0003
Taxes	0.0200
Vehicle Usage	0.0200

9 Urban Sprawl	
Fuel Costs	-0.0010
Population	0.0030
Veh Running Costs	-0.0040
Urban Sprawl	0.0030
Enviro Awarenese	-0.0030
Migration	0.0200
Economic Climate	0.0200
Bicycle Usage	-0.0070
Walking	-0.0020

10 Aging of Pop	
Population	0.0010
Aging of Pop	0.0200
Fitness Issues	-0.0050
Improved Techno	-0.0030

11 Decen Wrk Place	
Fuel Costs	0.0020
Veh Running Costs	0.0200
Urban Sprawl	0.0200
Decen Wrk Place	0.0050
Enviro Awareness	0.0300
Improved Techno	0.0002

12 Fitness Issues	
Fitness Issues	0.0100
Education	0.0200

13 Enviro Awareness	
Congestion	0.0002
Enviro Awareness	0.0080
Education	0.0400

14 Improved Techno	
Improved Techno	0.0200
Road Deaths	0.0050
Education	0.0300
Economic Climate	0.0300

15 Alternative Fuels	
Fuel Costs	0.0100
Fuel Availablity	-0.0004

Improved Techno 0.0100
Education 0.0006

16 Cost of Living
Fuel Costs 0.0010
Veh Purch Costs 0.0100
Veh Running Costs 0.0100
Decen Wrk Plac? -0.0100
Family Size 0.0300
Taxes 0.0200
Road Deaths 0.0030
Economic Climate -0.0080
Cars per Household 0.0100
Use of P/T -0.0060
P/T Fares0.0001
Vehicle Usage 0.0300
Bicycle Usage -0.0030
Walking -0.0006

17 Family Size
Enviro Awareness -0.0070
Cost of Living -0.0200
Family Size -0.0040
Education -0.0050
Economic Climate 0.0050

18 Taxes
Population 0.0200
Tourist -0.0050
Depend on SS 0.0200
Road Deaths 0.0200
Size Police Force 0.0006
Costs of P/T Ops 0.0200

19 Tourism
Enviro Awareness 0.0200
Cost of Living -0.0200
Tourism 0.0200
Size Police Force 0.0030
Economic Climate 0.0200

20 Migration
Enviro Awareness 0.0100
Cost of Living -0.0100
Migration 0.0080
Size Police Force 0.0003
Economic Climate 0.0100
21 Depend on SS
Population 0.0030
Aging of Pop 0.0040
Cost of Living 0.0200
Family Size 0.0060
Taxes 0.0200
Migration 0.0070
Education -0.0030
Economic Climate -0.0100

22 Road Deaths
Congestion 0.0020
Road Speeds 0.0200
Decen Wrk Place -0.0005
Size Police Force -0.0300
Cars per Household 0.0040
Use of P/T -0.0600
Vehicle Usage 0.0200
Bicycle Usage -0.0003
Walking -0.0030

23 Size Police Force
Road Speeds 0.0020
Population 0.0005
Urban Sprawl 0.0020
Road Deaths 0.0005

24 Education
Aging of Pop 0.0010

25 Economic Climate
Fuel Costs -0.0070
Fuel Availablity 0.0004
Veh Purch Costs -0.0050
Veh Running Costs -0.0070
Improved Techno 0.0100
Taxes -0.0040
Tourism 0.0200
Depend on SS -0.0040
Road Deaths -0.0700
Education 0.0200

26 Cars per Household
Road Supply 0.0003
Veh Purch Costs -0.0200
Veh Running Costs -0.0200
Urban Sprawl 0.0080
Aging of Pop -0.0030
Enviro Awareness -0.0150
Cost of Living -0.0200
Family Size 0.0040
Economic Climate 0.0060
Flexible Wrk Hours -0.0100

Vehicle Usage	0.0100
Bicycle Usage	-0.0004
Walking	-0.0010

27 Attract of P/T	
Congestion	0.0100
Road Speeds	-0.0002
Veh Running Costs	0.0200
Cost of Living	0.0100
Road Deaths	0.0300
P/T Availability	0.0080
P/T Fares	-0.0060

28 Use of P/T	
Population	0.0040
Urban Sprawl	-0.0100
Aging of Pop	0.0040
Enviro Awareness	0.0030
Cost of Living	0.0200
Tourism	0.0005
Depend on SS	0.0200
Economic Climate	-0.0050
Attract of P/T	0.0500
P/T Availability	0.0080

29 P/T Availability

30 Costs of P/T Ops	
Fuel Costs	0.0200
Veh Purch Costs	0.0005
Veh Running Costs	0.0200
Urban Sprawl	0.0050
Attract of P/T	0.0050
Use of P/T	0.0200
P/T Availability	0.0100

31 P/T Fares	
Costs of P/T Ops	0.0050

32 Flexible Wrk Hours	
Congestion	0.0100
Flexible Wrk Hours	0.0100

33 Vehicle Usage	
Congestion	-0.0200
Fuel Costs	-0.0030
Road Supply	0.0002
Road Speeds	0.0030
Population	0.0020
Veh Purch Costs	-0.0030
Veh Running Costs	-0.0300
Urban Sprawl	0.0100
Aging of Pop	-0.0005
Fitness Issues	-0.0030
Enviro Awareness	-0.0100
Cost of Living	-0.0200
Road Deaths	-0.0030
Cars per Household	0.0300
Use of P/T	-0.0300
Bicycle Usage	-0.0020

34 Bicycle Usage	
Congestion	0.0200
Population	0.0040
Veh Purch Costs	0.0002
Veh Running Costs	0.0020
Aging of Pop	0.0020
Fitness Issues	0.0200
Enviro Awareness	0.0040
Cost of Living	0.0050
Family Size	0.0100
Road Deaths	0.0200
Bicycle Usage	0.0004

35 Walking	
Congestion	0.0100
Population	0.0060
Veh Purch Costs	0.0003
Veh Running Costs	0.0040
Aging of Pop	0.0030
Fitness Issues	0.0230
Enviro Awareness	0.0050
Cost of Living	0.0060
Road Deaths	0.0100
Walking	0.0002

The following graphs were generated by the model tabulated above. You will note that there are a number of factors that appear to be trending in directions that could be considered as undesirable. These factors are: Congestion, Road Speeds, Urban Sprawl, Road Deaths and Size of Police Force.

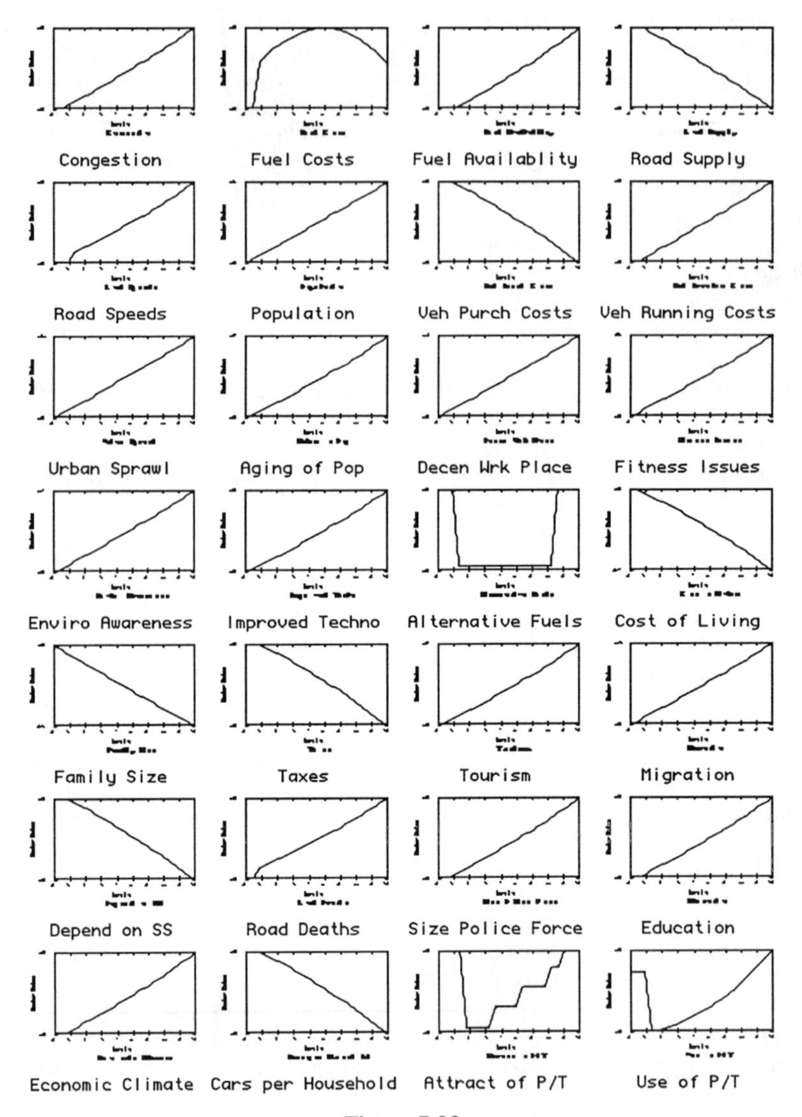

Figure 7.33

In an attempt to correct some of these problems it was decided to search for solutions to the model that had as its objectives a further increase in the use of public transport and to reduce the incidence of road death.

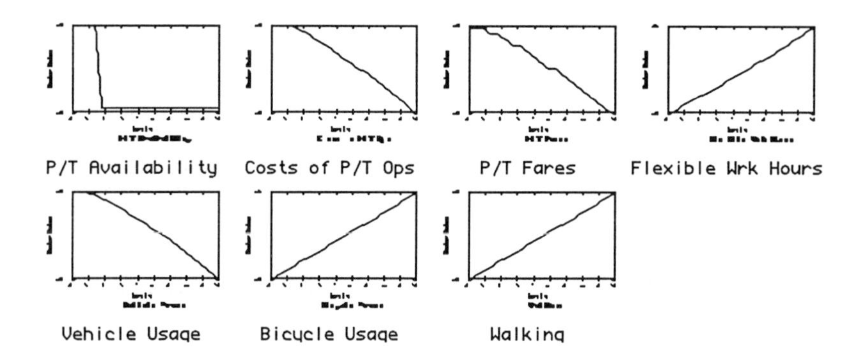

Figure 7.34

The GA suggested the following twenty possible solutions and recommended changes to the factors in the model that were considered to be under direct control of the decision maker:

Fitness	Road Supply	Urban Sprawl	Decen Wrk Place	Taxes	Size Police Force	Attract of P/T	P/T Avail	P/T Fares	Flexible Wrk Hours
0.0901	0.6383	0.2630	-0.4897	-0.5758	-0.6891	-0.0694	-0.2414	0.2942	0.2082
0.0464	0.1378	0.2786	-0.0850	-0.5758	-0.1887	-0.8358	0.7595	0.5249	0.4272
0.0792	0.6872	0.0264	-0.4643	-0.1065	-0.9550	-0.3470	-0.1163	-0.8377	-0.4565
0.0686	0.6852	0.0577	0.1457	-0.7771	-0.4819	-0.1359	-0.5015	-0.8886	0.1496
0.0774	0.6383	0.1535	-0.1163	-0.5758	-0.4409	-0.3196	-0.4917	0.4194	0.0518
0.0867	0.6774	0.0577	0.1261	-0.7771	-0.9824	-0.5640	-0.5054	-0.7791	0.3842
0.0456	0.7322	-0.9120	0.7243	-0.2766	-0.9824	-0.5660	-0.6931	-0.7654	0.0244
0.0627	0.6755	0.2610	-0.3314	-0.0147	-0.6520	-0.6129	0.0283	-0.8260	-0.3157
0.0668	0.6774	0.0616	0.1261	-0.7771	-0.9824	-0.4389	-0.5015	-0.7791	0.1496
0.0787	0.6305	0.2786	-0.9902	-0.5758	-0.7204	-0.0694	-0.9922	0.4174	0.2082
0.0827	0.6872	0.0264	-0.4487	-0.0714	0.0440	-0.4761	-0.0518	-0.8377	-0.4878
0.0716	0.6735	0.0616	0.3920	-0.8944	-0.9824	-0.3138	-0.5054	-0.5601	0.1496
0.0699	0.6852	0.0616	0.0010	-0.0890	-0.9824	-0.5640	-0.5953	-0.8886	0.1281
0.0737	0.6872	0.0459	0.1417	-0.7615	-0.4819	-0.1359	-0.5758	-0.7947	0.1535
0.0745	0.6872	0.0459	0.1417	-0.8788	-0.4467	-0.1359	-0.6383	-0.7869	-0.8006
0.0672	0.6813	0.0303	0.0186	-0.9570	-0.9198	-0.3314	-0.0068	-0.3099	0.1808
0.0769	0.6755	0.2610	-0.3627	-0.5073	-0.9648	-0.5484	-0.9726	-0.8260	-0.0655
0.0664	0.6852	0.0616	0.0010	-0.8084	-0.7322	-0.9707	-0.6109	-0.7654	0.1515
0.0443	0.6735	0.1554	-0.6070	-0.9570	-0.9198	-0.2571	-0.6950	-0.3099	0.0557
0.0245	0.2004	0.2942	-0.6168	-0.7654	-0.4858	-0.3353	0.3138	0.5249	0.3959

(Note that there are a number of factors that always has the same direction of suggested movement. These answers are dominant and every solution will have these factors moving in the same direction. These are Road Supply which should always increase and Taxes and the Attractiveness of Public Transport which should always decrease.)

The following table isolates answers that are significantly different from each other. Note that we have removed the dominating factors since they will always have the same suggested directions of effort.

Urban Sprawl	Decen Wrk Place	Size Police Force	P/T Availability	P/T Fares	Flexible Wrk Hours
0.2630	-0.4897	-0.6891	-0.2414	0.2942	0.2082
0.2942	-0.6168	-0.4858	0.3138	0.5249	0.3959
0.0264	-0.4643	-0.9550	-0.1163	-0.8377	-0.4565
0.0577	0.1457	-0.4819	-0.5015	-0.8886	0.1496
-0.9120	0.7243	-0.9824	-0.6931	-0.7654	0.0244
0.2610	-0.3314	-0.6520	0.0283	-0.8260	-0.3157
0.0264	-0.4487	0.0440	-0.0518	-0.8377	-0.4878
0.0459	0.1417	-0.4467	-0.6383	-0.7869	-0.8006
0.1554	-0.6070	-0.9198	-0.6950	-0.3099	0.0557

Since there are six factors that can take on either positive or negative values we, in theory, have an opportunity in this instance to discover $2^6 = 64$ possible 'families' of solutions, in fact the GA has found nine that appear to supply sensible results and adequate fitnesses, given the objectives we defined.

If the decision makers feel that it is not prudent to increase public transport fares then there are available seven other families of results, each of which allows for differing combinations in movement for the other factors. However, if it was decided to increase public transport fares then we can choose between increasing or decreasing public transport attractiveness, however; each of the other factors is now constrained. Flexibility of Work Hours and Urban Sprawl must increase and Decentralisation of Work Place and the Size of the Police Force must decrease. No other alternatives are possible given the model as constructed.

The following graphs were generated by the result with the best fitness.

Note the dramatic change in the forecasted future, as demonstrated by the graphs over the next two pages, if we accept the proposed answer. If we were to accept the answer selected as the most preferred then we would expect to undertake the following actions: increase road supply, allow urban sprawl to continue somewhat, to significantly bring effort to bear to reduce the centralisation of employment activity, significantly reduce taxes, reduce the size of the police forces, reduce the attractiveness of public transport but increase its availability and the cost of using public transport, and finally to increase the incidence of flexible working hours.

All these suggested movements can be understood within the context of the model developed. However, it is certainly possible for individuals to disagree with the model as structured. In these instances it is suggested that new models be constructed that concur with the understandings of the workings of the environment that these others may have.

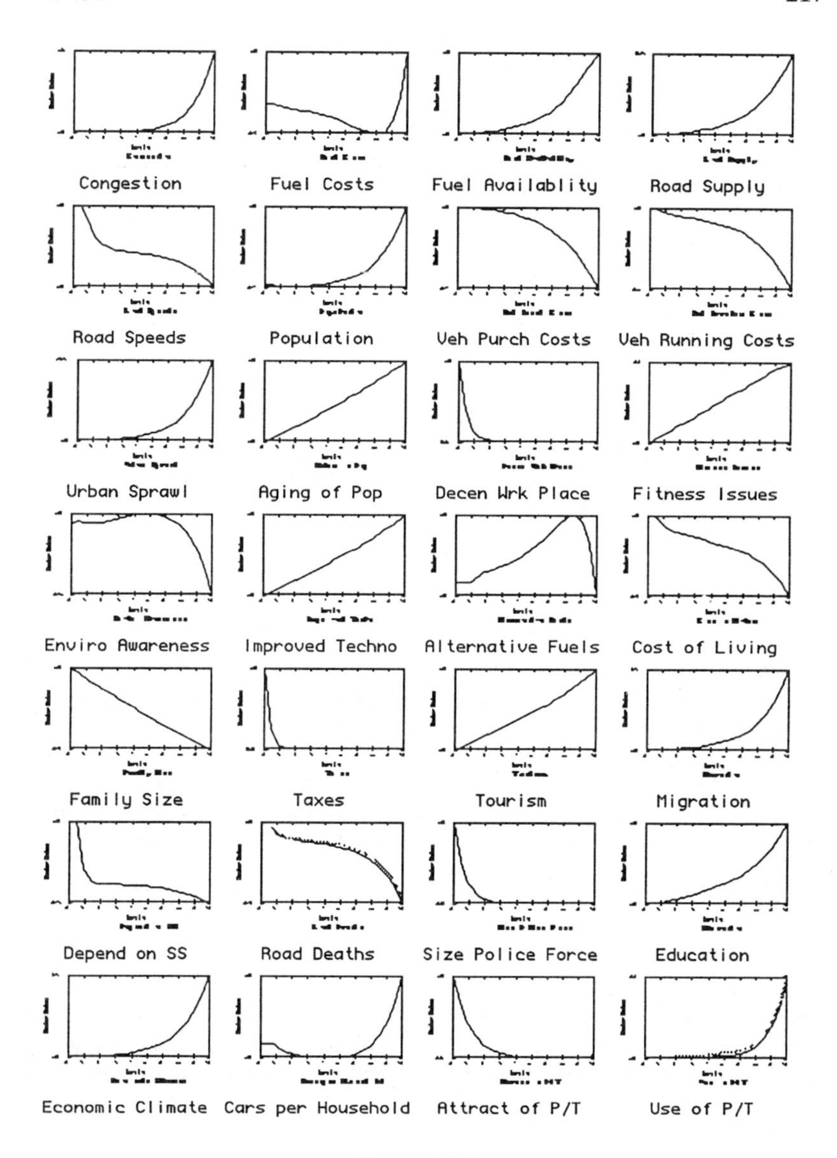

Figure 7.35

It then becomes a simple task to evaluate each of the models presented by the different players and to arrive at a single consensus model that each interested party agrees matches its understanding of the environment under consideration.

Once such a level of consensus is achieved then the task of actually running the model to generate pictures of the future and to elicit a set of solutions to probable adverse futures is a simple task. The implementation of the suggested 'best' solution is the subject of a totally different field of endeavor and is not included in this work.

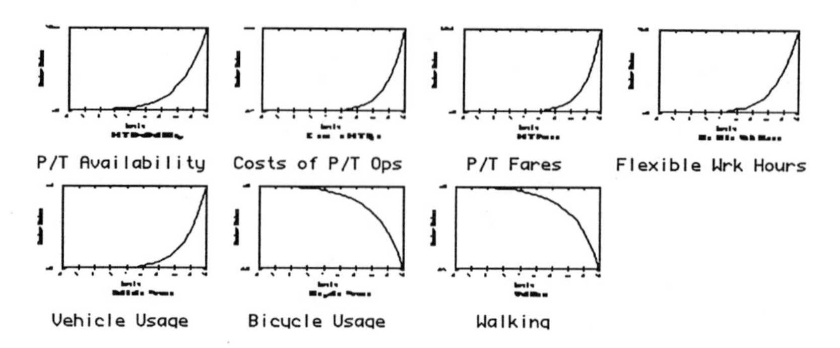

Figure 7.36

7.24 GAs as Assistors in Understanding Systems

One of the great values I find in using GAs is as mechanisms that assist in understanding the system under examination.

A useful proceedure is to save each member of the population, that is generated, and examine them as a complex data set after numerous runs of the same problem. The major advantage of GAs in the undertaking is that each 'citizen' of the populations generated are viable solutions to the problem and as the GA finds highly fit solutions the number of citizens close to good solutions increases (increasing the accuracy of interpretations around good solutions, see Figure 7.7).

With this information it becomes possible, at times and depending upon the problem at hand, to develop a conceptual understanding of the problem. Broad statements of effect can be determined and consistent effects can be observed.

For example, in the Gulf War scenario we observe that we should not increase rhetoric and our attempts to destroy the Iraqi War Capacity together. Any such move will precipitate war.

These analytical processes show us another value of GAs. In a number of real world areas it is not possible to discover a 'perfect' solution. What we seek is direction, not absolutes. GAs give us mechanisms to discover these directions. Statements such as; 'Don't increase rhetoric and destruction of Iraqi war capacity at the same time if you want to avoid war!' can be made by analysic of the phenotypic populations generated by GAs.

I agree that there are other methods and mechanisms that will help us discover the understandings I have described here, but GAs are one of the better tools for this undertaking.

Chapter 8

Dave Corne
Peter Ross
Hsiao-Lan Fang
Department of Artificial Intelligence,
University of Edinburgh,
80 South Bridge,
Edinburgh EH1 1HN, Scotland

{dave|peter|hsiaolan}@aisb.ed.ac.uk

Evolving Timetables

0-8493-2519-6/95/$0.00 + $.50

8.1 Introduction

Timetabling problems are rife in schools, colleges, universities, and many businesses. Solving them satisfactorily often presents great difficulties. Some researchers have begun to demonstrate successful results using genetic algorithms (GAs) on such problems [4, 18, 1]. In particular, [4] reports the successful and continuing use of a GA-based system for the timetabling of exams in a university department. In this chapter we will look at the basics of applying GAs to this sort of problem, consider various different ways of doing it, and look at a number of relevant and related issues.

The successful experiences reported so far, on quite varied kinds of timetabling problems, are good reason to be optimistic for the general business of using GAs on arbitrary timetabling problems. Then again, the range of possible kinds and sizes of such a problem is rather too great for such optimism to be anything but tentatively justified without delving further into the general issues involved, and doing lots of empirical work. In this chapter we attempt to do both, and leave the reader with some conclusions about GA based timetabling which we hope he or she will feel to be justified, and which we also hope will encourage him or her to pursue one or more of the many possible investigations and explorations around the issue which we suggest.

8.1.1 Overview

We begin in section 8.2 with a general description of what is usually meant by a 'timetabling problem.' This takes the form of abstracting the main features (e.g.: times, events, rooms, people), and the usual kinds of constraints that are imposed. It should be evident how this description covers examination timetabling and class/lecture timetabling problems, while also covering a very wide range of others. Section 8.3 has a terse introduction to Genetic Algorithms. Since we assume much familiarity with the basics of GAs there is no need for anything more comprehensive than this; *some* brief introduction is necessary and helpful however, since even if the reader is already highly familiar with GAs this section will help him or her to appreciate what we see as the main aspects of the GA to focus on in the context of timetabling applications. This section then looks at the chief issues involved, as we see them, in using a GA to address an optimisation problem, paying particular attention to the timetabling case. In section 8.4, we get down to illustrating and describing the main styles of a GA approach to timetabling that can be adopted and describe three basic ways in which genetic search can be applied to timetabling. Considering each in turn, their respective advantages and disadvantages are moot. Section 8.5 describes various simple experiments investigating the relative effectiveness of these methods on a test set of exam timetabling problems. We find that this confirms one of these methods, actually the simplest of them, to be best. We then take a look in section 8.6 at how the method fares on some real timetabling problems. The results seem most encouraging. In section 8.7 we look at a simple implementation issue which is easily missed, but makes the genetic algorithm much faster on problems with the sort of fitness function we use. This also gives us a more useful measure than 'number of evaluations' for measuring the time taken by the GA on timetabling problems. Using this measure, we take a step back to grasp what we can generally expect from GAs in timetabling in section 8.8, which looks at the question of how we can expect performance to vary across a very wide range of different kinds of timetabling problems. Results are reported which enable us to make some justified, positive claims in this regard. In particular, it seems that not only can we expect optimal or near-optimal performance across a wide range of problems typical of the size and difficulty of many real problems, but it seems we can also predict how long the GA will take in some cases. Toward the end, section 8.9 reminds us that we should always be careful about claiming too much for a GA application; specifically tailored methods may often be better, but it, nevertheless, usually seems possible to make an *even better* hybrid algorithm by using GA-style ideas carefully. Finally, section 8.5.1 summarizes the chapter, and provides some concluding discussion of some of the relevant things we have left out.

8.2 Timetabling Problems

A large class of timetabling problems can be described as follows. There is a finite set of events E = $\{e_1, e_2, ..., e_v\}$ (for example, exams, seminars, project meetings), a finite set of potential start-times for these events T = $\{t_1, t_2, ..., t_s\}$, a finite set of places in which the events can occur P = $\{p_1, p_2, ..., p_n\}$, and a finite set of *agents* which have some distinguished role to play in particular events (eg: lecturers, tutors, invigilators...) A = $(a_1, a_2, .., a_m\}$. Each event e_i can be regarded as an ordered pair $e_i = (e_i^l, e_i^s)$, where e_i^l is the length of event e_i (eg: in minutes), and e_i^s is its size (eg: if e_i is an examination, e_i^s might be the number of students attending that examination). Further, each place can be regarded as an ordered pair $p_i = (p_i^e, p_i^s)$ where p_i^e is the *event* capacity of the place (the number of different events that can occur concurrently in this place), and p_i^s is its size (eg: the total number of students it can hold). There is also an n x n matrix D of travel-times between each pair of places.

An *assignment* is an ordered 4-tuple (a, b, c, d), where a $\in$ E, b $\in$ T, c $\in$ P, and d $\in$ A. An assignment has the straightforward general interpretation: "event *a* starts at time *b* in place *c*, and with agent *d*". If, for example, the problem in hand was a lecture timetabling problem then a more natural interpretation would be: "lecture *a* starts at time *b* in room *c*, and is taught by lecturer *d*".

Given *E, T, P, A* and the matrix *D*, the problem involves producing a timetable which meets a large set of constraints *C*. A timetable is simply a collection of *v* assignments, one per event. How easy it is to produce a useful timetable in reasonable time depends crucially on the kind of constraints involved. In the rest of this section, we discuss what *C* may contain.

8.2.1 Timetabling Constraints

Different instances of timetabling problems are distinguished by the constraints and objectives involved, which typically make many (or even all) of the smn^v possible timetables poor or unacceptable. Each constraint may be hard (must be satisfied) or soft (should be satisfied if possible). Many conventional timetabling algorithms address this distinction inadequately: if they cannot solve a given problem they relax one or more constraints and restart, thus trying to *solve a different problem.* The kinds of constraints that normally arise can be conveniently classified as follows.

8.2.1.1 Unary Constraints

Unary constraints involve just one event. Examples include: "The science exam must take place on a tuesday", or "The Plenary talk must be in the main function suite". They naturally fall into two classes:

Exclusions: An event must not take place in a given room, must not start at a given time, or cannot be assigned to a certain agent.

Specifications: An event must take place at a given time, in a given place, or must be assigned to a given agent.

8.2.1.2 Binary constraints

A binary constraint involves restrictions on the assignments to a pair of events. These also fall conveniently into two classes:

Edge Constraints: These are the most common of all, and are the central difficulty in almost all timetabling problems arising because of the simple fact that people cannot be in two places (or doing two different things) at once. A general example is: "event x and event y must not overlap in time". It is computationally simple to check the status of such a constraint. The term 'edge' arises from a commonly employed abstraction of timetabling problems as graph colouring problems [19, 25].

Juxtaposition Constraints: This is a wide class of constraints in which the ordering and/or time gap between two events is restricted in some way. Examples include: "event x must finish at least 30 minutes before event y starts", and event x and event y must start at the same time".

Edge constraints are of course subsumed by juxtaposition constraints, but we single out the former because of their importance and ubiquity. Edge constraints appear in virtually all timetabling problems, and in some problems they may be the *only* constraints involved. We return to edge constraints later in this section, but for now we continue enumerating the other kinds of constraints that occur.

8.2.1.3 Capacity Constraints

Capacity constraints specify that some function of the given set of events occurring simultaneously at a certain place must not exceed a given maximum. Eg: in lecturing timetabling we must usually specify that a room can hold just one lecture at a time. In exam timetabling this capacity may be higher for many rooms, but in both cases we need also to consider the total student capacity of a room, eg, we may allow up to six examinations at once in a given hall, but only as long as that hall's maximum capacity of 200 candidates is not exceeded.

8.2.1.4 Event-Spread Constraints

Timetablers are usually concerned with the way that events are spread out in time. In exam timetabling, for example, there may be an overall constraint of the form."A candidate should not be expected to sit more than four exams in two days". In lecture timetabling, we may require that multiple lectures on the same topic should be spread out as evenly as possible (using some problem-specific definition of what that means) during the week. Event-spread constraints can turn a timetabling problem from one that can be solved simply by more familiar graph-colouring methods to one which requires general optimisation procedures and for which we can at best hope for a near optimal solution.

8.2.1.5 Agent Constraints

Agent constraints can involve restrictions on the total time assigned for an agent in the timetable, and restrictions and specifications on the events that each individual agent can be involved in. In addition to those already discussed (exclusions and specifications) we typically also need to deal with the following two kinds of agent constraints:

Agent Preference: event *ei* may be assigned to agent a_j with desirability d. This kind of constraint arises mainly in class timetabling problems where any of a set of teachers may teach certain courses, but have their individual preferences for particular courses. The value d might be taken from a symbolic range (eg: {highly desirable, desirable, indifferent, preferably not}, or some numeric range).

Agent Load: Agent load constraints come in two forms: "the total duration of events assigned to agent a_i must be (exactly/less than/more than) n (minutes/hours/...)", or "agent a_i must have n free (days/mornings/afternoons)".

8.2.2 The Importance of Edge Constraints

It is useful and instructive to consider the set of events E to be the vertices of a graph and the set of times T to be colours (or labels) that can be used to paint the vertices of this graph. We can represent any constraint of the form "e_i and e_j cannot share the same time-slot" as an edge between vertices e_i and e_j. The simple timetabling problem then becomes one of colouring every vertex in such a way that no two vertices connected by an edge are of the same colour. This is a well-known problem in graph theory; see for example [25]. It can easily be shown that the number of ways of vertex-colouring a simple graph[1] with u vertices (events) and c edges (constraints) using s colours (time-slots) is a polynomial $g(s)$ of degree v of the form:

$$g(s) = s^{v} - cs^{v-1} + \ldots + \alpha s \quad (8.1)$$

[1] A graph is called simple if no edge starts and ends at the same vertex and at most one edge joins any pair of vertices.

for some constant α, and that the coefficients alternate in sign. This result is at first glance a little surprising since it suggests that at least for large values of s, a proportion of at least (1 - c/s) of all timetables will be valid solutions. However, practical timetabling problems tend to involve a fairly small value of s and fairly large value of c so that not mnch can be said about the density of possible solutions. One further graph-theoretic result is worth mentioning: Brooks' theorem (see for example [25]) tells us that if not every pair of vertices is joined by an edge, and if no vertex has more than g edges attached to it, then the graph can be vertex-coloured using at most g colours. Thus, given an instance of the simple timetabling problem we can use Brooks' theorem to say how few time-slots are really necessary. Often, however, this number can be significantly larger than the minimum possible.

Graph-theoretic methods can be used to handle the simple timetabling problem. As we shall illustrate later in section 4, such methods consist of algorithms which assign events to slots one by one, employing a simple heuristic or deterministic choice of the slot to use in each case, always ensuring that this choice is legal. Unfortunately, real problems are hardly ever that simple. Transversal theory (again see [25]) can be applied to solve slightly harder timetabling problems such as finding some coherent set of assignments of events, time-slots and locations given a set of binary constraints between them. However, these techniques do not naturally extend to handling the range of nonbinary or nonsymmetric constraints that arise in many practical timetabling problems. More generally, timetabling problems are usually heavily complicated by soft constraints, which cannot all be satisfied. This means that effective optimisation techniques are required to address them, and that simple graph theoretic methods are no longer applicable. The rest of this section discusses examples of such constraints.

8.2.3 More on Juxtaposition Constraints

In simple timetabling problems without ordering constraints, any solution remains a solution when the time-slots are permuted. The polynomial in equation 1 counts these as being distinct; clearly, the presence of a significant number of ordering constraints will reduce the number of possible solutions dramatically, perhaps by a factor as large as $s(s - 1)...(s - v + 1)$. Graph vertex-colouring methods could be adapted to handle binary ordering constraints if there were few enough of them to allow perfect solutions to be possible. However, such ordering constraints often constrain the problem so much (especially in their interaction with other soft constraints) that we must use optimisation methods instead. It is also hard to see how graph-theoretic methods could be easily extended to handle nonbinary ordering constraints, for example of the form "at least one of the events in set S_1 must come before/after one/all of the events in set S_2". An instance of this might be that a weekly seminar session has to happen later than all relevant lectures.

8.2.4 Coping with it All

One of the great difficulties with conventional approaches to timetabling is that they make a distinction between hard constraints which must not be violated and soft constraints, or preferences, which it would be pleasing to satisfy but which can be violated if necessary. Talking to people who construct timetables by hand suggests that they may often begin by trying to solve a particular timetabling problem assuming certain sets of hard and soft constraints. If that effort fails, they may convert one or more hard constraints into soft ones, or even delete them altogether, after re-negotiating with all the other people involved. Thus they move on to trying to solve a new problem, for which the work done on the original problem may be largely unusable.

One of our GA-based approaches to timetabling typically avoids this distinction between hard and soft constraints; instead they award different levels of penalty for different kinds of constraint violation. Thus a timetable which turns out to have a large number of soft-constraint violations may be treated as worse than one with just one or two hard-constraint violations. The GA-based approach is thus addressing a whole spectrum of what would otherwise be regarded as differing timetabling problems, in a uniform way. This may often be what the user might really prefer - a system that takes some liberties with the problem formulation in the first place, rather than expecting the user to conduct a manual search through the space of possible timetabling problems.

8.3 Genetic Algorithms

8.3.1 Brief Description

A GA can be seen as an unusual kind of search strategy. In a GA, there is a set of candidate solutions to a problem; typically this set is initially filled with random possible solutions, not necessarily all distinct. Each candidate is typically (though not in all GAs) an ordered fixed-length array of values (called 'alleles') for attributes ('genes'). Each gene is regarded as atomic in what follows; the set of alleles for that gene is the set of values that the gene can possibly take. Thus, in building a GA for a specific problem the first task is to decide how to represent possible solutions. Assuming we have thus decided on such a representation, a GA usually proceeds in the following way:

- Initialisation: A set of candidate solutions is randomly generated. For example, if the problem is to maximise a function of x,y, and z then the initial step may be to generate a collection of random triples (x_i,y_i,z_i) if that is the chosen representation.

- Now iterate through the following steps, until some termination criterion is met (such as that all solutions in the population are the same, a perfect solution is found, or some other criterion). The process alters the set repeatedly; each set is commonly called a generation.

  1. Evaluation. Using some predefined problem-specific measure of fitness, we evaluate every member of the current set as to how good a solution to the problem it is. The measure is called the candidate's fitness, and the idea is that fitter candidates are in some way closer to being one of the solutions being sought. However, GAs do not require that fitness is a perfect measure of quality; they can to some modest extent tolerate a fitness measure in which the fitter of some pairs of candidates is also the poorer as a solution.

2. Selection. Select pairs of candidate solutions from the current generation to be used for breeding. This may be done entirely randomly, or stochastically based on fitness, or in other ways (but usually based on fitness, such that fitter individuals have more chance of being chosen).

3. Breeding. Produce new individuals by using genetic operators on the individuals chosen in the selection step. There are two main kinds of operators:
- Recombination: A new individual is produced by recombining features of a pair of 'parent' solutions.
- Mutation: A new individual is produced by slightly altering an existing one.

The idea of recombination is that useful components of the members of a breeding pair may combine successfully to produce an individual better than both parents; if the offspring is poor it will just have lower chance of selection later on. In any event, features of the parents appear in different combinations in the offspring. Mutation, on the other hand, serves to allow local hillclimbing, as well as introduce variation which cannot be introduced by recombination.

4. Population update. The set is altered, typically by choosing to remove some or all of the individuals in the existing generation (usually beginning with the least fit) and replacing these with individuals produced in the breeding step. The new population thus produced becomes the current generation.

8.3.2 Terminology

The GA research community is replete with terms such as 'chromosome', 'mutation', 'niche' and so on, naming aspects and structures of GAs after natural evolutionary analogues. Unfortunately the terminology is not quite standardised; different GA researchers sometimes use the same words to mean different things, or different words to mean the same thing. The rest of this section hopes to clear up any ambiguities which might otherwise remain in this article.

First off, unlike a few other researchers, we use the term 'GA' to refer to any of a wide variety of algorithms based, sometimes perhaps loosely, on Holland's original 'genetic plans' [15]. Some researchers restrict the use of GA to algorithms which use binary valued genes, and (more or less) call everything else an 'EA' (evolutionary algorithm). We do not make this distinction, preferring to call something a GA if, loosely speaking, the main search operator is recombination, and we call something an EA if the main operator is mutation (and the algorithm can trace its lineage back to Rechenberg's early work on evolution strategies, rather than Holland's on genetic plans). We are not saying that our use of these terms is somehow better than others, but just wish to explain our usage. Also, we call an individual member of a generation a 'chromosome'.

8.3.3 GA Variants

Different kinds of GA vary in the way that the selection and population update steps are performed, as well as in the choice of breeding methods. Other differences may lie in the flow of control, and/or the introduction of new steps for some reason (such as to promote the evolution of multiple distinct solutions, implement some way of controlling premature convergence, etc.). To partly illustrate some of these choices, we will outline here the options chosen for the GA used in the experiments described later on. The evaluation and variation aspects of the GA are tied intimately with the nature of the problem being solved, and hence these are discussed in a later section on the application of GAs to timetabling.

8.3.3.1 Selection

The kind of GA we use in various experiments reported later uses 'local mating', a form of spatial selection introduced in [2]. In spatial selection by local mating, the population occupies a two dimensional grid (actually, a toroid). A selection step involves first choosing a place *L*, somewhere in this grid. We then attempt to replace the chromosome at that position with a new one as follows: do a random walk of a given length *R* along the grid starting from location *L*. The selected chromosome is simply the fittest chromosome encountered during the walk. For crossover, we pick two parents via two different random walks from the same location, and crossover to produce one child. For mutation, we simply rotate the chromosome selected from one random walk. In both cases, we update the population by putting the child in the original location, *only if* that child is fitter than the chromosome currently residing there. In our GA, the initial choice of grid location is random; this hence resembles a steady state [23] form of local mating.

Local mating, in which individuals only compete within localised groups, is based on Wright's 'shifting balance' account of evolution [26, 27, 28, 29], rather than the more common *panmictic* style, where every individual in the population competes with every other, first promoted by Fisher [11]. The spatially-structured approach appears robust and highly competitive with other selection techniques across a range of fitness landscapes, although it may not always be the best choice.

8.3.3.2 Breeding

Given two parent chromosomes a single offspring, or two complementary offspring in some GAs, is produced by a *crossover* procedure. A new chromosome is generated by copying parts of the parents. In *one-point* crossover, a random 'cut' point is chosen at which to divide both chromosomes; the first parent is copied up to that point and the second thereafter. In *two-point* crossover, two such cut points are chosen and the first parent is copied before the first and after the second, the second parent is copied between them. *N-point* crossover is also sometimes used as is *uniform* crossover [21] in which, for each gene position, a random choice is made between the genes of the two parents at that position. Given two maximally dissimilar parents of length L it is worth noting that one point crossover can produce $(L + 1)$ possible children, two-point crossover can produce $L(L + 1)/2$ possible children and uniform crossover can produce all 2^L possible children.

If only one offspring is produced from two parents, certain alleles present in the parents will not be present in the child, and the genetic diversity of the next generation may well be smaller than that of the current generation. Even if two complementary offspring are produced, they displace others when installed in the new generation, so the diversity can still decrease. Typically, therefore, some method is sometimes needed to (re-)introduce diversity lest alleles necessary to a good solution have accidentally been lost from the entire set of chromosomes before they can show their worth. Normally this is handled by a mutation operator, which with small probability makes random changes to genes. There are many variants of mutation. In single-gene mutation, for example, one gene is randomly chosen and its allele randomly altered with some low probability (eg: 0.1). In multi-gene mutation, each gene in the chromosome is altered with a very low probability (eg: 0.005). In specific applications, some 'smart' mutation operator can usually be devised which exploits problem-specific knowledge to perform local hill-climbing on a chromosome.

8.3.3.3 Population Update

The choice of population update technique governs precisely how the population changes with time. The two extremes are *generational,* in which enough individuals are selected to replace the entire population at each time step, and *one-at-a-time* in which just one pair of chromosomes gets selected and their offspring gets installed in the current population, usually displacing the least fit if less fit than the offspring. This latter method was first introduced by Whitley [23] in his GENITOR system; comparative tests on a number of benchmark problems suggest that it is remarkably effective compared to a number of other widely-used GA variants [14]. Intermediates between these two extremes are also often used.

8.3.3.4 Reinitialisation

GAs often converge prematurely. Many researchers have found that this problem can be partially alleviated by employing a reinitialisation strategy. When the population has converged, a typical reinitialisation strategy is to retain one chromosome, and randomly reinitialise the rest of the population by using multi-gene mutation with a high mutation rate. In this way, the GA can often further improve on its best solution so far.

8.3.4 Applying Genetic Algorithms

As the previous section suggests, there are many choices to be made when designing a GA. Configuring a GA to a particular problem is an art, which relies on intuition and imperfect understanding of how a GA operates, as well as what makes a solution to the problem a good one. In this section we will briefly discuss some issues that need to be considered.

8.3.4.1 The Evaluation Function

Given a space P of candidate solutions to a problem, there are countless ways in which we can define a fitness function $f(p)$ for $p \in P$. We wish $f(p)$ to be a normally increasing function of the quality of p as a candidate solution of the problem, at least so that optimal solutions lie at the global maxima of f. We also want f to be a reasonably well-behaved function, so that its value conveys some information about the quality of p as a solution in most parts of the space. As p gets closer to being an optimal solution we would like $f(p)$ to change in some way that reflects this. For example, we could require that f was a continuous and increasing function of the quality of p. However, this is a very strong condition which is not essential to the success of a GA.

Notice that the quality of a solution p may not vary smoothly as the genes comprising p vary. For example, in some GA applications the chromosome might not directly represent a solution; instead it might represent a sequence of instructions for building a solution (see, for example, [10]). A trivial change to an early member of the sequence might result in the construction of a very significantly different candidate solution. In such cases one talks of the fitness of the *phenotype* – what the chromosome decodes to rather than the fitness of the *genotype,* based on the chromosome itself rather than what it represents.

The genetic operators such as crossover and mutation do not vary the gene values smoothly either. For example, if a chromosome is regarded as a vector of gene values, then one-point crossover produces a vector located at the intersection of two orthogonal subspace projections of the parents – a kind of quadrature motion through the space. This also correctly suggests that it is wise to discourage later breeding between a parent and its child, since they will lie in some shared projective subspace and their offspring will be confined to that subspace too if mutation does not shift them out of it.

Except for artificial test problems that are thoroughly understood, it is difficult or impossible to design a fitness function whose graph will be best explored by a given set of genetic operators. After all, if the problem is that well understood there would be no need to use a GA to solve it in the first place. However, there are usually some natural choices available for the fitness function. In timetabling problems one might choose to use something inversely proportional to the number of violated constraints; for instance, if $V(p)$ is the number of violated constraints in candidate p one could choose

$$f(p) = 1/(1 + V(p)) \tag{8.2}$$

so that the theoretical maximum of f is 1 and we can be clear about whether or not we have found a perfect solution. However, this function treats all constraints equally and conveys no information about the extent to which any is violated; a better choice can be made.

8.3.4.2 The Space to be Searched

Normally, it is wise to ensure that the sets of alleles for different genes in a chromosome are independent of one another. For example, suppose the first three genes in a certain chosen representation were all to be integers in the range 0 - 1000 but were to be made dependent on each other so that their sum was always 1000. The trouble with such a representation is that the genetic operators will almost always produce offspring that do not satisfy this representational constraint; the offspring will have to be forcibly mutated, perhaps drastically, into one that does. However, if the sets of valid alleles for different genes are independent then the space of possible chromosomes will, in vector terms, be a rectangular solid. Often, this will include regions that represent invalid candidate solutions to the original problem. In timetabling, if we represent a candidate as a list of u genes (one per event) each of which can take any value from 1 to s (representing the time-slot for that event), then many chromosomes will be invalid, for instance, chromosomes in which every gene has the same value so that every event is to happen in the same time-slot. It seems reasonable to design a representation that allows the space of possible chromosomes to include such infeasible regions; if the fitness function is appropriately designed, such infeasible regions may 'smooth' the transition between better and better feasible points. Recent GA research has tended to support this (eg see [20]). This is reassuring, since it can sometimes be difficult to design a representation such that the whole space of possible chromosomes is feasible for the original problem. Also, many problems will be so constrained that feasible solutions, those that satisfy all the hard constraints, are impossible. In such cases we have no choice but to represent infeasible regions, and use the GA to minimise the degree of violation of the hard constraints as well as the soft constraints.

8.3.4.3 Recombination Operators

Genetic search is primarily fueled by the recombination or 'mating' of previously generated individuals to produce new individuals. An underlying heuristic which motivates this process is that recombination should frequently lead to offspring which are fitter than their parents. Theoretical analysis of this process centres on the notion of *schemata;* these are just combinations of alleles. Eg: the schema ##a#b##...", in which '#' denotes "any allele", is contained by any chromosome whose third gene's allele is *a*, and whose fifth gene's allele is *b*. A schema has an associated 'order' and fitness. Its order is the number of defining alleles; ie: the number of non-#'s in the string used to define it. The fitness of a schema is usually taken to be the average fitness of the set of chromosomes that contain it, averaged over the current generation. Since recombination works by offspring taking some of their genes from one parent, and the rest from the other, schemata tend to be passed on from parent to child. Analysis [15, 12] sees genetic search as played out on the level of schemata: high-fitness schemata (or 'building blocks') propagate in the population, ousting lower-fitness schemata defined at the same positions. Selection tends to pick up chromosomes containing good building blocks, and recombination tends to produce new highly fit chromosomes by combining different building blocks from each parent. Recombination works in such a way that if a given building block is present in both selected parents it will be present in both offspring; this contributes to the successful propagation of good building blocks. A building block can often fail to be passed on though, as recombination may split it or mutation may disrupt it; this is less likely if it has low order; hence, low order high fitness building blocks tend to be propagated best from generation to generation.

Given this picture of how genetic search operates, we can see that a good representation for the GA is one in which what we consider to be good features of a solution can be represented by low order schemata, and hence constitute good building blocks. After first deciding what may constitute good building blocks, the chromosome representation can be chosen so as to maximise the degree to which the building blocks correspond to low-order schemata. The recombination operators are then chosen in such a way that the building blocks are effectively compounded by their repeated action in order to make up a good composite solution.

In practice, the business of applying a GA to a problem is often not attacked in the way sketched above. Typically, a chromosome representation almost 'chooses itself', being perhaps the only immediately conceivable way of representing the desired kind of solution structure as a fixed length string of attribute values. Similarly, a natural candidate fitness function may readily suggest itself, and a recombination operator is then chosen from a standard set of well-tried and tested ones; usually this is uniform crossover [21]. In many applications this will often suffice, and efficient and effective optimisation will ensue. Often though, the results obtained are rather weak; it is at this stage that the GA-engineer will go back to basics and consider the process more carefully, expending considerable effort on analysing what appear to be the important building blocks of a solution, how these may be represented more effectively, and/or how the recombination operator may be revised so as to more effectively traverse (or indeed transform) the fitness landscape.

8.4 Some Possible Methods for GA-Based Timetabling

8.4.1 Clash-Rich Spaces

The first GA approach we will consider is one which searches the space S of all timetables. We can satisfy most of the requirements discussed in the previous section by representing a timetable as a list of u numbers, each of which can take on any value from 1 to s, for example:

3621724235

which has the interpretation: e_1 is assigned to time-slot 3, e_2 is assigned to time-slot 6, e_3 is assigned to time-slot 2, and so on. If e_i is assigned to time-slot g_i we can immediately see that each edge constraint (e_i, e_j) translates into the constraint $g_i \neq g_j$. If there are many edge constraints, then a very large proportion of the space of timetables (and hence, with this choice of representation, the space of chromosomes) will have many edge-violations; we hence refer to the space searched in this approach as 'clash-rich'.

8.4.1.1 Fitness

If a timetable p violates v_{edge}(p) edge constraints, and $v_{ord}(p)$ ordering constraints, a possible choice of fitness function given the chosen representation is to have fitness inversely proportional to the number of violations. That is, fitness would be given by:

$$f(p) = 1/(1 + v_{edge}(p)+ v_{ord}(p)) \quad (8.3)$$

Notice that we may have a strictly infeasible timetable (say, in which just one edge constraint is violated) having near optimal fitness (since it has only one violation). But this is fine, because this timetable is genotypically very close to an optimal chromosome; perhaps only one gene (one of the two genes associated with events involved in the edge violation) needs a new allele. A problem with this fitness function, however, is that edge constraints are treated on a par with ordering constraints. If, as is typical, the ordering requirements are soft constraints, we would far prefer a timetable whose only violation is an ordering constraint to a timetable whose only violation is an edge constraint. This scheme fails to distinguish between these in terms of fitness, and hence selection will similarly fail to distinguish between them; the GA may be equally likely to converge to the edge-violation solution as to the ordering-violation solution. Indeed, if the ordering violation in question is particularly hard to satisfy, the GA will be more likely to converge on the edge-violating solution. Rather than have this situation, we would like the GA to make appropriate tradeoffs between timetables with differing degrees and kinds of constraint violation. For instance, a timetable with 5 edge-violations and 8 ordering violations is normally to be preferred to (and hence should be fitter than) one with 8 edge-violations and one ordering violation.

A straightforward way to achieve this is to penalise the different types of constraint violations in accordance with their relative importance; constraints we strongly need satisfied will carry heavier penalties. If we have k kinds of constraint, the penalty associated with constraint-type i is w_i, p is a timetable, and $v_i(p)$ is the number of violations of constraints of type i in p, then the fitness function becomes:

$$f(p) = 1/(1+\sum_{i=1}^{k} w_i u_i(p)) \tag{8.4}$$

This is actually a very flexible system; the main advantage is that we can incorporate any constraint we like into the fitness function, along with an appropriate penalty measure for it, and we can then expect the GA to take this constraint into account during its optimisation. An important consideration is that evaluation be fairly fast, which means here that it be computationally quick to check how many constraints of each type are violated in a given timetable chromosome. Fortunately, this is true for the kinds of constraints we have considered so far, which are indeed the most commonly occurring in timetabling problems. To check an edge constraint (e_i, e_j), for example, we simply need to check whether or not the value of the ith gene is equal to the value of the jth gene. If the chromosome is implemented as a simple array in C, this is a very fast operation.

It is unclear from the above description, however, whether this will actually work as intended. And, even if it works, it is unclear how sensitive performance might be to the particular relative choices of penalty values. Further work on these questions will be reported in the future, but initial evidence is very promising. Works [9] and [4], for example, report success with this technique using intuitively chosen penalty values.

The basic idea is that higher penalty settings for a constraint increase the artificial evolutionary pressure to remove that constraint from the population. Hence, we penalise the hard constraints more heavily than the soft constraints, since we *must* remove the hard constraints in order to get a feasible timetable, but can and must live with violations of the soft constraints. In practice, there may be more kinds of soft constraints, which we may penalise differentially; for instance, if we considered not having early morning exams as more important than avoiding consecutivity, then we might penalise the former with a. penalty of 0.2, and the former with a penalty of 0.03. This means that selection will more highly favour a timetable with seven consecutive exams and no early morning ones, than a timetable with one early rnorning exam. In this way, we can choose penalty settings according to our particular idea of how we would trade off the advantages and disadvantages of different solutions to the problem in hand. Penalties must be set with care, however. If the ratio between two penalties (say, ordering constraints vs. event spread constraints) is too high, then search will quickly concentrate on a region of the space low in violations of the more penalised constraint, but perhaps missing a less dense such region in which better tradeoffs could be found. If too low, then the capacity for the search to trade off between different objcctives is lost. Evidently, optimal penalty settings depend on many things, primarily involving the density of regions of the fitness landscape in relation to each constraint, as well as the subjective relative disadvantages of different constraint violations in the problem at hand. Forthcoming papers will explore this issue further. Related work includes the interesting approach in [20] (in the context of multiple objective facility layout problems), in which penalty settings are revised dynamically in accordance with the gradually discovered nature of the constrained fitness landscape.

A further style of alternative which should be mentioned is a method for constructing scalar functions for multi-objective problems presented in [8], and discussed in the context of GA optimisation in [16]. This is called Multi-Attribute Utility Analysis (MAUA). In MAUA, extensive questioning of an expert decision maker (in our case, an experienced timetable constructor) on example cases (pairs of distinct timetables) would lead to the construction of a nonlinear function M of the vector of summed constraint violations, designed so as to best match the judgement of the expert in that the ordering on timetables imposed by M optimally matches that imposed by the expert. The effort involved in performing MAUA, however, is unlikely to be rewarded with a function M which is significantly closer to the expert decision-maker's judgement than an essentially *ad-hoc* but intuitive linear weighted penalty function. Also, MAUA involves no attempt to structure M such that the fitness landscape is more helpful to the search process; so, on balance, we have not considered it a fruitful line of work to use MAUA in the context of timetabling.

Extensive experience so far suggests we need not be too careful in choosing the penalty settings for typical GELTPs, and can settle for a simple linear penalty-weighted sum of violations with an intuitive choice of penalty settings. This choice remains important in terms of the priority ordering it imposes on the kinds of constraint violation in question. Eg: if a juxtaposition constraint is ranked less important (and hence has a lower penalty term) than a room exclusion constraint, then the GA is more likely to evolve solutions with a single juxtaposition violation than one with a single room exclusion violation. Relative penalty settings should hence reflect this importance ordering, but the particular settings seem non-critical within this restriction on many problems.

Finally, we mention a further alternative which we have not yet explored in the timetabling case: this is to employ the GA to perform Pareto Optimisiation, attepting to evolve points on the Pareto surface. Proposed in [12] and later explored in [16], this seems a useful way of allowing us to grasp an idea of how the different constraints we impose interact (eg: the Pareto surface might tell us that room capacity constraints can only be satisfied at great cost to event-spread constraints), without us needing to specify distinct penalty terms; all we need is a priority ordering over the different constraints. For now, however, we have left this unexplored since there seems no great need to explore the Pareto surface in the timetabling problems we have addressed, because usually the GA finds a perfect solution anyway.

8.4.1.2 Operators

The choice of operators to use in breeding can be quite difficult. Let us consider recombination first. In a timetable, there are various possibilities as to what may constitute a good building block. Clearly any edge constraint (e_i, e_j) constrains the values assigned to the two events somewhat. If there exist many constraints between the members of some small subset of the set of events then there will be relatively few mutually compatible assignments for the members of the subset. Any such coherent set of assignments for these events might therefore be a useful building block, and it would be sensible to arrange for it also to be a short building block by arranging the chromosome so that those events occupied some contiguous slice of it.

Making such potential building blocks short would seem to be particularly useful when employing operators such as one-point and two-point crossover, since a short building block will be less likely to be cut and hence broken by the operator than a long building block. On the other hand this argument only applies when the building block already exists, either by chance in the initial population or by being created through the previous action of some breeding operator. If the initial population is not large this suggests that one-point or two-point crossover may take some while to form that good short building block in the first place.

Uniform crossover does not exhibit this dichotomy between production and destruction of good short building blocks. It explores the set of values that might compose a building block faster than one-point or two-point crossover. It also destroys a building block faster if *that building block is not present in both parents.* This suggests that uniform crossover is a good choice when the representation and fitness are such that short building blocks contribute an amount to the fitness which is significantly disproportionate to their size, so that selection and population update spread the building blocks through the population usefully faster than uniform crossover can destroy them. The idea of probabilistically selecting which of these types of crossover to use each time has been explored with considerable success in some simple GA applications; see [5].

One-point. two-point, uniform and similar crossover operators are very general; they can be applied in almost any GA. It is possible to design problem-specific operators too. For example, in timetabling problems one might use an operator that focuses on time-slots as follows. For each time-slot t, consider the set of events that are assigned to that slot in either parent. Divide the set into two disjoint subsets at random. Give the members of the first subset the value t in one child and a random value other than t in the other child. Give the members of the second set the value t in the second child and a random value other than t in the first child. This happens to be a variant of an operator investigated in [1] that is easy to implement with the simple timetable representation that we are considering in this section. They were particularly interested in exploring the possibilities for such 'period' (slot-based) building blocks which seem fairly natural to the task. Of course, one must always bear in mind that what seems natural may in fact be what makes the problem hard to solve in the first place; intuitive appeal may be a reason for trying something, but only experimentation or formal analysis can justify it.

Choice of a mutation operator is more straightforward. All that seems necessary is an operator which randomly changes the alleles of a small number of genes. Experience with GAs has shown, however, that 'smart' mutation can aid evolution greatly. We initially consider two such operators for use on timetabling problems in clash-rich spaces. Both involve the concept of the violation-score of an event: an event's violation score is simply the sum, weighted by the appropriate penalty values, of the constraint violations involving that event.

Violation-directed mutation Choose an event with a maximal violation score, and randomly alter its assigned time-slot.

Event-freeing mutation Choose an event with a maximal violation score. Then choose a slot which will maximally reduce this score, and assign the chosen event to the chosen slot.

Violation-directed mutation is simple and relatively cheap to implement. The set of events with a maximal violation score can easily be determined during evaluation. Mutating the assignment of this event would then seem to be directing mutation to where it is most needed. Event-fleeing mutation is a more expensive operation, because of the additional cost involved in searching for the best slots into which the chosen event can be placed. However, it promises to make up for this increase in complexity by more effective local improvement. Such hillclimbing or local-improvement mutation can be very useful and lead overall to significant speedup in the search, although it must not be overused; a high rate of hillclimbing will typically lead to getting stuck in local optima. On the other hand, such operators are particularly helpful in escaping from the kinds of local optima in which GAs often get stuck. As discussed earlier, GAs can converge quickly to a near optimal (near clash-free) timetable without being able to make the final steps to remove the remaining few clashes (or other violations). Event-freeing mutation, however, might manage to do this directly. There might still be cases in which the final clash or clashes simply cannot be removed without first altering the structure of other parts of the timetable which do not directly involve the events involved in the clash. Hopefully, the GA will have done its work well enough to make even this awkward case fairly easy to handle by some kind of non-GA-based local search.

8.4.2 Clash-Free and Clash-Sparse Spaces

As equation 8.1 suggests, if the number of time-slots s is large enough and the number of edge constraints c remains fixed, a very large proportion of possible timetables will be valid timetables. In these circumstances it ought to be possible to construct solutions directly via vertex-colouring methods. This suggests an alternative approach: consider the space of timetables that requires the same constraints between the u events but which use a varying number of time-slots: and get a GA to try to minimise the number used. If the number of slots used can be brought down to the s required by the original problem, or lower, the original problem will have been solved. In fact, even if clash-free timetables are very sparse in the space, such methods can often easily find them.

In this kind of approach a chromosome might be taken to represent some permutation of the set of events. Given such a chromosome, it can be decoded into a timetable by assigning time-slots to the events in the order given by the chromosome. This approach is discussed in [7]. For example. if there was no edge-constraint between the first and second of the events, they would be given the same time-slot. If there was an edge constraint between the second and the third but not between the first and the third, then the third event could be given the same time-slot as the first. The assignment of slots to events would proceed in this way, only introducing a further time-slot when the constraints prevented use of any existing time-slot. The fitness of the chromosome would be inversely proportional to the total number of time-slots needed by this procedure, so that if the GA maximised fitness it would be minimising the number of time-slots required. This method has been used in GA applications in scheduling, see for instance [30, 10]; the idea is discussed in more general terms in chapter six of [7].

In this approach the GA is searching the space of permutations of v events, of size $v!$, rather than a space of size s^v. This means that if the number of slots required (call it s^+) were $O(v)$ or worse it would be searching a larger space than in the previous approach. However, it is extremely unlikely that s^+ need be as large as v – that would correspond to putting every event in its own unique time-slot. Thus the expectation is that the GA will essentially be searching a rather smaller space, of timetables in which the number of slots required is much smaller than v.

The alert reader will notice that there seems to be a representational problem here. If two chromosomes are each just permutations of the numbers $1...v$ it is extremely unlikely that their offspring will be too. This has been handled in different ways by different authors. The crudest method is to forcibly mutate the offspring until it is a genuine permutation; however, this will often necessitate a great deal of mutation. [24] and others have explored the idea of using special operators which guarantee that the offspring are also genuine permutations. The one we prefer (and have used very successfully in [10]) is to make each chromosome be an indirect encoding of a permutation as follows. Let each chromosome be a sequence of v- 1 numbers, each in the range $1...v$, say $g_1 g_2 g_3$ .... A list of the events E (in simple numerical order) is maintained, and a permutation list P, which is initially empty. A permutation is retrieved from the chromosome by taking each gene g_i in turn, finding the g_ith member of the list E (treating E as circular), deleting this from E, and then adding it onto the end of P. Thus, for example, if $v = 6$ the following chromosome (which is thus of length 5):

45345

would decode to the permutation:

463125

A disadvantage of this is that many different chromosomes represent the same permutation. However, experience suggests that this drawback is more apparent than real; in the work reported in [10], the population typically converges quickly with the intial parts of chromosomes stabilising much more rapidly than the tail-end parts.

There is a further representational problem with the clash-free approach. We can only find an optimal timetable if indeed such a timetable can be represented in our chosen representation scheme. If we only consider optimality as meaning a clash-free timetable with minimal time-slots used, then it can indeed be shown that some permutation of the events will represent this timetable, in the sense that the clash-free decoding algorithm will produce this timetable given that permutation. When we add other kinds of constraints to the problem though, we can often no longer assume this. For example, we may wish to minimise the degree to which the timetable requires students to sit in consecutive exams. Aspects of the problem may immediately tell us that a timetable with no consecutive exams at all is infeasible, and so it therefore makes no sense to avoid consecutivity in the representation itself. We might therefore search the space of clash-free (but not consecutivity-free) timetables, while using the GA to optimise over the degree of consecutivity as well as minimisation of timeslots (if necessary). The problem, however, is that we cannot know for sure if optimal consecutivity timetables are represented. To see this, imagine a simple problem with two edge-constrained events and three consecutive slots. The entire search space for the clash-free approach comprises the chromosomes: e_1e_2 and e_2e_1, respectively representing the timetables $((e_1, s_1), (e_2, s_2))$ and $((e_2, s_1), (e_1, s_2))$. Notice that the timetables $((e_1, s_1), (e_2, s_3))$ and $((e_2, s_1), (e_1, s_3))$, both optimal in terms of consecutivity, are not reachable in this space.

This problem is avoided by an amendment to the representation and its interpretation. The key change is that instead of always assigning an event to its *first available slot* as the timetable is being developed, we assign an event to the n-th available slot, where n is a parameter provided by the representation. A subsidiary change is that we no longer search permutations of the events, but use a single fixed ordering. The representation thus looks the same as it did in the clash-rich approach, but the interpretation has changed. The decoding algorithm therefore takes as input an ordered list of u, values, $g_1, g_2, \ldots g_v$, each from 1... v,. This is decoded as follows. Starting at $i = 1$, assign the ith event to the g_ith available slot. A slot is 'available' to event i if it can be placed there without violating any edge constraints, and the list of available slots at any stage is treated as circular. If no slot is available to event i, then assign event i to the g_ith unavailable slot.

It is straightforward to show that every timetable represented in the clash-free approach can also be represented in this approach. More importantly, many other timetables can too. For example, the two consecutivity-optimal timetables in the simple example above can be represented by the chromosomes: s_1s_3, and s_3s_1. Indeed, every clash-free timetable occupying the available number of slots can be represented, irrespective of our a *priori* ordering on the events, and hence timetables which are optimal in respect of other constraints are certainly reachable in the space.

It is more difficult to devise 'smart' mutation operators for clash-sparse and clash-free spaces than in the clash-rich case. This is because of the implicit nature of the representation. A timetable represented via the clash-free approach is very 'directly' represented, in the sense that reading the chromosome immediately gives us the slot assigned to each event. If we wish to change the slot for any event to a particular new slot, we know precisely what to do. In the clash-sparse approach, however, the chromosome is opaque to such operations. Changing the slot of an event no longer corresponds directly to changing the allele in that event's place in the chromosome. This leads to a credit assignment problem: it can be expensive to distinguish which of a set of mutations is an improvement. We therefore do not use smart mutation operators in the clash-sparse and clash-free approaches in this paper, although it remains possible to devise useful such operators for the clash-sparse approach.

8.4.3 Further Issues

There are many ways of tweaking and tuning the above approaches, which may improve performance in certain conditions. This section just hints at the possibilities.

8.4.3.1 Minor Changes to the Representation

In section 4.1 we considered a representation consisting of v integers, the i-th giving the time-slot for event e_i. If there are many constraints that relate to days, such as keeping down the number of events that happen in the same day, then a representation which separates the day information from the slot within that day may work better. For example, one might try a representation containing $2v$ integers; in each pair, the first labels a day for event e_i and the second gives the slot within that day. This might help the GA to handle the day-related constraints more simply.

8.4.3.2 Speeding up Evaluation

In a large exam timetabling problem, the number of edge constraints is typically very high; this means that the evaluation stage of the GA will be slow (each edge-constraint must be considered for each chromosome). Considerable speedup might be attainable by only checking a subset of the edge-constraints for each chromosome (perhaps a randomly chosen subset). As time goes on, it is necessary to check the complete set of constraints (otherwise we will not know if a timetable with perfect fitness is actually a perfect timetable), hence there must be a carefully chosen regime by which the proportion of edge-constraints checked is gradually increased. Nevertheless, this 'noisy evaluation' technique could greatly speed up the early part of the search, without seeming to affect the quality of the early populations.

8.4.3.3 Solving in Stages

If there are ordering or event-spread constraints, the search may be carried out in two stages; first, evolve a timetable which satisfies the edge constraints, and second: evolve permutations of this timetable to find a permutation of the slots (within the already evolved assignments) which optimally satisfies the other constraints. Such approaches have been reported with some success using GAs in other areas [17]. In a sense the choice is between having the GA solve two problems at once, or one at a time. The former has the disadvantage of being slower, but the latter seems more likely to be trapped at local optima; eg: if we find the best clash-free timetable, and then find the best permutation of its slots to satisfy the ordering constraints, the end result may not be as good as starting the second stage with the second-best clash-free timetable.

8.4.4 Summary of Approaches

Three approaches to timetabling with GAs have been proposed. The first, 'clash-rich' involves a very simple representational scheme in which a chromosome *abc*... is interpreted as "assign event 1 to slot *a*, assign event 2 to slot *b*, assign event 3 to slot *c*,... ", and so on. The space searched by the GA is rich in violations of any constraints that may be imposed. In particular, it is rich in violations of edge-constraints (which are apparent in almost every timetabling problem). Nevertheless, as demonstrated in [4], effective evolution of suitable timetables for real problems can occur. Any assignment of events to the available slots is possible in this representation, and hence optimal timetables (however that is defined) are reachable in the space. Two problem-specific mutation operators are proposed for improving the speed of evolution in the clash-rich approach.

The second, 'clash-free', uses the GA to search the space of permutations of the events. Each permutation is decoded by a 'greedy' algorithm which takes events in the given order and assigns each to the earliest event in which it will fit without violating any constraints. This is the method discussed and used by Davis in chapter six of [7]. The space contains only clash-free timetables, some of which may use more slots than are allowed in the problem. Searching a permutation space with a GA requires either specialised operators (as used in [7]), or a method of representing permutations which allows standard operators to be used; we choose the latter method. Because there is no need for the GA to eliminate clashes from timetables, we might expect faster solutions (note: the increased complexity of decoding a chromosome is balanced somewhat by not needing to count clashes during evaluation), however, it is unclear how well the GA might do at reducing the number of extra slots used. Also, where difficult soft constraints are involved, this method fails to guarantee that optimal timetables are reachable in the space. Another possible difficulty from the point of view of the GA is that timetabling building blocks no longer map clearly onto short defining length schemata in the chromosome, although the GA has a way of dealing with this kind of situation.

The third, 'clash-sparse', involves a representation similar in appearance to the clash-rich approach, but decodes a chromosome via an algorithm which eliminates most (though not necessarily all) clashes. The GA should find clash-free timetables in this space far more quickly than in the clash-rich approach. Also, unlike the clash-free approach, timetables which are optimal in respect to the soft constraints will be reachable in the space. As with the clash-free approach, however, there are representational problems which somewhat hamper the ability of the GA to exploit good building blocks. For related reasons, it is also difficult to devise smart mutation operators for the clash-sparse and clash-free methods.

8.5 Some Investigation of the Three Approaches

Our current research in this area focuses on two questions: first, what can be said about the general behaviour of GAs on timetabling problems?; second, in applying a GA to a particular timetabling problem, what is the most appropriate choice of space, representation, operators, and so on? Here we focus more on the first of these. Considerations relating to the second are very important from the applications viewpoint, however, in the interests of keeping this article at a reasonable length, we prefer to refer readers to previous and forthcoming papers for such issues.

In this section, we will empirically test the different approaches described in previous sections, in an attempt both to test our intuitive expectations and to derive some ideas of their relative merit for general timetabling. We examine three versions of each of two test problems. Problem 1 involves 100 events which are to be organised around 10 slots, subject to 500 randomly generated edge constraints. The slots are interpreted to be in two groups of five, and ordered in time within each group such that slots 0 and 1, 1 and 2, 2 and 3,..., form consecutive pairs (but not slots 4 and 5, since these are in different groups (ie: on different days). Similarly, problem 2 involves 100 events and 15 slots (in three groups of five), subject to 1,000 randomly generated edge-constraints. Version 1 of each has the problem of finding a timetable free of edge-constraint violations. In version 2 of each problem, we additionally try to minimise the number of consecutive edge-constrained events. Version 3 is as version 2, with the addition of 10 (coherent) ordering constraints. With the clash-rich approach, we also examine the use, to varying degrees, of violation-directed mutation and event-fleeing mutation.

Version 3 of problem 1 is similar in magnitude, though rather more highly constrained, than the real university exam timetabling problems explored in [4, 9], and may be considered similar in difficulty to many realistic timetabling problems faced in schools and university departments. Problem 2 is rather more difficult, and so non-typical of most problems that would be faced in practice. However, we generally find it rather more enlightening to compare performance on difficult problem sets than on easier ones. This is because each method (clash-rich and clash-free in particular) tends to find perfect solutions quickly on less constrained problems, making it more difficult to tease out their relative differences in performance.

In each experiment we use a population size of 200, tournament selection with a tournament size of 5, a bit mutation rate of 0.02, uniform crossover, and one-at-a-time update. Tournament selection is used, because we have generally found it to be best for the timetabling problems we have looked at. A relatively low tournament size is chosen to combat premature convergence by reducing selection pressure. We have generally found the one-at-a-time population update technique to be most robust, as is also true of the choice of crossover operator and the use of bit mutation.

Each of the clash-rich, clash-free, and clash-sparse approaches were tried on each version of both problems, running (unless stated otherwise) for 50,000 generations. Each generation involved the production of one new chromosome, and hence one evaluation, unless it was a 'restart' generation, in which case it involved the production of 199 new chromosomes, and hence 199 evaluations. Typically only a handful of restarts occurred in each experiment.

Finally, the penalties for constraint violations were as follows:

- Violation of an edge constraint (ie: a clash): 20.
- Violation of a consecutivity constraint (ie: two edge-constrained events in consecutive slots): 1.
- Violation of an ordering constraint: 1.
- Penalty per additional slot used in the timetable: 20.

Only the first three of the above were used in the clash-rich and clash-sparse approaches, and the last three were appropriate in the clash-free experiments. The settings used here are informed but not necessarily optimal choices.

8.5.1 Results

Table 8.1 shows relative performance on version 1 of the two problems. Each entry in the first column simply refers to a number of edge-constraint violations, averaged over ten trial runs. The second column records the average number of evaluations until convergence. The occurrences of 'n/a' (not applicable) in this table are explained just below.

Table 8.1 tells a simple story. If we employ the GA on a problem defined by edge constraints only, then, first of all, solutions are readily evolved when we use the clash-rich approach. Secondly, as could be expected, speed reduces as the difficulty of the problem rises (equivalently, as the density of solutions falls). When it comes to using the clash-sparse and clash-free methods on these problems, however, it turns out that evolution is not necessary. Perfect timetables readily occur in the initial random populations, under the interpretation of either the clash-free or clash-sparse decoding algorithms.

	Averaged final best	Average evaluation before convergence

Problem 1		
Clash-Rich	0	2644
Clash-Sparse	0	n/a
Clash-Free	0	n/a
Problem 2		
Clash-Rich	0	4425
Clash-Sparse	0	n/a
Clash-Free	0	n/a

Table 8.1: Version 1: Edge constraints only

Table 8.2 shows relative performance on version 2 of the two problems. Each entry in the the first two columns is of the form (a, b), where a refers to clashes (in the clash-rich and clash-sparse columns) or additional slots (in the clash-free columns), while b refers to violations of the event-spread constraints; ie: b is the number of pairs edge-constrained events which appear consecutively in the best timetable found. Unlike the experiments in table 8.1, optima were not found in the experiments recorded in table 8.2, and in most of the experiments recorded in table 8.3, which address much harder optimisation problems. Hence, these experiments were run for a full 50,000 generations, using reinitialisation whenever the populations converged. For each such run we recorded the number of evaluations beyond which there was no improvement in fitness; the figures in the second column are the average of this number over the ten trial runs in each case.

For example: in the clash-rich approach on problem 2, the best timetable in the initial random generation had, on average, no clashes, 12.4 pairs of consecutive edge-constrained events, and was found after an average of 36,022 evaluations.

Table 8.2 demonstrates the effect of introducing event-spread constraints, and we can immediately see that these affect the performance of each method significantly. First, evolution is slowed down considerably in the clash-rich space, although feasible timetables (ie: timetables with no edge-constraint violations) are still reliably found. In the clash-sparse space, feasible timetables were again found each time, but on average significantly poorer in terms of event-spread problems. Finally, the 'best' found in the clash-free space was not always a feasible timetable. In one fifth of runs, the best-scoring chromosome found was a timetable which needed to use one extra timeslot. It so happens that many feasible timetables (using no extra timeslots) were found during the course of search in the clash-free space, but only with very poor performance in terms of the event-spread constraints.

	Averaged final best	Average evaluations before convergence
Problem 1:		
Clash-Rich	(0.0, 4.2)	35026
Clash-Sparse	(0.0, 16.2)	38062
Clash-Free	(0.6, 38.2)	18930
Problem 2:		

Clash-Rich	(0.0, 12.4)	36022
Clash-Sparse	(0.0, 38.6)	33300
Clash-Free	(0.2, 72.6)	14430

Table 8.2: Version 2: Edge and event-spread constraints

Table 8.3 shows relative performance on version 3 of the two problems. Entries in the first two columns are as in table 8.2, plus an additional number in each first column entry referring to violations of the ordering constraints. For example: in the clash-sparse approach on problem 2, the best timetable found had on average no clashes, 90.4 event-spread violations, edge-constrained events, and 4.6 violations of the ordering constraints.

	Averaged final best	Average evaluations before convergence
Problem 1:		
Clash-Rich	(0.0, 6.2, 1.8)	38763
Clash-Sparse	(0.0, 12.8, 1.8)	34484
Clash-Free	(1.8, 38.4: 3.8)	12250
Problem 2:		
Clash-Rich	(0.0, 10.6, 3.6)	36660
Clash-Sparse	(0.0, 90.4, 4.6)	40902
Clash-Free	(0.8, 78.8, 4.4)	14510

Table 8.3: Version 3: Edge, ordering, and event-spread constraints

The effect of ordering constraints in addition to edge-constraints and event-spread constraints seems complex. Evolution in the clash-rich space clearly remains the most effective overall method, while clash-free is worst in problem 1, and clash-free is worst in the more difficult problem 2. Note how the clash-free approach seems to get thoroughly trapped in local optima, rarely improving beyond 15,000 evaluations despite the fact that experiments continued for 50,000 generations.

	Averaged final best	Average evaluations before convergence
Problem 2 version 1:		
No smart mutation	0.0	4425
Violation-directed mutation	0.0	2140
Event-freeing mutation	0.0	980
Problem 2 version 2:		
No smart mutation	(0.0, 12.4)	36022
Violation-directed mutation	(0.0, 4.6)	44480
Event-freeing mutation	(0.0, 3.8)	53444
Problem 2 version 3:		
No smart mutation	(0.0, 3.6, 10.6)	36660
Violation-directed mutation	(0.0, 2.8, 3.6)	42020
Event-freeing mutation	(0.0, 2.2, 2.8)	46422

Table 8.4: Results for smart mutation

Table 8.4 investigates violation-directed mutation and event-freeing mutation. Results are shown for the clash-rich approach using violation directed mutation, event-freeing mutation, and neither (in which case the figures are simply borrowed from the earlier tables) on each version of problem 2. When either violation-directed mutation or event-freeing mutation was active, it was applied at a rate of 0.5. That is, each reproduction step either results in a standard selection of two parents with crossover and (standard) mutation, or it involves the selection of one parent to which smart mutation is applied. With a smart-mutation rate of 0.5, both are equally likely. Prior investigation into the effect of different rates found 0.5 close to optimal in each case; too high, and the algorithm degenerates to hillclimbing and gets stuck too easily at local optima; too low, and there is little significant speedup in the search.

In every case, event-freeing mutation found better solutions than violation-directed mutation, and both were better than results without smart mutation. In the edge constraints only version of the problem, there is an evident speedup of the search process with smart mutation active. When extra constraints are involved, however, smart mutation slowed down convergence. Additional results of these experiments were that the use of both event-freeing and violation-directed mutation sometimes led to optimal timetables for version 2 of the problem. Event-freeing mutation found a perfect solution in four of the ten runs; that is, a timetable with no consecutive events and no edge-constraint violations. Violation-directed mutation found such a timetable in two of the ten runs. Also, in version 3 of the problem, two of ten runs with event-freeing mutation found a solution with no edge-constraint violations, no event-spread violations, and only one ordering constraint violation.

The results hint at the following guidelines and generalisations:

1. If only edge-constraints are present, then simple graph-theory or heuristic based timetabling algorithms will probably suffice to find a perfect timetable.

2. When soft constraints are introduced, these can greatly alter the fitness landscape and necessitate the use of a powerful optimisation technique. It appears that GA-based search in a clash-rich space (ie: in which almost all of the space is composed of strictly infeasible timetables) is a good choice. Hybrids of the GA with simple heuristic timetable builders work up to a point, while searching completely or almost completely feasible spaces, but generally converge to poorer timetables.

3. Domain-specific mutation operators markedly improve evolution in clash-rich spaces.

8.6 Results on Some Real Problems

Elsewhere, success has been reported on the use of GAs on real examination and lecture timetabling problems [4, 1, 3, 18]. We provide just an example of such work here by illustrating the successful results obtained in our department on using GAs for timetabling lectures and tutorials. We will present this experience along the lines of a case study; that is, this is how we turned the technique described above into an application in one particular case.

8.6.1 A Specific Lecture Timetabling Problem

An MSc course in the Department of Artificial intelligence, University of Edinburgh (EDAI) involves eight taught course modules spread over two terms. The course is organised into 'themes', each involving a particular combination of 8 modules, of which some are compulsory and some optional. As well as choices from among the 30+ modules available in the AI Department, students may also choose modules from the Computer Science (CS) Department and others. A complicating factor here is that the CS Dept is an inconvenient bus ride away from the AI Dept.

The set T comprises 80 start times, 16 per day on each day of a five day week. Each day's slots are at half-hourly intervals from 9am to 4:30pm. The set E comprises a large collection of lectures, tutorials, and lab sessions, mostly an hour long, but sometimes two hours long. A student enrolled in a course must attend all the lectures in E involving the course, but only one of the tutorials or labs (E will include several tutorials or labs for each course). Separate courses are pre-assigned to either term 1 or term 2. Hence, in one academic year there is a separate lecture timetabling problem for each term involving roughly half of the modules available on the course as a whole.

In general, the full set of constraints which need to be faced are as follows.

Options: Student's options should be kept open as far as possible. No pair of lectures in the same theme should overlap in the timetable, and, in general, lectures x and y should not overlap if there is expectation that one or more students may wish to take both courses x and y.

Event Spread: The individual timetable for any student must be spread out fairly evenly. Eg: A student should not have to sit through four lectures in a single day. Rather, the events an individual student must attend should be evenly spaced out during the week. Also, different lectures on the same topic (eg: there may be 2 Prolog lectures per week) should occur on different days.

Travel Time: A student should have at least 30 minutes free for travel between events in the CS Dept and events in the AI Dept.

Slot Exclusions: CS lectures should occur in morning slots, and AI lectures in afternoon slots (this arises from an inter-departmental agreement). Also, lectures at lunchtime (starting at 1:00pm or 1:30pm) should be avoided if possible. In a similar vein, various constraints of the form "event e cannot start at time t" arise owing to other commitments of the staff involved.

Slot Specifications: Various constraints are given in the form "event e must start at time t", arising for various reasons.

Capacity: The size of a lecture or tutorial should not exceed the capacity of the room it occurs in. Also, a room can only cope with one event (lecture, tutorial, or lab) at a time (this is the main difference between lecture and examination timetabling).

Room Exclusions: Many constraints on room assignments for particular events can easily be derived from the capacity constraints, along with information about the expected sizes of events. In addition, however, there are other considerations which lead to several *a priori* constraints of the form "event e cannot occur in room r". For example, event e may demand disabled access, or certain audio-visual requirements unavailable in room r.

Room Specifications: Similarly, several constraints are apparent of the form "event e must occur in room r".

Juxtaposition: Preferably, all tutorials or laboratory sessions for any course should occur later in the week than the week's first lecture on that course. Sometimes this is particularly necessary, since a tutorial or lab session may be based on the lectures which were held (hopefully) earlier in the week. In other cases this is desirable but not vital.

8.6.2 Translating the Constraints into a Fitness Function

In this case, it so happens that every lecture's agent (ie: lecturer) is predetermined, and individual lecturers have already decided in advance (providing details in the style of exclusion constraints) for which slots they are not available. Tutors for tutorials and lab sessions are not predetermined in this way, but for these there is no point in incorporating them into the timetable at this stage (usually well in advance of term), since we simply do not know for sure who will be available and when. For this problem we therefore need not consider the set A, and hence can use chromosomes of length $2v$.

The constraints can all be incorporated into the fitness function as follows.

Keeping Options Open

The Options constraint is handled by interpreting it as a large collection of binary constraints, each involving a distinct pair of events expected to share students. For convenience, we derive a set of 'virtual' students, each of whom takes a distinct one of the set of possible four or five-module options for the term. This set of virtual students is then used to derive the collection of distinct binary constraints between events. Here, such a binary constraint occurs between every pair of distinct lectures taken by some virtual student.

A similar constraint also holds between distinct lectures on the same module, and between lectures and tutorials on the same module. No such constraint is needed between different tutorials or labs for the same course, or even between tutorials and labs on different courses. Such may be scheduled simultaneously, and often are; in due course this gives rise to constraints which affect each student's choice from the set of tutorial and/or lab sessions available for each module.

For each such constraint "e_1 must not overlap with e_2", the fitness function must check for its violation. This simply requires a quick computation using the time genes for events e_1 and e_2, along with their stored lengths.

Notice that we can incorporate the Travel Time constraint here. If e_1 and e_2 are timetabled with a break of, say, k minutes between them, but are assigned to rooms which take more than k minutes to travel between, then any Options constraint between them is effectively violated. Hence, the violation check for Options constraints can, simply via accessing the place genes for e_1 and e_2 and a given travel-time matrix for the places, also account for Travel Time constraints.

Event Spread

Event spread constraints can be handled in a number of ways. Eg: to even out the event spread for individual students we might explicitly calculate some measure or measures of event spread for each virtual student. Alternatively, we could reasonably approximate this by examining some measure or measures of the spread of the timetable as a whole. Both would seem to offer the same overall effect; the latter will typically be computationally cheaper, but the former approach would seem to offer more potential for control and tradeoff of different aspects of the event spread for individual students. More discussion of the choices available and their implications appears elsewhere [19].

The method used in the experiments detailed later is as follows: the fitness function notes, for each virtual student, the number of instances of the following two 'offenses':

a) four events are scheduled in one day for this virtual student;

b) five or more events are scheduled in one day for this virtual student. A different penalty term is associated with each, and the penalty weighted sum of instances of these offenses, summed over virtual students, makes up the contribution to (or, rather, detraction from...) fitness of the overall event spread constraint.

Finally, 'different-day' constraints can clearly be handled in the same way as Options constraints. For any pair of lectures on the same module, we simply check directly from the chromosome whether or not their assigned slots are on the same day. If they are, then the appropriate penalty is added.

Exclusions and Specifications

It is easy to see how exclusion constraints can be directly translated into one or more simple violation-check functions, given the chromosome representation in use. Notice, however, that we can just as simply pre-arrange it so that chromosomes never violate these constraints in the first place. We can doctor the allele range of each gene so that it is always the specified allele (if any), or only ranges over the non-excluded alleles.

Choosing between such pre-satisfaction of exclusion and specification constraints, and the option of penalising violations of them, is not always straightforward. If many such constraints exist, the 'pre-satisfied' space may well lack excellent timetables which violate a few exclusions, for example, but make up for this in other ways. On the other hand, pre-satisfaction speeds up evaluation and promises to speed up search via reducing the search space. The full ramifications of this choice are beyond the scope of this paper, but it suffices to point out here that either option should be available in a system which implements this technique. In the experiments discussed later, most exclusion and specification constraints were prespecified. The only 'penalised' such constraint was that for lunchtime lectures. According to the EDAI MSc course organisers, it is preferable to avoid these, but acceptable to trade these off against other constraint violations.

Capacity constraints
To check that a room's capacity isn't exceeded, we must first translate the overall room capacity constraint into constraints of the form 'room r should contain no more than $r_{c}ap$ students in timeslot t', for each room r and timeslot t, where $r_{c}ap$ is the student capacity of room r. During evaluation, the system simply pre-computes from the current candidate timetable the student load for each room in each slot, and then runs through this list of constraints checking each in turn and accumulating penalties for violations.

Juxtaposition Constraints
Finally, it is evident, how the ordering constraints between given groups of events can be incorporated. We first derive a collection of binary constraints from those given. "All lisp tutorials should occur later than the first lisp lecture of the week" is translated into a collection of binary constraints of the form "lisp_l must be before lisp_t3". Checking for violations of such constraints is then straightforward. Similarly, it should be clear how any juxtaposition constraint (eg: "there should be at least two days between event e_1 and event e_2") can be similarly handled.

8.6.3 Experiments

8.6.3.1 Experimental Setup

We addressed tile EDAI MSc lecture/tutorial problems for both terms of the academic years 92/93 and 93/94, respectively, involving 76, 73, 82, and 73 events. Other features of these problems are as described in section 6.1, and full details are available via anonymous FTP from ftp.dai.cd.ac.uk.

In all cases, the GA used here had a population size of 50, used uniform crossover, gene-by-gene mutation, and elitist generational reproduction. The crossover and mutation rates p_C and p_M were dynamically altered as follows. p_C started at 0.8 and was decreased by 0.001 after each generation (ie: after every 50 evaluations), with a lower limit of 0.6, while p_M started at 0.003 and increased by 0.0003 each generation, with an upper Iimit of 0.02. Each trial was run for 200 generations (10,000 evaluations). Separate experiments are recorded for each problem for each of three different selection strategies: fitness-proportionate (FIT), GENITOR-style rank-based with bias 2 (RANK) [23], and tournament selection with tournament size 10 (TOUR).

The result of a trial was a maximally fit timetable found during the trial. From this we record a vector of violations V = $\{c, j, p, l, e, d_{5+}, d_4, t)$, which, respectively, denote violations of Options constraints (clashes), *important* juxtaposition constraints, *desirable* juxtaposition constraints, lunchtime constraints, slot-exclusion constraints, cases where a 'virtual student' faced more than four events in a day, cases where a 'virtnal student' faced four events in a single day, and, finally, travel time constraints. Violations of all other constraints mentioned above (eg: room exclusions, capacity constraints) are not recorded, since they were fully satisfied in all cases.

Ten trials were run for each experiment, and results for each problem record the best *V* found overall, and the mean of *V* over the ten trials, for each of three selection schemes. We also present *V* for the timetables produced by the course organisers for each problem, and which were the actual timetables used (or in use), since the GA system itself was not yet in regular use. 'Best' means relative to the penalty-weighted sum of violations. The fixed penalty values used in these trials for the various violations were, in the order in which they appear in the tables: 500, 300, 30, 30, 10, 5, 1, 1. Hence, violations are listed in order of decreasing importance, as judged by the course organisers.

8.6.4 Results

Tables 8.5 and 8.6 clearly show that the GA approach leads to much better timetables in each case. One course-organiser produced timetable failed to keep all reasonable options open (ie: had clashes), while many failed to fully keep lectures and tutorials away from various restricted slots, and all failed constraints at least as important as the need to avoid lunchtime events. Problems caused involve lecturers and tutors being forced to work during timeslots previously designated for other things (eg: regular weekly seminars), forced to give up free afternoons or mornings designated for research, students facing excessively demanding days, and so on. On the other hand, for each problem, tournament selection found either a perfect timetable or one with a single travel-time violation in at least one of ten trial runs. For each selection scheme, mean and best results compared very favourably with the experts' efforts. Each GA trial was completed within 5 minutes on a SUN SPARC. Occasionally, a GA solution, was worse in terms of some attribute (eg: violations of desirable juxtaposition constraints) than the course organisers' solution, but better overall in terms of the penalty-weighted sum of violations. This suggests that the GA was more successful at trading off the relative occurrences of violations of different importance.

Problem	c	j	p	l	e	d_{5+}	d_4	t
92/93_term1								
Course Organisers	0	0	4	4	14	0	2	65
FIT/Best of 10	0	0	0	0	0	0	0	8
FIT/Mean of 10	0	0.1	0.2	1.2	0	0	0.9	16.6
RANK/Best of 10	0	0	0	0	0	0	0	17
RANK/Mean of 10	0	0	0	1.6	0	0	0.6	17.2
TOUR/Best of 10	0	0	0	0	0	0	0	0
TOUR/Mean of 10	0	0	0	1	0	0	0.4	0.7
92/93_term2								

Course Organisers	0	1	9	2	0	1	4	49
FIT/Best of 10	0	0	0	0	0	0	1	0
FIT/Mean of 10	0	0	0.1	0.5	0	0	2.1	8.2
RANK/Best of 10	0	0	0	0	0	0	3	9
RANK/Mean of 10	0	0	0.5	1	0	0	2.4	13.4
TOUR/Best of 10	0	0	0	0	0	0	0	0
TOUR/Mean of 10	0	0	0.3	0.6	0	0	0.7	0.1

Table 8.5: Comparative performance on the 92/93 problems

Problem	c	j	p	l	e	d_{5+}	d_4	t
93/94_term1								
Course Organisers	0	0	0	5	7	0	2	84
FIT/Best of 10	0	0	0	2	0	0	0	15
FIT/Mean of 10	0	0	2.6	2.3	0	0	0.7	28.6
RANK/Best of 10	0	0	1	2	0	0	0	23
RANK/Mean of 10	0	0	2.2	2.5	0	0.1	0.3	31.5
TOUR/Best of 10	0	0	0	0	0	0	0	1
TOUR/Mean of 10	0	0	0.6	1.4	0	0	0.2	0.6
93/94_term2								
Course Organisers	2	0	3	3	1	0	0	50
FIT/Best of 10	0	0	1	0	0	0	0	11
FIT/Mean of 10	0	0	0.2	1.4	0	0	1.4	14
RANK/Best of 10	0	0	0	1	0	0	1	13
RANK/Mean of 10	0	0	0.6	1.7	0	0	1.3	13.6
TOUR/Best of 10	0	0	0	0	0	0	0	0
TOUR/Mean of 10	0	0	0	0.7	0	0	0.2	0.2

Table 8.6: Comparative performance on the 93/94 problems

Tournament selection appears to be the best choice, with rank-based selection of the style used in [23] and fitness proportionate selection, in that order, being next best. This relative performance of different selection schemes cannot strictly be taken as read from these results without further experiments using different tournament sizes, biases, and so on; however, much GA literature backs up this ordering of relative performance.

8.6.5 Discussion

The results clearly indicate the benefits of using the clash-rich based GA approach on this problem, and by implication suggest similar utility for the same approach on similar problems. In designing the penalty-weighted fitness function itself for these experiments, several of the design decisions were *ad hoc*. For example, penalty values were chosen according to a rough judgement of relative importance. Also, there were several other possibilities, as discussed earlier, for dealing with the event-spread constraints. Even the underlying GA itself was far from optimal in terms of parameter settings and general configuration. Better choices of selection scheme, for example, are spatially-oriented schemes as presented in [2] and [6], as we have indicated before.

The 'rough-and-ready' aspect of the experimental configurations used, though, coupled with the good results reported and the case of implementing the approach suggest a promising future for both further research and also practical use of GAs on general timetabling problems. Naturally, there are a considerable number of theoretical and practical issues that need to be answered. How, for example, does this approach scale up to larger and more tightly constrained problems? The real problems addressed here, and similarly those addressed in [3, 18, 4], are similar in size or larger than a large proportion of the timetabling problems faced in many institutions. Hence the usefulness of this approach seems justified, inasmuch as we can expect the beneficial results displayed here to carry over to different timetabling problems of similar or smaller size. How the approach scales with increasing size and/or complexity is a harder question, which is the subject of continuing research. Initial indications in unpublished work are that the approach scales well, but suffers from a problem common to GA-based optimisation: that is, solutions near optimal regions are rapidly found on large complex problems, but further evolution toward optima becomes considerably slow, and may stop altogether. Fortunately this difficulty is readily aided by the use of smart hillclimbing mutation operators, as we have mentioned, and will show later.

8.7 Speeding Things Up: Delta Evaluation

The clash-rich approach looks so far like the best bet for general use. One helpful thing about the clash-sparse approach that we have seen is that, by virtue of the directness of the representation, it allows us to construct useful smart mutation operators. There is also another very promising thing about this approach: the potential for *delta evaluation.* As the alert reader may well have spotted, it is unnecessary for the objective function to check every constraint each time; if a similar timetable has already been evaluated, then all that needs to be checked is the set of constraints in which events are involved whose assignments differ between these two timetables. This type of approach is common in simulated annealing, for example, where changes are always small. With a GA, the changes between timetables will typically be more marked than in simulated annealing, but it seems safe to suppose that we might get significant speedup of the method if we take advantage of what we call *delta evaluation.*

8.7.1 Delta Evaluation

Consider two timetables, g and h, which differ only in the assignments made to some subset D of the events E. If C_X is the subset of C containing only those constraints which involve one or more events from the subset X of E, then we can say the following:

$$P(g) = \sum_{c \in C_{E/D}} w_c v(c,g) + \sum_{c \in C_D} w_c v(c,g)$$

and

$$P(h) = \sum_{c \in C_{E/D}} w_c v(c,h) + \sum_{c \in C_D} w_c v(c,h)$$

Since the assignments of events in E/D are the same in both g and h, we can go on to say:

$$P(h) = \sum_{c \in C_{E/D}} w_c v(c,g) + \sum_{c \in C_D} w_c v(c,h)$$

And further, using equation (8.2):

$$P(h) = P(g) - \sum_{c \in C_D} w_c v(c,g) + \sum_{c \in C_D} w_c v(c,h)$$

which expresses $P(h)$ solely in terms of violation checks involving the events in D. If the size of C_D is small in relation to C, then this promises to save much time. In general, if h differs from g in k places, then evaluating g needs $2k$ constraint-checks. This will obviously speed things considerably in later stages of a GA run, when new chromosomes are likely to be very similar to their progenitors. Very early on, it seems it won't save us much time.

Applying delta evaluation of timetable h in a GA needs a decision as to which already evaluated timetable to take as g. One possibility, for example, is to scan the population to find a previously evaluated timetable genotypically closest to h; we might even hold on to a limited number of 'dead' chromosomes in case of closer matches with these. A natural choice, which we use in the experiments we describe later, is to choose g as follows. First of all, each of the initial population of timetables is *fully* evaluated. There are then three cases, depending on the process which produced h, which was either mutation, crossover, or violation-directed mutation (see later). In the case of mutation, g was naturally taken to be the parent which was mutated to produce h; note that in this case we can expect C_D to be very small, and hence evaluation of mutants to be very quick. In the case of crossover, g was arbitrarily taken to be the first of the two parents selected which led to h. At early stages, the size of C_D after uniform crossover might typically be half the size of C, but since each member of C_D has to be checked *twice* in delta evaluation, once for g and once for h, there is no effective speedup. Clearly, however, the size of D after crossover reduces considerably with time as more and more alleles become fixed.

8.7.2 The `edai93` and `mis93` Exam Timetabling problems

For many of the experiments later in this chapter we use a real example of an examination timetabling problem. In this problem, called `edai93`, 44 events (exams) have to be arranged within a 9-day timetable with 4 slots per day. Data for 84 students gives rise to 414 edge constraints. The timetable must of course have no clashes; additionally, there is an event-spread constraint whereby students should have a minimum of 120 minutes between any two exams. In this problem, this comes down to specifying that edge-constrained events scheduled on the same day should either respectively be in the first and third, first and fourth, or second and fourth slots. In addition, `edai93` involved 480 instances of some event being excluded from appearing in some slot, but some of the time we will ignore this aspect and explore regions around the *exclusion-free* `edai93` problem.

A larger real examination timetabling problem which we call `mis93` is also used in some experiments; this involves 1034 events, 48 slots, and 8464 distinct edge constraints (as generated by the exam choices of some 2423 students). This problem involves a timetable structure of 24 days, with 2 slots per day, and has the event-spread constraint that no student should take more than one exam in a day.

`edai93` and `mis93` are hence taken to represent typical examination timetabling problems of the size and kind suffered by, respectively, a medium sized university department, and an entire university. In some ways this is misleading, but in most ways not. For example, a great many decisions and calculations and considerations typically go into producing, as a final result, the set of constraints which form part of the data for `edai93` or `mis93`. Then again, `edai93` and `mis93` as we present them above are precisely what the human timetable involved were presented with as their starting points. Each institution does their timetabling in their own way, and in the case of the institution involved in solving the `edai93` problem, for example, it so happens that it goes as follows:

1. Lecturers who need to be present at specific examinations provide details of times at which they cannot be available. This happens too for a minority of examination candidates.

2. A specific period (ie: number of days) is chosen which the eventual timetable will cover.

3. The information in (1) and (2) is used together with the student data to produce a feasible timetable which satisfies all the (1) and (2) constraints, has no clashes, and has as few as possible violations of event-spread constraints.

4. Given the timetable produced by (3), the timetabler assigns each event to a room (in this case, it so happens that a large room is available which can hold many exams at once).

The main job of the human timetabler, step (3), is what we ask of the Genetic Algorithm when asking it to solve `edai93`. Clearly, though, we can put the GA to work on the combined problem of (3) and (4), or we might try to involve (2) as well, in the sense of trying to minimise the total period which the timetable will cover. A key feature promised by the general GA approach is that this kind of thing seems to be possible, and institutions may be tempted, if informed of the possibility, to unload more and more of the timetabling problem on the computational tool in use.

8.7.3 GA Configuration

The GA employed in the delta-evaluation trials below had population size 400, used 'local mating' selection [2] with random walk size 5 (with the population on a 20 x 20 grid), and one-at-a-time reproduction with the following population-update procedure: at any reproduction step either select two parents and crossover to produce one child, or select one and mutate it. A mutation step happens with probability 0.4. During mutation, each gene takes a random new allele with probability 0.02. Selection begins with choosing a random grid location; either one (for mutation) or two (for crossover) random walks are performed from this location to collect a parent or parents. The single child resulting is inserted into the population at the chosen location only if it is strictly more fit than the current resident at that location.

This set of experiments involves 7 randomly generated problems. Each had 50 events, assumed the same 36 slot timetable structure as `edai93`, and imposed the same event-spread objective, and varied in number of (randomly generated distinct) edge constraints between 200 and 800.

It seems obvious that delta evaluation will immediately result in faster operation. But how much faster? For some ideas in this regard we begin with some experiments on randomly generated timetabling problems of a similar size to `edai93` (but with no exclusion constraints) and see how speedup changes as we make the problems more difficult by increasing the number of edge (and therefore event-spread) constraints. Figure 8.1 shows the results, showing how full evaluation compares in time with delta evaluation, where time is measured in number of constraint checks done by the GA before finding a perfect solution which violated no constraints. Evidently, speedup becomes more and more marked as the problem grows in difficulty. On a 500 constraint problem, similar in difficulty to the real `edai93` problem, speedup seems to be twofold; this quickly rises to fourfold on an 800 edge constraint problem. What is happening to cause this is in one sense clear; by counting constraint checks on a delta evaluation run we are actually measuring the total number of *allele changes* that occurred during the run before a global optimum was found, and hence measuring the number of alleles searched. It is fortunate that the structure of the typical timetabling fitness function enables us to do this. It is also intuitively clear why the number of alleles searched drops so apparently sharply as a proportion of number of timetables searched (ie: number of full evaluations); as problems get harder, the GA indeed needs to explore more regions of the space, but this exploration quickly gets contextualised via the fixing of several alleles which compose sound building blocks found early on.

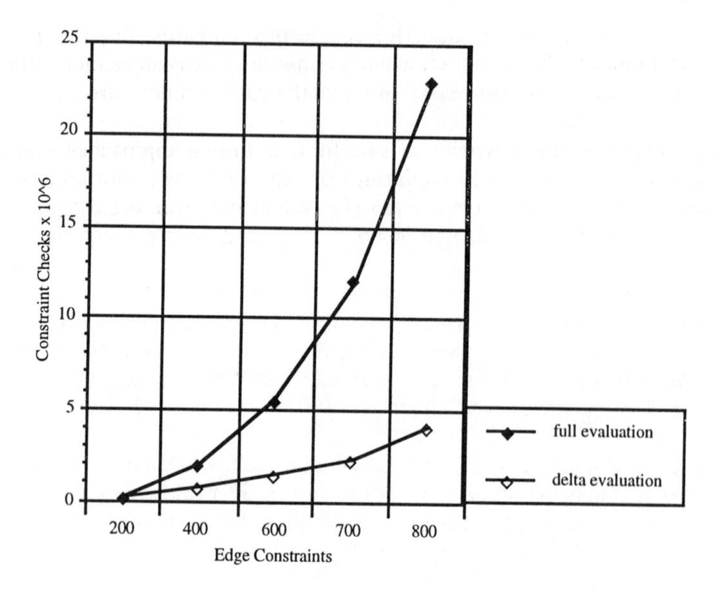

Figure 8.1: How delta-evaluation speeds up the GA on problems with increasing numbers of edge constraints

To test this further we collected results from 10 trial runs on the `edai93` problem and the `mis93` problem. A similarly encouraging result ensued. Delta evaluation proved 3.9 times quicker than full evaluation (2.0 million constraint checks vs. 7.8 million), on `edai93` and 8.2 times quicker (37.4 million constraint checks vs. 343 million) on `mis93`. Hence, it certainly seems to be the thing to do on large timetabling problem instances.

8.8 Investigating Further: Scope and Limitations

We already know something of the scope and limitations of GA's applied to timetabling. It is probable, for example, that GAs are not generally the best choice for applied solutions to edge-constraint-only problems (in particular, problems without event-spread considerations). Nevertheless, it is interesting and important from the point of view of GA theory that GAs still do well on such problems, in the sense that many such can be quickly solved in clash-rich spaces. In fact, if a problem is so highly constrained that there *must* be violated edge-constraints in a solution, and if (necessarily) some or all of these can be considered soft constraints, then the clash-rich GA-based method may be the most appropriate method to use. An example of such a problem is that of designing a tight lecture timetable, before knowing the students' module options, so as to allow as many combinations of modules as possible. For such a problem, we might have a high penalty for violating popular combinations of options and lower penalties for violating unlikely combinations.

Most practical timetabling problems require optimisation of many difficult soft constraints. This changes the nature of the problem considerably from the type that can be easily solved by graph theoretic or heuristic methods, to one requiring an effective optimisation procedure. Results here and in other published papers suggest that GAs are a very effective choice in such cases. In particular, our results here indicate that evolution within clash-rich spaces can be more effective than an approach which searches only for mainly feasible timetables.

From the application's viewpoint, it is useful both to find approaches which are effective and robust across a broad range of timetabling problems, and also to learn the extent to which different approaches can be constructed and optimised to best exploit particular details and features of specific problems. Results so far indicate that the clash-rich approach combined with smart mutation is likely to be an effective choice for a wide range of timetabling problems, and so is a candidate for a robust and general method. Much work is needed, however, to learn how to best customise a GA-based approach depending on particular aspects of the problem in hand.

Important groundwork for the latter endeavour is to understand how the GA copes as we change the problem in some way. We have shown that the GA can solve some real timetabling problems, but how does its behaviour change as we increase as we make the problem bigger, smaller, or more or less difficult in some fashion? We take a first look at this in the following.

8.8.1 Performance against Measures of Problem Size

The notion of how 'difficult' or 'large' a timetabling problem is seems hard to pin down. Immediate candidates for dimensions of variation are the number of events v, the number of slots s, and the number of edge constraints c. The size and/or difficulty of a given problem instance seems closely related to these three aspects. For convenience we will describe collections or classes of problem instances in terms of triples (e, s, c).

The structure we would like to understand is the four dimensional GA performance surface p as it varies across v, s, and c, where p is some measure of the average performance of the GA on timetabling problem instances of type (e, s, c). At the outset we can see that much of this space is undefined in p; for example, for a given value of v, there can be no more than $v(v - 1)/2$ edge constraints, and hence c is bounded above by $v(v - 1)/2$.

Some function of v, c, and s may be what we are looking for; a candidate for this is a measure of *solution density.* In any problem, a guide to this will be the proportion of possible timetables which satisfy' all the edge constraints. Remembering equation 1, it might seem that, we can find this in some cases. In general, however, we need to compute many more terms in this series before coming to a good estimate of solution density. Unfortunately, this computation is infeasible for problems of practiced size, so little can usually be discerned about the density of possible solutions.

Initially jarred by the complexity of the interaction of v, c, and s, we will begin our study of this performance surface by simply taking some 2-dimensional snapshots of it. This amounts to looking at GA performance with two of v, c, and s fixed while varying the other. Though having the benefit of simplicity, this might seem to lack usefulness. If we can, for example, gain clues to GA performance on problems in the class (v, s, c) as c varies, then it is hard to see to what extent this might generalise to different values of v, and s. There are three distinct replies: first, it seems intrinsically useful to find out if performance is indeed well-behaved within these 2D snapshots; second, these snapshots may help build a useful piecewise picture of large chunks of the performance surface; third, it may be that something useful can be found concerning how p varies with, say, c, across a *bounded range* of values for a and b. That is, we might eventually be able to give upper bounds on time to find an optimal solution as a function of c, for any problem with between, say, 200 and 300 events and between 50 and 60 timeslots.

8.8.2 Basic GA Performance Measures

If we know what the optimal fitness is in a particular case, we can look at the GA's *speed* and *robustness*. Speed is simply some measure of the time taken by the GA to find an optimal solution. We will use a number of constraint checks for the measure of time, since it is essentially more useful than number of evaluations. A constraint check has practically constant time between different problems, but an evaluation doesn't. Robustness is simply the proportion of trial runs in which the GA finds the optimum.

If in a particular problem instance the GA seems to *always* finds the global optimum, then we are lucky enough to be able to completely classify performance in terms of speed. The latter is the route we are able to take in this paper. In other words, (almost) all GA runs reported found optimal solutions. This is a very helpful aspect when it comes to presenting results. If we cannot expect 100% robustness, for example, then it becomes very difficult to decide on factors such as *when to stop a single GA trial.* This is a result of difficulty involved in deciding when a GA run has converged. We cannot simply define convergence as "all chromosomes the same", since this simply may not happen we may end up with a stable population containing differing chromosomes: which are more likely depends on the particular choice of population-update step. Also, independent of whether there is convergence to a stable homogeneous or inhomogeneous population, we cannot be sure whether further beneficial mutation will occur which starts things going again. Further still, using a reinitialisation strategy we might often turn a non-100% robust GA into a 100% robust one. This might depend on the particular reinitialisation strategy we use. Fortunately, however, we manage to avoid these problems with the experiments later in this chapter, because we almost always achieve optimal solutions from the GA, without even the need for reinitialisation.

This might seem essentially unhelpful though when generalising our observations to problems with an unknown optimal fitness, and/or when the GA is not 100% robust. The question is: given an arbitrary problem instance of type (a, b, c), how useful is a prediction of time to find an optimum, in the case where the prediction is based on instances of the same or similar type which all found optima, but where the GA is not 100% robust on the instance in question? The short answer is that we can just reinterpret the observations suggested by the experiments in this chapter as "time beyond which there will probably be no improvement in performance". The usefulness of our observations then depends on the degree to which the following is true: *time before no further improvement on problems where the GA is 100% robust is similar to time before no further improvement on problems where the GA occasionally settles on suboptima.* We feel this is true to a high extent, but cannot say for sure; ongoing work is attempting to find an answer to this.

In summary, we conjecture that a reasonable initial way to go about studying the four-dimensional performance surface defined by v, s, c, and p, where v is the number of events, s is the number of time-slots, c is the number of edges, and p is simply the number of constraint-checks until the GA found an optimum. The two-dimensional snapshot of this space which we will consider are planes parallel to the performance axis and other, so that variation in performance can be seen with the remaining two measures fixed.

8.8.3 The GA Configuration

The GA employed in the experiments described below has population size 400, uses 'local mating' selection [2] with random walk size 5 (with the population on a 20 x 20 grid), and uses *selective event-freeing mutation* SEFM. SEFM is a stochastic form of EFM which we will have rather more to say about later on. The GA uses one-at-a-time reproduction with the following population-update procedure: at any reproduction step either select two parents and crossover to produce one child, or select one and rotate it, or select one parent and do EFM. A mutation happens with probability 0.3, SEFM with probability 0.1. During mutation, each gene takes a random new allele with probability 0.001. Selection begins with choosing a random grid location; either one (for mutation) or two (for crossover) random walks are performed from this location to collect a parent or parents. The single child resulting is inserted into the population at the chosen location only if it is strictly more fit than the current resident at that location.

8.8.4 The `edai93` Problem Region

Taking exclusion-free `edai93` as arguably typical of problems continually faced by a medium sized university department, our experiments are designed so as to explore GA performance around the region of this problem, that is, the region of the class of problem instances (44, 36, 414). In practice, we took as our central point the class of instances (50, 36, 500), and varied one dimension in each experiment, each of which assumed the same 4 slots per day structure and the same event-spread constraint, The latter is an important point; these problems, and timetabling problems in general, are all far easier to solve without the event-spread constraints. Indeed, part of the reason why produced and used timetables tend to be of poor quality (and hence part of why many researchers are looking to GAs and other powerful optimisation methods), is because standard 'graph-colouring' style techniques used to produce them often simply ignore the problem of minimising violations of event spread constraints.

8.8.5 How Performance Varies with No. of Edges

The first set of experiments performed looks at how the time for finding an optimum varies against the number of edges. Eight problems in the class (50, 36, c) were randomly generated, for c from 200 to 900. The resulting plot of time (measured in constraint checks) against c is in figure 8.2. Each point plotted is the average of ten trial runs.

Variation in performance (the full line) takes on an intuitively reasonable shape. As c rises, the event-graph becomes more saturated with edges, and violation-free arrangements become harder and harder for the GA to find. The increase in difficulty accelerates as we get beyond 600 edges, and more so as we get beyond 800 edges. A (50, x, y) problem instance, for example, is impossible to solve anyway if $x < 50$ when y is 1225, without even considering the event spread constraints; so when x is 36, and considering the extra difficulty imposed by event-spread constraints, it seems reasonable to say that the regions of the graph to the right of the '600 edges' mark on the horizontal axis will rarely arise in practice as real problems. In one sense this is simply wrong, because problems of such difficulty can easily arise in the sense that a timetabler faced with a (50, x, y) problem instance where $y > 600$ may well first try to place these events in 36 or fewer slots; the point, however, is that the timetabler, through experience, will rarely expect to have much success if x is much smaller than 30, or y is much closer to 1225 (at which point the graph is fully saturated with edges).

Line L in figure 8.2 hence seems to indicate an upper bound on GA time *over a wide range of real potential problem instances.* This essentially seems to reveal that GA time on problems in the exclusion-free `edai93` region grows only linearly with the number of edges for a wide range until half-saturation. Noting too that the average number of constraint violations in the initial population grows perfectly linearly with the number of edges, it seems the GA is, in a sense, removing constraint violations at a constant rate in this region. To some extent, removing violations should become harder as candidate timetables become more violation-free, since changes which remove violations then tend to introduce others. But since 'violation-removal' (at least for less than half-saturation) appears to be happening at a constant rate, this suggests that the early implicit schema sampling work done by the GA smooths the way for a continued constant rate of violation removal at later stages.

More work is needed to analyse these effects properly. For now we will be content to feel encouraged by the essentially linear growth in time over a claimed wide range of problem instances, and continue with similar initial observations for some different two dimensional snapshots of the performance surface. Later, we will attempt to use these results to predict performance on `edai93`, and also later on `mis93`.

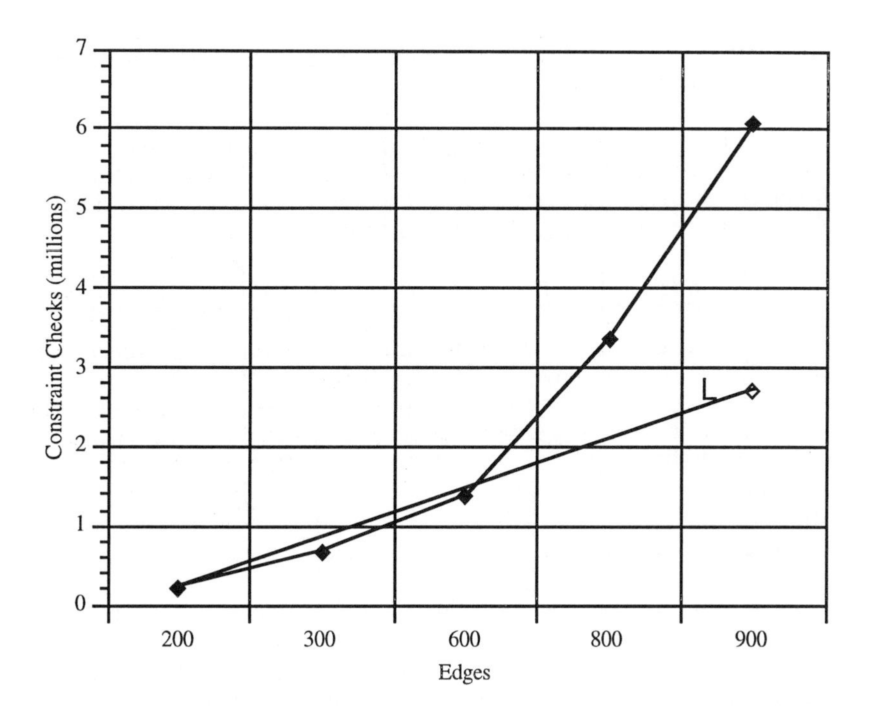

Figure 8.2: GA time against number of edge constraints

8.8.6 How performance Varies with No. of Slots

The next experiments look at how GA time varies with *s*. The set of events and edge constraints used was the same as in the randomly generated (50, 36, 500) problem used above, but ten distinct variants of this problem were addressed in which *s* varied from 50 down to 23. The resulting plot of time (measured in constraint-checks) against decreasing *s* is in figure 8.3. Each point plotted is the average of ten trial runs.

Figure 8.3 shows a rather more messy performance variation. There is a clear trend toward increasing time as s falls (that is, as the same timetabling problem is squeezed into fewer and fewer slots), but this is of course no surprise. The only result we can point to based on this series of experiments is that performance as a function of number of slots seems to exhibit considerable variance in the *easier* region, although we seem to have smooth degradation in performance as s falls from a point at which the problem is already similarly constrained to a problem like exclusion-free `edai93`. There thus seems to be smooth variation in the region of most practical interest.

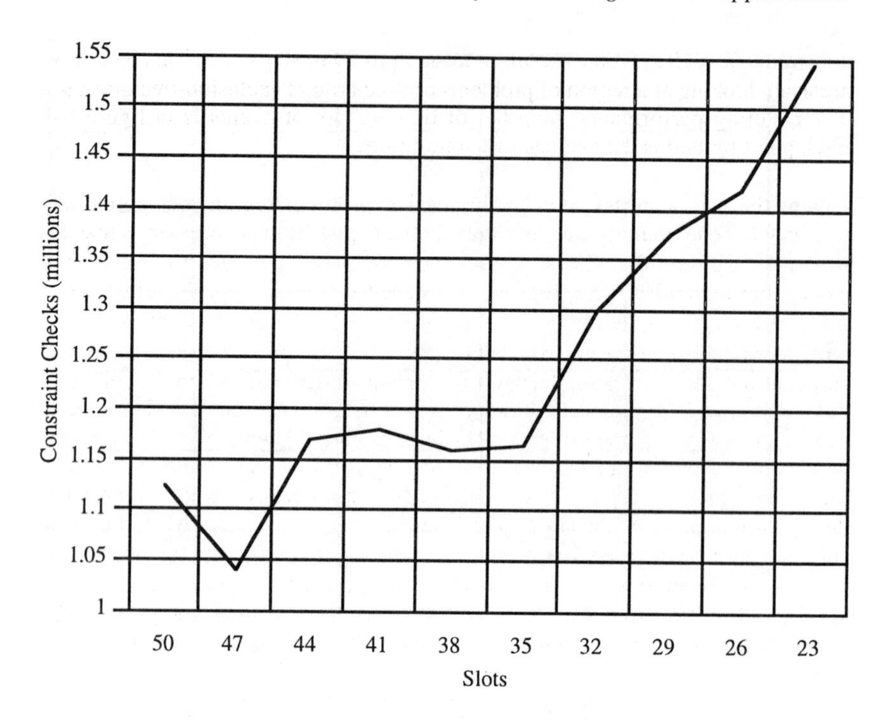

Figure 8.3: GA time against decreasing number of slots

Unclear variation in performance for larger s begs attention. We would expect it to be easier for the GA (or any method) to find solutions as s grows (and hence more scope for events to be placed apart from each other and avoid conflicts), but the curve suggests that in practice this is not uniformly or reliably so. A possible explanation lies in [13], which analyses the tendency for a GA with a k-ary allele alphabet to quickly settle on a smaller 'virtual' alphabet as evolution progresses. In our case, this means that the number of slots the GA is effectively using is smaller than we might expect. Early on, the GA picks out a subset of the available slots, governed by the distribution in the initial random population. Beyond a certain point, the number of slots the GA works with (ignoring for simplicity the occasional reintroduction of lost slots via mutation) will not increase significantly as the number of 'available' slots increases. Adding to this effect is the reduction in expected allele coverage [22] as s grows. In other words, the increase in s becomes less and less reflected in the increase in alleles present in the initial population (for a fixed population size); so although we present the GA with more slots to work with, it doesn't necessarily use them. A further confounding effect here is that the event-spread constraints, which concern the slots' temporal interrelationships, make the fitness landscape more sensitive to the particular slots in use.

8.8.7 How Performance Varies with No. of Events

Our final snapshot of the performance surface is built by fixing c and s, respectively, at 500 and 36 retaining the same slot structure and event-spread constraint `edai93`, while varying v from 35 to 70 in steps of 5. This means we are again looking at a region of problems centred around exclusion-free `edai93`. The resulting performance snapshot of time vs. no. of events is in figure 8.4. Each point plotted is the average of ten trial runs.

Again, there is a messy area in the 'easier' problem region, but potentially predictable behaviour appears as v falls. Some explanation is necessary since it is intuitively strange that the problems get harder as we reduce the number of events, but remember that c is *fixed;* this means that reducing v leaves us with a more and more saturated graph. In fact, when $v = 35$ the graph is 84% saturated with edges. Considering the extra imposition of event-spread constraints, it is no surprise that GA time rises sharply here. Again we can see that in the region of the graph at which graph saturation is between about 50% and 66%, between 50 events and 40 events, the GA's speed reduces linearly. Again, this seems to show predictable behaviour in a region of practical relevance, within which (we believe) many real timetabling problems lie. The more varied behaviour for larger v also deserves some attention. Intuition first suggests that the GA should find solutions more quickly as v rises, since the underlying graph becomes more and more sparsely connected, but speed does not seem to improve with any significant uniformity. A partial explanation is the increasing difficulty introduced by the fact that more events are being squeezed into a constant number of slots.

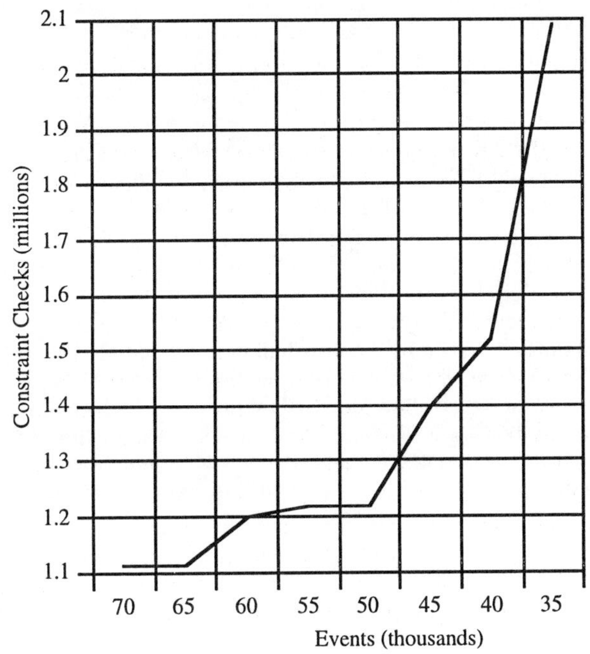

Figure 8.4: GA time against decreasing number of events

8.8.8 Attempted Predictions

To begin testing how successful the observations so far might be in characterising performance on real problems, we first turned to the exclusion-free `edai93` problem. Exclusion-free `edai93` is of type (44, 36, 414). The results which generated figure 8.4 would predict an expected mean time of about 780000 constraint checks for exclusion-free `edai93`. Ten trials were performed on exclusion-free `edai93`, resulting in a mean time (in constraint-checks) of 648629.8 with standard deviation 67460.9. By the normal distribution test it seems that this is too far away from the expected mean for us to have anything but a tiny (less than 1%) chance of expecting this result. It hence seems that exclusion-free `edai93` is significantly *easier* for the GA than our graphs would lead us to expect. Interestingly, this is almost certainly a product of the fact that the distribution of edges in `edai93` is nonrandom; as is quite typical, the exams in `edai93` fall into a small number of groups, with large interconnection within groups but sparse interconnection between; this greater regularity is very likely to help the GA in its task, since the problem amounts to solving a small handful of *smaller* loosely connected subproblems. It thus seems reasonable to say that predictions expected from our random-problem experiments are thus upper bounds for GA time on real problem instances; the closeness with which a real problem approaches these bounds will probably be a fraction of the degree of underlying modularity.

To generate a prediction for the complete `edai93` problem we first notice that the combination of the various exclusions and specifications amount to there being an average of 25.16 of the 36 slots available to each event. It therefore seems reasonable to expect the GA-difficulty of this real problem instance to be close to that at the 25-slots point in figure 8.3 The expected mean time at this point is 1.47 million constraint checks. What we found in practice, however, is that optima were found on average at the considerably longer time period of some 7.8 million constraint checks. This time, our prediction has erred significantly in the other direction. The cause seems certainly to be that we were not justified in supposing that the exclusion and specification constraints can be 'averaged away' in the way attempted. Evidently, the imposition of such extra constraints must be accounted for in some other way, and more study is needed to discern what this is. It seems there is a great difference in GA-difficulty between a problem for which the average number of slots available to any event is k, but *these are not all the same slots,* and a problem in which each event has access to the same k slots. The important consideration missed before our erred prediction is a more direct measure of the degree of freedom imposed by the particular slots available to each edge-constrained pair. Analysis of this degree-of-freedom measure for exclusion constrained problems will be pursued in forthcoming papers.

To test the idea of upper-bound prediction further, we considered `mis93`; recall, `mis93` involves 1034 events, 48 slots, and 8464 edge constraints, and involves a timetable structure of 24 days, with 2 slots per day, with the event-spread constraint that no student should take more than one exam in a day. We generated four random problems, each of type (1000, 48, x), with the same slot structure and event spread constraint as `mis93` but with x at, respectively, 1000, 1500, 2000, 2500. The result of this was the time vs no. of edges performance graph in figure 8.5[2]. A first point to note about this figure is that the full line, between 100 and 2500 edges, does appear to demonstrate linear upper bounds on GA time in this region. Since saturation of the event-graph arond here is small even for numbers of constraints as large as in `mis93`, it seems justified to expect performance to stay close to this line at the `mis93` point and for quite some way beyond (with v = saturation occurs at c = 250000). On this basis, illustrated by the dotted line in figure 8.5 extrapolation leads us to expect that `mis93` will be solved after some 58 million constraint checks.

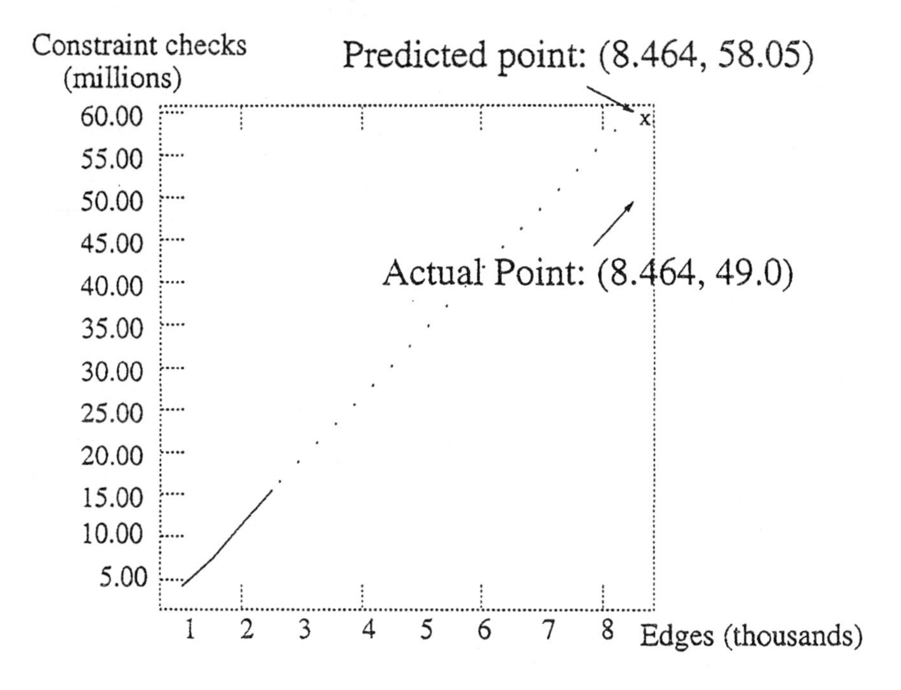

Figure 8.5: GA time against number of constraints on some random 1000-event problems

Results were as follow: five trials were run on the `mis93` problem; an optimal solution was found in 3 of the 5, while the remaining 2 trials converged to a solution which violated just one event-spread constraint (involving only one student). Number of constraint checks, averaged over the 5 trials, was 49.0 million, with a standard deviation of 19.99 million. Using the normal distribution test we find that the chance of being even more distant from the expected mean is 32%, so this result also agrees with the notion that performance on random problems as c varies generates reliable upper bounds for regions of low saturation. Additionally, the extrapolation appears reliable, although it must be pointed out that far too few trials have been made at the time of writing to

---

[2] Note: here the GA had a 30 x 30 population.

properly justify such a test. This observation also tallies with our result for the prediction for exclusion-free `edai93`; the prediction in both cases was correct as an upper bound on the time taken, and it seems safe to suggest that the main reason the GA found both problems easier than their random counterparts lies in the regularity of the edge distributions.

8.8.9 Discussion

The results and observations reported above represent initial steps in the endeavour to discover some general characteristics concerning GA performance on timetabling problems. One claim we attempt to justify is that, to a large extent, GA time taken on a large range of practically relevant instances of timetabling problems is linearly upper-bounded. This is encouraging for two main reasons: the *prediction* aspect, whereby we can take advantage of being able to predict expected time taken when designing a general GA-based timetabling system, or just considering, for example, at what level to set a 'max evaluations' parameter, or what particular regime to use for interpolated mutation and/or crossover rates. Second is the *result* aspect, whereby it seems we can expect optimal or near optimal importance on this wide range of problem instances; in particular, notice that the problems studied were all very highly constrained, considering in particular the imposition of event spread constraints.

It is clear from our experience with attempting to predict performance on the full `edai93` problem, however, that introduction of different types of constraints makes it very difficult to generate a reasonable prediction from the performance surface we were looking at. The obvious step is to see if a collection of problems which in some way *include* such extra kinds of constraints can combine to generate reliable predictions. It would be much better, however, if we were able to account for the effect of other kinds of constraints in some more principled way; work on these lines is ongoing, whereby we are looking at the interaction of different kinds of constraint in the fitness landscape, and also considering the effects of genetic operators tailored to specific kinds of constraint.

8.9 Strong Methods and Stronger GAs

Genetic Algorithms are a wonderful optimisation method. They, along with the more general class of evolutionary algorithms, are highly intriguing techniques of interest to anyone who is curious about evolution, genetics, or just nature in general; additionally, they are a powerful tool for addressing a very wide range of difficult problems that seem to see little consistent success from other techniques. An additional bonus is their natural parallelisability. What we must be sure not to forget, however, is that the GA is a *weak* method. A typical GA artificially evolves away quite happily on a given problem while ignoring any available problem or domain specific clues that other algorithms would hunger for. We can coax our GAs to use such information by hybridising them with other techniques, or adding a useful smart mutation operator, and sometimes we find this will improve things, but one must always remember to check for the possibility that the GA may not be necessary at all in this hybrid, and may indeed be slowing things down.

Consider the following experience we had when investigating potentially more useful forms of smart mutation.

8.9.1 Very Smart Mutation: Selective EFM

The information that can be amassed during evaluation concerning the *violation score* of each event can be put to use in various ways. Our earlier described methods, violation-directed mutation and event-freeing mutation, are just two. Among the other possibilities for an operator that makes use of this information is to introduce a biased stochastic element to the choice of event and the choice of its new slot assignment. A natural candidate for the stochastic aspect is to use a typial GA or EA selection operator. Initially, *select* some event for mutation, based on the relative violation scores, where selection pressure aims for events with higher violation scores. In the experiments described later using this method, we use tournament selection with a tournament size of 20 (for 50 events), and 100 (for 1000 events). Next, for the selected event, evaluate each of its possible choices of slot, calculating a separate violation score for this event in each slot. Finally, *select* one of these slots, again using a typical GA or EA selection operator, based on the relative fitnesses represented by the violation scores. Again, we use tournament selection at this step, with a tournament size of about $3\sqrt{s}$ where s is the number of slots. *Selective* event-freeing mutation (SEFM) thus attempts to put fairly troublesome events into fairly trouble free slots.

8.9.2 GA Configuration

The GA employed in the delta-evaluation trials below had population size 400, used 'local mating' selection [2] with random walk size 5 (with the population on a 20 x 20 grid), and one-at-a-time reproduction with the following population-update procedure: at any reproduction step either select two parents and crossover to produce one child, or select one and mutate it. A mutation step happens with probability 0.4. During mutation, each gene takes a random new allele with probability 0.02. Selection begins with choosing a random grid location; either one (for mutation) or two (for crossover) random walks are performed from this location to collect a parent or parents. The single child resulting is inserted into the population at the chosen location only if it is strictly more fit than the current resident at that location.

In the SEFM trials, population size was 900 (on a 30 x 30 grid), and all else was as above except that crossover and mutation steps happened in ratio 3/2 with a combined probability of $1 - r$, where r was the chance of SEFM. In the later described experiments on the `mis93` problem, all was as for the SEFM trials except that the genewise mutation rate was 0.001.

8.9.3 Random Test Problems

This set of experiments involves 7 randomly generated problems. Each had 50 events, assumed the same 36 slot timetable structure as `edai93`, and imposed the same event-spread objective, and varied in number of (randomly generated distinct) edge constraints between 200 and 800.

8.9.4 Speedup via SEFM on `edai93`

Figure 8.6 compares the effect of using selective event-freeing mutation on the edai93 problem for different rates from 0 to 1.0 in steps of 0.1. At a given rate of event-freeing mutation *r*, a reproduction step resulted in performing event-freeing mutation on a single selected parent with probability *r*, and resulted instead in crossover or ordinary mutation with probabilty (1 - *r*); crossover remained 1.5 times as likely as mutation throughout. This figure shows quite remarkable speedup with increasing use of event-freeing mutation. Most interestingly, improvement continues right up to the point where the event-freeing mutation rate is 1. At this point we are no longer using a genetic algorithm, since no recombination occurs. Instead, this belongs to the more general class of evolutionary algorithms, using a powerful domain specific mutation operator coupled with a robust selection procedure.

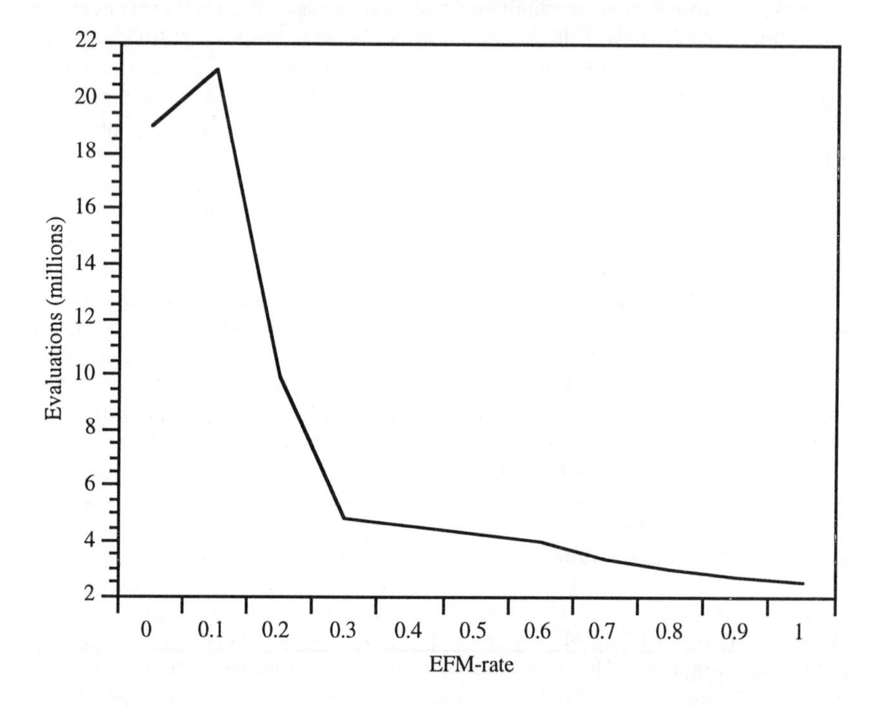

Figure 8.6: GA time on edai93 for varing rates of SEFM

A similar set of experiments performed on the larger mis93 problem showed a similar trend. In this case, 'average time to optimum' points could not be plotted for event-freeing mutation rates of 0 and 0.1 because the GA was not 100% robust over the ten trials at these points, but from a rate of 0.2 onward the GA consistently showed better performance as crossover became more and more replaced by selective event-freeing mutation until, again, the best version of the algorithm on this problem was the non-recombination version. These results are shown in figure 8.7.

8.9.5 Further Improvements: Smart Crossover

The results indicated above seem to tell us that the GA is outperformed by an SEFM alone on the problems addressed, and this surely means that SEFM will turn out to be best on a much wider range of problems too. There could be many explanations for this, and it will be interesting to see to what extent it is true on different kinds of timetabling problems. In the test problems addressed here, for example, we have successfully addressed two real, highly constrained problems of rather different size with, essentially, a domain-specific stochastic local hillclimbing algorithm. What is interesting is that this has performed significantly faster than the GA; in itself, this is no great surprise because SEFM exploits much inside knowledge of the nature of the problem, but it is interesting to reflect on why none of the 'hybrid' attempts (where SEFM happened at a rate between 0.1 and 0.9) were comparable with the full SEFM method. There are two quite distinct ways of looking at it: either these problems were *not difficult enough* for the GA to be able to truly exploit its prowess, or that we just happened to have found a better iterative stochastic solution to this kind of problem.

Several experiments not reported here in detail show indeed that SEFM alone still works out best when we constrain the problem more highly, even indeed if we construct a problem which has only one solution. Hence, it does not seem to be the case that the results shown here for SEFM arise because the problem was not difficult enough for the GA to show its prowess. Rather, we seem to have found a better way of solving many timetabling problems than the kind of GA we have considered so far. It seems that in the process of hybridising our original GA with a powerful domain-specific hillclimbing method, we have shown the hillclimbing method was better all along, to the extent that using it in conjunction with the GA (that is: employing a standard crossover operator in addition to the SEFM operator) *slows down* solution speed. It is instructive to consider why. When the SEFM rate is 1, the 'GA' configuration means that we are actually running 900 (the population size) individual and non-interacting trials in which an initial random timetable is undergoing repeated applications of SEFM, with the resulting mutant being kept or discarded (and hence the progenitor retained) only when it is fitter than its progenitor. Because of the stochastic nature of SEFM, we can expect the highly diverse random initial population to take on different, unrelated routes toward improvement. Effectively, this means that the population retains high diversity throughout when the SEFM-rate is high. A crossover operation would hence be expected to generate offspring from highly dissimilar parents, and thus achieve quite a low rate of success. This is especially so in our case. Consider for example two distinct individuals of high fitness: one may have, respectively, in slots s_1 and s_2, groups of events E_1 and E_2 such that there are no edges within groups, but high connection between them. The other individual may have these groups, respectively ,in slots s_2 and s_1. Crossover of these individuals may well mix these events up and, perhaps, place them all in s_1 or all in s_2. Standard recombination and SEFM therefore do not mix well.

It is, nevertheless, clear that genetic ideas should still be of help. SEFM as used above does not use any information regarding the population as a whole; there is simply a large collection of non-interacting iterative SEFM processes. It seems safe to consider, however, that some way of having these separate processes interact would improve things. An individual which happens to have been caught in a suboptimal chain of mutations might benefit from information that more successful individuals have implicitly gained. We have already seen that standard uniform crossover is not helpful in delivering such information. A more promising route, however, is quite simple to formulate: use the helpful information gleaned from violation-scores to inform the recombination procedure too.

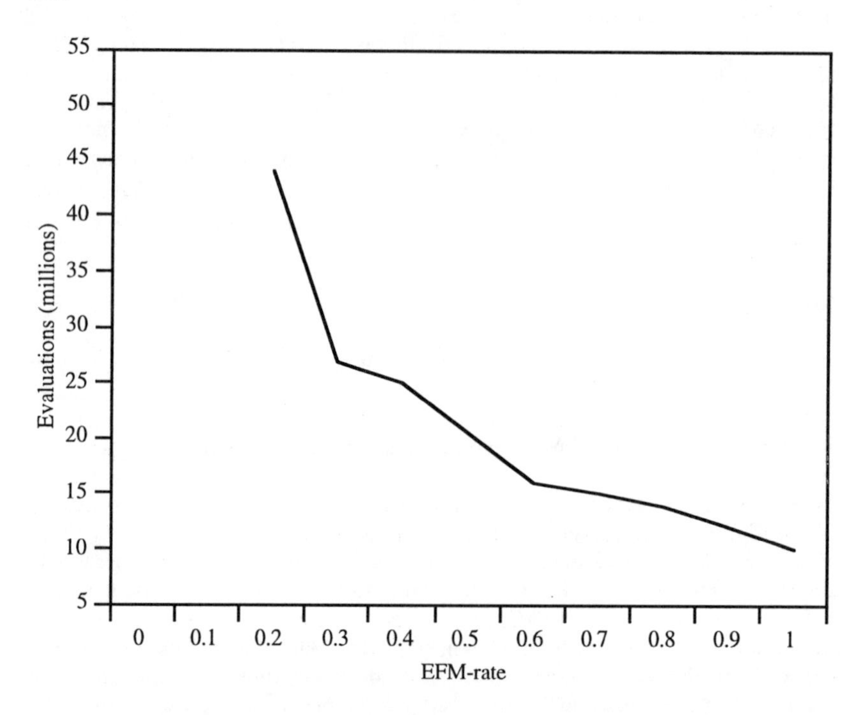

Figure 8.7: GA time on `mis93` for varing rates of SEFM

8.9.5.1 Violation-Directed Recombination

We have seen how making biased stochastic choices within mutation provide a very powerful operator for timetabling. This operator is incompatible with standard crossover, however, because the latter simply seems to destroy the good work done by the former. A route forward is to use the violation scores employed during SEFM to help us make crossover more effective. A way we can do this is as follows, in a process called *violation-directed recombination*, VDR.

Given two parents, parent1 and parent2, build a child as follows. Consider the violation $v(i)$ scores of the events i in parent1. Let $maxv$ and $minv$ be the maximum and minimum of these, respectively. For each event (gene position) i, calculate a swap probability $swap_i$, which is just $(v(i) - minv)/(1 + maxv - minv)$. Build the child by assigning its ith gene with the ith allele in parent1 with probability $1 - swap_i$, otherwise taking this allele from parent2.

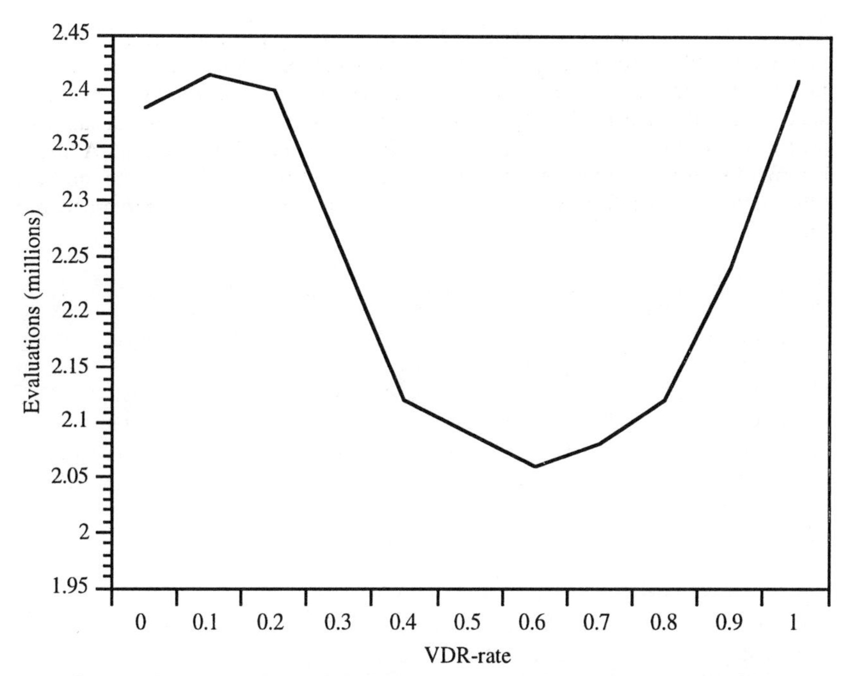

Figure 8.8: VDR+SEFM GA time on `edai93` for varing rates of VDR

This operator uses the violation-score information to make sure that, to a high extent, events with relatively small violation scores in the first parent retain their slots in the child. To some extent the low violation score of a particular event may be due to the particular slots assigned to other events which actually have high violation scores, so retaining the former but possibly changing the latter is certainly not guaranteed to help, but it seems safe to suppose that the general idea will work anyway. What VDR effectively does is to stochastically select a small number of events with generally high violation scores to have their slots changed, and then use the population (by virtue of using another selected parent) as a guide toward the choice of new slot. In this way it does half of the work of SEFM (choice of event), and then completes the job by choosing the new slot from another member of the population, rather than by a local search procedure which assesses candidate slots and then selects one. In other words, VDR uses the standard GA-style heuristic that an allele from another selected member of the population may be a good choice, saving time on the extra slot-choice work done by SEFM.

Does the GA formed by combination of sc vdr and SEFM yield an algorithm better than or comparable to SEFM alone? Figure 8.8 shows the results of some initial trials on the `edai93` problem where we combined VDR and SEFM; the horizontal axis shows the rate of application of VDR, which was always one minus the rate of application of SEFM. This figure indeed shows that the combination of VDR and SEFM, and hence a highly tailored domain specific GA, is even better than SEFM alone on the `edai93` problem.

8.10 Some Final Discussion

We started by presenting three styles of a GA approach to timetabling, and found that one of them looked promising enough for further investigation. This clash-free approach was then found to do very well on some real lecture timetabling problems. Indeed, the same basic approach has been used by a collection of researchers who have provided encouraging case studies on some real timetabling problems. We then noted an obvious way of making the implementation of this approach very fast, which we call delta-evaluation, and indeed found that it did speed things up by all order of magnitude on larger problems. We then looked at how we might expect this approach to perform in general over a wider range of problems. This investigation was understandably limited, considering the very large number of ways in which we could vary the statement of a timetabling problem, but we decided on some obvious candidates for dimension of variation which *all* timetabling problems share, and pursued some initial investigation of GA performance on them. What we found here was quite promising. On random problems of similar size and kind to the real `edai93`, the GA seems to maintain the ability to find optima; also, time to find optima across a large practically relevant range of problems seems to only rise linearly as we make the problem more difficult in either of the three dimensions.

Following this, we decided to look further at the idea of event-freeing mutation, and described a promising variant called *selective* event-freeing mutation, SEFM. We then gradually unveiled a story: SEFM was found to improve the GA's performance when we hybridised the two; but this went even further, to the extent that SEFM *alone* performed better than any of the hybrids. We were thus left with a powerful method of solving timetabling problems, which was better than any GA we had used so far. This is no great surprise: there is always one *best* method for any particular problem or class of problems, and it is equally certain that the best method for solving a problem is *not a* weak, domain-independent method like the GA. Indeed, it was always clear, and should be clear throughout this book, that much GA research focuses on the usefulness of the GA as an optimisation procedure which can be applied with high hopes for success on problems for which no other technique seems to work adequately. A further aspect of the GA, however, is that the basic ideas of genetic search are inherently very powerful, and are often likely to yield in improvement when hybridising with a strong problem-dependent algorithm as long as we make informed decisions on how to effect the hybridisation. We then indeed found that by introducing a smart recombination operator our results were further improved, yielding a powerful highly domain-tailored GA for solving general timetabling problems.

We have thus come quite far in this chapter, but it is, nevertheless, necessary to point out a large number of relevant issues which we simply have not addressed. For example, some common timetabling needs are not met by the approach as discussed here. Eg: timetablers may wish to generate several distinct timetables to choose from. There are several ways of using a GA to perform such *niching* or *speciation,* via which the final population is likely to contain several quite distinct solutions to the problem [12, 5, 16]. Indeed, the spatial selection method we tend to use [2] manages this, but we have not yet analysed the extent to which different methods yield a usefully large collection of distinct results, and which are individually *usefully distinct* from each other. These are difficult things to measure, and indeed it is not yet clear to what extent timetablers need such a collection.

Also, it is proper to point out that the general process involved in interpreting the constraints of a particular problem into a particular choice of fitness function for the GA can often be non-obvious. It is rarely apparent how best to do this, and further work is required to assess the various possibilities. Experience shows, however, that there is unlikely to be a major difference in performance between different such choices for problems of the size and type addressed here; hence, we feel that natural and/or arbitrary choices may be made with impunity for penalty settings, pre-specifications, event-spread constraint handling, and so on..., at least for problems of the size and type found in small or medium sized timetabling problems.

The topic of different variants of approach, in particular with respect to different representations, was also covered only barely. There are several more possible choices to apply. One, for example, is to make use of best-first search in the interpretation function. A timetable "*abcde*..." is interpreted as follows: "put event1 in the *a*th best choice of slot, put event2 in the *b*th best choice of slot, put event3 in the *c* best choice of slot, ..." and so on. This is similar to our clash-sparse approach, but may be more useful; 'best choice of slot' is taken to mean best in the context of the full penalty-weighted fitness function, rather than just in the context of avoiding clashes. This is evidently quite a time-consuming choice of interpretation function, but benefits may outweigh this. Note that using a GA on chromosomes thus interpreted amounts to a variant of best-first search, in which the GA searches choices of branch to take at all depths of the tree in parallel.

It is also interesting and important to compare the GA approach with other methods such as branch and bound search, simulated annealing, and so on. This endeavour is complicated by the differences between the techniques themselves. Eg, the promise of the GA-based approach is most strongly manifested in its robustness across a very wide range of different timetabling problems. Comparison with rule-based approaches to test this claim on the same variety of problems would then necessitate the lengthy and difficult development process of building rule-based systems with similarly wide applicability. Comparison with simulated annealing is a more likely prospect, and such is planned in due course.

Finally, we have not yet reported on the application of the technology to fully general problems which involve varied constraints on agents and rooms as well as slots, despite indicating early on that the general approach *applies* to such. This is partly because we have not yet obtained enough interesting data for such problems, and we prefer not to construct random examples of such because they may easily be very misrepresentative.

Readers are of course encouraged to try the techniqes we discuss on any problems they wish to address, either to make real use of the technique, or to perform some general research into the GA behavior. A program is available by anonymous

FTP called *GATT*, which will help in this regard. *GATT* is a general timetabling program which implements the clash-rich approach, and can optionally employ all of the other techniques we have discussed. It has a simple language for inputting data for a wide variety of problems, including exam and lecture timetabling, with or without constraints on rooms and/or agents. *GATT*, along with a collection of example timetabling problems, is available at the FTP site ftp.dai.ed.ac.uk.

References

[1] D. Abramson and J. Abela, 'A parallel genetic algorithm for solving the school timetabling problem', Technical report, Division of Information Technology, C.S.I.R.O., (April 1991).

[2] Robert J. Collins and David R. Jefferson, 'Selection in massively parallel genetic algorithms', in *Proceedings of the Fourth International Conference on Genetic Algorithms*, eds., R.K. Belew and L.B. Booker, pp. 249-256. San Mateo: Morgan Kaufmann. (1991).

[3] Alberto Colorni, Marco Dorigo, and Vitrorio Maniezzo, 'Genetic algorithms and highly constrained problems: The timetable case', in *Parallel Problem Solving from Nature*, eds., G. Goos and J. Hartmanis, 55-59, Springer-Verlag, (1990).

[4] Dave Corne, Hsiao-Lan Fang, and Chris Mellish, 'Solving the module exam scheduling problem with genetic algorithms', in *Proceedings of the Sixth International Conference in Industrial and Engineering Applications of Artificial Intelligence and Expert Systems*, eds. Paul W.H. Chung, Gillian Lovegrove, and Moonis Ali, 370-373, Gordon and Breach Science Publishers, (1993).

[5] J.C. Culberson, 'Genetic invariance: A new paradigm for genetic algorithm design', Technical Report TR92-02, Univerity of Alberta Dept of Computing Science, (1992).

[6] Yural Davidor, 'A naturally occurring niche and species phenomenon: The model and first results', in *Proceedings of the Fourth International Conference on Genetic Algorithms*, eds., R.K. Belew and L.B. Booker, pp. 257-263. San Mateo: Morgan Kaufmann, (1991).

[7] *Handbook of Genetic Algorithms*, ed., L. Davis, New York: Van Nostrand Reinhold, 1991.

[8] R. de Neufville, *Applied Systems Analysis: Engineering Planning and Technology Management*, McGraw-Hill Publishing Company, 1990.

[9] Hsiao-Lan Fang, *Investigating Genetic Algorithms in Scheduling*, Master's thesis, Department of Artificial Intelligence, University of Edinburgh, 1992.

[10] Hsiao-Lan Fang, Peter Ross, and Dave Corne, 'A promising genetic algorithm approach to job-shop scheduling, rescheduling, and open-shop scheduling problems', in *Proceedings of the Fifth International Conference on Genetic Algorithms*, ed., S. Forrest, 375-382, San Malco: Morgan Kaufmann, (1993).

[11] R.A. Fisher, *The Genetical Theory of Natural Selection,* Dover Books, 2nd edition, 1958.

[12] David E. Goldberg, *Genetic Algorithms in Search, Optimization & Machine Learning,* Reading: Addison Wesley, 1989.

[13] D.E. Goldberg, 'Real-coded genetic algorithms, virtual alphabets, and blocking', *Complex Systems,* 5(2), 139-167, (April 1991).

[14] V.S. Gordon and D. Whitley, 'Serial and parallel genetic algorithms as function optimizers', in *Proceedings of 5th ICGA,* ed., S. Forrest, pp. 177-183. Morgan Kaufmann, (1993).

[15] John H. Holland, *Adaptation in Natural and Artificial Systems,* Ann Arbor: The University of Michigan Press, 1975.

[16] Jeffrey Horn and Nicholas Nafpliotis, 'Multiobjective optimisation using the niched pareto genetic algorithm', Technical Report 93005, Illinois Genetic Algorithms Laboratory (IlliGAL), (July 1993).

[17] K. Juliff, 'A multi-chromosome genetic algorithm for pallet loading', in *Proceedings of 5th ICGA,* ed., S. Forrest, pp. 467-473. Morgan Kaufmann, (1993).

[18] Si-Eng Ling, 'Intergating genetic algorithms with a prolog assignment problem as a hybrid solution for a polytechnic timetable problem', in *Parallel Problem Solving from Nature, 2,* eds., R. Manner and B. Manderick, 321-329, Elsevier Science Publisher B.V., (1992).

[19] Peter Ross, Dave Corne, and Hsiao-Lan Fang, 'Timetabling by genetic algorithms: Issues and approaches', Technical Report AIGA-006-94, Department of Artificial Intelligence, University of Edinburgh, (1994). revised version to appear in Applied Intelligence.

[20] A.E. Smith and D. M. Tate, 'Genetic optimization using a penalty function', in *Proceedigns of the 5th ICGA,* ed., S. Forrest, pp. 499-505. Morgan Kaufmann, (1993).

[21] G. Syswerda, 'Uniform crossover in genetic algorithms', in *Proceedings of the Third International Conference on Genetic Algorithms and their Applications,* ed., J. D. Schaffer, 2-9, San Mateo: Morgan Kaufmann, (1989).

[22] David M. Tate and Alice E. Smith, 'Expected allele coverage and the role of mutation in genetic algorithms', in *Proceedings of the Fifth International Conference on Genetic Algorithms,* ed., S. Forrest, pp. 31-37. San Mateo: Morgan Kaufmann, (1993).

[23] Darrell Whitley, 'The GENITOR algorithm and selection pressure', in *Proceedings of the Third International Conference* on *Genetic Algorithms,* ed., J.D. Schaffer, 116-121, San Mateo: Morgan Kaufmann, (1989).

[24] Darrell Whitley, Timothy Starkweather, and D'Ann Fuquay, 'Scheduling problems and travelling salesmen: The genetic edge recombination operator', in *Proceedings of the Third International Conference on Genetic Algorithms,* ed., J.D. Schaffer, 133-140, San Mateo: Morgan Kaufmann, (1989).

[25] R.J. Wilson. *Introduction to Graph Theory,* Longman, London, 2nd ed., 1985.

[26] S. Wright, *Evolution and the Genetics of Populations, Volume 1: Genetic and Biometric Foundations,* volume 1, University of Chicago Press, 1968.

[27] S. Wright, *Evolution and the Genetics of Populations, Volume 2: The Theory of Gene Frequencies,* volume 1, University of Chicago Press, 1969.

[28] S. Wright, *Evolution and the Genetics of Populations, Volume 3: Experimental Suits and Evolutionary Deductions,* volume 1, University of Chicago Press, 1977.

[29] S. Wright, *Evolution and the Genetics of Populations, Volume 4: Variability within and among Natural Populations,* volume 1, University of Chicago Press, 1978.

[30] Takeshi Yamada and Ryohei Nakano, 'A genetic algorithm application to large-scale job-shop problems', in *Parallel Problem Solving from Nature, 2,* eds., R. Manner and B. Mandcrick, 281-290, Elsevier Science Publisher B.V., (1992).

Chapter 9

J.E. Everett
Department of Information Management and Marketing
The University of Western Australia
Nedlands, Western Australia 6009

jeverett@ecel.uwa.edu.au

Algorithms for Multidimensional Scaling

0-8493-2519-6/95/$0.00 + $.50

Abstract

In this chapter we will be looking at the potential for using genetic algorithms to map a set of objects in a multidimensional space. Genetic algorithms have a couple of advantages over the standard multidimensional scaling procedures that appear in many commercial computer packages. We will see that the most frequently cited advantage of genetic algorithms, the ability to avoid being trapped in a local optimum, applies in the case of multidimensional scaling. A further advantage of using a genetic algorithm, or at least a hybrid genetic algorithm, is the opportunity it offers to choose freely an appropriate objective function, without the restrictions of the commercial packages, where the objective function is usually a standard function chosen for its stability of convergence rather than for its applicability to the user's particular research problem. We will develop some genetic operators appropriate to this class of problem, and use them to build a genetic algorithm for multidimensional scaling with fitness functions that can be chosen by the user. We will test the algorithm on a realistic problem, and show that it converges to the global optimum in cases where a systematic hill-descending method becomes entrapped at a local optimum. We will also look at how considerable computation effort can be saved with no loss of accuracy by using a hybrid method, with the genetic algorithm being brought in to "fine tune" a solution which has first been obtained using standard multidimensional scaling methods. Finally, a full program description will be given allowing the reader to implement or modify the program in a C or C++ environment.

9.1 Introduction

9.1.1 Scope of this Chapter

In this chapter, we will be considering the nature and purpose of multidimensional scaling and the types of problems to which it can be applied. We shall see that multidimensional scaling techniques are susceptible to being trapped in local optima, and that it is important to use a measure of misfit that is statistically appropriate to the particular multidimensional scaling model being analysed. These factors will be shown to be a problem with the standard multidimensional scaling techniques available in commercial statistical packages, a problem that can be overcome by using a suitable genetic or hybrid algorithm. A number of suitable genetic operators will be discussed, differing from the more standard genetic operators because of the continuous nature of the parameters, and because of the ascription of these parameters to individual objects. The problem of evolving the best mapping of a number of interacting bodies is analogous to the evolution of a social organism with a joint fitness function. This analogy will be developed, and provides a rationale for using the alternative genetic operators suggested.

Some extensive test results on a realistic multidimensional scaling problem will be reported and examined.

The chapter ends with a program listing and a full description of each of its component parts. The program is written in C, using the simulation package Extend. Any reasonably proficient user of the language should be able to transfer the program to another C or C++ environment.

9.1.2 What is Multidimensional Scaling?

In many situations, we have data on the interrelationships between a set of objects. These interrelationships might be, for example:

- The distances or the travel times between cities;
- The perceived similarities between different brands of beer;
- The number of words shared between members of a group of languages;
- The frequency with which libraries lend items to each other;
- The frequency with which journals cite each others' papers;
- The mistakes that learners of the Morse code make, confusing one symbol for another;
- The similarities between shades of colours;
- The correlations between adjectives used to describe people.

In each of the cases listed above, the data take the form of a matrix **D**, whose components d_{ij} represent some measure of the similarity or dissimilarity between object 'i' and object 'j'. Each case is an example of a general and common situation where it would be useful to produce a mapping of the objects, and has been the subject of published research where multidimensional scaling has been used to produce such maps. On the map, the relative positions of the objects should provide a concise graphical representation of their interrelationships. Object 'i' will be mapped at a point having coordinates x_{if}, where f ranges from 1 up to however many dimensions are being mapped.

For example, in the case of the Morse code confusions, we would want a map where the symbols that get confused with each other most frequently appear close together, and those symbols which are rarely confused with each other map far apart.

Multidimensional scaling is a modelling technique which uses the matrix of interrelationships between a set of objects. These interrelationships could either be measures of similarity (such as the rate of confusion between symbols in Morse code) or of dissimilarity (such as the travel time between pairs of cities). Multidimensional scaling techniques attempt to find a set of coordinates for the objects, in a multidimensional space, so that the most similar objects are plotted close together and the most dissimilar objects are plotted furthest apart.

9.1.2.1 Metric and Non-Metric Multidimensional Scaling

In metric multidimensional scaling, the distances between objects are required to be proportional to the dissimilarities, or to some explicit function of the dissimilarities. In non-metric multidimensional scaling, this condition is relaxed so as to require only that the distances between the objects increase in the same order as the dissimilarities between the objects.

9.1.2.2 Choice of the Misfit or Stress Function

In a multidimensional scaling model, the parameters of the model are the coordinates at which we map the objects. These parameters, or coordinates, have to be chosen so as to minimise some measure of misfit, which will be a function of the differences between the observed inter-object data matrix, and a comparable matrix calculated from the model. As in most examples of model fitting, the fitness function to be optimised is actually a misfit function, requiring minimising. Multidimensional scaling texts tend to refer to this misfit function as 'stress'. Generalised treatments of nonlinear modelling commonly refer to it as 'loss'. For our purposes, fitness function, misfit, stress and loss will be treated as synonymous. In general, the fitness function to be minimised will here be referred to as misfit.

Standard multidimensional scaling procedures, commercially available in statistical computer packages such as SPSS, SAS and SYSTAT, use some convenient standard measure of misfit, chosen for its convergence properties, such as Kruskal's stress (Kruskal, 1964). However, the appropriate measure of misfit will be different for different problems, depending on the statistical nature of the model we are trying to fit. For example, when dissimilarities are distances, the misfit or stress may appropriately be some function of the squared errors between the computed and actual distances between objects. Depending on a knowledge of the way the data were gathered, and therefore of how errors might arise, it may be statistically more appropriate to use absolute error instead of squared errors, or to proportionate the error to the inter-object distance. For frequency data, such as the rate of confusions in the Morse code example, the maximum likelihood fit may be obtained by choosing parameters (coordinates) to minimise a chi-square function of the difference between observed and predicted frequencies. Statistical packages do not readily allow the user to tailor the misfit or stress function in these statistically appropriate ways.

9.1.2.3 Choice of the Number of Dimensions

There is no reason why a mapping should be in only two dimensions, but in general we would want to produce a map with as few dimensions as possible. It is not surprising that most published work in multidimensional scaling has produced two-dimensional (or at most, three-dimensional) solutions: mapping objects in one direction tends to be inadequate or trivial; more than three dimensions are impossible for us mere mortals to visualise, and more than two dimensions are unpopular with editors who like to publish figures on flat pages which can be easily understood.

In fitting a model to the data, for a given number of dimensions, the object coordinates will be chosen so as to minimise the residual misfit. Whatever function of misfit is used, it will be found that, unless a perfect fit has been obtained, the residual misfit can always be decreased by increasing the number of dimensions. In the limit, n objects can always be plotted perfectly in n-1 dimensions (although, in certain cases, some of the coordinates may be imaginary!). However, such a perfect fit may be entirely spurious, and an adequate fit may usually be obtained in fewer dimensions, with the residual misfit being ascribed to statistical error. Occam's razor (or its modern counterpart KISS), tells us that the preferred mapping model is one which represents the objects in as few dimensions as are needed to conform adequately to the inter-object data. Again, it is important to have a misfit function appropriate to the statistical properties of the model being fitted, before we can reasonably decide whether residual misfit is significant or not, and therefore decide whether the mapping requires more or fewer dimensions.

If the misfit function is statistically appropriate to the way the data were formed or gathered, the appropriate number of dimensions will be achieved when the residual misfit becomes small enough to be ascribed to random error. In practise, we may compromise on meaningfulness rather than statistical significance, and accept a simpler model, of fewer dimensions, that explains most of the original misfit, even if it leaves a statistically significant residue. This compromise between meaningfulness and significance has been discussed more fully in the earlier chapter on Modelling Techniques (Chapter 0), and is the essence of Occam's Razor.

9.1.2.4 Replicated Data Matrices

A further extension of multidimensional scaling occurs when the data consist of several matrices, one for each respondent. These replicated data matrices may be treated as repeat estimates of the same configuration, so that a single best fit map is produced. However, it may be reasonable to model each respondent as having a different map, with the same configuration, but stretched differently along the axes for different individual respondents. This refinement of multidimensional scaling is known as Indscal. For example, in the study of Morse code confusions by Shepard (1963), it was found that the symbols plotted on a two-dimensional map. One axis varied as the proportion of dashes in the symbol, so that Morse symbols containing predominantly dashes were at one extreme, and those containing predominantly dots were plotted at the other extreme. The second axis was found to relate to the number of items (dots or dashes) in the symbol, increasing from only one item to two, three, four and more item symbols. It could well be that if the data for individual operators had been analysed using Indscal, then individual respondents' maps would be expected to be elongated in the first dimension for those operators who were less confused by the dot/dash distinction than by the number of dots and dashes. Those operators who had more trouble distinguishing dots from dashes rather than identifying the number of dots and dashes would produce maps elongated along the second dimension.

9.1.2.5 Arbitrary Choice of Axes

In ascribing coordinates to objects, there are a number of arbitrary choices that do not affect the goodness of fit:

- Adding or subtracting a constant to coordinates of any particular dimension;
- Reversing the axis of any dimension;
- Rotating the entire set coordinate axes by any angle lying in any plane;
- Scaling the entire set of coordinates by any consistent factor.

Any of these operations will leave the misfit or stress function unaltered. To this extent, there are in theory an infinite number of global optima or, perhaps more appositely, an infinite number of representations of a single global solution. Some arbitrary rules have to be imposed to select which representation of the global solution to use. One set of rules used in the standard implementations is:

- Set the coordinates on each dimension to have zero mean;
- Make the first dimension be the one with greatest variance, and scale it to unit variance;
- Make each subsequent dimension be the remaining one of greatest variance.

An alternative set of rules, computationally easier to implement is to:

- Set the first object to have zero coordinates in all directions ($x_{1f} = 0$, all f);
- Establish the first dimension with ($x_{21} = 1$, $x_{2f} = 0$ for $f>1$);
- Establish further dimensions as needed with ($x_{nf} = 0$ for $f>n-1$).

In this approach, each successive object is used to introduce a new dimension.

In models where the inter-object data are specifically distances, then the scaling of the coordinates will be determined, although their origin, sense and rotation will still be arbitrary.

9.1.3 Standard Multidimensional Scaling Techniques

Several multidimensional scaling procedures are available in commercial statistical computer packages. Each package tends to offer a variety of procedures, dealing with metric and non-metric methods, and single or multiple data matrices. Among the most used procedures are Alscal, Indscal, KYST and Multiscale. Their development, methods and applications are well described by Schiffman et al. (1981), Kruskal and Wish (1978) and Davies and Coxon (1982). They are available to researchers in many major statistical computer packages, including SPSS (Norusis, 1990) and SYSTAT (Wilkinson et al., 1992).

9.1.3.1 Limitations of the Standard Techniques

Standard multidimensional scaling methods have two deficiencies:

- The dangers of being trapped in a local minimum, and
- The statistical inappropriateness of the function being optimised.

The standard multidimensional scaling methods use iterative optimisation that can lead to a local minimum being reported instead of the global minimum. The advantages of genetic algorithms in searching the whole feasible space and avoiding convergence on local minima have been discussed by many authors (see for example Goldberg, 1989, and Davis, 1991). This advantage of genetic algorithms makes them worthy of consideration for solving multidimensional scaling problems.

The second deficiency of the standard multidimensional scaling methods is perhaps more serious, though less generally recognised. They optimise a misfit or stress function which is a convenience function, chosen for its suitability for optimising by hill-descending iteration. The type of data and sampling conditions under which the data have been obtained may well dictate a maximum likelihood misfit function or other statistically appropriate function which differs from the stress functions used in standard multidimensional scaling procedures. One great potential advantage of a genetic algorithm approach is that it allows the user to specify any appropriate function for optimising.

The advantages that a genetic algorithm offers in overcoming these problems of the standard multidimensional scaling techniques will be discussed in greater detail in the next section.

9.2 Multidimensional Scaling Examined in More Detail

9.2.1 A Simple One-Dimensional Example

In multidimensional scaling problems, we refer to the dimensionality of the solution as the number of dimensions in which the objects are being mapped. For a set of n objects, this object dimensionality could be any integer up to n-1. The object dimensionality should not be confused with the dimensionality of the parameter space, with a much greater number of dimensions, one for each model parameter or coordinate, of the order of the number of objects multiplied by the number of object dimensions.

We will start by considering a simple problem, which in this particular case can be modelled perfectly with the objects lying in only one object dimension. The problem will seem quite trivial, but it exhibits more clearly some features that are essential to the treatment of more complicated and interesting problems of greater dimensionality.

Consider three objects, which we shall identify as 'Object n', for n = 1, 2 and 3. The distance d_{ij} has been measured between each pair of objects 'i' and 'j', and is shown in Table 9.1.

		i	
	1	**2**	**3**
j 1	0	10	20
2	10	0	10
3	20	10	0

Table 9.1: An Example Data Matrix of Inter-object Distances d_{ij}

The purpose is to map the objects in one dimension, with object 'i' located at 'x_i', so as to minimise the average proportionate error in each measurement. Thus a suitable misfit function to be minimised is:

$$Y = \Sigma|(|x_i - x_j| - d_{ij})|/d_{ij} \qquad (9.1)$$

With no loss of generality, we can constrain $x_1 = 0$, since a shifting of the entire configuration does not change the inter-object distances.

Using a spreadsheet program, such as Excel or Lotus, we can calculate the function Y over a range of values of x_2 and x_3, keeping x_1 zero. The three objects fit perfectly (with Y = 0) if $\underline{x} = (0, 10, 20)$, or its reflection $\underline{x} = (0, -10, -20)$. This global solution is drawn in Figure 9.1, with the three objects being the three solid spheres. However, if we move Object 3 to $x_3 = 0$, leaving the other two objects unmoved, we find a local minimum Y = 1 at $\underline{x} = (0, 10, 0)$. Small displacements of any single object from this local minimum cause Y initially to increase.

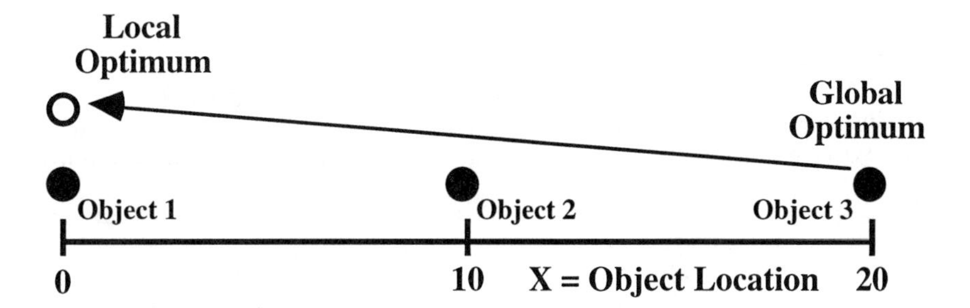

Figure 9.1: Global and Local Optima for the One-Dimensional Example

Figure 9.2 shows the misfit function values for the relevant range of values of x_2 and x_3. Values are shown on a grid interval of 2, for x_2 increasing vertically and for x_3 increasing horizontally. The global minima are surrounded by heavy bold circles, the local minima by light bold, and the saddle points are italicised inside ordinary circles.

x_2																							
	31	28	27	26	25	24	23	22	21	20	19	18	15	12	13	14	15	16	17	14	11	8	7
	27	24	23	22	21	20	19	18	17	16	15	14	11	12	13	14	15	16	13	10	7	4	3
+10	23	20	19	18	17	16	15	14	13	12	11	**10**	11	12	13	14	15	12	9	6	3	**0**	3
	23	20	19	18	17	16	15	14	13	12	11	14	15	16	17	18	15	12	9	6	3	4	7
	23	20	19	18	17	16	15	14	13	12	15	18	19	20	21	18	15	12	9	6	7	8	11
	23	20	19	18	17	16	15	14	13	16	19	22	23	24	21	18	15	12	9	10	11	12	15
	23	20	19	18	17	16	15	14	17	20	23	26	27	24	21	18	15	12	13	14	15	16	19
0	23	20	19	18	17	16	*15*	18	21	24	27	30	27	24	21	18	*15*	16	17	18	19	20	23
	19	16	15	14	13	12	15	18	21	24	27	26	23	20	17	14	15	16	17	18	19	20	23
	15	12	11	10	9	12	15	18	21	24	23	22	19	16	13	14	15	16	17	18	19	20	23
	11	8	7	6	9	12	15	18	21	20	19	18	15	12	13	14	15	16	17	18	19	20	23
	7	4	3	6	9	12	15	18	17	16	15	14	11	12	13	14	15	16	17	18	19	20	23
-10	3	**0**	3	6	9	12	15	14	13	12	11	**10**	11	12	13	14	15	16	17	18	19	20	23
	3	4	7	10	13	16	15	14	13	12	11	14	15	16	17	18	19	20	21	22	23	24	27
	7	8	11	14	17	16	15	14	13	12	15	18	19	20	21	22	23	24	25	26	27	28	31
		-20					-10					0					+10					+20	x_3

Figure 9.2: Misfit Function (Y) for the One-Dimensional Example

It is clear that a simple hill descending optimisation is in danger of getting entrapped not only at the local minimum of $\underline{x}$ = (0, 10, 0), and its reflection, but also at the saddle point $\underline{x}$ = (0, 0, 10) and its reflection. It can be seen that the axes of the saddle point are tilted, so a method that numerically evaluates the gradients along the axes will not find a direction for descending to a lower misfit value.

The problem we have considered, of fitting three objects in one dimension, had two parameters that could be adjusted to optimise the fit. It was therefore a comparatively straightforward task to explore the global and local minima and the saddle points in the two-dimensional parameter space. If we increase the number of dimensions and/or the number of objects, the dimensionality of the parameter space (not to be confused with the dimensionality of the object space) increases so as to preclude graphical representation, and to make analysis very difficult. The problem is especially severe if (as in our example) the misfit function is not universally differentiable.

We might expect the problem of local optima to diminish as we increase the number of dimensions and/or the number of objects, since there are more parameters available along which descent could take place. However, it is still possible that objects closely line up within the object space, or within a subset of it, generating local optima of the form we have just encountered. Without evaluating the misfit function over the entire feasible space, we cannot be entirely sure that a reported solution is not just a local optimum. This 'entrapment' problem therefore remains a real danger in multidimensional scaling problems, and cannot be ruled out without knowing the solution. Since entrapment may generate a false solution, the problem is analogous to locking oneself out of the house and not being able to get in without first getting in to fetch the key. Any optimisation method that has a danger of providing a local solution must be very suspect.

9.2.2 More than One Dimension

If we have 'n' objects to be mapped (i = 1, 2 ... n) in 'g' dimensions (f = 1, 2, ... g), then the data d_{ij} will comprise a matrix **D** measuring n by n, and the problem will require solution for g(2n-1-g)/2 coordinate parameters x_{if}, where f goes from 1 to g, and i goes from f+1 to n.

Because any translation or rotation of the solution will not alter the inter-object distances, we can arbitrarily shift or translate the whole set of objects so that the first object is zero on all coordinates. Rotations then allow us to make zero all but one of the second object's coordinates, all but two of the third objects coordinates, and so on. These operations are equivalent to setting x_{if} to zero when $i \leq f$.

The data matrix **D**, with elements d_{ij}, can be any appropriate measure of similarity or dissimilarity between the objects. At its simplest, it might just be measured inter-object distance, as in our one-dimensional example. In such a case, the diagonal of the data matrix will contain zeros, and the data matrix **D** will be symmetric ($d_{ij} = d_{ji}$), so there will be only n(n-1)/2 independent data observations.

For the case of a symmetric data matrix with zero diagonal, the number of coordinate parameters will be equal to the number of independent data observations when f, the number of dimensions f is equal to (n-1), one less than the number of objects. So such a symmetric zero diagonal data matrix can always be mapped into (n-1) or less dimensions. However, if the data matrix is not positive definite, the solution will not be real.

Multidimensional scaling methods are designed to find a solution in as few dimensions as possible that adequately fits the data matrix. For metric multidimensional scaling, this fit is done so that the inter-object distances are a ratio or interval transformation of the measured similarities or dissimilarities. For non-metric multidimensional scaling, the inter-object distances are a monotonic ordinal transformation of the measured similarities or dissimilarities, so that as far as possible the inter-object distances increase with decreasing similarity or increasing dissimilarity. In either case, we can refer to the transformed similarities or dissimilarities as 'disparities'. The usual approach with standard multidimensional scaling methods is to find an initial approximate solution in the desired number of dimensions, and then iterate in a hill-descending manner to minimise a misfit function, usually referred to as a 'stress' function. For example, the Alscal procedure (Schiffman et al., 1981, pp 347-402) begins by transforming the similarities matrix to a positive definite vector product matrix and then extracting the eigen vectors, by solving:

$$\text{Vector Product Transform of } \mathbf{D} = \mathbf{XX}' \tag{9.2}$$

In this decomposition, **X** is a matrix composed of n vectors giving the dimensions of the solution for the n objects coordinates, arranged in order of decreasing variance (as indicated by their eigen values). The nth coordinate will of course be comprised of zeroes since, as we have seen, the solution can be fitted with (n-1) dimensions. If, for example, a two dimensional solution is to be fitted, then the first two vectors of X are used as a starting solution, and iterated to minimise a stress function. The usual stress function minimised is 's-stress', which is computed as the root mean square value of the difference between the squares of the computed and data disparities, divided by the fourth power of the data disparities (see Schiffman et al., 1981, p. 355-357). The s-stress function is used because it has differentiable properties that help in the iteration towards an optimum.

9.2.3 Using Standard Multidimensional Scaling Methods

We have already seen, in the introduction to this chapter, that there are two major problems in the use of standard multidimensional scaling procedures to fit a multidimensional space to a matrix of observed inter-object similarities or dissimilarities.

The first shortcoming considered was the danger of a local minimum being reported as the solution. This problem is inherent in all hill-descending methods where iterative search is confined to improving upon the solution by following downward gradients. A number of writers (for example Goldberg, 1989, and Davis, 1991) have pointed out the advantage in this respect of using genetic algorithms, since they potentially search the entire feasible space, provided premature convergence is avoided.

The second and most serious shortcoming of standard multidimensional scaling procedures was seen to lie in the choice of the stress or misfit function. If we are trying to fit a multidimensional set of coordinates to some measured data which has been obtained with some inherent measurement error or randomness, then the misfit function should relate to the statistical properties of the data. The misfit functions used in standard multidimensional procedures cannot generally be chosen by the user, and have been adopted for ease of convergence rather than for statistical appropriateness. In particular, the formulation of s-stress, used in Alscal and described above, will often not be appropriate to the measured data.

For example, if the data consists of distances between Roman legion campsites measured by counting the paces marched, and we are fitting coordinates to give computed distances $d_{ij}*$ that best agree with the data distances d_{ij}, then sampling theory suggests that an appropriate measure of misfit to minimise is:

$$Y = \Sigma(d_{ij}*-d_{ij})^2 /d_{ij} \quad (9.3)$$

In other cases, the data may be measured as frequency of some sort of interaction between the objects, and the misfit function should more properly make use of the statistical properties of such frequency data. Kruskal and Wish (1978) describe a classic study by Rothkopf (1957), analysed by Shepard (1963). The data comprised a table of frequencies of confusions by novices in distinguishing between the 36 Morse code signals. The confusion frequencies were used as measures of similarities between the code signals, and were analysed using multidimensional scaling to generate an interpretable two-dimensional map of the Morse code signals, in which it was found that the complexity of the signals increased in one direction and the proportion of dashes (as opposed to dots) increased along the second dimension. However, instead of the standard stress function it would have been more appropriate to use a misfit measure that related the generation of confusions to a Poisson process, with the Poisson rate for each pair of Morse code signals depending upon their inter-object distance d_{ij}. Following Fienberg (1980, p. 40) a maximum likelihood solution could then be obtained by minimising the function:

$$Y = G^2 = 2\ \Sigma_{ij}\ F_{ij} \cdot \log(F_{ij} / E_{ij}) \quad (9.4)$$

where: F_{ij} = Observed confusion frequency, E_{ij} = Modelled confusion frequency, and:

$$E_{ij} = \exp(-d_{ij}) \quad (9.5)$$

The log-likelihood function defined in equation (4) has the fortunate property of being approximately a chi-squared distribution. The chi-square value can be partitioned, so that we can examine a series of hierarchical models step by step. We can fit the inter-object distances d_{ij} to models having successively increasing numbers of dimensions. Increasing the model dimensions uses an increasing number of parameters and therefore leaves a decreasing number of degrees of freedom. The improvement in the chi-square can be tested for significance against the decrease in the number of degrees of freedom, to determine the required number of dimensions, beyond which improvement in fit is not significant. This method could, for example, have provided a statistical test of whether the Morse code signals were adequately representable in the two dimensions, or whether a third dimension should have been included.

In some cases, the data matrix may not be symmetric. For example, Everett and Pecotich (1991) discuss the mapping of journals based on the frequency with which they cite each other. In their model, the frequency F_{ij} with which journal j cites journal i depends not only on their similarity S_{ij}, but also upon the source importance I_i of journal i, and the receptivity R_j of journal j. On their model, the expected citation frequencies E_{ij} are given by:

$$E_{ij} = I_i R_j S_{ij} \tag{9.6}$$

They use an iterative procedure to find the maximum likelihood solutions to I and R, then analysed the resulting symmetric matrix **S** using standard multidimensional scaling procedures, with the usual arbitrary rules applied to using the residual stress to judge how many dimensions to retain. They could instead have used the model:

$$E_{ij} = I_i R_j \exp(-d_{ij}) \tag{9.7}$$

It would have then been possible to evaluate the chi-square for a series of hierarchical models where d_{ij} has increasing dimensionality, to find the statistically significant number of dimensions in which the journals should be plotted.

The standard multidimensional scaling procedures available in statistical computing packages do not allow the user the opportunity to choose a statistically appropriate misfit function. This choice is not possible because the stress functions they do use have been designed to be differentiable and to facilitate convergence. On the other hand, genetic algorithms do not use the differential of the misfit function, but require only that the misfit function be calculable, so that it is not difficult for users to specify whatever function is statistically appropriate for the particular problem being solved.

We will now discuss the design of a genetic algorithm for solving multidimensional scaling problems, and report some preliminary test results.

9.3 A Genetic Algorithm for Multidimensional Scaling

Genetic algorithms, as described in many of the examples in this book, commonly use binary parameters, with each parameter being an integer encoded as a string of binary bits. The two most standard genetic operators of mutation and crossover have also been described in previous chapters.

In designing a genetic algorithm for multidimensional scaling, we will find some differences in the nature of the parameters, and in the genetic operators that are appropriate. The parameters in a multidimensional scaling model are the coordinates of the objects being mapped, so they are essentially continuous. The application of genetic algorithms to optimising continuous (or 'real') parameters has been discussed by Wright (1991).

In our multidimensional scaling case the situation is further enriched by some ambiguity as to whether the set of objects being mapped is best thought of as the optimisation of a single entity, or as optimisation of a community of interacting individuals. We shall see that the latter analogy, treating the set of objects as an interacting community of individuals, provides some insight triggering the design of purpose-built genetic operators.

9.3.1 Random Mutation Operators

In mutation, one parameter is randomly selected, and its value changed, generally by a randomly selected amount.

9.3.1.1 Binary and Real Parameter Representations

In the more familiar binary coding mutation randomly changes one or more bits in the parameter. One problem with binary coding is that increases and decreases are not symmetric: if a parameter has a value of 4 (coded as 100), then a single bit mutation can raise it to 5 (coded as 101), but the much more unlikely occurrence of all three bits changing simultaneously is needed to reduce it to 3 (coded as 011). This asymmetry can be avoided by using a modified form of binary coding, called Gray coding after its originator, in which each number's representation differs from each of its neighbours, above and below, by changing only one bit from '0' to '1' or vice versa.

In either standard binary or Gray coding of integers, if the parameter is a binary coded integer with maximum feasible value X_{max}, then changing a randomly selected bit from '0' to '1' or vice versa, the parameter value is equally likely to change by 1, 2, 4, ... (X_{max} /2) units.

This greater likelihood of small changes, while allowing any size of change, has obvious attractions, and can be mimicked for real parameters by setting the mutation amplitude to $+/-X_{max}/2^p$, where p is a randomly chosen integer in the range 1 to q, where $X_{max}/2^q$ is the smallest mutation increment to be considered, and the sign of the mutation is chosen randomly.

An alternative approach is to set the mutation to N(0, MutRad), a Gaussian distribution of zero mean and standard deviation 'MutRad', the desired 'mutation radius'. Again, with this form of mutation smaller mutation steps are more likely, but larger steps are possible, so that the entire feasible space is potentially attainable. In an evolving algorithm, the mutation radius can start by encompassing the entire feasible space, and shrink to encompass a smaller search space as convergence is approached.

Like Gray coding, mutation of continuous parameters avoids the asymmetry we noted for standard binary-coded integer parameters. With either of the continuous parameter mutation procedures just described, not only are small changes in parameter value more likely than large changes, but negative changes have the same probability as positive changes of the same magnitude.

9.3.1.2 Projected Mutation: A Hybrid Operator

A third way to specify the mutation amplitude provides a hybrid approach, making use of the local shape of the misfit function. The method can be applied only if the misfit function is locally continuous (though not necessarily differentiable).

Figure 9.3 shows how the suggested projection mutation operator works. The parameter to be mutated is still randomly selected (so that a randomly selected object is shifted along a randomly selected direction). However, the direction and amount of the projection is determined by evaluating the function three times, for the object at its present location (Y_1) and displaced small equal amounts ΔX in opposite directions, to yield values Y_0 and Y_2. A quadratic fit to these three values indicates whether the misfit function is locally concave upward along the chosen direction. If it is, the mutation sends the object to the computed minimum of the quadratic fit. Otherwise, the object is sent in the downhill direction by an amount equal and opposite to its distance from the computed maximum of the quadratic fit. In Figure 9.3, both situations are depicted, with the original location in each case being the middle of the three evaluated points, identified by small circles. In the first case, where the curvature is concave downward, the solution is projected downhill to the right by a horizontal amount equal but opposite to the distance of the fitted quadratic maximum. In the second case, where the curvature is concave upward, the solution is projected downhill to the left, to the fitted quadratic minimum.

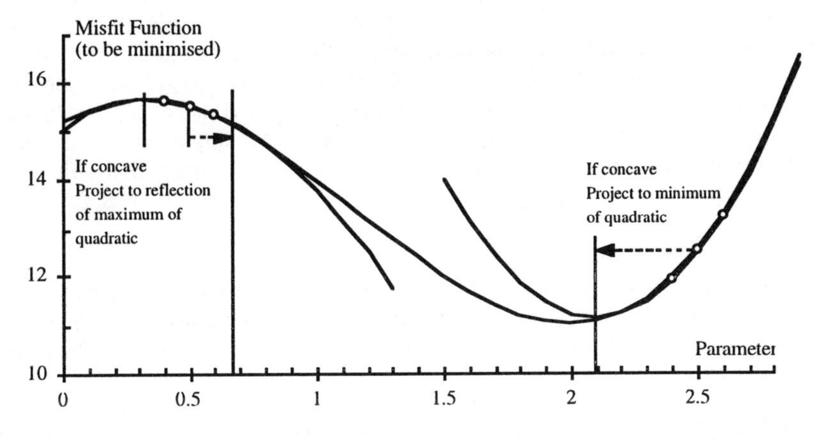

Figure 9.3: Projected Mutation

9.3.2 Crossover Operators

Crossover consists of the interchange of parameter values, generally between two parents, so that one offspring receives some of its parameter values from one parent, and some from the other. Generally, a second offspring is given the remaining parameter values from each parent.

Originally, a single crossover point was used (Goldberg, 1989). If the parameters were listed in order, an offspring would take all its parameters from one parent up to the crossover point (which could be in the middle of a parameter), and all the remaining parameters from the other parent. Under uniform crossover (Davis, 1991) each parameter (or even each bit of each parameter if they are binary coded) is equally likely to come from either parent. Uniform crossover can break up useful close encodings, but has the opportunity to bring together useful distant encodings. With continuous parameters, where the parameters have no natural ordering or association, an attractive compromise is to use uniform coding modified so that the offspring obtains each parameter at random from either parent.

In the multidimensional scaling, there is no *a priori* ordering of the objects. Suitable uniform crossover modifications would therefore be to get either:

- Each parameter (a coordinate on one dimension for one object) from a random parent, or
- Each object's full set of coordinates from a single random parent.

9.3.2.1 Inter-Object Crossover

A third, unorthodox, form of crossover that can be considered is to use only a single parent, and to create a single offspring by interchanging the coordinate sets of a randomly selected pair of objects. This postulated crossover variant has the attraction that it could be expected to help in situations of entrapment, where a local optimum prevents one object passing closely by another toward its globally optimum location.

We can consider the set of objects being mapped as a sub-population or group of individuals whose misfit function is evaluated for the group rather than for the individual. Using a biological analogy, a colony of social animals (such as a coral colony or a beehive) may be considered either as a collection of individuals or as a single individual. If we view the objects as a set of individuals, then each individual's parameter set comprises its identifier 'i' plus its set of coordinates. Inter-object crossover is then equivalent to a standard single point crossover, producing two new objects, each getting its identifier from one parent object and its coordinates from the other.

9.3.3 Selection Operators

We have considered how each generation may be created from parents, by various forms of mutation, crossover or combinations thereof. It remains to be considered how we should select which members of each generation to use as the basis for creating the following generation.

A fundamental principle of genetic algorithms is that the most fit members should breed. Many selection procedures have been implemented. It would appear preferable to avoid selection methods where a simple rescaling of the fitness function would greatly change the selection probabilities.

Procedures based on rank have the advantage of not being susceptible to the scaling problem. One approach is to assign a selection probability that descends linearly from the most fit member (with the smallest misfit value) to zero for the least fit member (with the largest misfit value). Tournament selection can achieve this effect without the need to sort or rank the members. A pair of members are selected at random, and the most fit of the pair is used for breeding. The pair is returned to the potential selection pool, a new pair selected at random, the best one used for breeding, and so on until enough breeders have been selected. The selection with replacement process ensures that a single individual can be selected multiple times. This procedure is equivalent to ranking the population and giving them selection probabilities linearly related to rank, as shown in the following proof:

- Consider 'm' members, ranking from $r = 1$ (least fit, with highest Y) to $r = m$ (most fit, with lowest Y)
- Each member has the same chance of selection for a tournament, a chance equal to $2/m$.
- But its chance of winning is equal to the chance that the other selected member has lower rank, a chance equal to $(r-1)/(m-1)$.
- So $P(win) = 2(r-1)/[m(m-1)]$, which is linear with rank.

In selecting members of the next generation, it would appear unwise to lose hold of the best solution found in the previous generation. For this reason, an 'elitist' selection procedure is often employed, with the 'best yet' member of each generation being passed on unaltered into the next generation (in addition to receiving its normal chance to be selected for breeding).

9.3.4 Design and Use of a Genetic Algorithm for Multidimensional Scaling

To investigate some of the issues that have been discussed, a genetic algorithm program was designed, using the simulation package Extend, which is written in C. The algorithm has been used to fit the inter-object distances of ten cities in the United States. This example has been chosen because it is also used as a worked example in the SPSS implementation of the standard multidimensional scaling procedure Alscal (Norusis, 1990, pp. 397-409). The data as given there are shown in Table 9.2.

	Atlanta	Chicago	Denver	Houston	L.A.	Miami	N.Y.	S.F.	Seattle	D.C.
Atlanta	0	587	1,212	701	1,936	604	748	2,139	2,182	543
Chicago	587	0	920	940	1,745	1,188	713	1,858	1,737	597
Denver	1,212	920	0	879	831	1,726	1,631	949	1,021	1,494
Houston	701	940	879	0	1,374	968	1,420	1,645	1,891	1,220
L. A.	1,936	1,745	831	1,374	0	2,339	2,451	347	959	2,300
Miami	604	1,188	1,726	968	2,339	0	1,092	2,594	2,734	923
N. Y.	748	713	1,631	1,420	2,451	1,092	0	2,571	2,408	205
S. F.	2,139	1,858	949	1,645	347	2,594	2,571	0	678	2,442
Seattle	2,182	1,737	1,021	1,891	959	2,734	2,408	678	0	2,329
D.C.	543	597	1,494	1,220	2,300	923	205	2,442	2,329	0

Table 9.2: Inter-City Flying Mileages (after Norusis, 1990, p. 399)

On the reasonable assumption that the expected variance of any measured distance is proportional to the magnitude of that distance, the misfit (or stress) function to be minimised was expressed as the average of the squared misfits, each divided by the measured inter-city distance. The elements d_{ij}^* representing the fitted distances and d_{ij} the measured distances:

$$\text{Misfit Function} = Y = \text{Average}[(d_{ij}^* - d_{ij})^2 / d_{ij}] \quad (9.8)$$

This is equivalent to the misfit function used in equation (3) above, but expressed as an average rather than as a sum, to aid interpretation.

The genetic algorithm in Extend was built with a control panel, as shown in Figure 9.4. It was designed so that the inter-object distances could be pasted into the panel, and the results copied from the panel. The control panel permits specification of how many objects and dimensions are to be used, and whether the optimisation is to be by systematic hill descent, or to use the genetic algorithm. If the genetic algorithm is being used, then the population size can be specified, together with how many members are to be subjected to each type of genetic operator. The allowed genetic operators, discussed in the previous sections, include:

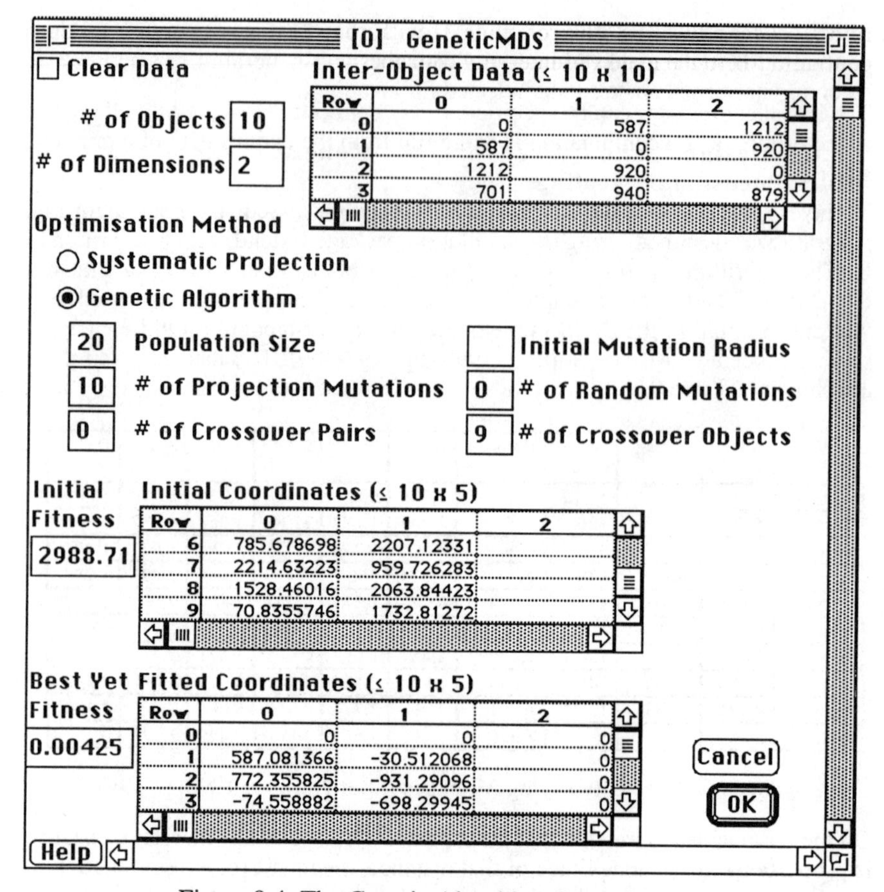

Figure 9.4: The Genetic Algorithm Control Panel

- 'Projection Mutation' of a randomly selected object along a randomly selected dimension, to the quadratic optimum, if the misfit function is upwardly concave for this locality and direction. If the function is downwardly concave, the projection is downhill to the reflection of the quadratic fit maximum, as shown in Figure 9.3.

- 'Random Mutation' of a randomly selected object along a randomly selected dimension, by an amount randomly selected from a Gaussian distribution. The Gaussian distribution has a zero mean, and a standard deviation set by a 'Mutation Radius', which shrinks in proportion to the root mean square misfit, as convergence is approached.

- Standard 'Crossover Pairing' where each offspring takes the coordinates of each object from one of its two parents (the source parent being selected at random for each object).

- 'Crossover Objects' where an offspring is created from a single parent by interchanging the coordinate set of a randomly selected pair of objects.

Figure 9.4 shows the control panel for a run, fitting an initial random configuration to the matrix of inter-city distances.

The initial coordinates can be specified, if it is not desired to start with all objects at the origin, or if a continuation is being run from the ending state of a previous run.

As the run progresses, the best fitting solution yet found (lowest Y value) is reported in the 'fitted coordinates' table. This solution is preserved as a member of the new generation. The parents of the new generation are selected by pairwise tournament selection, which we have seen is equivalent to ranking the population and giving them selection probabilities linearly related to rank.

The C language coding for the program is listed at the end of this chapter, and is supplied on the disk accompanying this book.

9.4 Experimental Results

9.4.1 Systematic Projection

The program was run first using systematic projection, with only a single population member, projected to the quadratic minimum once for each parameter in each iteration. Since the ten cities were being plotted in two dimensions, there were twenty projections each iteration. The fitting was repeated for ten different starting configurations, each randomly generated by selecting each coordinate from a uniform distribution in the range zero to 2,000 miles. The results for the ten runs are plotted in Figure 9.5.

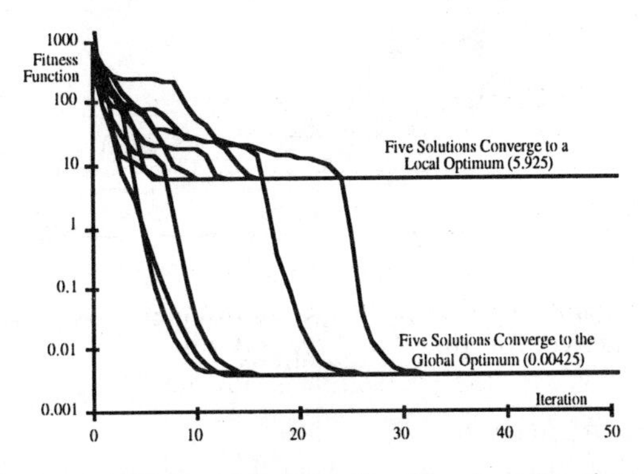

Figure 9.5: Systematic Projection from Ten Random Starting Configurations

It can be seen from Figure 9.5 that half the solutions converged to the global minimum, with the misfit function equal to 0.0045, but that the other five solutions became trapped on a local optimum, with the misfit function equal to 5.925.

Since the misfit function, Y, of equation (8), is the average of the squared error divided by the inter-city distance, the global minimum corresponds to a believable standard error of plus or minus one mile in a distance of 220 miles, or 2.1 miles in a thousand miles distance. The local optimum corresponds to an unbelievably high standard error of 77 miles in a thousand miles inter-city distance.

9.4.2 Using the Genetic Algorithm

The genetic algorithm was used on the same set of ten starting configurations. For the genetic algorithm (as shown in the control panel of Figure 9.4) a population size of twenty was used. An elitist policy was used, with the best member of the previous generation being retained unaltered in the next. Nineteen tournament selections were made from the previous generation for breeding each new generation. Ten new generation members were created from their parents by a projection mutation (along one randomly selected dimension for one randomly selected city), and for the remaining nine members, a randomly selected pair of cities were interchanged.

Figure 9.6 shows that the genetic algorithm brought all ten starting configurations to the global optimum, even in the five cases where the systematic projection had resulted in entrapment on a local optimum.

As is commonly the case with genetic algorithm solutions, the reliability of convergence on the global optimum is bought at the cost of a greater number of computations.

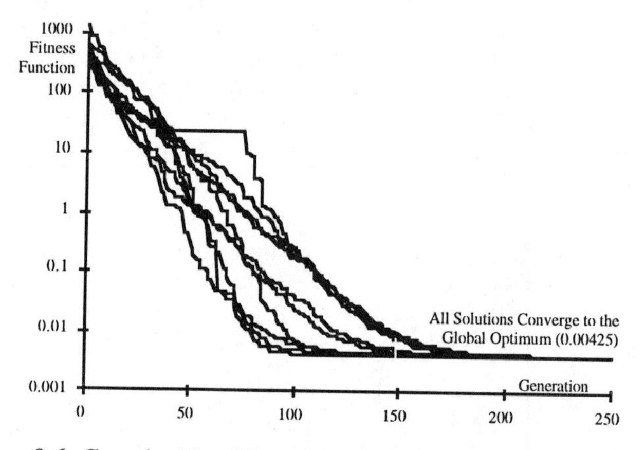

Figure 9.6: Genetic Algorithm Using the Same Ten Random Starting Configurations

9.4.3 A Hybrid Approach

A hybrid approach that can greatly reduce the computation effort is to use a starting configuration that has been obtained by a conventional method, and home in upon the global optimum using the genetic algorithm. This hybrid approach is illustrated in Figure 9.7.

The eigen values were extracted from the vector product transformation of **D**, as shown in equation (2) above. The vector product transformation is constructed by squaring the d_{ij} elements, subtracting the row and column means and adding the overall mean of this squared element matrix, and finally halving each element (see Schiffman et al., 1981, p. 350). Figure 9.7 shows that the genetic algorithm was able to converge the eigen solution to the global optimum in about 130 generations, only a moderate improvement upon the 150 to 200 needed for the random initial configurations. A much quicker convergence, in about 30 generations, was obtained using the Alscal solution in the SPSS computer package. It will be recalled, as discussed above, that the Alscal solution optimises a different misfit function, the s-stress, instead of the proportional error variance of equation (8). Consequently, it is to be expected that the two different misfit functions will have differing optimal solutions. The statistically inappropriate Alscal solution gives a convenient starting point for the genetic algorithm to approach the global optimum of the statistically appropriate misfit function.

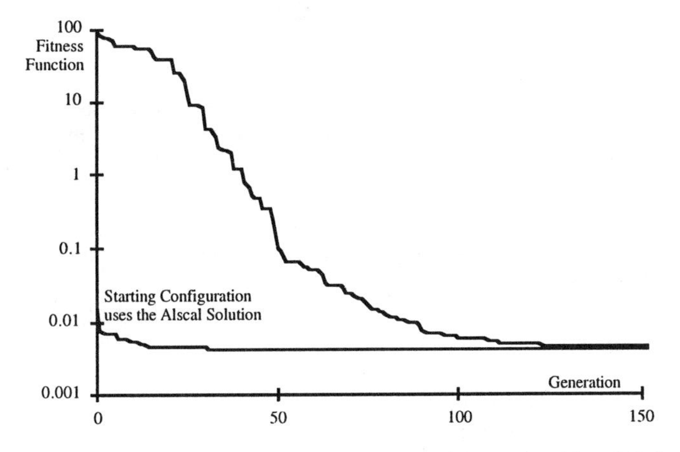

Figure 9.7: Starting from Eigen Vectors and from the Alscal Solution

Further investigations have been run, using standard genetic operators of random mutation and crossover of pairs of solutions, as described earlier. The same set of ten starting configurations were used as for Figures 9.5 and 9.6. The standard operators gave slower convergence than our projection mutation and object crossover operators, and exhibited entrapment on a local minima in the same cases as did the systematic downhill projection of Figure 9.5. However, the possibility remains that the most efficient algorithm may need to be built from a combination of our modified operators with the standard genetic operators. The interested reader is invited to experiment, using and adapting the computer software provided.

9.5 The Computer Program

9.4.1 The Extend Model

The computer program was written using the simulation package Extend, which is coded in a version of C. The Extend model and library files are included in the disk supplied with this volume. A text listing of the program block is also included. Users without Extend but some knowledge of C or C++ will be able to implement the program with little alteration.

Figure 9.8 shows the layout of the Extend model. It comprises a single program block, 'HybridMDS', connected to the standard library plotter, which collects and displays spreadsheet and graphical output for each computer run.

Each simulation step in Extend corresponds to one generation of the genetic algorithm.

Figure 9.8: The Extend Model

The program block can be double clicked to open up and display the control panel of Figure 9.4. As is standard with Extend, option-double-click on the program block displays the code listing of the block, in a form of C. This listing is given below.

9.5.2 Definition of Parameters and Variables

9.5.2.1 Within the Control Panel (Dialog Box)

A number of parameters are defined within the control panel or dialog box of Figure 9.4. These are:

- ClearData — Clicked if the control panel is to be cleared
- NumObj — The number of objects to be mapped
- NumDim — The number of dimensions to be mapped
- Data — Inter-object source data (NumObj by NumObj)
- Xopt — The number of dimensions to be mapped

You can choose to use systematic projection or the genetic algorithm by clicking one of:

- SystProj — To use systematic projection
- GenAlg — To use the genetic algorithm

If you choose to use the genetic algorithm, you should specify:

- NumPop — The number of population members in each generation
- NumRandProj — The number of members created by random projection
- NumCross — The number of pairs created by crossover
- MutRad — The initial mutation radius
- NumMut — The number of members created by random mutation
- NumCrossObj — The number of members created by object crossover

An initial configuration should be entered (random, eigen vectors or Alscal solution)

- Xinit — The initial coordinate configuration (NumObj by NumDim)

The program reports into the control panel:

- Avinit — The initial average misfit value

And at each generation, updates:

- Xopt The coordinate configuration of the best solution so far
- Avopt The average misfit value of the best solution so far

9.5.2.2 To the Library Plotter

The program block also has four connectors to the library plotter

- Con0Out The average misfit value of the best solution so far (Avopt)
- Con1Out =Y[0] = Best total misfit so far = Avopt x NumObj x NumObj
- Con2Out =Y[1] } Two more total misfit values from
- Con3Out =Y[2] } members of the current generation

9.5.2.3 Variables and constants set within the Program Listing

The following variables and constants are set within the program listing:

```
integer m, i, j, k, d, MaxObj, MaxDim, BlankRow, BlankCol,
        MaxPop, NumObjSq;

real Diff, Total, TotalSum, DX[20][20], X[][20][5], Y[],
        Xold[][20][5], Yold[];

real Yopt, Yinit, Y0, Y1, Y2, DelX, DelX2, Temp,
        LogSqData[20][20];

constant AllowObj is 10;  constant AllowDim is 5; constant
        Increment is 100;
```

9.5.3 The Main Program

The main program comprises three Extend calls.

The first is activated when the control panel is closed, and checks that the data are valid:

```
on DialogClose { CHECKVALIDATA();}
```

The second acts at the start of a simulation, checks for valid data, and initialises the simulation:

```
on InitSim{CHECKVALIDATA();TotalSum/=NumObj*NumObj;DelX=TotalSum
  /Increment;DelX2=2*DelX; INITIALISE();}
```

The third is activated at the each step of the simulation, and simulates one generation of the genetic algorithm (or one sequence of the systematic projection, if that is being used):

```
on Simulate {if(SystProj) {m=0; for i=0 to MaxObj for d=0 to
MaxDim DESCEND();} else

 {TOURNELITE(); m=1; for k=1 to NumRandProj RANDPROJ(); for k=1
  to NumCross CROSSOVER();
```

```
for k=1 to NumCrossObj CROSSOBJ(); MutRad=Sqrt(Avopt*1000);
for k=1 to NumMut MUTATE(); }

XYoptGET(); Avopt=Yopt/NumObjSq; Con0Out=Avopt; if(NumPop>1)
Con1Out=Y[0];

if(NumPop>2) Con2Out=Y[1]; if(NumPop>3) Con3Out=Y[2];}
```

9.5.4 Procedures and Functions

The main program calls upon several procedures. To make the program operation easier to follow, they will be listed here in the order in which they are called.

In the actual program listing, any procedure or function which is called must have already appeared in the listing, and therefore the listing order will not be the same as shown here.

9.5.4.1 CHECKVALIDATA()

Checks the input data for internal consistency.

```
Procedure CHECKVALIDATA()

{if(SystProj) NumPop=1; if(ClearData) {NumObj=0; NumDim=0; for
i=0 to AllowObj-1

{for j=0 to AllowObj-1 Data[i][j]=0; for d=0 to AllowDim-1
Xopt[i][d]=0; }

ClearData=0; usererror("Data Cleared: Object Data Needed");
abort;}

if((NumObj>AllowObj)OR(NumDim>Min2(AllowDim,NumObj-
1))OR(NumDim<1)OR(NumObj<2))

{usererror("Error: You must set the number of Objects in the
range 2 to "+AllowObj

+"  and the number of Dimensions in the range 1 to
"+AllowDim+", but less than the number of Objects"); abort;}

MaxObj=NumObj-1;  MaxDim=NumDim-1;  TotalSum=0; ** CHECK FOR
BLANK ROWS OR COLUMNS

BlankRow=0; for i=0 to MaxObj {Total=0; for j=0 to MaxObj
Total+=Data[i][j]; TotalSum+=Total; if(Total==0) BlankRow+=1;}

BlankCol=0;  for i=0 to MaxObj {Total=0; for j=0 to MaxObj
Total+=Data[j][i]; if(Total==0) BlankCol+=1;}

if(BlankRow+BlankCol>0) {usererror(BlankRow+" Rows Total Zero
"+BlankCol+" Columns Total Zero"); abort;}
```

```
Temp=0; for i=0 to MaxObj for d=0 to MaxDim
Temp+=realabs(Xinit[i][d]);

if (!((TotalSum>0)&&(NumObj>0)&&(NumDim>0)&&(NumPop>0)))
{usererror("Blank Data"); abort;};

if(GenAlg) if (!(MutRad>0)) {usererror("Set Mutation Radius");
abort;};

if (!SystProj) if(NumPop<=NumRandProj+2*NumCross+NumCrossObj
+NumMut) {usererror("Increase NumProj"); abort;}}
```

9.5.4.2 INITIALISE()

This initialises all the first generation to the initial configuration, entered in the control panel.

```
Procedure INITIALISE()

{for i=0 to MaxObj for j=i+1 to MaxObj LogSqData[i][j]
=2*Log(Data[i][j]);

MatCopy(Xopt,Xinit,NumObj,NumDim);

for m=0 to MaxPop for d=0 to MaxDim for i=0 to MaxObj
X[m][i][d]=Xinit[i][d];

m=0; Yinit=EVALUATE(); for m=0 to MaxPop Y[m]=Yinit; Yopt=Yinit;
NumObjSq=(NumObj*MaxObj)/2;

Avinit=Yinit/(NumObjSq); Avopt=Avinit; MakeArray (X,NumPop);
MakeArray (Y,NumPop);

MakeArray (Xold,NumPop); MakeArray (Yold,NumPop); MaxPop=NumPop-
1;}
```

9.5.4.3 EVALUATE()

This function evaluates the misfit Y for the mth population member. The misfit Y is the sum of the squared errors each divided by the inter-object distance, as defined in equation (9.3) above.

```
real EVALUATE() {integer ip, jp, dp; real Y; Y=0; for ip=0 to
MaxObj for jp=ip+1 to MaxObj

{DX[ip][jp]=0; for dp=0 to MaxDim DX[ip][jp]+=(X[m][ip][dp]-
X[m][jp][dp])^2; DX[ip][jp]=SQRT(DX[ip][jp]);

Y+=((DX[ip][jp]-Data[ip][jp])^2)/Data[ip][jp];} return(Y);}
```

9.5.4.4 DESCEND()

This procedure projects the solution Y[m] = fn(X) along the dth dimension of the ith object, to its quadratic minimum, as illustrated in Figure 9.3.

```
Procedure DESCEND(){Y1=Y[m]; X[m][i][d]-=DelX; Y0=EVALUATE();
X[m][i][d]+=DelX2; Y2=EVALUATE();

 X[m][i][d]-=DelX; Diff=(Y0+Y2)/2-Y1; if(Diff<>0) Temp=Delx*(Y0-
  Y2)/Realabs(4*Diff); X[m][i][d]+=Temp;

 Y[m]=EVALUATE(); if(Y[m]>Y0) {Y[m]=Y0; X[m][i][d]-=Temp;} }
```

9.5.4.5 TOURNELITE()

Here we preserve the best solution yet, then choose the rest of the breeders for the next generation by tournament contest of randomly selected pairs. A pair of the previous generation are chosen at random, and the better of the two is used for breeding.

```
Procedure TOURNELITE() {integer mp, mq, BestYet;

 for m=0 to MaxPop {Yold[m]=Y[m]; for i=0 to MaxObj for d=0 to
  MaxDim Xold[m][i][d]=X[m][i][d];}

 BestYet=0; for m=1 to MaxPop if (Y[m]<Y[BestYet]) BestYet=m;

  if (BestYet>0) for i=0 to MaxObj for d=0 to MaxDim
  {X[0][i][d]=X[BestYet][i][d]; Y[0]=Y[BestYet];}

 for m=1 to MaxPop {mp=Random(NumPop); mq=Random(NumPop); if
  (Yold[mq]<Yold[mp]) mp=mq;

    Y[m]=Yold[mp]; for i=0 to MaxObj for d=0 to MaxDim
  X[m][i][d]=Xold[mp][i][d];}}
```

9.5.4.6 RANDPROJ()

This procedure selects a random object, then projects it to the quadratic optimum along each coordinate.

```
Procedure RANDPROJ() {i=RANDOM(NumObj); for d=0 to MaxDim
DESCEND(); m++;}
```

9.5.4.7 CROSSOVER()

This procedure assigns the objects of two parent solutions at random to two members of the next generation.

```
Procedure CROSSOVER() {integer ip, dp; real Dum;

for ip=0 to MaxObj if(RANDOM(2)) for dp=0 to MaxDim

     {Dum=X[m][ip][dp]; X[m][ip][dp]=X[m+1][ip][dp];
  X[m+1][ip][dp]=Dum;}

  Y[m]=EVALUATE(); m++; Y[m]=EVALUATE(); m++; }
```

9.5.4.8 CROSSOBJ()

Coordinates of two randomly selected objects are interchanged for a member of the population.

```
Procedure CROSSOBJ() {integer ip, jp, dp; real Dum;

ip=RANDOM(NumObj); jp=RANDOM(NumObj); for dp=0 to MaxDim

  {Dum=X[m][ip][dp];X[m][ip][dp]=X[m][jp][dp]; X[m][jp][dp]=Dum;}

  Y[m]=EVALUATE(); m++;}
```

9.5.4.9 MUTATE()

One randomly selected object is mutated a random distance (normally distributed, with standard deviation equal to the mutation radius). The mutation radius is revised each generation in the 'on Simulate' call, so that it contracts at the same rate as the misfit function shrinks.

```
Procedure MUTATE()

   {i=RANDOM(NumObj);   for   d=0   to   MaxDim
  X[m][i][d]+=Gaussian(0.0,MutRad); Y[m]=EVALUATE(); m++;}
```

9.5.4.10 XYoptGET()

Copies the best yet solution (m=0) into Yopt and Xopt.

```
Procedure XYoptGET() {integer ip, dp; Yopt=Y[0]; for dp=0 to
MaxDim for ip=0 to MaxObj Xopt[ip][dp]=X[0][ip][dp];}
```

9.5.5 Adapting the Program for C or C++

If Extend is not available to the user, then the following revisions are needed to implement the program in C or C++. The program can also be translated into Pascal without major change.

9.5.5.1 Substitution for the Control Panel Input and Output

The following parameters (described in Section 9.4.2.1) should be read in, either from an input file such as a text spreadsheet, or by prompted request on the screen:

```
NumObj, NumDim, Data, Xopt, SystProj, GenAlg, NumPop,
NumRandProj, NumCross        , MutRad, NumMut, NumCrossObj, Xinit
```

Avinit can be reported to the screen at the start of the run.

Xopt, Avopt can be output for each generation, to a text file.

9.5.5.2 Substitution for the Library Plotter

Con0Out, Con1Out, Con2Out, Con3Out can be output for each generation, to a text file.

9.5.5.3 Changes to the Main Program

The commands within the three Extend calls will need to be incorporated into a main program.

on DialogClose and on InitSim commands would be unaltered;

on Simulate commands would be placed within a do-loop, with each iteration corresponding to one generation. The extent of the do-loop would be set to the number of generations required.

9.6 Using the EXTEND Program

To use the Extend program supplied requires version 2 or later of Extend (Imagine That, Inc.) The model itself (as shown in Figure 9.8) is in the file 'GeneticMDS'. It requires the library file 'GeneticMDSLib' to supply the two blocks 'HybridMDS' and the modified plotting routine 'Plotter, I/O'.

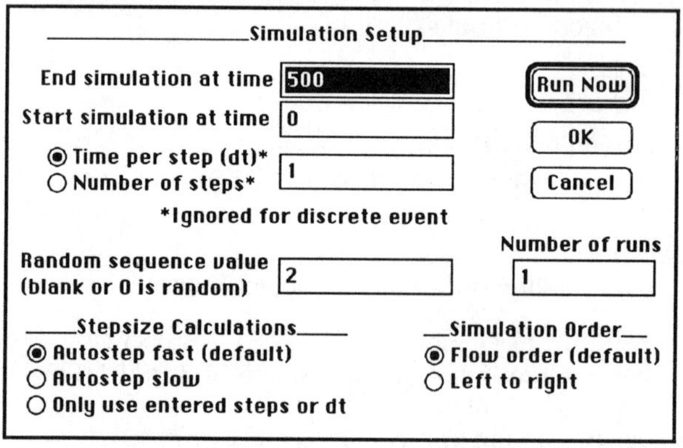

Simulation Setup
End simulation at time 500
Start simulation at time 0
Time per step (dt)* 1
Number of steps*
*Ignored for discrete event
Run Now
OK
Cancel
Random sequence value (blank or 0 is random) 2
Number of runs 1
Stepsize Calculations
Autostep fast (default)
Autostep slow
Only use entered steps or dt
Simulation Order
Flow order (default)
Left to right

Figure 9.9: The Extend Simulation Setup Screen

Once the model file 'GeneticMDS' has been opened, the control panel (as seen in Figure 9.4) can be opened by double-clicking on the 'HybridMDS' block. Starting parameters can then be entered into the control panel. This can be done by directly typing them in. Other parameters, such as the inter-object distances or the initial solution may be better entered by copying and pasting them from a spreadsheet or other source file.

The Simulation Setup panel shown in Figure 9.9 can be brought down from the Run menu in Extend. The parameters should be entered as shown in Figure 9.9, except that the 'End simulation at time' box can be changed if more or less than 500 generations are required.

The program can then be run. The 'Plotter, I/O' block can be double-clicked to monitor progress of the simulation run.

References

Davies, P.M. & Coxon, A.P.M., (eds) (1982). *Key Texts in Multidimensional Scaling.* London: Heinemann.

Davis, L.(ed.) (1991). *Handbook of Genetic Algorithms.* New York: Van Nostrand Reinhold.

Everett, J.E. & Pecotich, A. (1991). A combined loglinear/MDS model for mapping journals by citation analysis. *Journal of the American Society for Information Science, 42,* 405-413.

Fienberg, S.E. (1980). *The Analysis of Cross-Classified Categorical Data.* 2nd ed. Cambridge, Mass.: The MIT Press.

Goldberg, D.E. (1989). *Genetic Algorithms in Search, Optimization and Machine Learning.* Beverly Hills: Sage.

Kruskal, J.B. (1964). Nonmetric multidimensional scaling: A numerical method. *Psychometrika, 29,* 115-129.

Kruskal, J.B. & Wish, M. (1978). *Multidimensional Scaling.* Beverly Hills: Sage.

Norusis, M.J. (1990). *SPSS® Base System User's Guide.* Chicago: SPSS.

Rothkopf, E.Z. (1957). A measure of stimulus similarity and errors in some paired-associate learning tasks. *Journal of Experimental Psychology, 53,* 94-101.

Schiffman, S., Reynolds, M.L. & Young, F.W. (1981). *Introduction to Multidimensional Scaling.* New York: Academic Press.

Shepard, R.N. (1963). Analysis of proximities as a technique for the study of information processing in man. *Human Factors, 5,* 33-48.

Wilkinson, L. Hill, M. & Vang, E. (1992). SYSTAT: Statistics. Evanston, IL: SYSTAT.

Wright, A.H. (1991). Genetic algorithms for real parameter optimization, in Rawlins G.J.E., (ed.) *Foundations of Genetic Algorithms.* San Mateo, CA: Morgan Kaufman, 205-218.

Chapter 10

A.E. Eiben
P.-E. Raué
Zs. Ruttkay

Artificial Intelligence Group
Department of Mathematics and Computer Science
Vrije Universiteit Amsterdam
De Boelelaan 1081a
1081HV Amsterdam, The Netherlands

{gusz, peraue, zsofi}@cs.vu.nl

How to Apply Genetic Algorithms to Constrained Problems

0-8493-2519-6/95/$0.00 + $.50

Abstract

Traditional GAs are primarily applicable for problems where there is no interaction (epistasis) between the variables of a candidate solution. However, the most natural formulation of many practical problems involves constraints. To solve these problems using a GA both the problem and the (traditional) GA have to be adapted. In this chapter we divide search problems into three classes: free optimisation problems, constrained optimisation problems and constraint satisfaction problems. We discuss how to transform a search problem to a GA-suited counterpart problem by incorporating some constraints in an objective function and leaving others as constraints. In general, the resulting problem is a constrained optimisation problem, hence we consider different possibilities for solving constrained optimisation problems by GAs. We review different representations (types of chromosomes), genetic operators and suitable ways to define the fitness function. A crucial issue is to guarantee that genetic operators preserve the constraints that are not incorporated in the fitness function. Additionally, the operators should produce children that tend to satisfy those constraints that are in the fitness function, thus not maintained by the representation and the genetic operators. To this end, the random mechanism common in classical operators can be replaced by some heuristic-based construction mechanism. Search heuristics applied in traditional constraint satisfaction techniques can be used for this purpose. We elaborate and demonstrate a GA-based approach to 4 constraint satisfaction problems, the traffic-lights, N-queens, graph 10.3 colouring and the Zebra problem, and one constrained optimisation problem, the travelling salesman. In the appendix we give the C code of a simple GA environment which can be used to define the non-classical GAs discussed in the chapter.

10.1 Introduction

The main goal of this chapter is to describe the applicability of genetic algorithms (GAs) to solve constraint satisfaction problems (CSPs) and constrained optimisation problems (COPs). Our main question is whether GAs can be tailored to solve constrained problems and whether GAs provide a competitive technique for constraint satisfaction. Our original motivation for this study comes from both the CSP and the GA field. Since a general CSP is known to be NP-complete [Mac77], there is a lot of interest in new, non-classical stochastic methods which can be used to solve difficult CSPs. On the other hand, since problems where constraints play an essential role have the common reputation of being GA-hard [DavL85, DavY91], the modification of classical GAs so that they can cope with such problems is a significant extension of the scope of applicability of the evolutionary principle.

The basic idea of our approach is to use genetic operators that take constraints into account. Strictly speaking an operator can take constraints into account in two ways:

(i) it ensures that some of the prescribed constraints always hold, and

(ii) it directs the search by applying heuristics in such a way that the rest of the constraints will be satisfied.

Note that an operator need not have both features. Inclusion of heuristics in genetic operators can be done based on information which can be elicited by statistical analysis on the available candidates or by including search heuristics common in traditional CSP solving methods. In both cases we obtain *heuristic operators*. GAs applying such heuristic operators will be called *heuristic GAs* (HGAs). The mechanism which chooses values for the variables of a child-candidate in a heuristic genetic operator can be either deterministic or probabilistic, making the choice based on some evaluation of the possibilities. An interesting possibility is to take the structure of the network of constraints into account and to treat values not individually, but according to certain *masks* representing the structure of tightly constrained subproblems. In genetic terminology this means that certain variables within a candidate are grouped and treated together. The masking principle can be applied alone, or in combination with some heuristics.

In HGAs the other essential components of classical GAs, mutation, the probabilistic selection of parents and survivors and other components, are not changed essentially. The random and heuristic-based components are used to counterbalance each others deficiencies. That is, the application of heuristics can improve the performance of the traditional random mechanism and the random components can compensate the strong bias introduced by the heuristics. A natural expectation is that once good heuristics are incorporated, an HGA will outperform traditional GAs. A natural question from the CSP side is if GAs will outperform dedicated, classical heuristic search methods on certain constrained problems. We will give examples and explanations of different outcomes when discussing the case studies. In general, our test results justify our basic idea of HGAs, benefiting from heuristic genetic operators, genetic diversity and fitness-based selection.

In section 10.2 an introduction of CSPs, the limitations of classical deterministic solution methods and an overview of recent, non-deterministic solution methods are given. In section 10.3 we give a classification of search problems and discuss the abstract applicability of GAs to these problems. In section 10.4 the main issues of defining GAs for solving constrained problems are discussed: different representations, (heuristic) genetic operators and the definition of the evaluation function. This part includes classification schemes of different genetic operators. In section 10.5 we discuss the five case studies one by one, quoting the most important results of the experiment. We summarise our work in section 10.6. The appendix (at end of chapter) contains the C-code to a GA which was used to run some of the experiments described in this chapter. This C-code can be obtained by anonymous ftp from ftp.cs.vu.nl, /pub/papers/ai/GA/.

10.2 A CSP perspective

This section contains a general introduction into CSPs, for a more complete overview see [Mes89].

10.2.1 Concepts and terminology

In general, a constraint satisfaction problem is stated as the problem of finding an instantiation of a set of variables in such a way that prescribed relations, i.e. constraints, on (some of) the variables hold. We assume that the variables involved can only take on a finite number of different values. In specific papers this problem is referred to as the *consistent labelling problem*, and the term CSP is kept for any, general case. In our theoretical discussion we will only consider CSPs where all constraints are binary, meaning that each constraint refers to two variables. It was shown that for any non-binary CSP an equivalent binary CSP can be constructed [Nud83], hence our assumption is not restrictive.

We will denote a CSP with **N** variables and **K** binary constraints as $\mathbf{C} = \langle [D_1, ..., D_N], [c_{i_1 j_1}, ..., c_{i_K j_K}] \rangle$ where $\mathbf{D_i}$ is a finite set of positive integers, called the *domain* of the i-th variable. The constraint $\mathbf{c_{ij}}$ refers to the **i**-th and the **j**-th variables. For our discussion it is sufficient to assume that each constraint is given as a binary relation, i.e. $c_{ij} \subseteq D_i \times D_j$. For the sake of convenience, we will use the term constraint for the relation itself (the pairs of values for which it is satisfied) as well as for the Boolean function that corresponds to it. Finding a solution for a given **C** means finding an instantiation $[v_1, ..., v_N] \in D_1 \times ... \times D_N$ so that all constraints are satisfied. Note that a (binary) constraint $\mathbf{c_{ij}}$ restricts only the value on the **i**-th and **j**-th positions. For each pair of values **[p, q]** satisfying $\mathbf{c_{ij}}$, a subset of the search space can be assigned, containing all candidates which have **p** on the **i**-th and **q** on the **j**-th position, that is the cylindrical set $D_1 \times ... \times D_{i-1} \times \{p\} \times D_{i+1} \times ... \times D_{j-1} \times \{q\} \times D_{j+1} \times ... \times D_N$. The set of all the candidates satisfying a given constraint is:

$$C_{ij} = \cup\{D_1 \times ... \times D_{i-1} \times \{p\} \times D_{i+1} \times ... \times D_{j-1} \times \{q\} \times D_{j+1} \times ... \times D_N \mid [p, q] \in c_{ij}\}.$$

The assumptions on binary constraints are made only to simplify the discussion. The concepts we will discuss can easily be generalised for any CSP with finite domains.

The majority of CSP solving algorithms, which we will refer to as *classical* ones, are *deterministic* and *constructive* search algorithms [Mes89]. That is, a member of $D_1 \times ... \times D_N$ which is a solution is constructed by step-by-step specifying a value for a still uninstantiated variable in such a way that all the constraints which can be evaluated, are satisfied. If there is no such value for the current variable to be instantiated, then previous instantiations are revised (backtracking). The selection of the variable to be instantiated and of the value to be assigned to it are done in a deterministic way, based either on the representation (e.g. variables are instantiated in the order in which they are enumerated) or based on some evaluation of the uninstantiated variables and their possible values. In all cases, the evaluation-based variable and value selection is meant to arrive at a solution in fewer search steps (extensions and backtrackings) than blind, uninformed search. The algorithms differ in the evaluations applied as the basis for heuristics for variable and value selection. The heuristics are based on evaluations of some characteristics of the given CSP (e.g. the number of unsatisfied constraints referring to a variable, the connectivity of the graph representing the still unsolved subproblem, the number of values for other variables excluded by a possible value for the current variable, etc.).

10.2.2 Limitations of classical search methods

It is known that the general CSP is NP-complete [Mac77, Fre78], hence it is unlikely that a classical search algorithm exists which is capable of finding a solution for any CSP in an acceptable time. Much effort has been spent on inventing and investigating different algorithms for effectively solving CSPs with specific characteristics [Dec88, Dec89, Fox89, Fre82, Fre85, Har80]. In spite of the relatively long and fruitful history of CSP solving, there is still an urgent need from demanding application fields (e.g. in the domain of scheduling and design) for effective solution methods, and intensive work is carried out to invent and evaluate new algorithms and heuristics.

Theoretical results on finding the best heuristics for variable and value selection in the above scheme exist [Nud83]. However, these characterisations are only probabilistic, not ensuring that using the suggested heuristics a given problem can be solved efficiently. Some algorithms work well for certain easy CSPs. The different criteria of easiness are rather restrictive, given in terms of the structure of the problem (e.g. lack of cycles in the constraint graph) and/or the lack of implicit constraints (arc-, path-, k-consistency). As CSPs modelling real problems are usually big, have many cycles and implicit constraints, it is matter of trial-and-error to find an efficient CSP solving algorithm for a problem. Hence, generic and robust problem-solving methods (based on some principle other than classical search) are of great theoretical and practical interest.

10.2.3 Heuristic GAs as a promising paradigm

The application of GAs introduces a new way of solving CSPs. In contrast to classical search, the possibilities are not explored in an exhaustive way in some order, but a current set of candidate solutions (candidates) is improved by stochastically selecting one or more parents and using them to generate one or more children by applying a genetic operator. The advantage of a GA is that of genetic diversity: each step in the search process a set of candidates, instead of a single candidate, is considered and more candidates are simultaneously involved in the creation of new candidates. Furthermore, the selection of candidates that are involved in the creation of children is done on the basis of their quality. Hence, a GA explores and exploits the relationship between the inner structure and the quality of candidate solutions automatically. As the so-called *building block hypothesis* states: a GA detects simple patterns within candidates, the so-called building blocks, that contribute to the quality, and composes candidate solutions consisting of high quality building blocks. In theory there is only a limited guarantee that a GA will approximate or find a solution, [Hol92, Eib91], but in practice they show a good performance.

Traditional GAs based on uniform random mechanisms are primarily suited to problems where no interaction between the variables occurs [DavL85, DavY91]. Hence, it is inevitable that classical GAs must be modified to be able to cope with CSPs. The essential modification is that of the representation and the child generation mechanism, the genetic operator. Our approach is to replace the uniform random mechanism of the genetic operator with a mechanism which takes (some of) the prescribed constraints into account. In defining such genetic operators we will adopt heuristics, some of which are also used by classical search methods. Because of the presence of stochastic elements and heuristics within an HGA we hope to have the best of two worlds. In other words, the random and heuristic steps are presumed to counterbalance each other's deficiencies. Although an HGA executes a directed search, it is able to discover promising candidates outside the scope of the search heuristics due to the occasional random steps.

Let us note that in principle a constructive, exhaustive search method will always find a solution (if one exists), whereas GAs cannot guarantee this. Moreover, using a GA it is not possible to conclude that a CSP does not have a solution.

10.2.4 Related works in the CSP field

In the extensive CSP field almost exclusively constructive and deterministic algorithms have been considered. Recently, attempts have been made to apply probabilistic approaches. The first encouraging result is that an algorithm based on random selection performs better on the N-queens problem than depth-first search, where the selections take place according to a fixed order of the variables and values (which can be seen as a bad heuristic) [Bra88]. This result suggests that when not enough information about the search space is available (serving as the basis for some good search heuristics), it is more appropriate to rely on random selection than on poor heuristics. This principle has been proven in a few practical applications, where big and complex CSPs had to be solved. In [Ado90] a stochastic network was used to solve a scheduling problem. Another direction in stochastic CSP solving methods is the application of simulated annealing [Joh91].

In [Min92] they applied a *generate and repair method* in order to solve CSPs by implementing the ideas employed in the network by [Ado90]. This repair method performed much better than traditional, constructive methods. E.g. on the N-queens problem, it could find a solution for any number of queens up to one million very quickly [Min90].

We intend to apply a new combination of classical search and stochastic methods. Some research into exploiting the benefits of the alteration of the two techniques has been done, in both cases for constrained optimisation problems [Par92, Tsa90]. GAs have been applied to the subset sum problems and the minimum tardy task [Khu94a] as well as the zero/one multiple knapsack problem [Khu94b] and to the N-queens problem [Cra92, Hom92]. The incorporation of heuristic information into GAs has already been proposed and applied, [Suh87], but merging general CSP search heuristics is a new possibility devised to overcome the limitations of the single approaches.

10.2.5 Related works on GAs for solving COPs

The COPs that have been investigated the most in the GA literature are the so-called *order-based problems* where individuals are encoded using permutations, used to model routing or scheduling problems. The best known example is the TSP. It is possible to treat the TSP by GAs by designing a 'tricky' binary representation resulting in genotypes that all denote valid phenotypes [DavL85]. A more common approach is that of using order-based representation and defining permutation preserving crossovers [Oli87, Fox91, Sta91]. These studies can generally be applied to any permutation optimisation problem, e.g. sorting problems [Cor93]. The results on TSP can also be applied to solve scheduling problems [Bie90, Cle89, Sys91a, Whi89b, Whi91], although good performance of a certain crossover on the TSP does not guarantee good performance for scheduling. Colorni et al. treat time table problems by using matrix representation and applying row and column manipulations [Col91]. In [Tsa90] COPs are solved by GAs, where forward checking during the crossover guarantees that offspring are allowed. In [Par92] scheduling is treated by allowing wild cards in individuals, and using look ahead techniques. For the case of numerical linear (in)equalities spanning a convex search space of reals special crossovers were defined and applied in [Mic92]. The book also presents a comparison between three approaches based on, respectively, penalty functions, repairing and the special crossovers.

10.3 A GA point of view

GAs have proven to be satisfactory in solving a wide scale of optimisation problems, but in the meanwhile so-called GA-hard problems (where GAs are not particularly successful) are also reported [Gol89, DavL91]. In practice GAs are primarily used *as* function optimisers but it should not be inferred that GAs *are* function optimisers [DeJ93]. Somewhat deviating from DeJong's intended message, we interpret this fact in the following way. Notice, that although GAs are indeed pursuing fit candidates, fitness is not necessarily derived from an objective function to be optimised. What is especially interesting for us is that any heuristic measure to evaluate candidate solutions can be adapted as fitness, if the fittest elements in the search space correspond to solutions of the original problem to be solved, and vice versa.

10.3.1 Classification of search problems and their relation to GAs

In the forthcoming discussion of problem types the term *free search space* stands for a Cartesian product, i.e. in general a free search space is $\mathbf{S} = \mathbf{D_1}$ ¥ ... ¥ $\mathbf{D_N}$, where $\mathbf{D_i}$ are finite sets (domains). A *free optimisation problem* (FOP) is given by a free search space $\mathbf{S}$ and a real valued function $\mathbf{f}$ on $\mathbf{S}$; a solution of an FOP is an element $\underline{\mathbf{x}}$ Œ $\mathbf{S}$, such that $\mathbf{f}(\underline{\mathbf{x}})$ is optimal (minimal or maximal).

As we have discussed in Section 2, in case of a CSP, the problem is to find an instantiation of given variables with finite domains so that all of the constraints are satisfied. In this case we have a search space $\mathbf{S} = \mathbf{D_1}$ ¥ ... ¥ $\mathbf{D_N}$ and constraints $[c_{i_1 j_1}, ..., c_{i_K j_K}]$ to be satisfied, but no function to be optimised. By a *feasible element* we mean an element of the search space that satisfies all constraints, the conjunction of the constraints is therefore called the *feasibility condition.* In these terms an $\underline{\mathbf{x}}$ Œ $\mathbf{S}$ is a solution of a CSP if it is feasible.

A *constrained optimisation problem* (COP) is a mixture of an FOP and a CSP. It is given by a free search space $\mathbf{D_1}$ ¥ ... ¥ $\mathbf{D_N}$, a function $\mathbf{f}$ to be optimised and also by some constraints $[c_{i_1 j_1}, ..., c_{i_K j_K}]$ to be satisfied. We use the terms feasible element and feasibility condition in the same way as for CSPs. A solution of a COP is an element that is feasible (satisfies all constraints) and has an optimal objective function value.

Observe, that each of the problem types described above can be characterised as a triple **(S, j, f)**, where **S** is the free search space, **j** the feasibility condition and **f** a numeric function on **S** called the *objective function.* FOPs, COPs and CSPs are, respectively, characterised by the following schemes: **(S, ., f)**, **(S, j, f)** and **(S, j, .)**, a dot meaning the absence of the given component. In general, it is possible that a problem can be modelled by different problems of the same type or even by search problems of different types. However, we are not interested in this modelling aspect here. The only factor important to us is that a GA needs a fitness function, that is a function to be optimised. This implies that FOPs and COPs are *GA-suited* in the most elementary sense: they have an objective function. Solving a CSP using GAs will require a transformation of the CSP into a *GA-suited counterpart*, i.e. into an FOP or a COP so that solving the FOP (or COP) will yield a solution to the original CSP.

We say that a search problem A with a search space **S** is *equivalent* to a search problem B with a search space **S**, if it holds that for each $\underline{\mathbf{x}}$ **Œ S** which is a solution of A, $\underline{\mathbf{x}}$ is also a solution of B, and vice versa. For a CSP of the form **(S, j, .)**, we can define an equivalent COP **(S, j', f)**, if **j'** is a relaxation of **j** (often a subset of the set of constraints in **j**) and **f** reflects the violation of the constraints not included in **j'**. The objective function **f** can be defined in several ways. The most commonly used technique is that of using so-called penalties, where the violation of each constraint is penalized by a certain weight. According to this approach the objective function $\mathbf{f}(\underline{\mathbf{x}})$ of the COP equals the *penalty function*, that is the sum of the penalties of those constraints that are violated by $\underline{\mathbf{x}}$ **Œ S**:

$$\mathbf{f}(\underline{\mathbf{x}}) = \sum_{\mathbf{k=1}}^{\mathbf{K}} \mathbf{w}_{\mathbf{i_k j_k}} \text{ ¥ } \mathbf{d}_{\mathbf{i_k j_k}}(\underline{\mathbf{x}}),$$

where $\mathbf{d}_{\mathbf{i_k j_k}}(\underline{\mathbf{x}})$ is 1 if $\underline{\mathbf{x}}$ violates constraint $\mathbf{c}_{\mathbf{i_k j_k}}$, otherwise it is 0, the weights of all $\mathbf{w}_{\mathbf{i_k j_k}}$ are non-negative and all weights of the constraints not included in **j'** are positive. Then it will trivially hold that $\underline{\mathbf{x}}$ is a solution of **(S, j, .)** if and only if $\underline{\mathbf{x}}$ is a solution of the COP **(S, j', f)**, provided the CSP has a solution.

It is obvious that for one particular CSP several equivalent COPs can be defined by choosing the subset of constraints incorporated in **j'** and/or defining the objective function that measures the violation of the remaining constraints differently. This means that the GA-suited counterpart of a CSP is not unique. Notice, that FOPs can be perceived as special COPs where there are no constraints. Hence the transformation scheme discussed above also covers transformation of CSPs into equivalent FOPs. Obviously, in the penalty function of such an FOP all the weights are positive.

After transforming a CSP into a GA-suited counterpart (a COP or an FOP) we have to define a GA that can handle this optimisation problem. We say that GA is *applicable for an FOP* (**S**, **.**, **f**) if:

(i) the individuals are elements of the free search space;

(ii) the fitness function is (a monotonic transformation of) the objective function f;

(iii) the genetic operators do not lead out of the search space.

A GA is *applicable for a COP* (**S**, **j**, **f**) if:

(i) the individuals are elements of the free search space;

(ii) the fitness function is (a monotonic transformation of) the objective function f;

(iii) the genetic operators do not lead out of the search space;

(iv) if the n-th population consists of allowed candidates then so does the (n+1)-th.

The basic difference between an FOP and a COP from a GA point of view is that in an FOP every individual (i.e. element of the search space) is allowed, thus classical crossover can be applied freely (see also section 4.1). For COPs this is not the case. What makes it difficult to define an applicable GA for a COP, is clearly (iv): the classical genetic operators satisfy (iii) but do not ensure that if a constraint is not violated by the parents, then it will not be violated by the child(ren).

Let us note that by 'GA-suited' we only mean the fact that an objective function is present in a search problem. In this sense every COP is GA-suited, which does not imply that an applicable GA can be (easily) defined for it. If the constraints in a COP (**S**, **j**, **f**) cannot be maintained by the genetic operators we have at our disposal, then transforming (**S**, **j**, **f**) to an equivalent COP (**S**, **j'**, **f'**) may be needed before we can find a solution for (**S**, **j**, **f**) using a GA. This equivalent COP (**S**, **j'**, **f'**) can be defined by relaxing **j** into **j'**, penalising constraints from **j** \ **j'** and defining a new objective function which includes this penalty function as well as **f**. For the sake of clarity, in the rest of the text we distinguish the feasibility condition used in the context of the original constrained problem (a CSP or a COP) and the constraints in the GA-suited counterpart (if it is a COP). The latter will be called *allowability* and candidates satisfying the allowability condition are called *allowed*. With this terminology the above considerations can be summarised as shown in figure 10.1.

Example 10.3.1.1
The well-known N-queens problem can be formulated as a CSP with $\mathbf{S}=(\mathbf{1}, \ldots, \mathbf{N})^{\mathbf{N}}$ being the set of all configurations of N queens on the chessboard, where $\mathbf{v_i}$,

the i-th value in a string, indicates that there is a queen in the i-th column and on the v_i-th row. The feasibility condition $\{v_i \neq v_j, |v_i - v_j| \neq i - j, 1 \leq i < j \leq N\}$ expresses that there are no two queens on the same row or diagonal. This CSP is equivalent to the COP (**S, j', f'**), where the allowability $j' = \{v_i \neq v_j, 1 \leq i < j \leq N\}$ expresses that a candidate $\underline{\mathbf{x}}$ **Œ S** needs to be a permutation, and **f'** is the total number of diagonal attacks.

Note that applicability of a GA is a formal requirement which says nothing about the performance of a GA, in particular whether a solution will be found or not. In practice we obviously want an applicable GA which performs well in finding a solution for a given CSP. As for the performance, different measures can be maintained. *Effectivity* concerns the quality of the best individual obtained by the GA, *efficiency* indicates the time requirement for finding solutions (speed), and *robustness* concerns the domain of applicability, i.e. the scope of problems that can be solved by a given method.

Original problem	GA-suited counterpart (a COP or an FOP)
CSP - feasibility	**COP /FOP** - objective function = measure of constraint violation[1] - allowability
COP - objective function **f** - feasibility	**new COP/FOP** - obj. function **f'** = **a** ¥ **f** + **b** ¥ measure of constraint violation[2] - allowability

1) For instance, a penalty function as discussed above
2) Here are **a** and **b** non-negative coefficients

Figure 10.1

10.3.2 Basic ideas of maintaining constraints by GAs

From the discussion above it is clear that from a GA point of view there are two basically different possibilities when we transform a given constrained problem into a GA-suited counterpart:

(i) an equivalent FOP is defined;

(ii) an equivalent COP is defined.

The first possibility amounts to putting every constraint into the penalty function. Using this approach when solving COPs has important limitations, [DavL87, Ric89, Mic92]. Intuitively, if more constraints are expressed by one single objective function, then we lose information with regard to the search space because the evaluation of the single elements in the search space is less informative. For instance, we do not know if a candidate is nearly optimal with regard to the original objective function but violates some constraints or is allowed but scores poorly on the original objective function. Besides, the search space to be explored increases as explicit constraints have been discarded. Hence, in case (i) the GA would spend more time on bad candidates than in case (ii), which is expected to reduce its performance. This is justified by our experiments as well as ones reported in [Mic92]. These deficiencies can be reduced if (some) constraints are maintained by the representation and the genetic operators, i.e. option (ii) is applied.

If we represent a constrained problem by a COP as a GA-suited counterpart, then the original set of constraints that define feasibility is divided into two subsets: constraints to be maintained indirectly, respectively directly. The first group is incorporated in the objective function of the GA-suited counterpart. Thus, the fitness function used by the GA expresses the quality of the individuals with respect to the satisfaction of constraints of the first kind. The second group forms the allowability condition of the GA-suited counterpart and to enforce these constraints we need child generation mechanisms preserving allowability. To this end there are three basic possibilities. One could use:

filtering: use 'reckless' genetic operators, check the allowability of each generated child and eliminate children that are not allowed;

repairing: use 'reckless' genetic operators and modify candidates that are not allowed, i.e. post process them in order to make them allowed;

preserving: use such specific genetic operators which produce allowed children from allowed parents.

Filtering is obviously very inefficient because many children are produced which are not allowed. We do not advocate repairing either, because it requires the introduction of an extra repairing operation, [Col91, Müh89]. Nevertheless, as our case study on the Zebra problem shows adding a (simple) repair mechanism to a GA can improve performance. In the rest of the text we put the emphasis on the preserving approach, as also suggested in [Mic91, Mic92].

Summarising, our general idea is to transform a constrained problem (a CSP or a COP) into an appropriate COP for which an applicable GA can be constructed. The difficulty of constructing a GA that is applicable to a COP depends on the constraints. Hence, from the great variety of equivalent representations we will prefer those for which an applicable GA can be defined easily, that is where appropriate genetic operators (preserving the allowability of individuals) are at our disposal or can be easily defined.

10.4 Representations, operators and fitness

The two fundamental problems we face when solving a constrained problem using a GA are to find an equivalent COP and to define a GA which is applicable, effective and efficient on this COP. Applicability is primarily a question of child creation mechanisms, effectivity and efficiency involve questions of different GA models, selection schemes, mutation rates, etc. [Bäc92, Gol91, Sch89, Sys91b, Whi89a]. In this chapter we restrict ourselves to the most crucial issue, that of genetic operators extended with some considerations on the fitness function.

10.4.1 Representations

In the main body of GA literature bit-string representation is used, where the elements of the search space are from the set $\{0,1\}^N$, N being the chromosome length. In principle, 'any' problem can be coded to bit-strings but there are some obvious disadvantages to this. If a certain variable has a domain with a cardinality 2^k, then each value has a unique binary representation. However, in general, when the domain has an arbitrary cardinality we would either have redundancy (i.e. more bit-strings denoting the same value) or meaningless chromosomes (i.e. a genotype having no corresponding phenotype). In the latter case we would need some constraints to specify meaningful individuals and operators respecting these constraints. These constraints, however, do not originate from the problem, but from the representation, thus looking for a more appropriate representation is a natural idea.

The most straightforward way to represent a search space $D_1 \times ... \times D_N$ is to define individuals of length N and require that at the i-th position there are $|D_i|$ possible values. (For the sake of convenience we can assume that the elements of D_i are used as values.) This *domain-based representation* is more transparent than bit-string representation: there is no redundancy in the coding and there are no meaningless individuals to be monitored and eliminated. Furthermore, the classical 1-point (and n-point) and uniform crossover do not lead out of the search space, that is if the parents are from $D_1 \times ... \times D_N$ then so are the children. Let us remark that a search space belonging to a CSP as defined in Section 2 is always finite. In general, however, the concept of domain-based representation does not require that the cardinality of D_i is restricted. The floating point representation in [Mic92] is in fact a domain-based representation with infinite domains and it has been shown to be superior to binary representation for function optimisation. (Let us note that for CSPs with continuous domains the heuristic techniques we discuss in the upcoming sections are not applicable.)

Observe that bit-string and domain-based representations are not derived from constrained problems, they are just different ways in which to represent a free search space. The so called *order-based representation*, however, does originate from studying COPs, in particular routing and scheduling problems, [Oli87, Fox91, Sta91]. In this representation every individual is a permutation of a finite set, e.g. $\{1, ..., K\}$ denoting cities. This can be seen as a domain-based representation with $D_i = \{1, ..., K\}$ extended by an allowability condition requiring that every $j \in \{1, ..., K\}$ occurs exactly once in every individual. During the intensive study of order-based representation quite a few order-based crossovers have been invented (see next section). This offers an advantage: if one decides to apply order-based representation, a whole machinery of crossovers is at one's disposal 'for free'.

Let us note that the essence of order-based representation is its applicability to sequencing. This gives us a special way of applying permutations to constrained problems. We can look upon a permutation as representing the order in which the variables of an individual are assigned values. For instance $\{3,2,1,4\}$ would prescribe that first x_3 is given a possible value then depending on the value of x_3 we assign a value to x_2, etc. Obviously, according to this approach we need a mechanism, called a *decoder*, assigning values to variables and it is the combination of (order-based) candidates and this decoder that connect genotypes to phenotypes. Such a decoder is obviously problem dependent, but there are some general principles which can be used to assign values to the variables. For example, taking the first value which can be assigned to the variable in such a way that there is no conflict with earlier assigned values (if such a value does not exist, one can take the first value from the listed domain). A possible danger of such a rigid, deterministic decoder is, however, that some phenotypes might never be generated. Summarizing, we can say that if a good decoder can be designed then order-based representation can be used even if the problem at hand has nothing to do with sequencing. Although this approach results in an extra computational overhead it is often successfully used for scheduling and, as we discuss in section 5, it resulted in superior GA performance for graph colouring.

As the fourth possibility let us mention *matrix representation* which differs from the previous representations by having non-linear, two-dimensional chromosomes. As discussed at the end of the introduction, in some cases the variables can be divided into groups, so-called masks, and variables within a group can be handled together. If the groups are disjoint and have the same number of elements then it is very natural to represent a candidate as a matrix. For instance, for scheduling or time tabling this is a natural form, [Col91]. It could also be very well applied for our Zebra problem and gave better results then the other approaches we have tested. Obviously, every matrix can be 'linearized' to a string, but using a matrix representation offers advantages:

(i) the concepts of rows and columns make terminology and formalism transparent (e.g. constraints on rows can be easily formulated);

(ii) the matrix view suggests ideas that would not be so natural in a linearized from, e.g. row and column swap operators (see section 5).

To this end let us note that operators working on strings can be applied easily to (the rows, respectively, columns of) matrices too. If there are extra allowability constraints on rows, respectively, columns, then operators suited to that allowability can be used. For instance, if each row must be a permutation then the order-based operators can be applied to the rows of the matrix.

Finally, let us mention that the notion of a constrained problem does depend on the notion of a free search space. Our basic idea was that deciding on $\underline{\mathbf{x}} \in \mathbf{S}$ should be possible by testing $\underline{\mathbf{x}}$ pointwise, i.e. the membership relation of a search space should be simply composed from elementary membership relations and nothing else. Thus, we have defined a free search space in general as a Cartesian product $\mathbf{D_1 \times \ldots \times D_N}$. As a consequence, domain-based representation only reflects a free search space, it has nothing to do with constraints. In the works of Michalewicz, however, chromosomes with a (possibly different) domain for each variable are seen as constrained and genetic operators that do not lead out of $\mathbf{D_1 \times \ldots \times D_N}$ are seen as constraint preserving operators. Such a vision leads to more constrained problems then that of ours. Radcliffe, [Rad92], represents the opposite view when he suggests that the TSP can be seen as an unconstrained search problem in which the search space is the set of all permutations of city labels. His motivation comes from distinguishing constraints coming from the problem and those that are artifacts of the representation, a distinction that we have also made. However, it is hard to put the border between the two kinds of constraints without a rigorous definition of what belongs to the problem. In our framework there are formal definitions of CSPs, COPs, FOPs, yielding the conclusion that TSP is constrained and optimising a function on $\mathbf{D_1}$ ¥ ... ¥ $\mathbf{D_N}$ is not.

10.4.2 Heuristic and multi-parent genetic operators

In this section we introduce our special operators in detail. As mentioned before, we try to boost the performance of a GA by incorporating heuristics in operators and by involving more than two parents in the genetic operator.

10.4.2.1 Sub-individual fitness measures

When solving CSPs using classical search techniques one usually applies measures on variables and possible values for variables to guide the search. Following the standard GA approach complete individuals are estimated by the fitness function. Thus, from a GA point of view such a local measure can been seen as a sub-individual fitness measure giving information on a particular piece within an individual. Next we discuss some of the most frequently used measures in characterising CSPs.

Measures on constraints

The *tightness of a constraint* is the proportion of the number of those instantiations of the variables referred to by the constraint which satisfy the constraint and the total number of instantiations of these variables. Formally,

$$\mathbf{T}(\mathbf{c_{ij}}) := \frac{|\mathbf{c_{ij}}|}{|\mathbf{D_i}| \times |\mathbf{D_j}|}$$

If a partial instantiation $\underline{\mathbf{x}}$ is given, then based on the uninstantiated variables of the constraint, the *tightness of a constraint with respect to a given instantiation* can be defined, and the notation $\mathbf{T}(\mathbf{c_{ij}}, \underline{\mathbf{x}})$ will be used.

Measures on CSPs

In order to estimate the difficulty of a CSP $\mathbf{C}$ = · $[\mathbf{D_1}, ..., \mathbf{D_N}], [\mathbf{c_{i_1 j_1}}, ..., \mathbf{c_{i_k j_k}}]$ Ò, the following measures are common:

- The simplest estimation is the *total number of the constraints*, **TN := K**.

- More informative estimations are the *total tightness of the constraints*:

$$\mathbf{TT} := \sum_{k=1}^{K} \mathbf{T}(\mathbf{c_{i_k j_k}}),$$

• and the *average tightness of the constraints*,

$$\mathbf{TA} := \frac{\mathbf{TT}}{\mathbf{K}}.$$

Another kind of estimation is based on the structure of the problem. One such estimation is the *number of cycles in the constraint graph,* **TC**.

Each measure on CSPs can be defined in a straightforward way for any subset of the constraints for a particular CSP. Typically, in classical constructive methods these measures are defined to estimate the difficulty of the subset of those constraints which are not satisfied. In this case, the *total* and *average tightness of constraints with respect to a given instantiation* can be defined in a straightforward way.

Next we list some common measures used by classical CSP solving techniques for selecting a variable and its possible value, with or without respect to a particular instantiation of the variables. In general, a measure on variables will be denoted by a function **F(i, x̲)** or **F(i)**, where **i** identifies the variable in question, and **x̲** the current instantiation. A measure on values will be denoted by a function **G(i, v, x̲)** or **G(i, v)**, where **i** identifies the variable in question, **v** the value to be evaluated, and **x̲** the current (partial or complete) evaluation.

Measures on variables

An estimation of the difficulty of the i-th variable can be based on the *number of constraints referring to the variable*: **NC(i)**. If a (partial or complete) instantiation **x̲** is given, then the *number of unsatisfied constraints referring to the variable*: **NC(i, x̲)** can be used.

More subtle measures are the *average* or *minimum tightness of the constraints referring to the variable*: **AC(i)**, respectively, **MC(i)**. For any partial instantiation **x̲** the *average* or *minimum tightness of the constraints with respect to the instantiation* can be defined based on the still unsatisfied constraints. These will be denoted by **AC(i, x̲)** and **MC(i, x̲)**, respectively.

Another measure is the *number of possible values* for the variable **PV(i)**, and with respect to a given instantiation of other variables **PV(i, x̲)**:

$$\mathbf{PV(i)} := |\mathbf{D_i}|.$$

A more informative measure on the possible values for a variable is the *number of arc-consistent values* for the variable (also with respect to a particular partial instantiation of other variables): **AV(i)**, **AV(i, x̲)**.

Measures on values for variables

The estimation of a how good a certain value **v** for the i-th variable is can be based on the *number of satisfied constraints* which refer to the variable (also with respect to a particular instantiation of other variables, **SN(i, v, x)**):

$$SN(i, v) = | \{ j \mid c_{ij}(v, v_{ij}) \text{) holds for some } v_{ij} \} |$$

A more complex measure is the total number of possibilities for fulfilling the constraints referring to the variable:

$$ST(i, v) = \prod_{j=1}^{N} | \{ w \mid c_{ij}(v, w) \text{ holds} \} |$$

Another similar measure is the number of possibilities for fulfilling the tightest one of the constraints referring to the variable:

$$SW(i, v) = \min \{ \, | \{ w \mid c_{ij}(v, w) \text{ holds} \} | \, \}$$

10.4.2.2 Heuristic unary operators

Below we define genetic operators based on the idea of local repair: improving an individual by changing the value at one or more of those positions which correspond to variables referred to by unsatisfied constraints. Such a genetic operator applies to one individual and produces one new individual. If the selection of the value to be modified and the modification itself is done by a uniform random mechanism, then we perceive the operation as a mutation. If there is a non-uniform mechanism in either the selection of the value to be modified and/or the modification itself, then we will call it an *asexual* or *unary genetic operator.*

The number of modified values, the criteria for identifying the values to be modified, and the criteria for defining the new values are the defining parameters of these operators. Therefore, such an operator will be denoted by the triple **(n, p, v)** standing for (number, position, value). In our implementation **n** is either 1, 2 or #, indicating the number of modified values, where # means that the number of values to be modified is selected randomly (but in our implementation at most 1/4 of all positions). The attribute **p** denotes the way in which the positions of values to be modified are selected, **v** denotes the way in which the new value is defined. For both variable and value selection we distinguish uniform random and heuristic-based methods. Therefore, the attributes **p** and **v** can take their values from the set **{r, b}**, where **r** indicates uniform random selection, **b** indicates some heuristics-based selection. Thus, basically we have four possibilities to combine position selection with value selection mechanisms. These possibilities are multiplied by the cases corresponding to a different number of modified genes. This means that we define 12 genetic operators within this framework. Three of them, namely **(1, r, r)**, **(2, r, r)** and **(#, r, r)**, are non-heuristic ones, the other 9 are heuristic genetic operators. For example, the **(2, r, b)** operator is defined in such a way, that on two randomly selected positions the value of the parent is changed on the basis of some heuristic evaluation of the possible values. Of course, another multiplying factor is the range of applicable measures (listed in 4.2.1) for heuristic position and value selection. The most straightforward measure for position selection is the number of non-satisfied constraints referring to the appropriate variable, **NC(i, x)**; while for the new value selection it is the number of constraints which refer to the position and which are satisfied, **SN(i, v, x)**.

When defining a genetic operator, the allowability of the child should be ensured. That is, in addition to the selection of new gene values for some positions, the allowability of the child has to be checked. If the child is not allowed, then some correction has to be performed, still as a part of the genetic operation. This is a quite critical issue, which cannot be addressed without knowing the allowability criteria. For example, if the allowability criterion requires that individuals should be permutations of the numbers 1, 2, ..., N, then a swap has to be performed: if on a position the new gene value, **x**, is different from the one in the parent, **y**, then **x** occurs in the parent in a different position. In the child on that position, the new value will be **y** (see figure 10.2).

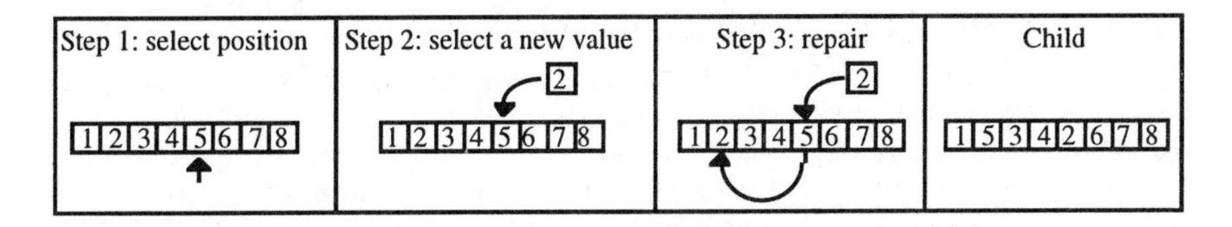

Figure 10.2: Repairing as a means of preserving allowability

10.4.2.3 Multi-parent operators

The operator we discuss in detail is called *gene scanning* which can be seen as a generalisation of uniform crossover extended by special mechanisms. The idea behind scanning is to examine (to put a marker on) the positions of the parents consecutively and choose one of the values on the marked positions for the child. Choosing from the marked genes can be done by a uniform random mechanism. This version will be called *uniform scanning*; the generalisation of the uniform crossover for n > 2 parents. The choice of a value for the child can also be based on the analysis of the values present in the parents. For instance, *occurrence-based scanning* chooses that value that occurs the most among the marked genes of the parents (see figure 10.3); this is somewhat similar to the p-sexual voting technique reported in [Müh89]. The choice can also be made by a random mechanism biased by the fitness of the parents, we call this *fitness-based scanning*. Note that each of these mechanisms are problem independent and represent a different level of bias, uniform scanning having the slightest bias and fitness-based scanning the strongest.

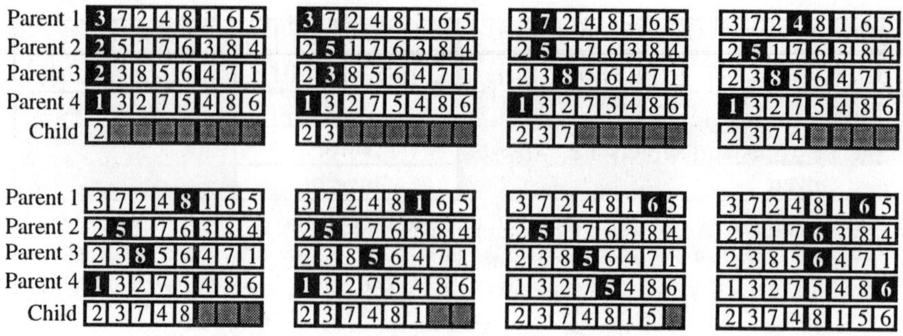

Figure 10.3: Occurrence-based scanning for order-based representation

Note, that above we tacitly assumed that the markers are updated by always shifting them one position to the right. In this case scanning does not work for order-based representations. Nevertheless, the marker update mechanism can be modified very easily so that in each parent the marker is shifted to the first value that does not yet occur in the child. This results in an order-based scanning mechanism that can be uniform, occurrence-based or fitness-based. The figure below shows occurrence-based scanning for order-based representation. The marked genes are those ones shown in reverse.

Scanning can also be enriched in a natural way using heuristics relying on extra information. For instance, for routing problems we can consider edge length and choose that city label to be inserted at the n-th position in the child that denotes the city closest to the (n-1)-th.

Let us briefly mention another multi-parent operator which is a generalisation of n-point crossover, which is in turn the generalisation of 1-point crossover. The idea is that n parents create n children by choosing n-1 crossover points and creating the children along the diagonals. Figure 10.4 illustrates how diagonal crossover works for n = 3.

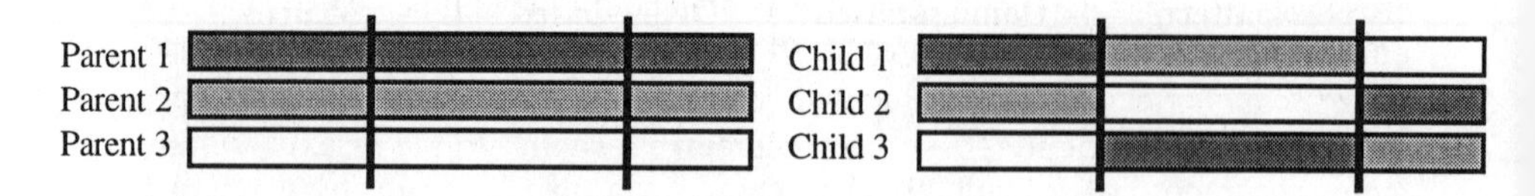

Figure 10.4: Diagonal crossover

Among the test cases we discuss in this chapter we only applied diagonal crossover to the Zebra problem; results of applying diagonal crossover to numerical optimisation are subject to forthcoming publications.

10.4.3 Overview of genetic operators belonging to different representations

In this section we sum up several genetic operators grouped by the representation they are suited for. We divide them by their arity, i.e. the number of parents that are involved in creating the child(ren).

Bit-pattern	Domain-based	Order-based	Matrix
(n, {r,b}, {r,b})[1]	(n, {r,b}, {r,b})	(n, {r,b}, {r,b})[2]	(n, {r,b}, {r,b})[3]
		swap	column/row swap
		shift	
invert		invert	

1) The standard mutation operator is equivalent to (1,r,r).
2) We used this operator with a repair mechanism as discussed in 4.2.2.
3) Applied in such a way that the units {r,b} refers to are columns or rows, see also section 5.4.2.

Figure 10.5: Unary operators

Bit-pattern	Domain-based	Order-based	Matrix
n-point	n-point	order1[1]	ATEX[2]
uniform	uniform	UOX	uniform
		order2	
		PMX	
		cycle	
		position	
		edge-recombin.[3]	
			CPX[2]
			IREX[2]

1) Orderl can be seen as a permutation based 2-point crossover, where the alleles from the second parent are not inherited position-wise, but by their relative order in which they are encountered when parsing from left-to-right.
2) ATEX (adapted tail-exchange), CPX (column permutation) and IREX (informed row exchange) are explained in detail in section 5.4.2
3) The edge recombination crossover is defined specifically for the TSP, [Whi91].

Figure 10.6: Binary operators

Bit-pattern	Domain-based	Order-based	Matrix
U-Scan	U-Scan	U-Scan[1]	
OB-Scan	OB-Scan	OB-Scan[1]	
FB-Scan	FB-Scan	FB-Scan[1]	
diagonal	diagonal		diagonal[2]
		U-ABC[3]	
		OB-ABC[3]	
		FB-ABC[3]	

1) The scanning techniques use a special marker update mechanism to ensure that the children will also be permutations as explained in section 4.2.2.
2) The diagonal crossover we applied works on rows.
3) The ABC (adjacency based crossover) is defined specifically for the TSP.

Figure 10.7: N-ary operators

10.4.4 Classification of operators

One obvious way to classify genetic operators is distinguishing them by their arity and the representation they are suited for, as we have done in the previous section. In this section we present another classification scheme that distinguishes operators by how specific they are, respectively, what kind of heuristics they apply.

First of all we can distinguish standard and non-standard genetic operators. The *standard genetic operators* apply to binary strings, use uniform random mechanisms and apply to either 2 parents and 2 children or 1 parent and 1 child. We call all other genetic operators *non-standard*, they apply to other representation forms, rely on random mechanisms with a non-uniform distribution and/or use more than 2 parents. This distinction means that a genetic operator can be both standard (when applied to binary strings) and non-standard (when applied to other representations) as is shown in figure 10.8. A non-standard genetic operator is called *heuristic* if it relies on non-uniform mechanisms, otherwise it is called *non-heuristic*. For example, PMX and order1 are non-standard, non-heuristic genetic operators used for order-based problems.

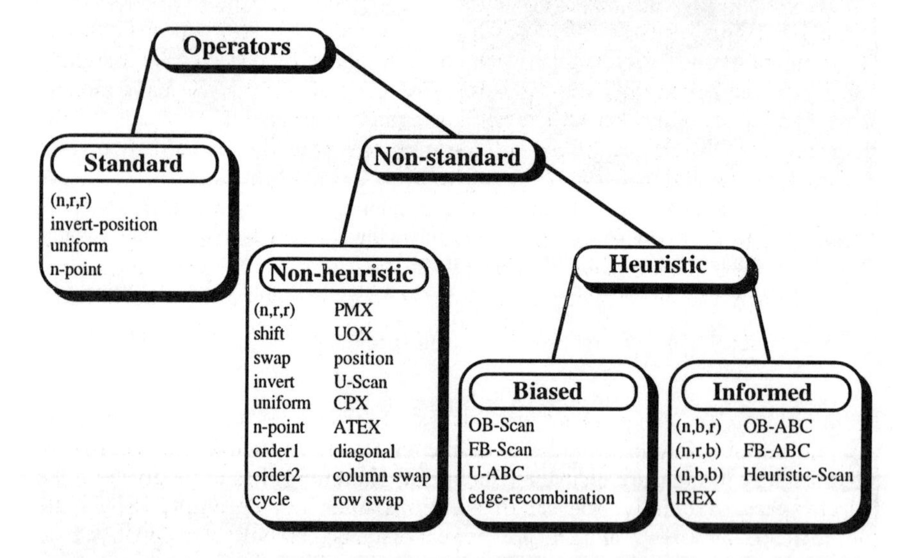

Figure 10.8: Classification of genetic operators

Among the heuristic genetic operators we distinguish two types. *Biased genetic operators* are based on some statistical analysis of a sample of the search space, e.g. the parents, such as occurrence-based and fitness-based scanning. In such genetic operators the uniform random mechanism is replaced by a mechanism so that gene patterns occurring more in (fit) individuals in the considered sample are propagated with a high probability. *Informed genetic operators* are based on some estimation of the fitness of different possible children of the parent(s) in question. That is, the genetic operator is defined in such a way that from all the possible children the 'most promising' one(s) will actually be generated. The criteria for evaluation of the possible children, incorporated into the genetic operator, may correspond to heuristics used by classical search algorithms or problem specific knowledge. Both types of heuristic genetic operators, i.e. biased and informed, can be deterministic or probabilistic when choosing the alleles of the child(ren). In the first case, the best choice according to the given heuristics is used, in the latter case good choices are made with a higher probability than poor ones.

10.4.5 Fitness function

The fitness function should express how good a candidate is from the point of view of satisfying the constraints not represented in the allowability criteria. Obviously, candidates which satisfy all the constraints should be the fittest elements. For the sake of convenience, we can assume that candidates have a fitness value in the range $[0..\infty]$, the best fitness value being 0. There are a number of ways in which to define the fitness for candidates. Here we discuss the constraint-based, the value-based and the search-space based possibilities.

A common approach we have discussed in section 10.3.1 is the use of penalty functions. In its simplest form a penalty function equals the number of violated constraints and this is the fitness value of a candidate in case of CSPs, see also figure 10.1. This measure, however, does not distinguish between difficult and easy constraints. A difference between constraints can be reflected by assigning weights to the constraints which leads to the general formula we have already presented in section 10.3.1. The weights can be given on the basis of the measures used in classical CSP solving methods to evaluate the difficulty of constraints (see section 10.4.2.1 for an overview of possibilities).

Another definition of fitness can be based on the evaluation of the 'incorrect' positions, that is positions where the value violates at least one constraint. In the simplest form, the number of the positions with incorrect values is used. More sophisticated definitions can be given on the basis of the evaluations of the positions (see 10.4.2.1).

A third possibility is to consider the fitness as a kind of distance of the candidate from the set of solutions. In order to define the fitness as such a distance, first of all some way to measure distance (Euclidean, Hamming) has to be given on the search space. Secondly, the set of solutions must not be empty. Since the solutions are not known in advance, the real distance can only be estimated. One such an estimation is based on the idea of using the distance of a candidate from the cylindrical sets represented by the constraints (see section 10.1). Since the solutions are elements of the intersection of all sets corresponding to the constraints, it is a natural idea to consider the total sum or the maximum of the distances of an individual in question from the cylindrical sets as the fitness of the individual.

10.5 Case Studies

We have studied the performance of GAs on 13 problems. We have chosen to describe 5 of them in this chapter, namely: the traffic lights problem (CSP), the N-queens problem (CSP), graph 3 colouring (CSP), the Zebra problem (CSP) and the travelling salesman problem (COP). These problems have been selected to illustrate specific aspects of applying GAs to constrained problems. For more results see [Eib93, Eib94].

The majority of the experiments described in this chapter were run using a library of GA routines (LibGA) written by A.L. Corcoran and R.L. Wainwright [Cor93]. The other experiments were performed using the GA described in the appendix.

All of the experiments described below were run using the generational model, roulette-wheel parent selection and elitism. The mutation-rate denotes the chance that a newly generated candidate is mutated, not the chance that a value is mutated.

10.5.1 The Traffic Lights Problem

This problem is defined as [How93]: find the safe light combinations for 8 traffic lights, four of which are vehicle lights having four possible colours (red, yellow/red, yellow and green) and the other four are pedestrian lights having only two colours (red and green).

When v_i denotes the colour of vehicle light i and p_i the colour of pedestrian light i, the constraints for this problem can be described as:

$$C_{v_i p_i v_{1+i \bmod 4} p_{1+i \bmod 4}} = \{ (r,r,g,g), (ry,r,y,r), (g,g,r,r), (y,r,ry,r) \}$$

This problem has four solutions.

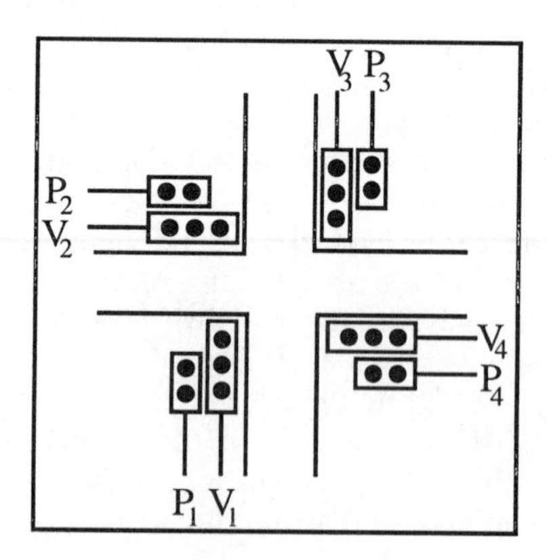

Figure 10.9 The traffic lights problem

10.5.1.1 Using domain-based representation (FOP)

Although it would be possible to encode this problem using bit-pattern representation (12 bits, 2 bits for each of the vehicle lights and 1 bit for each of the pedestrian lights) we opted to represent a candidate solution using domain-based representation. The candidates are strings of length 8 where the first four positions denote the vehicle lights and thus have a domain of [1...4], and the last four positions denote the pedestrian lights and have a domain of [1...2]. This results in a new representation for the GA which can be initialised as is shown in figure 10.10. (Take note: in the constraint tuples, sets of lights, vehicle and pedestrian, are grouped, in the representation all vehicle lights are grouped together as are the pedestrian lights.)

```
void GenerateTrafficLightsCandidate( GA *ga, Candidate *candidate
)
{
  int x;

  for(x = 0; x<4; x++) candidate->String[x] = RandomDomain(1,4);
  for(x = 4; x<8; x++) candidate->String[x] = RandomDomain(1,2);
}
```

Figure 10.10: Generating new candidate for the traffic lights problem

Fitness

The fitness is defined as S $d(T_i, C_{v_i p_i v_{1+imod4} p_{1+imod4}})$, where T_i is (v_i, p_i, $v_{1+imod4}$, $p_{1+imod4}$) and $d(T_i, C_{v_i p_i v_{1+imod4} p_{1+imod4}})$ is defined as the Hamming distance from T_i to the constraint-tuple with matching v_i, i.e. (v_i, ., ., .). For example:

$d($ (r, g, r, r), $C_{v_i p_i v_{1+imod4} p_{1+imod4}}) = d($ (r, g, r, r), (r, r, g, g)) = 3.

Genetic operators

Because we use domain-based representation we have been able to use the standard crossovers (1-point and uniform). The mutation operator for this problem was random which means choosing a random position in the candidate and generating a random value for this position (not necessarily different from the value currently at this position. The code for this mutation operator is shown below.

```
void TrafficLightsMutation( GA *ga, Candidate *candidate )
{
    in pos = RandomDomain( 0, ga->CandidateLength - 1 );

    if( pos < 4 ) candidate->String[pos] = RandomDomain( 1, 4 );
    else          candidate->String[pos] = RandomDomain( 1, 2 );
}
```

Figure 10.11: Random mutation for the traffic lights problem.

Results

Using this simple implementation we achieved very good results. A poolsize of 100 led to 93-100% solutions (100 runs of the GA) regardless of the crossover (both 1-point and uniform) or mutation rate.

The problem is not finding one solution, but finding all solutions. By running the GA a number of times (usually 20 or less) with different initial pools (guaranteed by a changing random seed) we were able to find all solutions.

To ensure that different solutions will be found in subsequent runs of the GA, those solutions that have already been found are penalised. For example if the first run of the GA found a solution where v_1,p_1,v_3 and p_3 set to green, v_2, p_2, v_4 and p_4 set to red, candidates approaching this solution are penalised in the next run of the GA by increasing their fitness as shown in figure 10.12. Upon running tests, we found that the average number of experiments needed to find all four solutions was 5.24 when using this penalty fitness function (poolsize 50, mutation-rate 0.5, max. generations 20, traffic-lights-mutation and 1-point crossover).

```
#define V_GREEN      1
#define V_YELLOW     2
#define V_YELLOW_RED 3
#define V_RED        4
#define P_GREEN      1
#define P_RED        2
int AlreadyFound[4][8] = {0};
double Fitness( GA *ga, Candidate *candidate)
{
    int vi, pi, vj, pj, x, y, diff, fitness = 0;
    for( x = 0; x < 4; x++ )
    {
        vi = candidate->String[x];
        pi = candidate->String[x+4];
        vj = candidate->String[(x+1)%4];
        pj = candidate->String[(x+1)%4 + 4];
        if( vi == V_RED)
 fitness += (pi!=P_RED  )+(vj!=V_GREEN)+(pj!=P_GREEN);
        if( vi == V_YELLOW_RED )
 fitness += (pi!=P_RED  )+(vj!=V_YELLOW)+(pj!=P_RED  );
        if( vi == V_GREEN)
 fitness += (pi!=P_GREEN)+(vj!=V_RED)+(pj!=P_RED  );
        if( vi == V_YELLOW)
 fitness += (pi!=P_RED  )+(vj!=V_YELLOW_RED)+(pj!=P_RED  );
        for( diff=y=0; y < 8; y++ ) diff += (candidate->String[y]
!= AlreadyFound[x][y]);
        if(      diff == 0                 ) fitness += 10;
        else if( diff == 1 && fitness != 0 ) fitness +=  2;
        else if( diff == 2 && fitness != 0 ) fitness +=  1;
    }
    return( candidate->Fitness = fitness );
}
main()
    GA  *ga        = AllocateGA( Fitness, NULL );
    int  solutions = 0, x;
    AddRepresentation(ga,"traffic-lights",GenerateTrafficLights
                                                    Candidate );
    AddMutationOperator(ga,"traffic-lights-mutation",Traffic
                                              LightsMutation );
    ConfigureGA( ga, "config-file" );
    while( solutions < 4 )
    {
        RunGA(ga);
        if( ga->Parents->Best->Fitness == 0)
        {
```

```
            for( x=0; x<8; x++ ) AlreadyFound[solutions][x] = ga-
                                         >Parents->Best->String[x];
            solutions++;
        }
    }
```

Figure 10.12: Finding all four solutions to the traffic lights problem[1]

5.2 The N-queens problem

The N-queens problem is defined as: place N queens on a chessboard of size N¥N so that no two queens attack each other.

If you represent each queen by one variable x_i (i indicates the column and the value of x_i indicates the row in which the queen is placed) then the constraints are:

$x_i \neq x_j$ for $i \neq j$ (row constraint)
$|x_i - x_j| \neq |i - j|$ for $i \neq j$ (diagonal constraint)

This problem is of theoretical interest only since Minton has been able to solve one million queens [Min92] and Sosic [Sos91] three million queens much quicker than we could even solve our maximum case of 10.000 queens. The N-queens problem has previously been solved using GAs as reported in [Cra92, Hom92]. Our GAs performed much better than the results reported by Crawfords especially with regard to the time needed to find solutions. We have been unable to find the [Hom92] publication.

10.5.2.1. Using domain-based representation (FOP)

A natural way to encode this problem is using a string of length N where each variable has a domain of [1, ..., N] (domain-based representation).

Fitness

The fitness of a candidate solution is defined as the number of constraints the candidate violates. For each pair of queens in the same row or diagonal 1 is added to the fitness value of the candidate. This means that a candidate with a fitness value of 0 is a solution to the CSP.

Genetic operators

We used the standard crossover operators. We also tested the asexual heuristic operators (n,r,b), (n,b,r) and (n,b,b) (see also section 5.2.2). The mutation operator was random.

Results

1 To run the GA include the code in figures 10.11 and 10.12.

Using this representation we were able to solve the problem in only 70-80% of all cases even when using the heuristic genetic operator (#,b,b). Obviously, either the genetic operators or the representation were not optimal.

10.5.2.2 Using order-based representation (COP)

When taking into account that each of the queens must be placed, not only in a different column, but also in a different row, a second representation becomes obvious.

In the order-based representation, candidates are permutations of [1, 2, ..., N]. This means that the row constraints are satisfied automatically.

Fitness

The fitness function was not changed from the one described in 5.2.1, except for the fact that we no longer need to check the row constraints.

Genetic operators

We ran tests using all order-based operators (see section 4.3). The mutation operator was swapped, which swaps the values in two randomly chosen (not necessarily different) positions. Of all genetic operators the asexual heuristic (n,p,v) operators (of which the framework for order-based representation is shown in figure 10.13) performed the best.

```
void AsexualOperator( GA *ga, Candidate *parent, Candidate
                                                *child, int n )
{
    int x, y, position, value, old_value;

    CopyCandidate( ga, parent, child );

    for( x = 0; x < n; x++ )
    {
        position = SelectPosition( ga, child );
        old_value                   = child->String[position];
        child->String[position] = SelectValue( ga, child,
                                                        position );

        for( y = 0; y < ga->CandidateLength; y++ )
            if( y != position && child->String[y] == child-
                                                >String[position] )
            {
                child->String[y] = old_value;
                break;
            }
    }
}
```

Figure 10.13: The framework for the (order-based) asexual operators

Heuristic selection of the position for the operators is done by identifying the number of constraint violations per position and choosing that position which adds the most constraint violations. The heuristic choice of a new value for the selected position is done by exhaustively checking all values and setting the position to that value which results in the least constraint violations. In order for this not to disturb the allowability we need to introduce a repair step (the last for() loop in the code) which ensures that the generated child will still be a permutation.

Results

Using this new representation we obtained very good results. Every operator (even mutation alone) was always able to find a solution for the problem (100% solution rate). We tested 25, 50, 75, 100 and 250 queens for all operators using poolsizes of 100, 200, 300, 400 and 500 respectively. Optimal mutation rates varied, but were mostly 100% for all binary and 10% for all unary operators. Of the operators we tested (#,b,b) proved to be the best, conventional order-based crossovers turned out to be inferior to the heuristic ones.

A natural question is whether the poolsize has a substantial effect on the performance of an asexual operator. Upon running tests with different poolsizes for 25 queens we found that the poolsize did have a (small) effect on the performance. Increasing the poolsize up to about 25 meant that the number of generations needed to find a solution decreased, increasing the poolsize further had (almost) no effect. This decrease was more pronounced for smaller numbers of queens. Running the same tests for 100 queens we found the decrease was still present, but it was much smaller. This can be explained using the study on the density of solution/equilibrium points which states that when the number of queens is increased the density of solution/equilibrium points approaches 1 [Mor92], meaning that a solution is more likely to be found for larger problems.

When increasing the number of queens we found that the number of generations needed to reach a solution increases approximately linear while the time needed to reach a solution increases quadratically for most genetic operators. For the asexual operators, in particular (n,b,b), the increase was less pronounced, see figure 10.14.

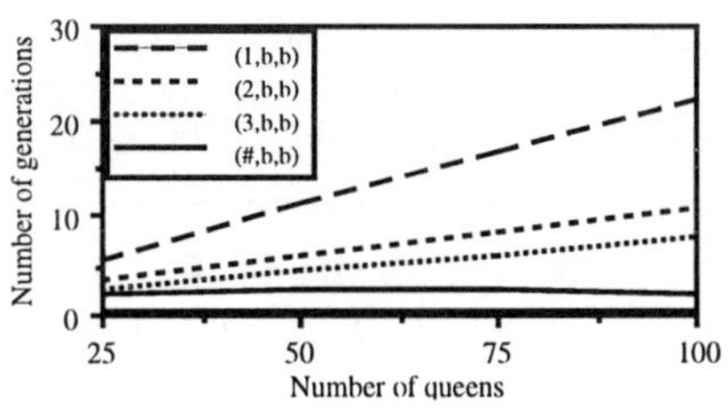

Figure 10.14: Performance of the asexual operators[2]

---

2 The number of generations for (#,b,b) does not increase because the number of modified positions increases when the number of queens increases.

Since the poolsize did not seem to have a significant effect on the performance we kept the poolsize stable at 10 candidates and ran tests with up to 10.000 queens using the (1,b,b) and (#,b,b) operators. We found that increasing the number of modifications for the asexual heuristic operators meant that the increase in generations (as well as time) was smaller, as is shown above for 25 through 100 queens (poolsize 10, mutation rate 10%).

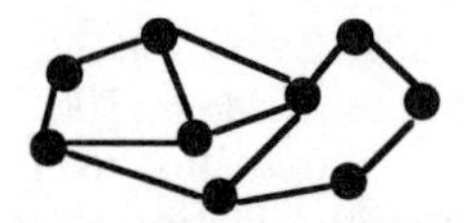

10.5.3 Graph 3 colouring

The graph 3 colouring problem is defined as: colour the nodes of a graph using three colours in such a way that no two nodes connected by an edge are coloured using the same colour. This has proven to be a very difficult CSP for specific edge-densities (see below). For our experiments we have restricted ourselves to graphs which are colourable.

This means that if n_i indicates the colour of node i and e_{ij} indicates whether node i and j are connected by an edge the following constraint must hold:

$n_i \neq n_j \quad \text{for } e_{ij} = 1$

When running experiments with different numbers of edges per node we found that the most difficult graphs had on average 5 edges per node as is shown in figure 10.15. This has also been found by Cheeseman [Che91] who used a deterministic algorithm to solve graph colouring problems.

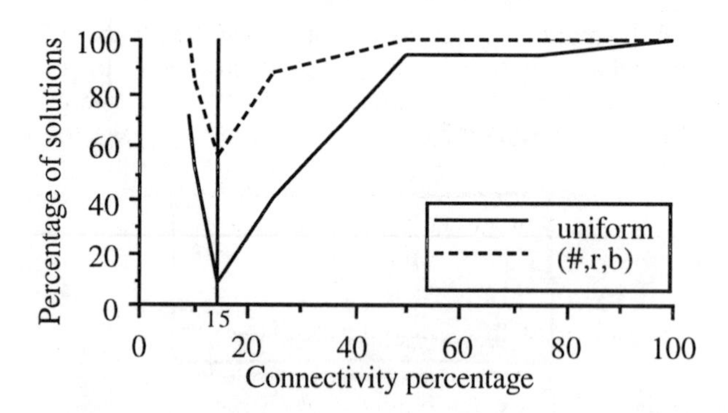

Figure 10.15: Performance for different connectivity percentages[3]

10.5.3.1 Using domain-based representation (FOP)

---

3 The connectivity percentage is the (average) number of edges per node divided by the maximum number of edges per node. The graph used in these experiments had 50 nodes divided into three groups (one group for each color). This means that a connectivity percentage of 15 indicates: 0.15*(50*2/3)=5 edges per node.

The most natural way in which to represent candidate solutions is using a string of length N (N being the number of nodes in the graph) where each position has a domain of [1...3] indicating the colour of that particular node. In [Min92] this representation was used in the generate-and-repair method (MC Backtrack) on graphs with on average 4 edges per node.

Fitness

The fitness is defined as the number of constraint violations.

Genetic operators

We tested the performance of the traditional crossover operators as well as the scanning techniques, the mutation operator was random.

Results

Using this representation we found that the maximum number of solutions for a graph with 50 nodes and 5 edges per node (on average) was approximately 56% for the (#,r,b) operator (with a poolsize of 500 and 75% mutation). When running tests we found that the poolsize had an approximately linear effect on the number of solutions as is shown in figure 10.16 (since we ran only 25 experiments per poolsize to keep running times down, the irregularities are — probably — caused by the low number of experiments).

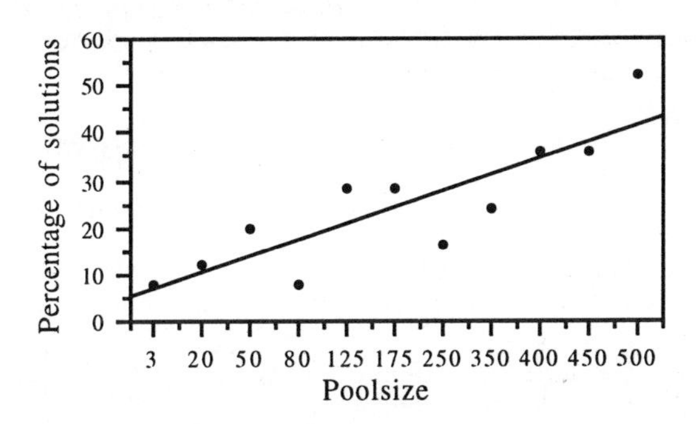

Figure 10.16 The effect of the poolsize on (#,r,b)

To improve performance we implemented a different GA using order-based representation as described below.

<u>10.5.3.2 Using order-based representation with a decoder (COP)</u>

When using order-based representation for the graph 3 colouring problem, each node is assigned a number and the order in which these numbers appear in the candidate solution denotes the order in which the nodes are coloured. When traversing the candidate solution from left to right each node is coloured using the first available colour. If all colours have been used for its neighbours a node is left uncoloured. A similar scheme was using in [DavL91] for weighted graph colouring with two colours.

Fitness

The fitness value of a candidate is defined as the number of uncoloured nodes. The decoder is included in the fitness function shown below[4].

```
double Fitness( GA *ga, Candidate * candidate )
{
    int           x, y, neighbour, uncoloured_nodes = 0;
    static int *colours = NULL;

    if( !colours && !( colours = (int *)calloc( ga->Candidate
                                    Length, sizeof(int) ) ) )
        Error( "Not enough memory." );

    memset( colours, 0, ga->CandidateLength * sizeof(int) );

    for( x = 0; x < ga->CandidateLength; x++ )
    {
        int node = candidate->String[x] - 1, used[3] = { 0, 0, 0
};

        for( y = 0; y < ga->CandidateLength && neighbour =
                          Connectivity[node][y]; y++ )
            used[colours[neighbour - 1] - 1] += colours[neighbour
                                                            - 1];

        for( y = 0; y < 3; y++ ) if( !used[y] ) break;

        y == 3 ? uncoloured_nodes++ : colours[gene] = y + 1;
    }

    return( candidate->Fitness = uncoloured_nodes );
}
```

Figure 10.17: The fitness function for the order-based implementation

Genetic operators

For the order-based representation we applied the order-based crossovers as well as the scanning procedures discussed in 4.2.3. The mutation operator was swapped.

```
void X_FB_Scan( GA *ga, Candidate *parents, Candidate *children )
{
    static int *used = NULL,
    int     markers[MAX_PARENTS]      = { 0 }, child_marker, i, x;
```

4 Before running this procedure the graph must have been read into the matrix "Connectivity" where row x is filled with the neighbours of node x, and a 0 indicates that there a no more neighours. For example, if row 1 is [2,4,5,8,0,0,0,0] (for a graph with 8 nodes) it means that node 2, 4, 5 and 8 are neighbours of node 1.

```
    double probability[MAX_PARENTS] = { 0 }, total = 0, sum = 0,
                                                            prob;

    if( !used )
        if(!( used = calloc( ga->CandidateLength, sizeof(int))))
            Error( "Not enough memory." );

    memset( used, 0, ga->CandidateLength * sizeof(int) );

    for( i = 0; i < ga->NrOfParents; i++ ) total += parents[i]
                                                   .Fitness;

    if( ga->Minimize )
    {
        for( i = 0; i < ga->NrOfParents; i++ )
        {
            probability[i] = total / parents[i].Fitness;
            sum            += probability[i];
        }

        for( i = 0; i < ga->NrOfParents; i++ ) probability[i] /=
                                                            sum;
    }
    else
        for( i = 0; i < ga->NrOfParents; i++ ) probability[i] =
                                    parents[i].Fitness / total;

    for( child_marker = 0; child_marker < ga->CandidateLength;
                                           child_marker++ )
    {
        for( x = 0; x < ga->NrOfParents; x++ )
            while( used[parents[x].String[markers[x]] - 1] )
                                                   markers[x]++;

        x    = -1;
        prob = RandomFraction();

        do
            prob -= probability[++x];
        while( prob > 0 && x < ga->NrOfParents );

        children[0].String[child_marker] = parents[x].String
                                                 [markers[x]];
        used[children[0].String[child_marker] - 1] = 1;
    }
}
```

Figure 10.18: Fitness based scanning for order-based representation

Results

The best results were obtained using an asexual operator, (1,r,r). Most operators were able to solve the graphs described by Minton in 100% of all cases up to 60 nodes, and although the performance dropped a little after this point our performance was still much better than Minton's (as shown in figure 10.19).

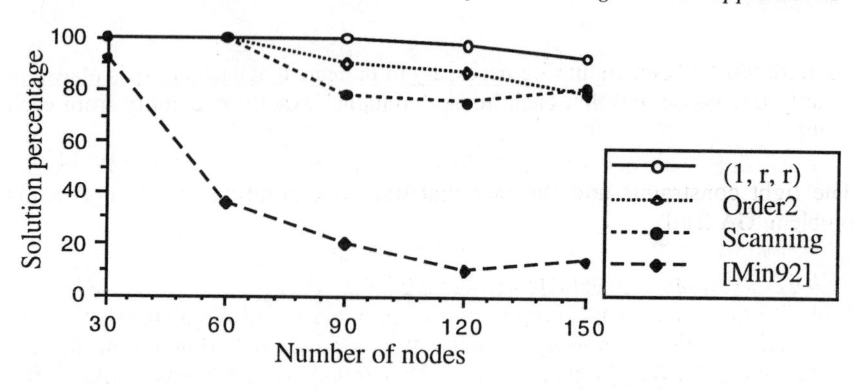

Figure 10.19: (Order-based) GAs versus Minton MC Backtrack

10.5.4 The Zebra Problem

The Zebra problem was offered by [Dec90] as a very challenging problem with tight constraints and only one single solution. Five persons with different nationalities, five colours, five beverages, five pets and five brands of cigarettes are given. This means that in total 25 entities distributed over 5 groups are given:

Nationalities : Norwegian, Ukrainian, Japanese, Spaniard, Englishman.
Colours : blue, red, green, ivory, yellow.
Beverages : coffee, tea, orange-juice, milk, water.
Pets : zebra, dog, horse, snails, fox.
Cigarettes : Old-Gold, Chesterfield, Lucky Strike, Parliament, Kools.

The problem is defined as: determine which person lives in which house, what is the colour of the house, what is his/her pet, favourite beverage and what brand of cigarette does (s)he smoke. The constraints given below must hold for a solution.

C_1	: HouseIs(Milk, 3)	C_2	: HouseIs(Norwegian, 1)
C_3	: SameHouse(Englishman, Red)	C_4	: SameHouse(Spaniard, Dog)
C_5	: SameHouse(Coffee, Green)	C_6	: SameHouse(Ukrainian, Tea)
C_7	: SameHouse(OldGold, Snails)	C_8	: SameHouse(Kools, Yellow)
C_9	: SameHouse(Lucky, Orange Juice)	C_{10}	: SameHouse(Japanese, Parliament)
C_{11}	: NearHouse(Chesterfield, Fox)	C_{12}	: NearHouse(Kools, Horse)
C_{13}	: NearHouse(Norwegian, Blue)	C_{14}	: NextHouse(Ivory, Green)

Figure 10.20: Constraints for the Zebra problem

The houses are numbered from 1 to 5 and each of the entities is assigned a variable. If the value of a variable denotes the number of the house in which the entity is placed then the constraints mean:

SameHouse(X, Y) fi $X = Y$ NearHouse(X, Y) fi $|X - Y| = 1$
HouseIs(X, C) fi $X = C$ NextHouse(X, Y) fi $Y - X = 1$

50 more (binary) constraints are necessary to indicate that each entity is placed in exactly one house and that each house 'contains' exactly one entity from each group.

The tight constraints and the fact that only one solutions exists make this problem 'GA-hard'.

10.5.4.1 Using order-based representation (COP)

Probably the easiest way to represent this problem is to have a string of length 25 (each position denoting a different entity) with a domain of [1...5]. Nevertheless, for the N-queens problem (where unicity conditions similar to the Zebra problem must hold) this representation did not work well. Therefore, we encoded candidate solutions using order-based representation. The entities were labelled by numbers from 1 to 25 and each permutation of [1, ..., 25] denoted a candidate solution. The order in which the numbers from a particular group occur was used to assign the houses. The first occurrence of a number from a group was assigned house 1, the second was assigned house 2 and so on. For example, if Norwegian, Ukrainian, Japanese, Spaniard and Englishman are assigned 1, 2, 3, 4 and 5, respectively, and they occur in the order 5, 4, 1, 3, 2 in a permutation then the Englishman lives in house 1, the Spaniard in house 2, the Norwegian in house 3 and so on.

Fitness

```
int Weights[17] = { 1,1,1,1,1,1,1,1,1,1,1,1,1,1,1,1,1 };
#define SatisfiedIsHouse(GROUP1, ENTITY1, HOUSE)(house[GROUP1]
                                            [ENTITY1] == HOUSE)
#define SatisfiedNextHouse( GROUP1, ENTITY1, GROUP2, ENTITY2) \
        ( house[GROUP2][ENTITY2] == house[GROUP1][ENTITY1] + 1)
#define SatisfiedSameHouse( GROUP1, ENTITY1, GROUP2, ENTITY2) \
        ( house[GROUP1][ENTITY1] == house[GROUP2][ENTITY2])
#define SatisfiedNearHouse( GROUP1, ENTITY1, GROUP2, ENTITY2) \
        (abs(house[GROUP1][ENTITY1]-house[GROUP2][ENTITY2]) == 1)
double FitnessFunction( GA *ga, Candidate *candidate)
{
  int x, house[5][5], fit = 0;
  for(x=0; x<25; x++) house[x/5][candidate->String[x]-1] = x % 5;
    if( !SatisfiedIsHouse( DRINKS, MILK, HOUSE_3 ) )
                                    fit+=Weights[0];
    if( !SatisfiedIsHouse( NATIONALITY, NORWEGIAN, HOUSE_1 ) )
                                    fit+=Weights[1];
    if( !SatisfiedSameHouse( NATIONALITY, ENGLISHMAN, COLOURS,
                                    RED )fit+=Weights[2];
    if( !SatisfiedSameHouse( NATIONALITY, SPANIARD, PETS, DOG ))
                                    fit+=Weights[3];
    if( !SatisfiedSameHouse( DRINKS, COFFEE, COLOURS, GREEN ) )
                                    fit+=Weights[4];
    if( !SatisfiedSameHouse( NATIONALITY,UKRAINIAN,DRINKS, TEA))
                                    fit+=Weights[5];
    if( !SatisfiedSameHouse( SMOKES, OLD_GOLD, PETS, SNAILS ) )
                                    fit+=Weights[6];
    if( !SatisfiedSameHouse( SMOKES, KOOLS, COLOURS, YELLOW ) )
                                    fit+=Weights[7];
```

```
    if( !SatisfiedSameHouse( SMOKES,LUCKY,DRINKS,ORANGE_JUICE))
                                fit+=Weights[8];
    if( !SatisfiedSameHouse( NATIONALITY, JAPANESE, SMOKES,
                            PARLIAMENT ) ) fit+=Weights[9];
    if( !SatisfiedNearHouse( SMOKES, CHESTERFIELDS, PETS, FOX ))
                                fit+=Weights[10];
    if( !SatisfiedNearHouse( SMOKES, KOOLS, PETS, HORSE ))
                                fit+=Weights[11];
    if( !SatisfiedNearHouse( NATIONALITY, NORWEGIAN, COLOURS,
                                BLUE ) ) fit+=Weights[12];
    if( !SatisfiedNextHouse( COLOURS, IVORY, COLOURS, GREEN ) )
                                fit+=Weights[13];
    return( candidate->Fitness = fit );
}
```

Figure 10.21: The fitness function for the Zebra (order-based) problem

The fitness is defined as the number of violated constraints. The way in which this is calculated is shown below. (Each entity is given a value from [0..4] which is used to access the house matrix.)

Genetic operators
We applied all order-based crossover operators to this problem, the mutation operator was swapped.

Results
For all crossovers and mutation rates we found a solution percentage not exceeding 4%. Clearly the problem cannot be solved efficiently using this implementation.

10.5.4.2 Using matrix representation (COP)

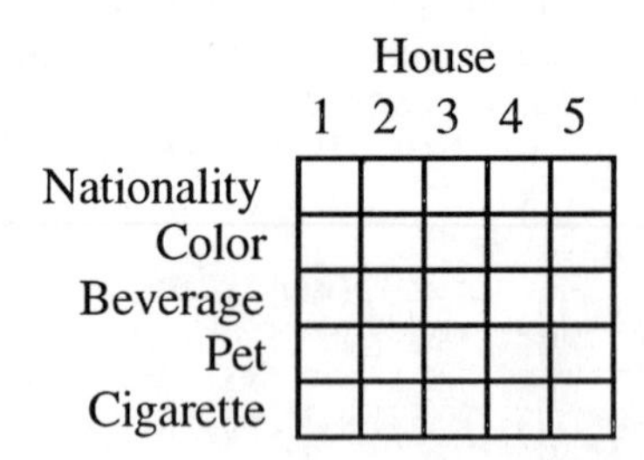

Figure 10.22 The matrix encoding

Since order-based representation did not work very well we decided to implement the Zebra problem using a matrix encoding scheme that allows us to keep entities from the same group together. In this representation each column in the 5 ¥ 5 matrix denotes one house, while each row in the matrix denotes one group, as is shown in figure 10.22.

As a consequence of the unicity conditions, each row in the matrix must be a permutation of [1, ..., 5] so that all entities in the group occur exactly once. This means that the 50 extra constraints indicating that each variable in a group must have a unique value are automatically satisfied by the representation.

Fitness

As was the case for the order-based encoding of candidates, the fitness is still the number of violated constraints. The only change in the fitness function is the first for() loop which decoded the order-based representation.

Genetic operators

For this encoding we have defined a number of genetic operators:

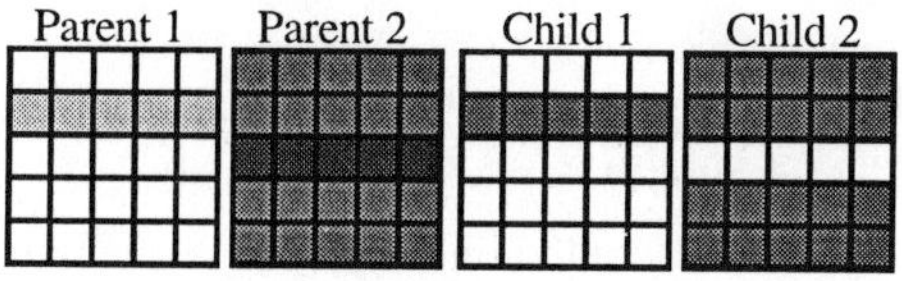

• **informed row exchange** (IREX): Create two children from two parents by replacing the worst row (the row which adds the most constraint violations) in the first parent with the matching row in the second parent and vice versa. (In the figure above, the "worst" row in each parent is shaded slightly darker.)

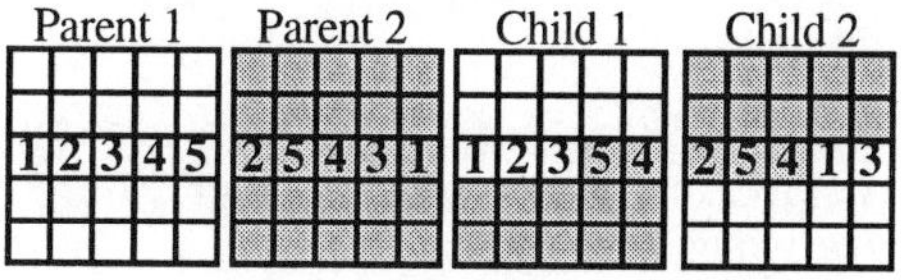

• **adapted tail exchange** (ATEX): Create two children from two parents. The first (second) child inherits the top part of the matrix from the first (second) parent. The bottom rows are inherited from the alternate parent. The row containing the crossover point is inherited up to the crossover point, the remaining positions in this row are filled with the unused values from the alternate parent in the order in which they appear in that parent.

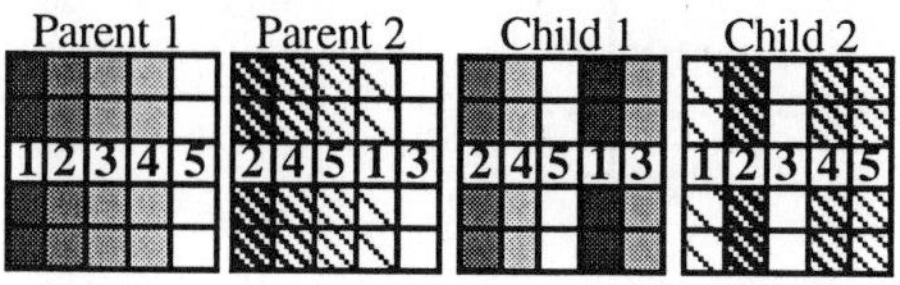

• **column permutation** (CPX): Create two children from two parents reordering their columns by a permutation of [1, ..., 5]. The permutation applied for the first parent, respectively, the second parent, equals a randomly selected row in the second, respectively, first, parent (the same row in each parent).

• **(1,b,r)**: Create one child from one parent by replacing the worst row (the row which adds the most constraint violations) in the parent by a random permutation.

• **row-swap**: Create a child by swapping two randomly chosen rows in the parent.

• **column-swap**: Create a child by swapping two randomly chosen columns in the parent.

Notice that the operators IREX and (1,b,r) are heuristic because they modify the worst row. This row is defined by a heuristic estimation of quality.

Results
When running experiments using this new representation we found that the results had improved. Instead of only 4% solutions we could find up to 11% solutions.

10.5.4.3 Improving the results
Considering that the results were still not very good we decided to try to change the implementation in such a way that a solution was more likely to be found. To achieve this we introduced five changes, namely:

Increasing the poolsize: We increased the poolsize from 100 to 300.

Weighing the constraints: Each (type of) constraint is given a separate weight to express the difference in difficulty. In our case we tried four different sets of weights (HouseIs,SameHouse,NearHouse,NextHouse): *constant* (1,1,1,1), *ranked* (4,3,2,1), *tightness-based* (10,6,5,1) and *exponential* (1000,100,10,1).

Incorporating implicit constraints in the fitness: A number of implicit constraints can be added explicitly in the fitness function. For example, the constraint HouseIs(Blue, 2) can be derived from the constraints HouseIs(Norwegian, 1) and NearHouse(Norwegian, Blue). The pre-processing was done by hand and the following constraints were derived and added:

- C_{15}: HouseIs(Blue,2)
- C_{16}: NearHouse(Yellow,Horse)
- C_{17}: NextHouse(Ivory,Coffee)

Fixing certain values (repairing): Another improvement was realised by noticing that the HouseIs() constraints (including the manually derived HouseIs(Blue,2) constraint) explicitly prescribe the values for certain variables, i.e. a part of the solution is known in advance. We introduced a repair step before checking the fitness of an individual so that all the variables in these HouseIs() constraints are given their correct value. In order to avoid changing the implementation of the GA, the fitness function called the repair routine in our implementation.

Automatic learning of constraint weights: It is rather obvious that the constraint weights applied have an important impact on the behaviour of the GA.

Therefore, we extended our implementation with a feature which allows the GA to learn from its own failures and to adapt constraint weights after each run. In particular, the weights of those constraints that remained unsatisfied by the best individual were increased after each run and the modified weights were used in the next evolution.

Using a poolsize of 300, a mutation-rate of 1.0, ranked weights, the addition of implicit constraints and repair we were able to find 42% solutions for the column-swap operator.

Although it is possible to estimate the weights of the four types of constraints in advance, this is very difficult to do for the individual constraints. Instead of calculating the weights of the individual constraints by hand, we implemented a GA which was able to learn from its mistakes. The GA started with an ad hoc set of weights (all constraint weights set to 1) and learned the weights of the individual constraints by increasing those constraint weights that were not satisfied by the best individual at the end of the evolution. The adapted constraints were then used as a starting point for the next evolution. We found that using the learning feature we could improve upon the results achieved without learning. The results are summarised in figure 10.23 (poolsize 100, mutation-rate 0.5, 100 runs on the GA).

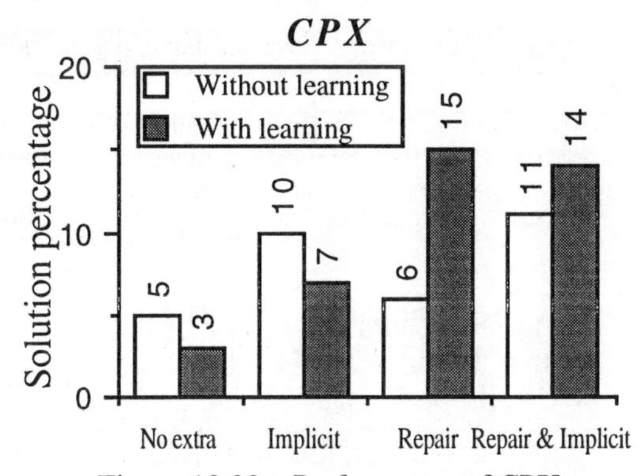

Figure 10.23a: Performance of CPX

Especially the IREX operator benefits from the learning of constraint weights. This is conformed to our expectations. Since this operator actually uses constraint weights when it determines the “worst” row, it is natural that its performance becomes better when using the newly learned weights.

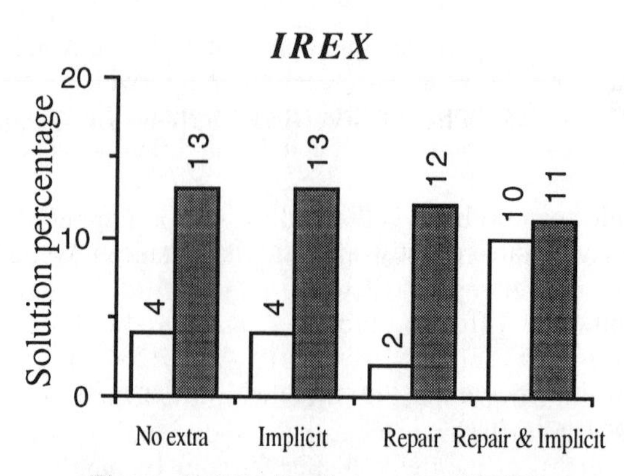

Figure 10.23b: Performance of IREX

The tests indicate that combined application of repair and the inclusion of implicit constraints yields the highest performance. Attempting to improve performance further we experimented with increasing the poolsize from 100 to 300 for the most promising configuration: extra constraints and repair. We found that the percentage of cases in which solutions are found increases significantly when using constraint weight learning, see figure 10.24.

	(1,b,r)	ATEX	Column-Swap	IREX	CPX
Without learning	10	12	15	17	22
With learning	22	17	24	26	31

Figure 10.24: Learning, extra constraints and repair with 300 poolsize

Note that the configuration with the learning facility starts with an ad hoc set of weights (all weights set to 1) and during the 100 runs the GA is learning a more suitable weight distribution. It can be expected that starting with the constraints weights learned by the GA, the performance would improve. We ran tests for the IREX and CPX operators using the constraint weights learned in 6 sessions of 100 runs each, see figure 10.25.

Using a poolsize of 100 this resulted in 22% solutions for both IREX (was 11%) and CPX (was 14%). Increasing the poolsize to 300 led to performances of 24% for IREX (was 26%) and 36% for CPX (was 31%).

C_1	C_2	C_3	C_4	C_5	C_6	C_7	C_8	C_9
1	1	10.5	12.2	8	12.8	11.8	11.3	12.8
C_{10}	C_{11}	C_{12}	C_{13}	C_{14}	C_{15}	C_{16}	C_{17}	
12.2	9.7	5	1	13.8	1	7.5	7.8	

Figure 10.25: The constraint weights learned after 6 sessions of 100 runs[5]

[5] The weights for the C_1, C_2 and C_{15} constraints are 1 because these are the IsHouse() constraints and are thus always set correctly due to the repair

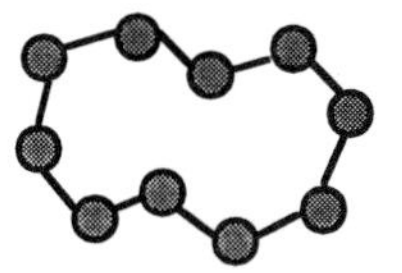

10.5.5 The Travelling Salesman Problem

The travelling salesman problem is defined as: visit a number of cities in such a way that all cities are visited exactly once and the distance travelled is minimal.

We tested a number of different problem sizes, namely: 30 cities (taken from [Oli87]), 51 cities (lin51), 105 cities (lin105) and 226 cities (pr226) (the last three taken from the ftp site softlib.cs.rice.edu in the /pub directory under the name tsplib.sh or tsplib.shar).

Order-based representation (COP).
The most natural representation for this problem is order-based since any tour can be easily represented as a permutation of cities. This is not surprising considering that order-based representation was originally introduced to deal with scheduling and TSP. Each city is assigned a number and the order in which the numbers appear in the candidate solution is the order in which they are visited. This ensures that all cities are visited exactly once.

Fitness
The fitness of a candidate is defined as the length of the tour encoded by the candidate.

Genetic operators
We tested all the order-based crossover operators on this problem. Adjacency based crossover (ABC) was designed specifically for the TSP. It is based on the premise that the absolute position in the string does not matter as much as the relative position. This crossover takes any number of parents 2 or over and generates one child by inheriting the first position from the first parent and inheriting values for subsequent positions by choosing from among the (first unused) successor to the previously inherited value in each of the parents. Occurrence-based ABC is shown below (take note: our tests were performed with a slightly different implementation of ABC).

```
void X_OB_ABC( GA *ga, Candidate *parents, Candidate *children )
{
   static int *used = NULL, **after = NULL, choices[MAX_PARENTS] = { 0 };
   int marker, x, y, next_city, last_city, nr, best, best_nr;

   if( !used )
   {
      if(!( used  = calloc( ga->CandidateLength, sizeof(int) ) ) ||
         !( after = (int **)calloc( ga->NrOfParents, sizeof(int *) ) ) )
           Error( "Not enough memory." );
```

---

mechanism. The constraint C_{13} is 1 because this is the implicit IsHouse(Blue,2) constraint.

```
        for( x = 0; x < ga->NrOfParents; x++ )
          if(!( after[x] = (int *)calloc( ga->CandidateLength, sizeof(int) ) ) )
              Error( "Not enough memory." );
    }

    memset( used, 0, ga->CandidateLength * sizeof(int) );

    for( marker = 0; marker < ga->CandidateLength - 1; marker++ )
       for( x = 0; x < ga->NrOfParents; x++ )
          after[x][parents[x].String[marker] - 1] = parents[x].String[marker + 1];

    for( x = 0; x < ga->NrOfParents; x++ )
       after[x][parents[x].String[ga->CandidateLength-1]-1] = parents[x].String[0];

    children[0].String[0] = last_city = parents[0].String[0];
    used[last_city - 1]  = 1;
    for( marker = 1; marker < ga->CandidateLength; marker++ )
    {
       for( x = 0; x < ga->NrOfParents; x++ )
       {
          int t = last_city - 1;

          while( used[after[x][t] - 1] ) t = after[x][t] - 1;

          choices[x] = after[x][t];
       }

       for( x = 0; x < ga->NrOfParents; x++ )
       {
          for( nr = 1, best = best_nr = 0, y = x + 1; y < ga->NrOfParents; y++ )
             nr += ( choices[x] == choices[y] );

          if( nr > best_nr ) best_nr = nr, best = x;
       }

       children[0].String[marker] = last_city = choices[best];
       used[choices[best] - 1]   = 1;
    }
}
```

Figure 10.26: Occurence-based ABC

Results

The results for the smaller numbers of cities were very encouraging. It was possible to find a solution to the 30 city problem in all cases when using a poolsize of 500 for the order1 crossover. When increasing the number of cities in the tour, however, we found that the best tourlength at the end of the evolution increased dramatically. The results for the order1 and the FB-ABC (with 4 parents) crossover are shown below (the poolsizes were, respectively, 100, 200, 400 and 500, mutation-rates were 20% for order1 and 100% for FB-ABC).

	Number of cities			
	30	51	105	226
Optimum	423.741	429.980	14383	Unknown
ABC	463.909	528.765	23479	181075
order1	470.160	539.458	17775	119999

Figure 10.27: Solving TSP using order1 and FB-ABC

When looking at the evolution we found that FB-ABC was (in general) a better crossover in the first part of the evolution because it was able to find better (shorter) tours much quicker than order1. However, at a certain point in the evolution (depending on the poolsize) the performance of the order1 crossover overtakes that of ABC because the genetic diversity has been lost in the case of ABC (see figure 10.28).

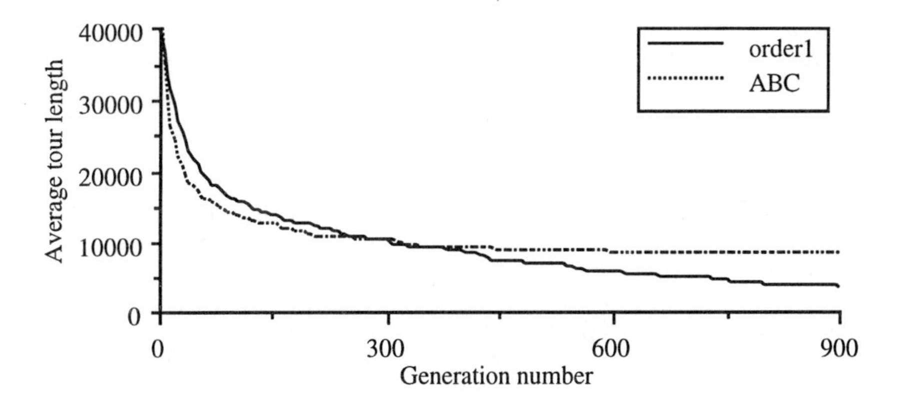

Figure 10.28: The evolution for order1 and ABC

A number of mechanisms have been tried to retain genetic diversity longer, and although we have not been able to run experiments using all these mechanisms we did find that a combination of order1 and ABC retained genetic diversity longer than ABC alone and was faster than order1 (at least at the start of the evolution).

The experiments for ABC shown above were performed using four parents. We tested different numbers of parents to see what the effect would be on the performance of the GA and found that increasing the number of parents did have a significant effect for the occurrence-based version of ABC (OB-ABC) as is shown in figure 10.29.

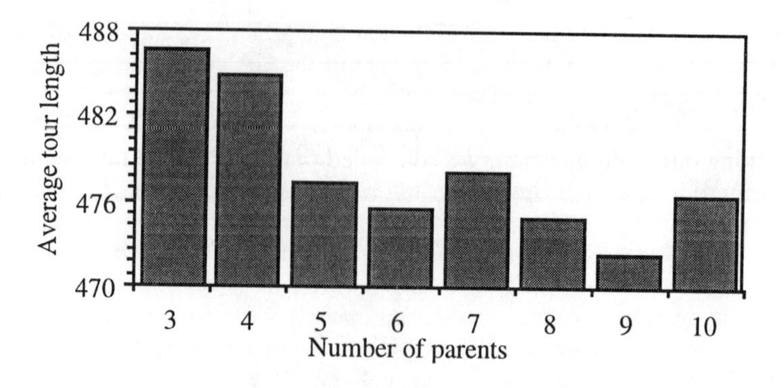

Figure 10.29: Effect of the number of parents for OB-ABC

10.6 Conclusions

We have considered the application of genetic algorithms for constrained problems. Our main question was whether GAs can be tailored to solve constrained problems and whether GAs provide a competitive technique for constraint satisfaction. First we have analysed the notion of a constrained problem and presented a framework where a problem in its original (i.e. GA independent) formulation and its GA-suited counterpart counterparts are distinguished. In its original form a search problem can be an FOP, a COP or a CSP, while a GA-suited counterpart is either an FOP or a COP. The case of a COP Æ FOP transformation (i.e. an FOP being a GA-suited counterpart of a COP) has been given some attention in the GA literature, the experiences are rather negative. We investigated the CSP Æ FOP transformation for the N-queens problem and gained similar experiences: it did not work well. Thus, the most promising, and also the most challenging, case from a GA point of view is when the GA-suited counterpart is a constrained optimisation problem. In this case one has to guarantee that the allowability constraints on candidates are always satisfied and that the genetic operators produce children that tend to be better and better along the evolution. The latter requirement means a double aim:

(i) children tend to satisfy those constraints of the original problem that are not included in the allowability but incorporated in the fitness function (in case the original problem is a CSP or COP);

(ii) children tend to score better on the original objective function in case the original problem is a COP.

For both purposes, the random mechanism common in classical operators can be replaced by some heuristic-based construction mechanism. For aim (i) heuristics from classical CSP solving search techniques can be adopted. We have applied this approach successfully for the N-queens, graph colouring and the Zebra problem, where our heuristic operators outperformed the applicable non-heuristic ones. As for aim (ii) we tested the TSP and concluded that the ABC operator based on problem dependent heuristics increased fitness more rapidly then the best conventional order-based crossover. As a negative side effect, however, the ABC operator lost genetic diversity faster. This raises the dilemma whether to prefer hare-type or turtle-type operators. Note that trying to achieve aim (ii) by domain dependent heuristics is a long existing idea in the GA community, but applying CSP heuristics for a) is a novel one.

Concerning our main question, we concluded that GAs can be tailored to solve constrained problems. For the graph colouring and the traffic light the evolutionary approach was superior to the corresponding classical methods. Classical analysis relates the density and distribution of the solutions to the easiness of the problem (similarly to the significance of the fitness landscape). N-queens is not a very hard problem in this sense, hence designing a good GA was rather easy. For the same reason, however, we could not reach the performance of the best classical techniques. The Zebra problem with difficult constraints and only one single solution turned out to be hard. Nevertheless, our experiments illustrate a number of generally applicable concepts to improve GA performance: incorporating implicit constraints, fixing certain values and a self-learning facility, where the GA learns from its own failures. One could ask whether we can suggest a 'generic' GA to solve CSPs? The answer is no; in our experiments we considered only permutation as allowability, even in the two dimensional chromosomes for the Zebra problem this was the kernel constraint. It requires further research to define operators for different allowability conditions, however, once a GA shell is given, it is easy to implement different variants of GAs.

Further work will concern comparative analysis of GAs with different fitness. An especially interesting issue is the study of COPs (in the original formulation), where the GA-suited counterpart is a COP. For these problems the aims a) and b) above may influence each other and applying more genetic operators in one evolution can be a successful approach. Also the optimal balance between heuristic based and random components in genetic operators is a subject of further investigations.

Literature

[Ado90] Adorf, H. M. and Johnston, M. D., A discrete stochastic neural network algorithm for constraint satisfaction problems, Proc. International Joint Conference on Neural Networks, San Diego, CA (1990).

[Bäc92] Bäck, T., The Interaction of Mutation Rate, Selection, and Self-Adaptation Within a Genetic Algorithm, Proc. of PPSN-92, Elsevier Science Publishers, 1992, pp. 85-94.

[Bie90] Biegel, J.E. and Davern, J.J., Genetic Algorithms and Job Shop Scheduling, Computers and Industrial Engineering, Vol. 19, Nos 1-4, 1990, pp. 81-91.

[Bra88] Brassard, G. and Bratley, P., Algorithmics — Theory and Practice, Prentice Hall, Englewood Cliffs, NJ, 1988.

[Che91] P. Cheeseman, B. Kenefsky and W.M. Taylor, Where the really hard problems are, Proc. of IJCAI-91, 1991, pp. 331-337.

[Cle89] Cleveland, G.A. and Smith S.F., Using Genetic Algorithms to Schedule Flow Shop Releases, Proc. of ICGA-89, Morgan Kaufmann, 1989, pp. 160-169.

[Col91] Colorni, A., Dorigo, M. and Maniezzo, V., Genetic Algorithms and Highly Constrained Problems, The Time-Table Case, Proc. of PPSN-90, Springer-Verlag, 1991, pp. 55-59.

[Cor93] Corcoran, A.L. and Wainwright, R.L., LibGA: A User Friendly Workbench for Order-Based Genetic Algorithm Research, Proc. of Applied Computing: Sates of the Art and Practice-1993, Indianapolis, IN, (1993), pp. 111-117.

[Cra92] Crawford, K.D., Solving the N-Queens Problem using Genetic Algorithms, Proc. ACM/SIGAPP Symposium on Applied Computing, Kansas City, Missouri, March 1-3, 1992, pp. 1039-1047.

[DavL85] Davis, L., Applying Adaptive Algoritms to Epistatic Domains, Proc. of IJCA-85, Lawrence Erlbaum Associates, 1985, pp. 162-164.

[DavL87] Davis, L. and Steenstrup, M., Genetic Algorithms and Simulated Annealing: An Overview, in L. Davis (ed.) Genetic Algorithms and Simulated Annealing, Los Altos, CA, Morgan Kaufmann, 1987, pp. 1-11.

[DavL91] Davis, L., Handbook of Genetic Algorithms, Van Nostrand Reinhold, New York 1991.

[DavY91] Davidor, Y., Epistasis Variance: A View-point on GA-Hardness, Proc. of FOGA-90, Morgan Kaufmann, 1991, pp. 23-35.

[Dec88] Dechter, R. and Pearl, J., Network-Based Heuristic for Constraint-Satisfaction Problems, Artificial Intelligence 34, 1988, pp. 1-38.

[Dec89] Dechter, R. and Pearl, J., Tree clustering for constraints networks, Artificial Intelligence 38, 1989, pp. 353-366.

[Dec90] Dechter, R., Enhancement schemes for constraint processing: Backjumping, learning, and cutset decomposition, Artificial Intelligence 41,. 1990, pp. 273-312.

[DeJ93] DeJong, K.A., Are genetic algorithms function optimizers?, Proc. of PPSN-92, Elsevier Science Publishers, 1993, pp. 3-13.

[Eib91] Eiben, A.E., Aarts, E.H.L. and van Hee, K.M., Global Convergence of Genetic Algorithms: a Markov Chain Analysis, Proc. of PPSN-90, Springer-Verlag, 1991, pp. 4-12.

[Eib93] Eiben, A.E., Raué, P.-E. and Ruttkay, Zs., Heuristic Genetic Algorithms for Constrained Problems, Working Papers of the Dutch AI Conference 1993, Twente, pp. 241-252.

Available from ftp.cs.vu.nl, /pub/papers/ai/NAIC.EibenRaueRutt kay.ps.Z

[Eib94] Eiben, A.E., Raué, P.-E. and Ruttkay, Zs., Solving Constraint Satisfaction Problems Using Genetic Algorithms, Proc. of the IEEE World Conference on Evolutionary Computation, to appear in 1994. Available from ftp.cs.vu.nl, /pub/papers/ai/WCCI.EibenRaueRutt kay.ps.Z

[Fox89] Fox, M. S., Sadeh, N. and Baykan, C., Constrained Heuristic Search, Proc. of 11th IJCAI, Detroit, 1989, pp. 309-315.

[Fox91] Fox, B.R. and McMahon, M.B., Genetic Operators for Sequencing Problems, Proc. of FOGA-90, Morgan Kaufmann, 1991, pp. 284-300.

[Fre78] Freuder, E.C., Synthesizing Constraint Expressions, Communications of the ACM Vol. 21, no. 11, 1978, pp. 958-966.

[Fre82] Freuder, E.C., A Sufficient Condition for Backtrack-Free Search, Journal of the ACM 29, 1 (1982), pp. 24-32.

[Fre85] Freuder, E.C., A Sufficient Condition for Backtrack-Bounded Search, Journal of the ACM 32, 4 (1985), pp. 775-761.

[Gol89] Goldberg, D.E., Genetic Algorithms in Search, Optimization and Machine Learning, Addison-Wesley, 1989.

[Gol91] Goldberg, D.E. and Deb, K., A Comparative Analysis of Selection Schemes Used in Genetic Algorithms, Proc. of FOGA-90, Morgan Kaufmann, 1991, pp. 69-93.

[Har80] Haralick, R.M. and Elliot, G.L., Increasing Tree Search Efficiency for Constraint Satisfaction Problems, Artificial Intelligence 14, 1980, pp. 263-313.

[Hol92] Holland, J.H., Adaptation in Natural and Artificial Systems, Ann Arbor: University of Michigan Press, 1992, second edition.

[Hom92] Homaifar, A., Turner, J. and Ali, S., The N-Queens Problem and Genetic Algorithms, Proc. IEEE Southeast Conf. 1992, Vol. 1, pp. 262-267.

[How93] Hower, W. and Jaboci, S., Parallel distributed constraint satisfaction, Proc. of the of the Second International Workshop on Parallel Processing for Artificial Intelligence, IJCAI-93, Chambery, France 1993, pp. 65-68

[Joh91] Johnson, D.S., Aragon, C.R., McGeoch, L.A. and Schevon, C., Optimization by simulated annealing: an experimental evaluation, Part II, J. Oper. Res., 39 (3) (1991), pp. 378-406.

[Khu94a] Khuri, S., Bäck, T. and Heitkötter, J., An Evolutionary Approach to Combinatorial Optimization Problems, to appear in the proceedings of CSC'94, Phoenix Arizona, March 8-10, 1994 (ACM Press).

[Khu94b] Khuri, S., Bäck, T. and Heitkötter, J., The Zero/One Multiple Knapsack Problem and Genetic Algorithms, to appear in the ACM Symposium of Applied Computation (SAC'94) proceedings (ACM Press).

[Mac77] Mackworth, A.K., Concistency in Networks of Relations, Artificial Intelligence Vol. 8 (1977), pp. 99-118.

[Mes89] Meseguer, P., Constraint Satisfaction Problems: An Overview, AICOM, Vol. 2, no. 1, March 1989, pp. 3-17.

[Mic91] Michalewicz, Z. and Janikow, C.Z., Handling constraints in genetic algorithms, Proc. of ICGA-91, Morgan Kaufmann, 1991.

[Mic92] Michalewicz, Z., Genetic Algorithms + Data Structures = Evolution Programs, Springer-Verlag, 1992.

[Min90] Minton, S., Johnston, M. D., Philips, A. and Laird, P., Solving large scale constraint staisfaction and scheduling problems using a heuristic repair method, Proc. of AAAI-90, Boston, MA (1990), pp. 17-24.

[Min92] Minton, S., Johnston, M. D., Philips, A. and Laird, P., Minimizing conflicts: a heuristic repair method for constraint satisfaction and scheduling problems, Artificial Intelligence 58 (1992), pp. 161-205.

[Mor92] Morris, P., On the density of solutions in equilibrium points for the queens problem, Proc. AAAI-92, San Jose, CA (1992) pp. 428-433.

[Müh89] Mühlenbein, H., Parallel Genetic Algorithms, Population Genetics and Combinatorial Optimization, Proc. of ICGA-89, Morgan Kaufmann, 1989, pp. 416-421.

[Nud83] Nudel, B., Consistent-Labeling Problems and their Algorithms: Expected Complexities and Theory Based Heuristics, Artificial Intelligence 21, 1983, pp. 135-178.

[Oli87] Olivier, I.M., Smith, D.J. and Holland, J.C.R., A Study of Permutation Crossover Operators on the Travelling Salesman Problem, Proc. of ICGA-87, Lawrence Erlbaum Associates, 1987, pp. 224-230.

[Par92] Paredis, J., Exploiting constraints as background knowledge for a case-study for scheduling, Proc. of PPSN-92, Elsevier Science Publishers, 1992, pp. 229-238.

Also as: Technical Report 92-60 of the Research Institute for Knowledge Systems, Maastricht, 1992.

[Rad92] Radcliffe, N.J., Non-Linear Genetic Representations, Proc. of PPSN-92, Elsevier Science Publishers, 1992, pp. 259-268.

[Ric89] Richardson, J.T., Plamer, M.R., Liepins, G.E. and Hilliard, M., Some guidelines for genetic algorithms with penalty functions, Proc. of ICGA-89, Morgan Kaufmann, 1989, pp. 191-197.

[Sch89] Schaffer, J.D., Caruana, R.A., Eshelman, L.J. and Das, R., A Study of Control Parameters Affecting Online Performance of Genetic Algorithms for Function Optimization, Proc. of ICGA-89, Morgan Kaufmann, 1989, pp. 51-60.

[Sta91] Starkweather, T., McDaniel, S., Mathias, K. and Whitley D., A Comparison of Genetic Sequencing Operators, Proc. of ICGA-91, Morgan Kaufmann, 1991, pp. 69-76.

[Sos91] Sosic, R. & Gu, J., 3.000.000 queens in less than a minute, SIGART Bulletin No. 2, vol. 2, pp. 22-24.

[Suh87] Suh, J.Y. and van Gucht, D., Incorporating Heuristic Information into Genetic Search, Proc. of the ICGA-87, Lawrence Erlbaum Associate, 1987, pp. 100-107.

[Sys91a] Syswerda, G., Schedule Optimization Using Genetic Algorithms, pp. 332-349 in [DavL91].

[Sys91b] Syswerda, G., A Study of Reproduction in Generational and Steady State Genetic Algorithms, Proc. of FOGA-90, Morgan Kaufmann, 1991, pp. 94-101.

[Tsa90] Tsang, E.P.K. and Warwick, T., Applying genetic algorithms to constraint satisfaction optimization problems, in: Arello, L. C. (eds): Proc. of ECAI-90, Pitman Publishing, 1990, pp. 649-654.

[Whi89a] Whitley, D., The GENITOR Algorithm and Selective Pressure, Proc. of ICGA-89, Morgan Kaufmann, 1989, pp. 116-121.

[Whi89b] Whitley, D., Starkweather, T. and Fuquay, D., Scheduling Problems and Taveling Salesmen: The Genetic Edge Recombination Operator, Proc. of ICGA-89, Morgan Kaufmann, 1989, pp. 133-140.

[Whi91] Whitley, D., Starkweather, T. and Shanner, D., The Traveling Salesman and Sequence Scheduling: Quality Solutions Using Genetic Edge Recombination, pp. 350-372 in [DavL91].

Appendix: Source code for a genetic algorithm

GA.h

```
#include <stdio.h>
#include <stdlib.h>
#include <string.h>
#include <time.h>
#include <limits.h>

#define MAX_PARENTS         20
#define MAX_CHILDREN        20
#define NO_OPTIMAL_FITNESS  -9E99

#define Error( message )          { fprintf( stderr, "%s\n", message ); exit( 0 ); }
#define RandomFraction()          ( (double)rand() / INT_MAX )
#define RandomDomain( low, high ) (int)( RandomFraction() * ( (high)-(low) +
0.999999 ) + (low) )
#define Better( ga, cand1, cand2 ) ( ga->Minimize ? \
                               (cand1)->Fitness < (cand2)->Fitness : \
                               (cand1)->Fitness > (cand2)->Fitness     )
#define Worse( ga, cand1, cand2 )  !Better( ga, cand1, cand2 )

struct _GA;

typedef struct _Candidate
{
   int    *String;
   double  Fitness, FitnessPercentage;
} Candidate;

typedef struct _Pool
{
   Candidate *Candidates, *Best;
   int        Size;
   double     TotalFitness;
} Pool;

typedef struct _Node
{
   struct _Node *Next;
   char          Name[64];
} Node;

typedef struct _MuOperator
{
   Node  Node;
   void (*MutationFunction)( struct _GA *, Candidate * );
} MuOperator;
```

```
typedef struct _XOperator
{
   Node  Node;
   void (*XOverFunction)( struct _GA *, Candidate *, Candidate * );
} XOperator;

typedef struct _Representation
{
   Node  Node;
   void (*GenerateCandidate)( struct _GA *, Candidate * );
} Representation;

typedef struct _GA
{
   Pool           *Parents, *Children;
   int            Poolsize, MaxNrOfGenerations, CandidateLength, NrOfParents,
NrOfChildren,
                         Minimize, SteadyState, Elitism, Eon, ReportInterval,
Generation;
   double         CrossoverRate, MutationRate, OptimalFitnessValue;
   double         (*FitnessFunction)( struct _GA *, Candidate * );
   void           (*EonFunction)( struct _GA * );
   Representation *Representations, *Representation;
   XOperator      *CrossoverOperators, *Crossover;
   MuOperator     *MutationOperators, *Mutation;
} GA;

void Error( char *message );
void CopyCandidate( GA *, Candidate *, Candidate * );
void RunGA( GA * );
void AddMutationOperator( GA *, char *, void (*mu_fun)( GA *, Candidate * )
);
void AddRepresentation( GA *, char *, void (*gen_fun)( GA *, Candidate * ) );
void AddCrossover( GA *, char *, void (*x_fun)( GA *, Candidate *, Candidate *
) );
void ConfigureGA( GA *, char * );
GA *AllocateGA( double (*fitness)( GA *, Candidate * ), void (*eonfun)( GA * )
);
```

Representation.c

```
#include "GA.h"

void GenBitCandidate( GA *ga, Candidate *candidate )
{
   int y;

   for( y = 0; y < ga->CandidateLength; y++ )
      candidate->String[y] = RandomFraction() >= 0.5 ? 1 : 0;
}
```

```
void InitializeRepresentations( GA *ga )
{
   AddRepresentation( ga, "bit-pattern", GenBitCandidate  );
}
```

Mutation.c

```
#include "GA.h"

void M_InvertPosition( GA *ga, Candidate *candidate )
{
   int x = RandomDomain( 0, ga->CandidateLength - 1 );

   candidate->String[x] = !candidate->String[x];
}

void InitializeMutationOperators( GA *ga )
{
   AddMutationOperator( ga, "invert-position", M_InvertPosition );
}
```

Crossover.c

```
#include "GA.h"

void X_TailExchange( GA *ga, Candidate *parents, Candidate *children )
{
   int x, x_point = RandomDomain( 0, ga->CandidateLength - 1 );

   CopyCandidate( ga, &parents[0], &children[0] );
   CopyCandidate( ga, &parents[1], &children[1] );

   for( x = x_point; x < ga->CandidateLength; x++ )
   {
      children[0].String[x] = parents[1].String[x];
      children[1].String[x] = parents[0].String[x];
   }
}

void X_Uniform( GA *ga, Candidate *parents, Candidate *children )
{
   int x;

   for( x = 0; x < ga->CandidateLength; x++ )
      if( RandomFraction() >= 0.5 )
      {
         children[0].String[x] = parents[0].String[x];
         children[1].String[x] = parents[1].String[x];
      }
```

```
        else
        {
           children[0].String[x] = parents[1].String[x];
           children[1].String[x] = parents[0].String[x];
        }
}

void InitializeCrossovers( GA *ga )
{
   AddCrossover( ga, "tail-exchange", X_TailExchange );
   AddCrossover( ga, "uniform"      , X_Uniform      );
}
```

GA.c

```
#include "GA.h"

extern void InitializeCrossovers( GA * );
extern void InitializeMutationOperators( GA * );
extern void InitializeRepresentations( GA * );

GA DefaultGA = { NULL, NULL, 100, 100, 10, 2, 2, 1, 0, 1, 0, 1, 1.0, 0.0,
NO_OPTIMAL_FITNESS };

void SetPercentages( GA *ga, Pool *pool )
{
   int        x;
   double     total = 0;
   Candidate *can   = pool->Candidates;

   if( ga->Minimize )
   {
      for( x = 0; x < pool->Size; x++ )
         total += can[x].FitnessPercentage = pool->TotalFitness / can[x].Fitness;

      for( x = 0; x < pool->Size; x++ )
         can[x].FitnessPercentage /= total;
   }
   else
      for( x = 0; x < pool->Size; x++ )
         can[x].FitnessPercentage = can[x].Fitness / pool->TotalFitness;
}

Candidate *SelectParent( GA *ga )
{
   double percentage = RandomFraction();
   int    x          = 0;

   while( percentage > 0 && x < ga->Parents->Size )
      percentage -= ga->Parents->Candidates[x++].FitnessPercentage;
```

```
    return( &ga->Parents->Candidates[x - 1] );
}

void InitParentPool( GA *ga )
{
    int x;

    for( x = 0; x < ga->Poolsize; x++ )
    {
        ga->Representation->GenerateCandidate( ga, &ga->Parents->Candidates[x]
);

        ga->Parents->TotalFitness += ga->FitnessFunction( ga, &ga->Parents-
>Candidates[x] );

        if( !ga->Parents->Best || Better( ga, &ga->Parents->Candidates[x], ga-
>Parents->Best ) )
            ga->Parents->Best = &ga->Parents->Candidates[x];
    }

    ga->Parents->Size = ga->Poolsize;
}

void CopyCandidate( GA *ga, Candidate *src, Candidate *dst )
{
    int *tmp = dst->String;

    memcpy( dst->String, src->String, ga->CandidateLength * sizeof(int) );
    memcpy( dst, src, sizeof(Candidate) );

    dst->String = tmp;
}

Candidate *AllocateCandidates( GA *ga, int nr_of_candidates )
{
    Candidate *candidates;
    int       x;

    if( candidates = calloc( nr_of_candidates, sizeof(Candidate) ) )
        for( x = 0; x < nr_of_candidates; x++ )
            if(!( candidates[x].String = calloc( ga->CandidateLength, sizeof(int) ) ) )
                return( NULL );

    return( candidates );
}

Pool *AllocatePool( GA *ga )
{
    Pool *p = calloc( 1, sizeof(Pool) );
```

```
        if( p && ( p->Candidates = AllocateCandidates( ga, ga-
>Poolsize+MAX_PARENTS+MAX_CHILDREN ) ) )
      return( p );
   else
      return( NULL );
}

void SelectSurvivors(GA *ga,Candidate *parents,int nr_of_parnts,Candidate
*children,int nr_of_children)
{
   int x, y, z;

   for( x = 0; x < nr_of_parnts; x++ )
   {
      z = 0;

      for( y = 1; y < nr_of_children; y++ )
        if( Worse( ga, &children[y], &children[z] ) )
             z = y;

      if( Better( ga, &parents[x], &children[z] ) )
        CopyCandidate( ga, &parents[x], &children[z] );
   }
}

void AddToPool( GA *ga, Pool *pool, Candidate *candidates, int
nr_of_candidates )
{
   int x;

   for( x = 0; x < nr_of_candidates; x++, pool->Size++ )
   {
     CopyCandidate( ga, &candidates[x], &pool->Candidates[pool->Size] );

      pool->TotalFitness += candidates[x].Fitness;

      if( !pool->Best || Better( ga, &candidates[x], pool->Best ) )
         pool->Best = &pool->Candidates[pool->Size];
   }
}

void ReplaceInPool( GA *ga, Pool *pool, Candidate *candidates, int
nr_of_candidates )
{
   int x, y;

   for( x = 0; x < nr_of_candidates; x++ )
       for( y = 0; y < pool->Size; y++ )
         if( Worse( ga, &pool->Candidates[y], &candidates[x] ) )
```

```
        {
                    pool->TotalFitness += candidates[x].Fitness - pool-
>Candidates[y].Fitness;

            CopyCandidate( ga, &candidates[x], &pool->Candidates[y] );

            if( Better( ga, &pool->Candidates[y], pool->Best ) )
               pool->Best = &pool->Candidates[y];

            break;
        }
}

int GenerateNewIndividuals( GA *ga, Candidate *parents, Candidate *children )
{
   int x, crossover = 1, nr_of_parents = ga->NrOfParents, nr_of_children = ga-
                                                            >NrOfChildren;

   if(!(crossover = (RandomFraction()<=ga->CrossoverRate))) nr_of_parents=nr_
                                                            of_children = 1;

   for( x = 0; x < nr_of_parents; x++ ) CopyCandidate( ga, SelectParent( ga ),
                                                            &parents[x] );

   if( crossover ) ga->Crossover->XOverFunction( ga, parents, children );
   else         CopyCandidate( ga, &parents[0], &children[0] );

   for( x = 0; x < nr_of_children; x++ )
      if( RandomFraction() <= ga->MutationRate )
         ga->Mutation->MutationFunction( ga, &children[x] );

   for( x = 0; x < nr_of_children; x++ ) ga->FitnessFunction( ga, &children[x] );

   if( ga->Elitism ) SelectSurvivors( ga, parents, nr_of_parents, children,
                                                            nr_of_children );

   return( nr_of_children );
}

void SteadyState( GA *ga, Candidate *parents, Candidate *children )
{
   int nr_of_children = GenerateNewIndividuals( ga, parents, children );

   ReplaceInPool( ga, ga->Parents, children, nr_of_children );
}

void Generational( GA *ga, Candidate *parents, Candidate *children )
{
   Pool *tmp;
   int  nr_of_children;
```

```
  ga->Children->Size = ga->Children->TotalFitness = 0;
  ga->Children->Best = NULL;

  if( ga->Elitism ) AddToPool( ga, ga->Children, ga->Parents->Best, 1 );

  while( ga->Children->Size < ga->Poolsize )
  {
    nr_of_children = GenerateNewIndividuals( ga, parents, children );

    AddToPool( ga, ga->Children, children, nr_of_children );
  }

  tmp         = ga->Parents;
  ga->Parents  = ga->Children;
  ga->Children = tmp;
}

void RunGA( GA *ga )
{
  int              x;
  static Candidate *parents = NULL, *children = NULL;

  if( !ga->Crossover || !ga->Mutation || !ga->Representation )
    Error( "Not set: crossover, mutation or representation." );

  if( !parents )
  {
    parents     = AllocateCandidates( ga, MAX_PARENTS  ),
    children    = AllocateCandidates( ga, MAX_CHILDREN );
    ga->Parents  = AllocatePool( ga );
    ga->Children = AllocatePool( ga );
  }

  if( !ga->Parents || !ga->Children || !parents || !children ) Error( "Not enough
memory." );

  InitParentPool( ga );

  if( ga->Eon && ga->EonFunction ) ga->EonFunction( ga );

  for( ga->Generation = 1; ga->Generation <= ga->MaxNrOfGenerations; ga-
>Generation++ )
  {
    if( !ga->Generation || !( ga->Generation % ga->ReportInterval ) )
      printf( "(%5d) Best element found: %G, Average: %G\n",
              ga->Generation, ga->Parents->Best->Fitness,
              ga->Parents->TotalFitness / ga->Parents->Size );
```

```
            if( (   ga->Minimize && ga->Parents->Best->Fitness <= ga->OptimalFitnessValue ) ||
                ( !ga->Minimize && ga->Parents->Best->Fitness >= ga->OptimalFitnessValue )   )
          break;

       if( ga->Eon && !( ga->Generation % ga->Eon ) && ga->EonFunction ) ga->EonFunction( ga );

       SetPercentages( ga, ga->Parents );

      if( ga->SteadyState ) SteadyState( ga, parents, children );
      else                  Generational( ga, parents, children );
   }

    printf( "Best element found in gen %d (%G): ", ga->Generation-1, ga->Parents->Best->Fitness );

   for( x = 0; x < ga->CandidateLength; x++ ) printf( "%d ", ga->Parents->Best->String[x] );

   printf( "\n" );
}

void *AddNode( int size, char *name, void *next )
{
   Node *node = calloc( 1, size );

   if( !node ) Error( "Not enough memory." );

   strcpy( node->Name, name );
   node->Next = next;

   return( node );
}

void AddMutationOperator( GA *ga, char *name, void (*mu_fun)( GA *, Candidate * ) )
{
      MuOperator *mu_op = AddNode( sizeof(MuOperator), name, ga->MutationOperators );

   mu_op->MutationFunction = mu_fun;
   ga->MutationOperators   = mu_op;
}

void AddRepresentation( GA *ga, char *name, void (*gen_fun)( GA *, Candidate * ) )
{
```

```
    Representation *rep = AddNode( sizeof(Representation), name, ga->Representations );

  rep->GenerateCandidate = gen_fun;
  ga->Representations    = rep;
}

void AddCrossover( GA *ga, char *name, void (*x_fun)( GA *, Candidate *, Candidate * ) )
{
    XOperator *x_op = AddNode( sizeof(XOperator), name, ga->CrossoverOperators );

  x_op->XOverFunction    = x_fun;
  ga->CrossoverOperators = x_op;
}

void *FindName( void *list, char *name )
{
  Node   *t;

  for( t = (Node *)list; t; t = t->Next )
    if( strcmp( t->Name, name ) == 0 ) return( t );

  fprintf( stderr, "Unknown: %s\nPossible choices:\n", name );

  for( t = (Node *)list; t; t = t->Next ) printf("%s\n", t->Name );

  exit( 0 );
}

GA *AllocateGA( double (*fitness)( GA *, Candidate * ), void (*eonfun)( GA * ) )
{
  GA *ga;

  if( ga = calloc( 1, sizeof(GA) ) )
  {
    memcpy( ga, &DefaultGA, sizeof(GA) );

    InitializeCrossovers( ga );
    InitializeMutationOperators( ga );
    InitializeRepresentations( ga );

    ga->FitnessFunction = fitness;
    ga->EonFunction     = eonfun;
  }

  return( ga );
}
```

```
char _Line[256];

void ConfigureGA( GA *ga, char *configfile )
{
  FILE *file;
  char  argument[64];

  srand( time( NULL ) );

  if(!( file = fopen( configfile, "r" ) ) ) Error( "Can't open configuration file." );

  while( fgets( _Line, 255, file ) )
  {
    int parsed_line = 1;

    if( sscanf( _Line, "crossoverrate %lf", &ga->CrossoverRate ) ) ;
    else if( sscanf( _Line, "mutationrate %lf", &ga->MutationRate ) ) ;
    else if( sscanf( _Line, "mutation %s", argument ) )
      ga->Mutation = FindName( ga->MutationOperators, argument );
    else if( sscanf( _Line, "crossover %s", argument ) )
      ga->Crossover = FindName( ga->CrossoverOperators, argument );
    else if( sscanf( _Line, "minimize %s", argument ) )
      ga->Minimize = ( strcmp( argument, "true" ) == 0 );
    else if( sscanf( _Line, "ga-model %s", argument ) )
      ga->SteadyState = ( strcmp( argument, "steady-state" ) == 0 );
    else if( sscanf( _Line, "representation %s", argument ) )
      ga->Representation = FindName( ga->Representations, argument );
    else if( sscanf( _Line, "elitism %s", argument ) )
      ga->Elitism = ( strcmp( argument, "true" ) == 0 );
    else
      parsed_line = sscanf( _Line, "#%c", &argument[0] );

    parsed_line += sscanf( _Line, "nr-of-parents %d" ,&ga->NrOfParents);
    parsed_line += sscanf( _Line, "nr-of-children %d" ,&ga->NrOfChildren);
    parsed_line += sscanf( _Line, "optimal-fitness %lf", &ga-
                                          >OptimalFitnessValue );
    parsed_line += sscanf( _Line, "candidate-length %d", &ga-
                                          >CandidateLength     );
    parsed_line += sscanf( _Line, "poolsize %d"     , &ga->Poolsize          );
    parsed_line += sscanf( _Line, "max-generations  %d" , &ga-
                                          >MaxNrOfGenerations  );
    parsed_line += sscanf( _Line, "generations-per-eon %d", &ga->Eon );
    parsed_line += sscanf( _Line, "report-interval %d", &ga->ReportInterval );

    if( !parsed_line ) printf( "IGNORED: %s", _Line );
  }

  fclose( file );
```

```
        if( ga->NrOfParents >MAX_PARENTS || ga->NrOfChildren >
MAX_CHILDREN )
      Error( "Too many parents and/or children." );
}
```

Example GA

```
#include "GA.h"

double Fitness( GA *ga, Candidate *candidate )
{
        int      x;
        candidate->Fitness = 0;
        for( x = 0; x < ga->CandidateLength; x++ )
                candidate->Fitness += candidate->String[x];
        return( candidate->Fitness );
}

main()
{
        GA *ga = AllocateGA( Fitness, NULL );

        ConfigureGA( ga, "config-file" );
        RunGA( ga );
}
```

config-file

```
mutation invert-position        #
crossover tail-exchange         #
minimize true                   # true false
ga-model generational           # steady-state generational
representation bit-pattern      #
crossoverrate 1.0               # 0.0 - 1.0
mutationrate 1.0                # 0.0 - 1.0
nr-of-parents 2                 # 0 - MAX_PARENTS
nr-of-children 2                # 0 - MAX_CHILDREN
optimal-fitness 0.0             # 0 - Inf.
candidate-length 100            # 1 - Inf.
poolsize 100                    # 1 - Inf.
max-generations 100             # 0 - Inf.
elitism true                    # true false
report-interval 1               # 1 - Inf.
generations-per-eon 0           # 0 - Inf.
```

Chapter 11

Simon Ronald
School of Computer and Information Science
The University of South Australia
The Levels
S.A. 5095, Australia

maspr@lux.levels.unisa.edu.au

Routing and Scheduling Problems

0-8493-2519-6/95/$0.00 + $.50

11.1 Scheduling Genetic Algorithms

GAs have been used successfully to solve continuous functional optimization problems. We typically use a binary encoding to map a GA chromosome to a single point in the problem space. In many functional problems we must solve an n-dimensional problem. To do this we use a parameter set which is partitioned into bit-strings. Each bit-string encodes a single parameter. The classic operators which act on these bit-strings are *crossover* and *mutation.* Crossover typically involves exchanging randomly selected bit-string chunks between two parents to create two children genotypes. Mutation involves scanning each bit in the bit-string, and with a low probability, toggling that bit. However, the genetic encoding and these genetic operators are not suitable for routing type problems. With routing problems we must find an optimal ordering of a list of objects in order to solve the problem. For example, in a Travelling Salesperson Problem (TSP) we must find an optimal ordering of towns (objects) in order to minimize travelled distance (see section 11.2). In the real world we can find many examples of these hard-to-solve routing and scheduling problems. These include timetable optimization, train scheduling, resource scheduling, vehicle re-routing and job shop scheduling.

In well studied problems such as the travelling salesperson problems, there are problem specific ways of finding exact solutions to smaller sized problems [EJKe85]. However, often these techniques require a deep insight into the nature of the specific problem and the techniques are often not portable to other scheduling type problems. It is well accepted that the basic GA algorithm can be used to optimize a very broad class of problem. If the genetic representation of two different scheduling problems is identical, then it is usually possible to optimize both problems while making few changes to the underlying GA. In this case we may only need to tune the operational parameters of the GA and write a new way of calculating the cost of the solution (the objective function). Genetic algorithms that use a standard genetic encoding scheme and off-the-shelf genetic operators may be used to solve a wide range of routing problems. However, we will see in Section 11.5, that a number of real-world problems use problem-domain specific genetic representations and use operators augmented with problem related knowledge. These techniques may be required to model a complex real-world problem, or simply to improve on the results that can be attrained with a conventional routing GA.

In this chapter we discuss three different scheduling problems, namely the travelling salesperson problem, the job shop scheduling problem and the open shop scheduling problem. We define and discuss some practical applications of each problem. We then describe the C++ code required to calculate the fitness of an input genotype (fitness function). We discuss the C++ classes relevant to each problem and describe, with code fragments, how each member function operates. If the reader does not wish to become bogged down with implementation details, then the code sections can be skipped without losing the thread of the discussion. We follow on to describe, in detail, the genetic operators suited to each problem type. For each genetic operator, namely Edge Recombination, and Linear Order Crossover, we discuss the theoretical basis for the operator and present a C++ implementation of each. All of the code in this chapter (and more) can be downloaded from the University of South Australia without charge using ftp on the Internet service[1]. This account also contains a complete GA implementation with other features and test problems not discussed in this text. In the final section (Section 11.5) we discuss a selection of a number of routing and scheduling applications to illustrate the diversity of uses and the impact that GAs have made in solving notoriously complex real-world problems.

11.2 The Travelling Salesperson Problem

11.2.1 Introduction

The Travelling Salesperson Problem (TSP) is a classic scheduling problem in which N towns are distributed around a two dimensional Euclidean plane. A salesperson starts at an initial town, and travels to each other town and returns back to the initial town. Each town is visited once. The salesperson may travel from any town, to any other unvisited town. The goal of this problem is to find a circuit of all towns which has a minimal total distance. The distance between two towns is calculated as the Euclidean distance between the coordinates of each town, and the cost of a circuit is the sum of all of the distances between adjacent towns comprising the circuit. This problem can be mapped to the general graph optimization problem 'Find the optimal length Hamiltonian circuit of a fully connected weighted graph of order N as described in [GMJ79], and is known to be NP-hard. No polynomial time algorithm is known to solve the general problem of N towns. However, heuristic and nondeterministic methods have been used to find good, and sometimes optimal solutions in reasonable amounts of time ([EJKe85, EM92]). In the TSP, the different towns, which may be represented by numerical labels as 0, 1,..., N - 1 represent the problem related object list. The objects in the list are towns. The fitness of a circuit depends on the *order* of the towns which comprise that circuit, or the order of the objects in the object list.

11.2.2 Applications of the TSP

There are many applications which can be modelled as a TSP. These applications include: routing security surveillance visits to locations, courier pickup routing, and the design of electricity supply networks. Foulds [Fou92] gives a good example of a problem which often arises in the pharmaceutical industry. In the example cited therein, *N* drugs are manufactured in a single reaction vessel one at a time. Each drug is manufactured one by one, in a specific order. After the last (*N*th) drug is manufactured the first drug is then manufactured again so manufacture proceeds this way in a cyclic manner.

If drug d_i is manufactured, and then is followed by the manufacture of drug d_j, there is a cleaning cost that is incurred which depends on the two drugs d_i and d_j. This problem can be directly mapped from a TSP if we use the cleaning cost c_{ij} in place of distances, and the drugs $d_1 \ldots d_N$ in place of towns. The goal of the problem then becomes 'find the order of manufacture of drugs which minimizes the total cleaning cost.'

---

[1] ftp to genopop.levels.unisa.edu.au. Use the user name of 'anonymous' and enter your email address as a password. Change to the directory/pub/genopop. Get the README file first and read it before downloading other files.

Another practical example of the TSP was cited in [EM92]. Evans discussed a printed circuit board manufacturing environment where N holes must be drilled in printed circuit boards. Between holes, the machine moves the work on the bench a measure of distance sufficient to align the drill to the next hole. The objective is how we find an optimum drilling sequence for the programmable drilling machine that minimizes the total amount of distance the machine needs to move the PCB on the bench. This problem can be modelled as a TSP if we consider each hole identifier to be a town identifier, and the distance between holes to be the distances between the towns.

The TSP has been extensively studied and there are dozens of good heuristic and exact techniques for finding good solutions, to smal1 problems (less than 1000 towns), in achievable amounts of time. However, the general problem for large numbers of towns is still intractable. When applying GAs to the TSP we must always keep in mind that the result is not guaranteed to be the best possible solution. However, GAs typically find good solutions to the TSP, especially when used with suitable genetic operators such as the edge recombination operator [WSF89]. We can learn a good deal about scheduling genetic algorithms by examining how we solve the TSP which is a classic hard scheduling problem. We can gain a good deal of insight into how scheduling GAs work even if the scheduling problem we wish to solve is different to that of a TSP.

11.2.3 A Sample TSP Problem

Before trying out our GA on our own TSP problem, it is a good idea to test our GA implementation against a number of well know problems where we are given the value of the optimal cycle length. We may then confirm that our GA solves these test problem instances efficiently before tackling our own problem. Knowledge of the best possible problem solution can help us tune the parameters of the GA, such as: the population size, the selection scheme, the selection bias, the type of operator, the mating scheme, and the reproduction technique. Often we must play with these and other such parameters to get a feel for how they affect the quality of the results obtained.

The Harrell 30 town TSP problem is one of the more popular problems used as a benchmark in GA research.

The following list of coordinate pairs is ordered according to the optimal circuit which has a length of 420 distance units[2].

H_{30} = ((64 60), (68 58), (71 44), (83 46), (91 38), (82 7), (62 32), (58 35), (45 21), (41 26), (44 35), (25 38), (24 42), (18 40), (13 40), (4 50), (18 54), (22 60), (25 62), (7 64), (2 99), (41 94), (37 84), (54 67), (54 62), (58 69), (71 71), (74 78), (87 76), (83 69), (64 60))

---

2 All town distances are rounded to the nearest integer, see Equation 17.2.4.2 in Section 17.2.4

H_{30} above contains 31 (X Y) coordinate pairs. Each coordinate pair describes the location of a town on a 2D plane. The optimal tour of these 30 towns starts at location (64 60), proceeds to location (68 58) etc. and 28 towns later finally finishes back at the starting point of location (64 60). This optimal tour can help us to determine whether our GA is performing correctly.

We know that a GA consists of a population of genotypes where each genotype encodes a point in the solution space [Gol89, Dav91]. To encode a genotype to a valid TSP cycle, the genotype is arranged as a list of integer values where each integer encodes a town identifier. In an example problem with a total of four towns the genotype might consist of the integer list (1, 2, 0, 3). This is interpreted as a TSP cycle that begins with town id. 1, followed by town id. 2, to 0, to 3, and then back to town id. 1. Obviously such a list must not contain any absent or duplicate town identifiers or else the resulting cycle will not be Hamiltonian (i.e. pass through each town once). Such anomolies are avoided as the initial random population is generated by creating each genotype by shuffling a complete list of gene values. From that point on, the genetic operator, which is responsible for creating new children, will only generate legal TSP circuits.

To illustrate a test problem we configured a steady state genetic algorithm. Other than the genetic operator and the implementation of the fitness function, we will not describe the implementation of this steady state GA in detail, rather we refer the reader to a general reference about steady state GAs. See [Dav91].

We ran our GA with pool-size of 1000 genotypes, with a linear selection bias of 1.9. The Edge Recombination Operator was used to create new children from selected parents (see Section 11.2.7.1). We reached the best known solution before 10,000 generations. In this example 37 population-best genotypes (champions) were produced before the final solution was reached. We can examine the progression of the solutions in the GA by graphing the fitness of the population-best genotype champions and noting the generation that each champion was produced. We do this by picking selected circuit files which are produced by the *circuit_dump* member function in the code (see Section 11.2.4). Figures 11.1 through 11.3 illustrate a selection of six such champion TSP circuits taken at generations 0, 152, 3652, 5121, 6627, and finally generation 9672.

11.2.3.1 The Initial Population

We can observe in Figure 11.1 the poor quality (fitness is 0.32967) of the best population member at generation 0. This in not surprising as all 1000 initial population genotypes were generated using a random gene shuffling algorithm. Random schedules are often used to start GA populations. They ensure a high degree of genetic diversity which means the genetic pool contains most, if not all possible edge relationships from the complete graph defining the TSP. Despite this useful diversity, a random population such as the one in this example relies on the GA to perform a good deal of unnecessary optimization to reach a reasonable level of fitness which occurs well into the population. It has been shown ([Bra91]) that larger TSP problems (N >> 50) can be solved faster, with better final solutions attained, if the initial population is created using a greedy heuristic algorithm coupled with a tour improvement algorithm. Braun stresses that his greedy initialization algorithm was specially designed to ensure not only reasonably good starting genotypes, but also a diverse population composition.

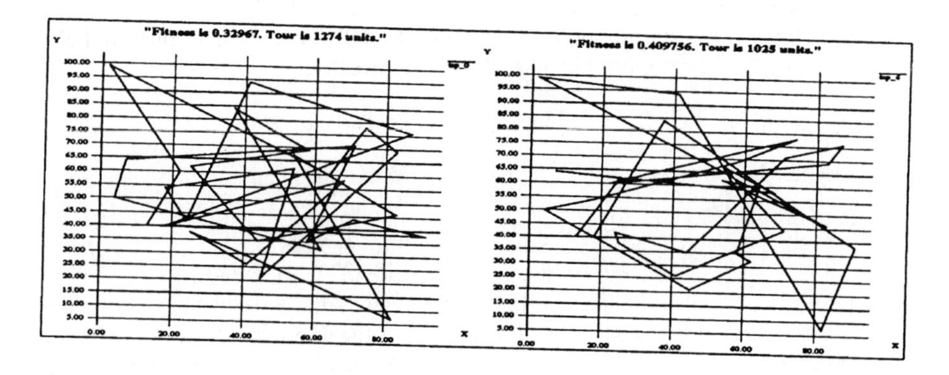

Figure 11.1: The Best Individual in the Initial Random Population (left). Champion at generation 152(right).

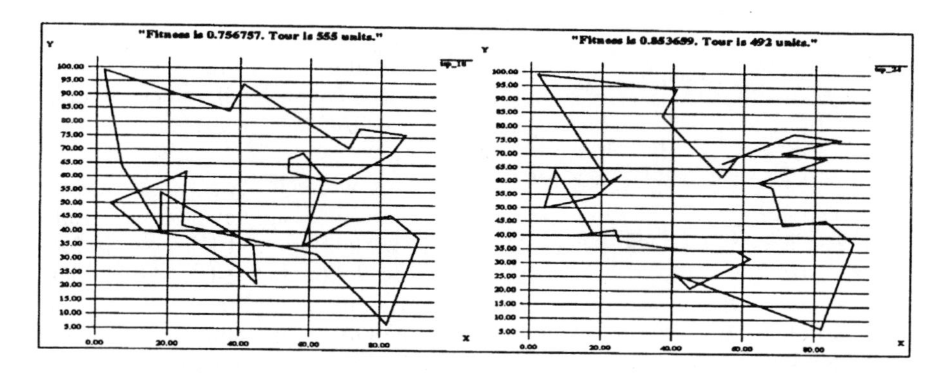

Figure 11.2: Champion at generation 3652 (left). Champion at generation 5121 (right).

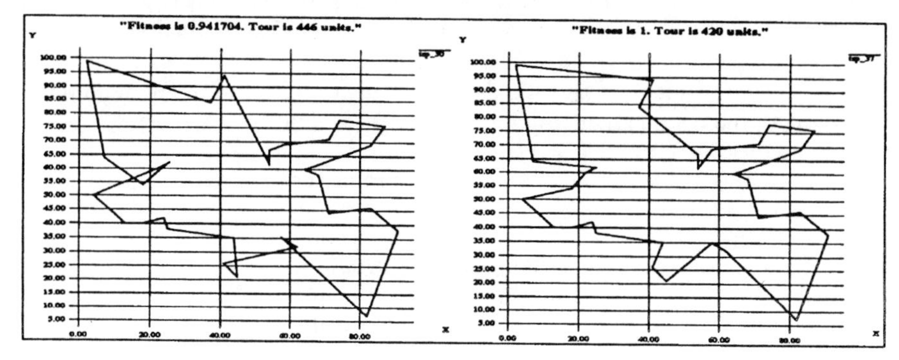

Figure 11.3: Champion at generation 6627 (left). Optimal Solution reached at generation 9672 (right).

11.2.4 Implementing the TSP

In a GA, a genotype encodes a point in the problem space, and for scheduling type problems consists of a list of N genes. In the case of the TSP, if we have a problem size of 30 cities, then there will be 30 genes in the genotype. A single gene stores a number which identifies a town. Towns are identified by 0, for the first town, 1 for the second town, through to (N - 1) for the last encoded town. In our GA implementation we use genotypes which are represented by a class called *individual.* The *individual* class stores other information in addition to the genotype such as references to the parents of that individual, and the generation that the individual was created. We do not delve into the implementation of the *individual* class, however, our code examples will use two member functions called *value* and *set.* To determine the value of the gene encoded at gene location x, we use a member function called *value* of the *individual* class. For example *genotype_in-->value(5)* gives us the value of the gene in the 6th position in the *individual* object called *genotype_in*[3]. To set the value of a gene v at position w in the individual called *genotype_in,* we would use the expression *genotype_in->set(w - 1, v).*

11.2.4.1 The TSP Class

The *tsp_class* defines both data and methods that allow the fitness of a provided genotype (called *genotype_in)* to be calculated.

The following list summarizes the main data items of the *tsp_class.* We will then discuss each member function in detail.

i) The *town_coords* data item is a two dimensional array. The first dimension specifies the town identifier (from 0 to *GENES_PER_GENOTYPE*1), and the second dimension specifies the ordinate (0=X and 1=Y). When this array is indexed with a town id. and an ordinate (X or Y), the array gives us the numerical value of the ordinate of the position of the town.

ii) The *town_dist* array is a two dimensional array which, when each dimension is set to two different town ids., gives us the value of distance between these two towns.

iii) The *dmin* variable stores the theoretical lower bound to the minimum tour length, or the optimal tour length if it is supplied in the TSP data file.

```
const X=0, Y=1;

class tsp_class {
  private:
    int town_coords[GENES_PEX_GENOTYPE][2];
    int town_dist[GENES_PER_GENOTYPE][GENES_PER_GENOTYPE];
```

---

[3] Note that *genotype_in->value(0)* gives us the gene value in the first position in the genotype called *genotype_in.*

```
    int dmin;

    void towns_load();
    void town_distances_build();
    void establieh_dmin();
    void tour_dump(float fitness, individual *genotype_in,
    float tour_length);
  public:
    tsp_class();
    float objective_calculate(individual *genotype_in, float
    objective_best);
};
```

11.2.4.2 The TSP Member Functions

When the GA program is run, the global object *tsp* is created as an instance of the previously described *tsp_class.* The constructor, described next, then performs the preliminary setting up of the TSP problem. When the GA needs to calculate the fitness of a new child genotype it will make a call to the function called *objfunct.* The *objfunct* function is responsible for measuring and returning a value of objective worth of the input individual *indiv.* This function in turn calls the *objective_calculate* member function of the TSP class. The objective worth (or fitness) of the individual it then calculated with the *objective_calculate* member function.

```
    tsp_class tsp;

    float objfunct(individual *indiv, int problem_id, float all_objective_best)
    {
  return tsp.objective_calculate(indiv, all_objective_best);
    }
```

The *tsp_class* constructor is called when the program begins and the *tsp* object is created. The constructor loads the towns in from the town file *(towns_load),* calculates the distance between all possible combinations of towns *(town_distances_build)* and then sets the lowest possible distance attainable for any individual *(establish_drain).* These functions update the relevant class variables of the newly created *tsp* object.

The *all_objective_best* input parameter contains the fitness of the best population member and is passed through to the *objective_calculate* member function. This value is used so that the *tsp* class knows when a champion is produced so a profile of the champion can be recorded for later inspection (see the *objective_calculate* member function).

```
    tsp_class::tsp_class()
  towns,load();
  town_distances_build();
  establish_dmin();
    }
```

The *towns_load* member function reads in the town coordinates from a filename which is a string returned by a call to the *get_pararn* function. The *get_pararn* function allows us to specify GA specific information, such as the population size, the number of generations, the selection scheme, and in this case, the objective function data file name, in a single configuration file. This strategy allows us to set our GA specific parameters in a single file without having to recompile our code. This parameter configuration file system is also used in the Genitor GA [Whi89]. We will not delve into the implementation of the *get_param* function.

The file format defining a TSP problem is similar to the file format stipulated by the TSPLIB package of TSP problems [Rei91]. The following small example file is for a four town TSP.

```
Progr Name:                 test4
Comment:                    This is an example simple 4 town TBP
Problem Type:               Complete Euclidean 2D TSP Graph
Towns:
Edge Weights:               4
Optimal length:             Derive from coordinates
City Listing:               0
0        5        89
1        42       66
2        7        64
3        18       80
EOF
```

The test file above contains six lines of descriptive information which describes the problem, the size of the problem, and how distance is to be measured. In the 'Optimal Length' field the value of zero indicates that an optimal value of distance is not known for this problem. If an optimal distance value is known, then this would be set to the length of the shortest circuit. If the value is zero then the member function *establish_dmin* will calculate a theoretical lower bound value of distance for the best Hamiltonian circuit. The coordinates for the towns then appear in the following lines of the data file. Each line is numbered from 0 to (N-1) where N is the number of towns in the problem. The line number is ignored and the following two fields which contain the X and Y coordinates are stored in the array *town_coords.* Note that the order in which these coordinate values appear in the data file is not important. Remember the GA will optimize the ordering of the towns represented by the coordinates. The data file is terminated with the three characters 'EOF'.

```
        void tsp_class::towns_load() {
        int junk;
        char jstr[1001];
        char filename[100];

extern int got_param(char *name, char *param);          //get the filename of
        the TSP problem file
assert(get_param("ob_file", filename));                 //from the GA parameter
        configuration file
```

```
ifstream in_file(filename); //using the get_paramfunction
        assert(!in_file.bad());

        for (int i=0; i< 5; i++)
    (void) in_file.getline(jstr, 1000);
    (void) in_file.get(jstr, 1000, ':');        // go to the colon first
    (void) in_file.get(jstr, 2);                 // move to the optimal value field
    in.file >> dmin;
    (void) in_file.getline(jstr, 1000);         // read in tho dmin value
    (void) in_file.gotline(jstr, 1000);         // and advance two more line
    for (int j=O; j< GENES_PER_GENOTYPE; j++)
        in_file >> junk >> town_coords[j][X] >> town.coords[j][Y];
    in_file >> jstr;
    assert(!strcmp(jstr, "EOF"));
}
```

Before we measure the length of any TSP circuits, we precalculate the distance between all pairs of towns. This calculation is done at the start of the program in the constructor of the class *tsp_class*. The constructor calls the *town_distances_build* function. We use a Euclidean measure of distance which is rounded up or down to the nearest integer value to define the distance between two towns c_i and c_j. The formula is:

$$d_{ij} = \left\lfloor \sqrt{\left(c_i(X) - c_j(X)\right)^2 + \left(c_i(Y) - c_j(Y)\right)^2} + 0.5 \right\rfloor \qquad (11.1)$$

This way of calculating the distance between towns c_i and c_j, d_{ij}, is standard and is used extensively in the research community for the TSP [Rei91]. By converting the distance, which may well be a natural number, to an integer saves a great deal of floating point arithmetic in subsequent computation. This results in a faster algorithm. The resulting quantization error, which is due to the rounding of the distances between town pairs to the nearest integer, in most cases will not prevent the GA from reaching the optimal true solution to the TSP.

Obviously from the equation above $d_{ij} = d_{ji}$. We can use this fact to halve the number of distance measurements that need to be done. Since $d_{ii} = 0$, we only need to perform $\frac{N(N-1)}{2}$ distance calculations.

```
        void tsp_class::town_distances_build() {
    int first, second;
    for (first=O; first< GENES_PER_GENOTYPE; first++) {
        town.dist[first][first]=0;
        for (second=0; second(first; second++)                          {
            town_dist[first][second] =
                (int)  (sqrt(  (double)  pow(town_coords[first][X]-
                    town_coords[second][X], 2) + pow(town_coords [first][Y]-
                    town_coords[second][Y],2)) + 0.5);
```

```
            town_dist[second][first] = town_dist[first][secoad];
        } // for second
    } //for first
        } //funct
```

If the minimum solution is not already known for this problem, then this is indicated by a 0 *dmin* value in the Optimal Value field of the TSP data file (see the *towns_load* member function). If the optimal distance is not known, in order to convert our simulated Hamiltonian cycle tour length to an objective value between 0 and 1, we must calculate a theoretical lower bound of the Hamiltonian cycle length. We calculate this bound using the equation

$$d\min = \frac{1}{2}\sum_{i=1}^{N}(\min\{d_{ij}\} + \min\{d_{ik}\})\ \forall j,k \in \{1,\ldots,N\}\ and\ i \neq j \neq k \qquad (11.2)$$

This is to say, for every town i we find the two smallest distances connecting that town with two other towns j and k. We sum these two distances d_{ij} and d_{ik}, and repeat the process for all other values of i (where i is a town id.). We will then have N distance sums, which we sum together and divide by two. This gives as a theoretical lower bound to the Hamiltonian cycle distance, in which in most difficult problems will not be attainable. In a few simulations of problems of less than 100 towns we have found this value of *dmin* to be approximately 20% shorter than the optimal solution to the problem. This suggests that if we have an objective value for an individual of 0.8 in a TSP problem where we do not know the optimal solution in advance (which is normally true), the GA is achieving results in the correct ballpark.

```
        void tsp_class::establish_dmin() {
    if (dmin > 0)  //the optimum distance was supplied in the configuration file
        return;      //so we won't calculate a lower bound, just return
    int dsum=0;
    int min2, minl;
    for (int i=0; i< GENES_PER_GENOTYPE; i++) {
        min2=min1=LARGE_INT;
        for (int j=0; j < GENES_PER_GENOTYPE; j++) {
            if (i==j)
                continue;
            if (town_dist[i][j] < minl && minl > min2)
                min1 = town_dist[i][j];
            else
                if (town_dist[i][j] < min2)
                    min2 = town_dist[i][j];
        }//int j
        dsum += min1 + min2;
    } //for
    dmin = (int) ( (dsum + 0.5)/2);
    cout << "The minimum calculated bound on circuit distance is "<< din";
        }
```

The *objective_calculate* member function is used to calculate the objective worth of the genotype passed in as a parameter. We start by taking the first two gene values in the genotype *genotype_in,* and we use the distance array *town_dist* to determine the distance between the towns represented by these two values. We then take the second and third gene values and calculate the distance between these two towns. We continue this process and we keep a running sum of the distances found. The final step of the algorithm adds the component of distance between the town represented by the last gene value, and the town represented by the first gene value. This last step ensures that we have calculated the distance of a closed loop, or circuit, of towns. Having calculated the distance represented by the individual *genotype_in,* we then convert the distance value into an objective value which lies between 0 (minimum worth) to 1 (maximum worth - most optimal). We achieve this by the calculation $objective = \frac{dmin}{tour\_length}$ where *dmin* is our minimum distance value calculated in the *establish_dmin* member function. If this new objective value is greater than the objective value of the best population member (which is contained in an input parameter variable *alLobjective_best)* then we call the *tour_dump* member function to write this new tour to a data file; this allows us to maintain a history of champion solutions. The final step of this function returns the objective value back to the objective function *objfunct.*

```
float tsp_class::objective_calculate(individual, *genotype_in, float
    all_objective_best) {
    int tour_length=0;
    for(int i=0; i< GENES_PER_GENOTYPE-l; i++)
        tour_length+=town_dist[genotype_in->value(i)][genotype_in
         ->value(i+1)]

    //now close loop and form Hamiltonian Cycle
    tour_length+=town_dist[genotypa_in->value(0)][genotype_in->value
    (GENES_PER_GENOTYPE-1)];

    float objective = dmin*1.O/tour_length;

    if (objective> all_objective_best)
        tour_dump(objective, genotype_in, tour_length);

    return objective;
        }
```

11.2.4.3 Reporting

The *tour_dump* member function is called when we have a tour which is better than the current best population tour fitness which is defined by the input parameter called *alLobjective_best.* This function is responsible for dumping the x, It coordinates of the new best population champion in a file called *tsp_x* in a directory called tsp. Note that the first file dumped is *tsp_0,* the second file dumped is *tsp_1* etc. At the end of our GA run we will have a number of files (perhaps between 20-40) that records the chronology of champion TSP solutions generated by the genetic algorithm. In this implementation this member function dumps the ordered list of coordinates to a file format readable by *xgraph,* a public domain, UNIX, graphing program. Obviously, such a member function can be adapted to dump coordinate information into any file format which may be more relevant. Note that this reporting can be switched on by defining a compiler directive *REPORT* in the header file.

```
    void tasp_class::tour_dump(float fitness, individual ,genotype_in, float
    tour_length) {
$if def REPORT
    static dump_count = 0; //variable to keep track of the file number
    char file_name[20];
    sprintf(file_name, "tsp/tsp_%i", dump_count++);
    ofstream tour_dump_file(file_name); //open the next consecutive file

    tour_dump_file << "TitleText: \"Fitness is" << fitness <<
     ". Tour is " << tour_length << " units." << "\"\n\n";

    for (int i=O; i< GENES_PER_GENOTYPE; i++)
        tour_dump_file << town_coords[genotype_in->value(i)][0] <<
        " " << town_coords[genotype_in->value(0)][1] << "\n";

    //complete cycle
    tour_dump_file << tomn_coords[genotype_in->value(0)][0] <<
    " " << town_coords[genotype_in->value(0)][1] << "\n";
$endif
    }
```

11.2.5 Variations from the TSP Problem

11.2.5.1 The Non-Symmetric TSP

The TSP, in graph theoretic terms, equates to the problem of finding the lowest cost simple cycle in a weighted graph with N vertices where the cycle visits all vertices in the graph. A simple cycle which visits all vertices is known as a Hamiltonian cycle. We assume that the problem is symmetric which is to say that the graph of vertices (towns) and edges (distances) is an undirected graph. This means that for all vertex identifiers $i,j \in \{1,..., N\}$, $d_{ij} = d_{ji}$ where d_{ij} is the weight of edge (i,j) (or in the TSP analogy d_{ij} is the distance between towns i and j). However, in some practical problems this symmetric property may not hold. For example, if we were planning a ring road system to connect a number of geographic locations with a daisychain type service road, then it is possible that we may connect two adjacent locations with two one way roads having traffic flows in opposing directions. It is conceivable that the two roads have different lengths i.e. $d_{ij} \neq d_{ji}$. In [HGL93] Homaifar et al. cite an application of a non-symmetric TSP where an optimal air route of towns is required in the presence of a wind factor. To model this situation correctly we would need to model the TSP as a directed graph. This variation on the TSP could easily be implemented in our genetic algorithm. We could simply modify our objective function to read in a precalculated non-symmetric distance file. The genetic encoding of the problem would not require any modification. This is because two neighbouring genes g_i and g_{i+1} not only encode the information about which two towns are adjacent, they also encode the order in which the two towns are visited in the tour. The objective function could then exploit this ordering to calculate d_{ij} or d_{ji}.

11.2.5.2 The Triangle Inequality Does Not Hold

In the standard GA implementation of the TSP, the triangle inequality relates the three distance measures d_{ij}, d_{jk} and d_{ik} by the inequality $\sqrt{d_{ij}^2 + d_{jk}^2} \geq d_{ik}$. This inequality holds because the coordinates are considered to lie on a 2D Euclidean plane, and all of the distance functions are calculated to be Euclidean. However, in a practical problem, such as road planning, these locations may not be connected with straight roads. Even if the roads were straight lines when projected in two dimensions, one of the roads could have an extra component of length if it ran through mountainous terrain (albeit in a straight line when looking from above) between two towns. In the more general case of the TSP, the distances are not calculated as Euclidean distances from a set of town (X, Y) coordinates. The distances, or weightings, are provided by a distance matrix, which may, or may not be, symmetric (see Section 11.2.5.1). In our TSP objective function, we would only need to modify our *tsp_class* object constructor to read in an explicit list of distances. No other GA changes would be required.

11.2.5.3 The Graph is Not Complete

The TSP problem is modelled on a complete graph, i.e. there exists a defined value of distance $\exists d_{ij}$, $\forall i,j | i \neq j$ and $i,j \in \{1,...,N\}$. However, in many practical problems it is likely that there cannot exist a path between certain pairs of towns. In graph theoretic terms the graph may be incomplete and contain less than $\frac{N(N-1)}{2}$ edges. We could deal with such a problem by extensively reworking our GA to ensure that the genes only encoded legal Hamiltonian circuits of the incomplete graph. However, let's consider the problem of generating an initial population. We ideally wish to generate P pseudo-random Hamiltonian circuits in our initial population, where P is the number of genotypes in our population. The problem of finding a Hamiltonian circuit in an incomplete graph is an NP complete problem [GMJ79]. That is to say that no efficient polynomial time algorithm is known to do this. We could use heuristics, or a branch and bound technique to find the P initial Hamiltonian circuits. However, generating an initial population in this way could adversely affect the genetic diversity of the starting population. We will explore this further.

In a regular TSP problem, to find a random Hamiltonian circuit, we simply create a list of N towns and apply a shuffle algorithm to the list. The resulting circuit is obtained by reading the resulting shuffled list from the first list position to the Nth list position (then returning from the last town in the list back to the first). The gene value in the jth position determines the jth town that is visited. If our shuffle algorithm generates high-quality random lists of genes that encode TSP towns, then we will have generated random initial Hamiltonian circuits. The resulting population will then consist of Hamiltonian circuits in which any specific Hamiltonian circuit is likely to appear in the initial population as any other Hamiltonian circuit. This gives us excellent genetic diversity, and a good starting point for our GA. However, in an incomplete graph, when we apply heuristics, or use a branch and bound technique to generate initial Hamiltonian circuits it would be difficult to generate an initial population with the kind of diversity that a GA requires. Many experiments suggest that a good random genetic starting pool helps the GA to converge to an optimal or near optimal solution.

The problem of finding a diverse initial, and legal, population is only the start of our problems when modelling a TSP problem in an incomplete graph. The genetic operators responsible for generating new children must also generate legal Hamiltonian circuits. Let us consider the genetic edge recombination operator [WSF89]. In many cases, the ERO would generate legal children, as the ERO will try to preserve the edge relationships that exist in the two parents when producing a child. If the ERO manages to produce a child genotype with no new town adjacencies, then the child will be a legal child. However, the ERO will often be forced to introduce new town adjacencies. Experiments with small problem sizes (<80 towns) indicate that approximately one out of every three children (Section 11.2.7.1) generated by the ERO have at least one pair of adjacent towns which was not inherited from either parent. In this case, the child will have one or more new edge relationships. If a new edge relationship say (i,j) is not a member of the set of edges in the problem TSP graph (because d_{ij} is undefined), then this new child will contain an illegal edge. We could disregard this illegal child and choose two new parents. Alternatively we could attempt to repair the child with some repair heuristic.

A different approach is to model our incomplete TSP graph G_i by a complete TSP graph G_c. G_i would be a subgraph of G_c. We will consider the set of edges I which exist in G_c but not in G_i. This set I is the infeasible edge set, and we would assign weights to each edge in I according to a penalty system which relates to the problem we are modelling. If all edges in I are equally infeasible in our problem, then we would assign each edge in I some weight greater then the maximum weight of any edges in graph G_i. If it makes sense in our problem to set the weights in I according to the cost detriment of any final solution containing these edges, then we would penalize the weights of these edges accordingly. This is a common GA technique [RPLH89] where we penalize infeasible, or unattractive solution components in the hope that through the course of evolution that these components will be bred away. If we penalize an infeasible edge in I by too much, then the edge will be bred out early in the evolution run. This may well lead to a suboptimal solution from forming. We may also find that this practice could lead to a very high infant mortality rate, where children which have these edges are instantly the populations worst members.

If we do not penalize *E* enough, then members of *E* might well form as part of our optimum solution. Richardson et al. remark that 'Care must be taken to write an evaluator which balances the preservation of information with the pressure for feasibility'. With this in mind the correct penalty settings should be determined empirically. Therefore the degree that we should penalize an infeasible edge in *I* is best found by experimenting with different values.

The ideal strategy is to find a genetic encoding of our problem which not only naturally incorporates all constraints, but allows genetic operators to work effectively. In situations such as a TSP with an incomplete graph, when such a strategy is not effective, penalty strategies can be useful. However, the determination of ideal penalty settings for such problems is still an open area of research in the GA field.

11.2.5.4 Hamiltonian Paths

A common variation on the TSP is the evaluation of Hamiltonian Path rather than a Hamiltonjan Circuit. A Hamiltonian Path, in a graph order *N* is a simple path which begins with vertex v1 and ends with vertex vn, and visits each vertex in the graph. ($v_1 \neq v_n$). Like the TSP, finding the least-cost Hamiltonian path in a weighted graph is an intractable problem and is a member of the NP-hard family. We can easily implement this variation by simply changing our objective function so that we do not add a component of distance from the last town back to the first town, which would normally be added into the length of the tour when calculating the length of the Hamiltonian circuit.

11.2.5.5 Other Variations of the TSP

We could consider a number of other variations to the TSP problem. However, we should take care when attempting to solve a problem which is a variation on another problem as the resulting problem may be easily solved using an exact method. Many common graph theoretic problems have efficient algorithmic solutions, for example, finding the shortest path between two vertices, or finding a minimal spanning tree of a graph. Many network type problems can be modelled as integer programming problems and can be solved efficiently. When considering an interesting problem, a genetic algorithm implementation should be the last resort and only used when all possibilities of better, exact, methods are exhausted.

11.2.6 The Edge Recombination Operator

11.2.6.1 How it Works

We have described the implementation of the fitness function of the TSP. The next important task is to implement a useful operator which can be used to create a child TSP genotype from two parent genotypes. A number of different operators for the TSP have surfaced in the GA literature over the past few years, and these include those known as Partially Mapped Crossover (PMX), Position Crossover, Cycle Crossover (CX), and Order Crossover (OX). Each of these operators creates a child from two parents and qualities are transferred from both parents to the child. In this way these operators satisfy Holland's requirement that fit schemata (or short solution components) are transferred into the child from the parents [Hol75]. These operators produce scheduling type children, or object lists where there are no duplicate or absent object values encoded in the genes. However, Whitley, Starkweather and Fuquay [WSF89] presented a new operator called the Edge Recombination Operator (ERO) and presented a performance comparison with these other scheduling type operators in [SMM+91]. Whitley et al. showed that ERO had the fastest convergence and converged on better solutions than these other operators when optimizing a number of TSP problems of sizes 30 to 76 towns. They attributed the success of this operator to the way that the ERO transfers adjacent town relationships from both parents into the new child, and the small probability in which new foreign edges are introduced into the child.

Starkweather et al. also showed in [SMM+91] that ERO is not the operator of choice for all scheduling type problems. They described a beer production scheduling system where a genotype encoded a list of customer order numbers, with each gene encoding a single order number. The production schedule would fill each of the customer orders in the order encoded by the first gene through to the order encoded in the last gene in the genotype. They argued that for this kind of genotype the relative order of the genes was more important than the adjacency of order numbers and their experiments showed that the ERO was the worst performer compared to the other scheduling operators.

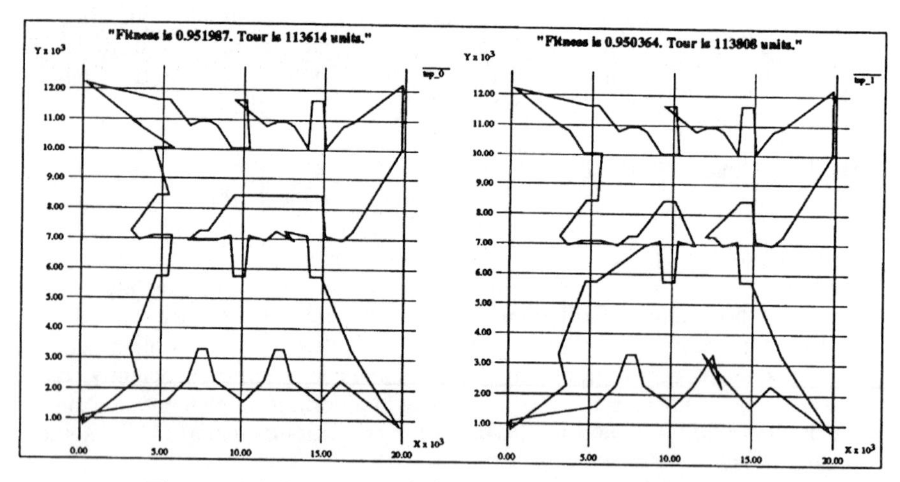

Figure 11.4: Parent 1 *(P1)(left)*. Parent 2 *(P2) (right)*.

The ERO builds a child genotype based on the combined set of edge relationships from both parent genotypes. The ERO can be used for TSP problems or other problems where adjacency between gene values is the critical factor in solution quality. To illustrate the ERO at work we ran a 76 town TSP and recorded the parents and child which where involved in a particularly successful recombination. Figure 11.4 shows both parents p1 and operator applied to parents p1 and p2. It is interesting to see how short path components of the child circuit are inherited from both parents. In this example no new edges, apart from those belonging to the parents were introduced into the child. In this instance the fitness of the child was approximately 3% better than both parents. The following discussion and implementation of the ERO is based on the ideas presented in [WSF89].

To describe how the ERO algorithms work we must first introduce the edge map. An edge map can be created from a genotype by examining the adjacent neighbours of each gene. For example, the genotype (p1) equal to

$$p_1 = (2, 3, 5, 4, 1, 0) \qquad (11.3)$$

means that the genotype p_1 encodes a TSP cycle (2, 3, 5, 4, 1, 0, 2) where each number in the cycle relates to a town identifier. Genotype p_1 has an edge map

Gene Value	*Neighbours*
0	1,2
1	0,4
2	0,3
3	2,5
4	1,5
5	3,4

(11.4)

The edge map above tells us that gene value 0 has two neighbouring edges 1 and 2 (remembering that the genotype for the TSP is considered cyclic). Gene value 1 has two neighbouring edges 0 and 4. Hence the edge map allows us to quickly determine the neighhours of a particular gene value.

In the example above, we have added the edge relationships from one parent p1 to the edge map. The ERO begins by adding edge relationships from both parents to the single edge map. For example, if our second parent genotype p_2 is

$$p_2 = (2,1,3,0,5,4) \qquad (11.5)$$

Then the combined edge map (M_{12}) of both parents p_1 and P_2 becomes

Gene Value	*E.M. from p_1*	*E.M. from p_2*	*Combined E.M. M12*
0	1,2	3,5	1,2,3,5
1	0,4	2,3	0,2,3,4
2	0,3	1,4	0,1,3,4
3	2,5	0,1	0,1,2,5
4	1,5	2,5	1,2,5
5	3,4	0,4	0,3,4

(11.6)

From the table above (Table 11.6), the last column represents the combined edge map. This new combined edge map M_{12} contains the combined edge relationships from both parents. We will called the combined edge relationships for a particular gene value g, the adjacent edge set $E(g)$. From the example above the adjacent edge set for gene value 3 is $E(3) = \{0, 1, 2, 5\}$. The number of elements in $E(3)$ is denoted by $|E(3)| = 4$.

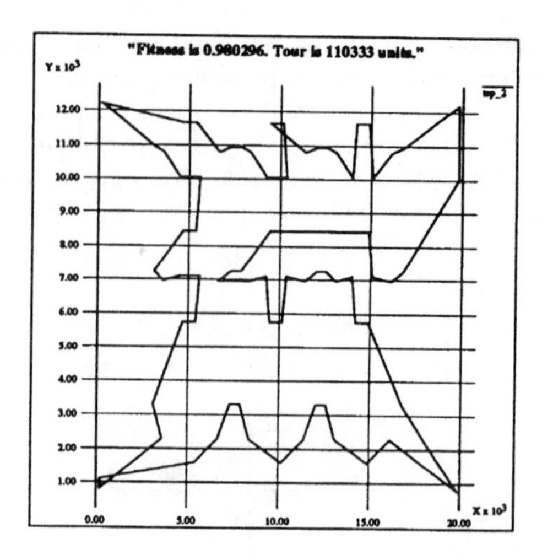

Figure 11.5: New Child produced from parents P_1 and P_2.

The following algorithm describes how the ERO works. Note that in the following terminology, $p_1(0)$ means the first gene value in genotype p_1.

<u>11.2.6.2 Algorithm Edge Recombination Operator</u>

Inputs: Genotypes p_1 and p_2 (parents). The combined edge map M_{12}. The number of genes per genotype N.

Output: c_1 the child genotype.

1) Copy the gene g from first position in p_1 to the gene in the first position in c_1. Remove this gene value g from the edge map.

2) $i \leftarrow c_1(0)$

3) If all N gene values have been added to child c_1 then exit and return c_1.

4) If $|E(i)| = 0$ then randomly choose a gene value g to be added be added to c_1 where g has not already been added to c_1. Add g to the next consecutive gene location in c_1, remove g from the edge map M_{12}, set $i \leftarrow g$ and go to step 3.

5) Choose the element $j \in E(i)$ where $|E(j)|$ is the minimum. if $a_1, a_2,..., a_n \in E(i)$ and $a_1, a_2,..., a_n$ have the equal smallest $|E|$, then choose j from $a_1, a_2,..., a_n$ randomly. Once we have chosen j we then add j to c_1 at the next consecutive gene location. We then remove j from the edge map M_{12}, assign j to i and go to step 3.

This ERO algorithm is used to generate one child c_1 from two parents p_1 and p_2. However, once we have created a combined edge map M_{12} we have enough information to create a second child in addition to the first. A simple technique to obtain the second child is to exchange p_1 with p_2 and run the ERO algorithm, and this time we consider the new child to be c_2. In practice this technique is quite efficient, as it takes a reasonable amount of computation to create the combined edge map. We can save on time if we use the combined edge map, as is, to create the second child.

We will apply our example parent genotypes p_1 and p_2 to the example above. Step 1; We copy across the first gene value in p_1 over to the first gene value in the child c_1. The child becomes c_1 = (2,_,_,_,_,_) where the character '_' represents an undecided gene value. We then remove the gene value 2 from the edge map, which becomes

Gene Value	*Neighbours*
0	1,3,5
1	0,3,4
2	1,3
3	0,1,5
4	1,5
5	0,3,4

(11.7)

We assign the value 2 to *i* and proceed to step 4 (condition in step 3 is false). Now $E(2) = \{1,3\}$ and $|E(2)| \neq 0$ therefore we proceed to step 5. Step 5 requires that we choose *j* where $j \in E(2)$ and $|E(j)|$ is minimum. We know $|E(1)| = 3$ and that $|E(3)| = 3$, therefore we choose between the values 1 or 3 randomly. We will choose 1. We add the gene value 1 to c_1 which now becomes c_1 = (2,1,_,_,_,_). We then remove the gene value 1 from the edge map, which becomes

Gene Value	*Neighbours*
0	3,5
1	0,3,4
2	3
3	0,5
4	5
5	0,3,4

(11.8)

We then assign *i* the value of 1 and proceed to step 3. We have only added 2 out of the 6 possible gene values in c_1 so we proceed to step 4. Now $|E(1)| = 3 \neq 0$ so we proceed to step 5. Now the elements of $E(1) = \{0,3,4\}$ and $|E(0)| = 2$, $|E(3)| = 2$ and $|E(4)| = 1$. We must choose *j* when $|E(j)||$ is minimum so we choose gene value 4. We add 4 to c_1 which becomes c_1 = (2,1,4,_,_,_). We remove 4 from the edge map which becomes

Gene Value	*Neighhours*
0	3
1	0,3
2	3
3	0,5

(11.9)

4	5
5	0,3

We assign $i = 4$ and continue to step 3. We pass over steps 3 and 4 (both conditions are false) and arrive again at step 5. This time we have only single value from which to choose j from, so $j = 5$. We add this to c_1, so now $c_1 = (2,1,4,5,\_,\_)$. We proceed in this way through the ERO algorithm until we obtain the final value of c_1. Depending the random choice in subsequent steps of the algorithm, c_1 will be (2, 1,4,5,0,3) or (2,1,4,5,3,0).

It is important to realize why, at step 5 of the ERO algorithm, we visit those towns with fewer adjacent edges earlier on in the town selection process. A town x with fewer adjacent edges is likely to cause difficulties if x is chosen later in the selection process because at this late stage it is likely that towns adjacent to x have been eliminated from the edge map, leaving x as a dead-end town. A dead-end town x introduces mutation as a random choice must be made for the next city following town x. This situation (mutation) occurs in step 4 of the algorithm when a chosen gene value x has $|E(x)| = 0$. This de&d-end situation means that we have already exploited the edge relationships relative to gene value x so the only solution is to randomly pick the next gene value in the new child from the set of unchosen gene values. Therefore to minimize these events we choose a value of x which has minimum $|E(x)|$. We make this choice because by using this rule, we minimize the chance that we introduce new foreign edge relationships into the child (mutations) which do not exist in either parent at later steps in the algorithm.

We know at step 5 in the ERO algorithm, that if there is more than one value to choose j from, then we choose randomly between the candidates. The choice that is made at this step introduces an element of controlled random exploration into the ERO. Typically for a genotype of reasonable length, we will make a number of random choices in the process of producing a single child. These choices represent the different ways in which we combine the parents without introducing any new, foreign, edge relationships into the child. It is a good way of intelligently exploring a much reduced and promising part of the solution space. In this way, depending on the random choices made at step 5 in the algorithm, many different children could result from the same set of parents.

11.2.7 Implementing the Edge Recombination Operator

11.2.7.1 Data Items of The *edge_class*

The defined constant *EM_WIDTH* relates to the maximum number of adjacent towns that a town could have when two parents are combined in an edge map, and is equal to four. The defined constant *EM_CYCLIC* determines whether we are modelling Hamiltonian cycles or Hamiltonian paths, and is described in detail when we discuss the member functions. The data item central to the *edge_class* is the *map* array. Each element of array *map* is a structure which describes the details of a single gene value in the edge map. When this array is indexed by gene value x the structure attributes give us the edge information for gene value x. In this structure the *edges* attribute tells us how many available towns are adjacent to *x*. The *visited* attribute is set to 1 if the gene value *x* has already been incorporated into the child's circuit. The *edge* array attribute gives us the actual adjacent gene values to *x*. The *edges* attribute tells us how many elements of this array are relevant to the gene value *x*. For example if *map[5].edges* = 4, then gene value, or the town with id. 5 has four adjacent towns. The first adjacent town is *map[5].edge[0],* and the last adjacent town is *map[5].edge[3].* We will discuss each of the member functions in turn in the subsequent sections.

```
    $define EM_WIDTH 4
    $define EM_CYCLIC 1
    class edge_class {
private:
    struct
    int edges;
    int edge[EM_WIDTH];
    char visited;
    } map[PARAMS];
    void remove_visited_towns(int town, int edge_to_go);
    void edge_insert(int cur_town, int new_town);
    int dead_end_next_town(int town_stuck);
    int choose_next,town(int current_town);
    int width;

public:
    void print();
    void reset();
    void edge_generate(individual ,parent);
    void tour_make(individual *parent, individual *child);
    }
```

11.2.7.2 The Edge Recombination Operator Function

Before we examine the member functions of the edge class in detail we will look at the *edge_recombination_operator* function. The GA calls this function after the two parent genotypes *parent1* and *parent2* have been chosen for crossover. This function is the high level function which delegates most of the specific implementation details to the *edge_class* class. The children to be created are defined as *childl* and *child2.* These four genotypes are passed into the function as parameters. The code for this function follows.

```
void edge_recombination_operator(individual *parent1, individual *parent2,
            individual *child1, individual *child2)
    {
edge_class edge_map;                  // define a new edge_map class
edge_map.reset();                     // clear out the contents of the edge map
edge_map.edge_generate(parent1);      // put edges from parent1 into the edge map
edge_map.edgs_generate(parent2);      // put edges from parent2 into the edge map
edge_clase edge_copy = edge_map;      // create copy of the edge map
edge_map.tour_make(parent1, child1);      // create childl from edge map
edge_copy.tour_make(parent2, child2);     // create child2 from edge map
```

```
}
```

The function first declares a local class variable called *edge_map*. The new edge map class variable is then initialized with the *reset* member function. The function then adds edge relationships from both parents with two separate calls to the *edge_generate* member function. We then take a copy of the newly created edge map and store the copy in the class variable called *edge_copy*. The function will use this copy later when creating the second child. The next step involves creating a tour for the first child, and this is achieved with the *tour_make* member function of the *edge_class*. We then create the tour for the second child using the copy of the edge map. Once this function has completed both children *childl* and *child2* will have been created. We will describe the specific member functions for the *edge_class* class next.

11.2.7.3 The *edge_class* Member Functions

The reset member function clears the edge map. This function is called prior to the *edge_generate* function, which will add edge relationships to the edge map from the cleared state.

```
    void edge_class::reset() {
  for(int i=0; i < GENES_IN_GENOTYPE; i++) {
    map[i].edges=0;
  }
}
```

The *edge_generate* member function adds the town relationships that exist in the *parent* genotype to the current edge map. Note that this function does not clear out the edge map array before adding the new town relationships from the *parent* parameter. This allows us to build up a combined edge map that is a union of two parent genotypes. The function establishes a loop to step through each position in the input genotype. For each position *i* the gene values to the right and the left of this position are added to the edge map as neighbours to the gene value at position i. In this way if two neighbouring genes have the values (7, 5) then gene value 7 will be added to the edge map as an adjacent edge to gene value 5, and vice versa. A special case occurs when i = 0; what is considered the left hand neighbouring gene of the first gene in the genotype? If the *EM_CYCLIC* constant is set to 1, then the genotype is considered to be cyclic and the left hand neighbour of position 0 will be the rightmost gene in the genotype. Hence, the *EM_CYCLIC* constant should be set to 1 when solving cyclic problems such as finding the shortest Hamiltonian Cycle (TSP). In problems such as the shortest Hamiltonian path problem where the genotype is not considered cyclic, then the constant *EM_CYCLIC* should be set to 0.

```
    void edge_class::edge_generate(individual *parent) {
  int i, prev_edge, cur_town;

  for(i=O; i< GENES_IN_GENOTYPE; i++) {
    cur_town = parent->value(i);
```

```
// is next gene in edge map

if ( !(!EM_CYCLIC && i== GENES_IN_GENOTYPE-1))
   edge_insert(cur_town, parent->value( (i+l) % GENES_IN_
      GENOTYPE));

// is prev gene in the edge map?
if ( !(!EM_CYCLIC && i==0)) }
   prev_edge = i-1;
   if (prev_edge == -1)
      prev_edge = GENES_IN_GENOTYPE - 1;
      edge_insert(cur_town, parent->value (prev_edge));
} // if
} // for i
   }
```

The *tour_make* function is responsible for creating a new child genotype. The child genotype is represented by the *child* pargreeter. As is stipulated by the first step in the ERO algorithm in Section 11.2.6.2, the gene value in the first position of the first parent genotype is assigned to the first town of the child genotype. The function then sets the visited attribute for all gene values in the *map* array to zero. The function then establishes a loop which will terminate when all gene values have been added to the *child* genotype. Within this loop the following steps are executed:

i) The function first removes any references to the previous town in the combined edge map *(map)* with a call to the *remove_visited_towns* function.

ii) If there is one or more connecting edge from the previous town to another town in the edge map, the the next town is determined by a call to the *choose_next_town* member function. Otherwise a foreign edge is to be introduced into the edge map and the new, randomly chosen town is selected by a call to the *dead_end_next_town* function.

iii) The function then adds this next town into the next consecutive position in *child* genotype. This is achieved with a call to the *set* member function of the child.

iv) The function then sets the visited flag of the previous town entry in the edge map to 1 and assigns the *next_town* attribute to prev_town.

At the end of this loop the function checks to see if one or more foreign edges (disruptions) were added into the child, and if so, the global variable *op_ero_disruptions* is incremented. This variable can be useful when investigating the performance of the ERO.

```
void edge_class::tour_make(individual *parent, individual *child) {
int i,j, prev_town, next_town, prov_edges;
int disruption=0;
```

```
        prev_town = paront->value(0);
        child->set(O, prev_town);

  for(i=0; i< GENES_IN_GENOTYPE; map[i++].visited=O); // clear town visited
                                                            flag
  for (i=1; i < GENES_IN_GENOTYPE; i++) {

     for(j=0; j< map[prev_town].edges; j++)
        remove_visited_towns(map[prev_town].edge[j], prev_town);
     prev_edges = map[prev_town].edges;

     // determine which next town to choose
     if ( prev_edgos > 0)
        next_town = choose.next_town(prev_town);
     // this new town has no links left to other towns, ERO will fail
     else {
        if (i != GENES_IN_GENOTYPE-1) //last edge being empty is not a
                                                            disruption
           disruption=1;
        next_town = dead_end_next_town(prev_town);
        }

        child->set(i, next_town);                 // add now town to child
        map[prev_town].visittd-l;                 // set prev town to visited

        prev_town = next_town;
                  // advance to next town

        print();
     } /*for i*/

     if (disruption)
        op_ero_disruptions++;
        }
```

The *edge_insert* function is responsible for adding a new edge to to the edge map. The function receives two parameters *cur_town* and *new_town.* The function first looks through the list of the towns which are adjacent to *cur_town* and if the function finds that the town *new_town* already exists then it returns control back to the calling function. If the edge does not already exist in the edge map, then the function finds the next position in the adjacency array *map* which is indexed by the *current_town,* and inserts the new town into this position. Note that this function only inserts edge (cur_town) into the edge map. When constructing an edge map from a genotype this function is called twice to insert both edge orientations (a, b) and (b, a) where a and b are gene values which appear in a genotype.

```
     void edge_class::edge_insert(int cur_town, int new.town) {
  for(int j=0; j<map[cur_town].edges; j++)
```

```
        if (map[cur_town].edge[j] == new_town)
            //the gene value is already present in the edge map
            return;
        } //if

    // insert next edge
int edge_position = map[cur_town].edges++; // select next free spot in E.M.
map[cur_town].edge[ edge_position ] = new_town; // assign new town to this
                                                            spot.
}
```

This *remove_visited_towns* function removes any reference to the *town* town from the edge map. This removal from the edge map is essential, as we cannot choose the same town twice. By removing the town from the edge map, we update the edge map to represent the current state of the algorithm. This updated information is essential to facilitate the workings of the algorithm to help produce children which inherit all edge relationships from their parents without introducing new, unwanted edges.

```
        void edge_class::remove_visited_towns(int town, int town_to_go) {
    int town_edges,k,j;
    town_edgee = map[town].edges;

    for (j=0; j<town,edges; j++)
        if (map[town].edge[j] == town_to_go) {
            if (town_edges == 1) {
                town_edges=0;
                break;
            }
            for (k=j+l; k<town_edges; k++) {
                // remove hole from edge map
                map[town].edge[k-1]= map[town].edge[k];
                }
            --town_edges;
            --j;
        ) //if
    map[town].edgee = town_edges;
        }
```

The *dead_end_next_town* function is called when the next town for the child genotype must be selected at random from the unvisited towns in the edge map. This is achieved by first counting the number of unvisited towns. The function does this and stores this result in the variable *free_towns*. Then a random number r is selected in the range 0 to *free_towns-1*. The function then scans through the edge map and selects the rth occurrence of a free town.

The position of this randomly selected, unvisited, town is then returned back to the calling function.

```
int edge_class::deed_end_next_town(int town_stuck) {
int towns_free = 0;
for (int i=0; i< GENES_In_GENOTYPE; i++)
    if (!map[i].visited && i != town_stuck)
        towns_free++;
int random_town_no = randobj.random_get(towns_free);
for (i=0; i < GENES_IN_GENOTYPE; i++)
    if (!map[i].visited && i!= town_stuck) {
        towns_free--;
        if (towns_free == rtndo-_town_no)
            return i;
    }
}
exit(l);        // error here
return 0;       // to stop warning
    }
```

The *choose_next_town* function is called when there are one or more connecting towns from the previously selected town remaining in the edge map *map.* The function is responsible for choosing the next town for the child genotype. The function examines each forwarding edge from the edge map in turn. The function then chooses the town from the adjacent town set of the current town which has the least number of adjacent towns in its adjacent town set. If there are a number of towns which have an equal least number of adjacent towns, then we choose randomly between this tied set of towns. In the case of a tie, we use the array *towns_least_tied* to represent all of towns involved in this tied set, and we choose a random integer between zero and the length of this array, less one, to choose the next town.

```
    int edge_class:: choose_next_town(int current_town) {
int towns_least_tied[EM_WIDTH];

int least_edges=EM_WIDTN+l;
int greatest_number = 1;
int index = 0;
int number_j, town_j;

for (int j=0; j< map[current_town].edges; j++) {
    town_j = map[current_town].edge[j];

    number_j = 1;
    int edges_j = map[town_j].edges;

    if (edges_j==least_edges && number_j == greatest_number)
        towns_least_tied[index++]=town_j;
    else if (number_j > greatest_number) {
        greatest_number = number_j;
        towns_least_tied[0] = town_j;
        index = 1;
        least_edges = edges_j;
```

```
        }
      else if (edges_j < least_edges && number_j == greatest_number) {
            towns_least_tied[0] = town.j;
            index = 1;
            least_edges = edgee_j;
        }//if edges
    }//for j

    int tied_index;
    if (! index)
        tied_index = 0;
    else
        tied_index = randobj.random_get(index);

    return towns_least_tied[tied_index];
        }
```

11.2.7.4 Reporting Edge Maps

The *print* member function of the *edge_class* prints out the edge map contained in the *map* array. The function prints the gene value, followed by the character 'V' if that gene value has been included into the child's circuit. The function then prints a list of adjacent gene values. This process is printed for all gene values. The following example report illustrates a 6 city TSP.

```
Edge Map Report
0       15
1       05
2   V  53  0  1
3   V  45
4   V  10
5       01
```

In this example the towns with identifiers 2, 3, and 4 have already been included into the child genotype (i.e. they have been visited). Note that once a town is visited, its adjacency list is not maintained from that point onward.

```
    void edge_class::print() {
#if REPORT
    cout << "Edge Map Report\n";
    for (int i=0; i< GENES_IN_GENOYYPE; i++) {
        cout << setw(3) << i << " ";
        if (map[i].visited)
                cout << "V ";
            else
                cout << " ";

        for (int j=0; j< map[i].edges; j++)
            cout << setw(3) << map[i].edge[j];
        cout <<"\n";
```

```
    }
#endif
    }
```

11.3 Job Shop and Open Shop Scheduling Problems

The Job Shop Scheduling problem (JSS) is a classic scheduling problem where we have *j* jobs which must be assembled or processed by a set of m machines. A particular job must be processed by the m machines in a specific, predetermined order. In real problems these precedence constraints often correspond to technological requirements such as one machine drilling a hole and then another machine mounting a component into it.

As an example of a JSS problem we will consider a metalworking factory which has four machines; a drilling machine, a cutting machine, a folding machine, and a welding machine. We wish to process ten different jobs. Five of these jobs need processing in the machine order of cutting, drilling, folding, and welding. The time required on each machine depends on the job being executed. The other five jobs require a subset of these machines (i.e. drilling not required) in five different orderings. This simple example represents a problem where many jobs share the same machine precedence constraints. Obviously, in this example the cutting and drilling machines will be high demand in the early parts of the schedule.

The JSS optimization problem requires that we find an optimal schedule that specifies which machines are to process which jobs, and when, in order to minimize some cost function such as the time required to machine all jobs. This problem is very hard to solve when we are dealing with problems which involve hundreds of machines and jobs. The general JSS problem belongs to the NP hard set of problems [GMJ79]. The JSS problem has been the focus of many researchers and numerous exact and heuristic algorithms have been presented. A good summary is available in [AC91]. A genetic algorithm approach to the JSS has shown promising results in a number of sample problems. However, simple GAs to date have not out-performed some of the better JSS heuristics such as the Shifting Bottleneck heuristic [ABZ88] and the S.B.-Shuffle heuristic [AC91]. However, a recent hybrid-GA in [DP93] uses a good deal of problem specific knowledge and a genetic strategy for sequencing the way in which the single machine scheduling sub-problems are solved in the Shifting Bottleneck Procedure. This combination of a genetic search and the Shifting Bottleneck Procedure was found to be superior to the Shifting Bottleneck Procedure alone, especially for larger problem sizes. A genetic approach to the JSS is beginning to look more promising, especially for solving problems with multiple objective fitness functions, or problems which incorporate additional problem-specific complexity, factors which could easily complicate, or invalidate the use of other heuristic techniques.

11.3.1 Classifying Scheduling Problems

Scheduling problems can be classified according to the ESR classification scheme as was proposed in [BW91]. We discuss each of these categories in turn in relation to the JSS and OSS problems.

11.3.1.1 The E or Event Category

The first category, E, relates to the event. When considering the JSS problem, an event corresponds to a job. A job *j* may be independent which is to say that once a machine processes that job the machine is not constrained from executing that job by any other factors. Alternatively, an event may be preemptive where the processing of a job *j* may be interrupted by another job, and job *j* then resumes on the machine at a later stage. Events may have precedence constraints where a job *j* depends upon prior processing by a set of machines before it can be processed by machine *m*. In the most general sense we could model the precedence constraints for a job by a state chart. However, in the classic JSS problem we only use a simple set of precedence constraints where the execution of job *j* on machine m is dependent on the prior execution of *j* on a single other machine *n*. The E category also defines any event communication facilities which may be necessary. Communication between two events implies a communication delay and some implied match in state between the two events (communication is only allowed when two events are in a certain state). Event communication normally relates to computer process scheduling on processor units and is not normally relevant to JSS problems.

Finally, the E category specifies the arrival of events. Events might all commence at time 0, be available to commence after an initial delay, or events might arrive stochastically according a set of arrival probability distributions. In the case of the JSS problem all jobs are ready to begin at time 0.

11.3.1.2 The S or Environment Category

The S category relates to the scheduling environment. This category defines the number of classes of resources where each class contains a set of similar machines. The physical qualities of each class relates, in the case of the JSS problem, to the machine processing time requirements for each job. The S class also defines the communication mechanism; i.e. can a job be physically moved from one machine to another machine? If so, what is the delay required to achieve this move? Do machines require setup time for each job or for classes of jobs?

11.3.1.3 The R or Objective Function Category

The R category relates to the objective function of the scheduling problem. Objective functions can relate to a single parameter which must be optimized, or can relate to multiple objectives. We will explore objective functions relating to the JSS problem in detail in Section 11.3.8

11.3.1.4 Issues Relevant In Real Job Shop Scheduling Problems

Faulkenauer [FB91] discusses a number of practical problem constraints relevant to JSS problems in the Belgian Metalworking Industry. Apart from precedence constraints, these include:

i) *Temporal Constraints.* Jobs may be assigned a starting time t_s and a finishing time t_f. t_s is considered a hard constraint and is due to the absence of ordered materials required before the job can begin. t_f relates to the delivery time required to meet a client's order. t_f may be exceeded, but at cost (see Equation 11.3.4.1).

ii) *Capacity Constraints.* Machine resources are finite and have a limited capacity. Some machines may have insufficient capacity to be used on certain jobs, so other, higher capacity machines must be scheduled in their place.

iii) *Pre-emption Constraints.* These constraints occur when certain machines are scheduled for repair or maintenance during the JSS simulation. An algorithm which considers this constraint should work around the scheduled repair time window with the minimal impact on cost.

As we can see, the general scheduling problem can take many forms. The ESR classification begins to formalize some of the many possible aspects. The JSS problem, however, was not designed to model all of these constraints, but rather, was intended to represent a simple to specify problem which is very hard to solve in practice (as are all NP-hard problems). The precedence constraints in a JSS problem make it hard to solve as a GA must utilize a genetic encoding which does not constantly violate these constraints. For this reason it is a good exercise to apply a GA to the JSS problem. From there we can incorporate more and more of the aforementioned constraints and variations. To date, no GA has been devised which models all of the possibilities that the ESR classification suggests.

11.3.2 The Open Shop Scheduling Problem

The Open Shop Scheduling problem (OSS) is similar to the JSS, however, the order in which the machines process each job is not important, as long as each job is processed once for the correct duration by each machine. The OSS problem models a factory where each machine performs a measure of work on a job which is not dependent on other machines.

An example of an OSS problem is a Printed Circuit Board (PCB) component mounting factory. We will consider an example OSS factory with four machines. Machine A mounts resistors and capacitors, machine B mounts large transformers, machine C mounts integrated circuits, and machine D mounts ribbon connectors to a printed circuit board (PCB). The purpose of this factory is to produce j different assembled PCBs. Each PCB must be processed for a time by each machine A, B, C, and D and the ordering of the machine processings is not important because for a given job machine x does not depend on the prior processing of another machine.

11.3.3 Aspects Common to the JSS and OSS

Both the OSS and JSS problems have many similarities. The first similarity is that a machine may only process a single job at a time. If a job arrives at a machine, and that machine is busy, then the job must wait in the input buffer of that machine along with any other waiting jobs. When simulating either an OSS or JSS plan we often find that certain machines are in high demand at key points in the simulation. When jobs are scheduled to use a busy machine these jobs must wait until the machine becomes free. This waiting period introduces idle time into the schedule. When simulating a randomly generated schedule we often find a considerable amount of this idle time. In both JSS and OSS problems, e~ch (machine, job) double has a corresponding processing time. We describe this with a j x rn matrix of associated machine processing times, which we shall call the time matrix *tim*. In the following sections we discuss aspects common to both problems unless we note otherwise.

11.3.4 The Fitness Function

In a GA implementation of either the JSS or OSS we design a fitness function which determines the quality of a given schedule. There are a number of different ways that we could decide this measure of quality. GAs are flexible in this regard as they allow us to arbitrarily specify an objective function. This feature compares favourable with certain heuristic methods such as the Shifting Bottleneck Procedure [ABZ88] which has a single unchangeable objective (minimizing makespan). We consider two ways to calculate fitness, the first, relates to finishing all jobs before a given deadline, and the second relates to minimizing the time that the last job takes to complete (makespan).

11.3.4.1 Finishing all Jobs Before a Deadline

In a commercial situation it might be important that we manufacture all, or almost all, jobs before a given deadline time (d) so as to be able to meet a order due at this nominated time. In this situation we wish to minimize the number of jobs that are unfinished after time d. We consider a fitness function in which the variable e_h is the time that job h finishes earlier than d, a value which is 0 if the job finishes later that this deadline. The variable l_i is the time that job i finishes later than d (0 if it finishes earlier). A fitness function f was used in the experiments of [FB91], and was determined by the formula

$$f = c\sum_{h=1}^{j} e_h - \sum_{i=1}^{j} l_i^2 \qquad (11.10)$$

This function awarded a penalty according to the square of job tardiness (with respect to our deadline time d). The function awarded fitness proportionally with respect to advances, or the time that jobs finished earlier than the deadline. This fitness function f was designed to penalize job tardiness and when used in a GA, will encourage the GA to converge on results in which many jobs complete around the deadline time, rather than a few jobs being well advanced and other jobs being tardy[4].

11.3.4.2 Minimizing Makespan

Another popular cost function relates to makespan, which is the simulation time when the last job finishes in the schedule. We used makespan as our fitness function in our sample code (Section 11.3.8).

We may calculate a theoretical lower bound l on the value of makespan. We do this so we can map a simulated value of makespan to a fitness value in the range

---

[4] This fitness function f determines the fitness of a schedule relative to single deadline d. It would not be difficult to amend this fitness function so that each job has its own deadline time d_x where x is the job number.

0 to 1, as was done in the TSP problem in Section 11.2. We calculate l by considering how long each job takes to process with the condition that there is no idle time when a job is simulated. With this condition in place, we then choose the job that takes the longest to process, and this time forms our lower bound on makespan time. Formally, this lower bound l is given by

$$l = \max\{\sum_{i=1}^{m} t_{ai} : a = 1 \ldots j\} \qquad (11.11)$$

In our coded example we define the fitness f of our schedule to be

$$f = \frac{l}{M} \qquad (11.12)$$

where M is the simulated makespan of the schedule under test.

This value of fitness f tells the GA how close the makespan of the test genotype has evolved to the theoretically best makespan l. Of course, the best fitness value would be $f = 1$ but is unlikely to occur with most practical and hard problems. We use this value of l to map our simulated makespan of the problem schedule into a objective value between 0 and 1[5].

11.3.5 An Example Problem

The following time matrix represents the processing time requirements for the popular 10 machine, 10 job problem defined by Muth and Thompson [MT63], which we shall refer to as MT10.

$$t_{jm} = \begin{vmatrix} 29 & 78 & 9 & 36 & 49 & 11 & 62 & 56 & 44 & 21 \\ 43 & 28 & 90 & 69 & 75 & 46 & 46 & 72 & 30 & 11 \\ 85 & 91 & 74 & 39 & 33 & 10 & 89 & 12 & 90 & 45 \\ 71 & 81 & 95 & 98 & 99 & 43 & 9 & 85 & 52 & 22 \\ 6 & 22 & 14 & 26 & 69 & 61 & 53 & 49 & 21 & 72 \\ 47 & 2 & 84 & 95 & 6 & 52 & 65 & 25 & 48 & 72 \\ 37 & 46 & 13 & 61 & 55 & 21 & 32 & 30 & 89 & 32 \\ 86 & 46 & 31 & 79 & 32 & 74 & 88 & 36 & 19 & 48 \\ 76 & 69 & 85 & 76 & 26 & 51 & 40 & 89 & 74 & 11 \\ 13 & 85 & 61 & 52 & 90 & 47 & 7 & 45 & 64 & 76 \end{vmatrix} \qquad (11.13)$$

This matrix indicates that for the first job, the first machine requires 29 units of processing time, the second machine requires 78, the third 9, etc. For the JSS problem, the precedence matrix is relevant and this is defined by:

$$\begin{vmatrix} 0\,1\,2\,3\,4\,5\,6\,7\,8\,9 \\ 0\,2\,4\,9\,3\,1\,6\,5\,7\,8 \\ 1\,0\,3\,2\,8\,5\,7\,6\,9\,4 \end{vmatrix}$$

---

[5] Of course, in practice we will not know in advance the best possible solution to our problem.

$$p_{jx} = \begin{vmatrix} 1\,2\,0\,4\,6\,8\,7\,3\,9\,5 \\ 2\,0\,1\,5\,3\,4\,8\,7\,9\,6 \\ 2\,1\,5\,3\,8\,9\,0\,6\,4\,7 \\ 1\,0\,3\,2\,6\,5\,9\,8\,7\,4 \\ 2\,0\,1\,5\,4\,6\,8\,9\,7\,3 \\ 0\,1\,3\,5\,2\,9\,6\,7\,4\,8 \\ 1\,0\,2\,6\,8\,9\,5\,3\,4\,7 \end{vmatrix} \quad (11.14)$$

This precedence matrix indicates that the first job must be first machined by a machine with an id. of 0, followed by machines with ids. 1, 2, 3, 4, 5, 6, 7, 8, and 9. The precedence matrix is not relevant for OSS type problems where no such constraints apply.

For the MT10 problem, the optimal solution, obtained with exact methods is known to be 930 units. So for JSS simulations we calculated the fitness as $f = 930/M$ where M is the makespan of the schedule.

Using the formula 11.11, the lower makespan bound l for this problem is 655 units. So for the case of the OSS schedule, if we did not know the value of the optimal solution, we would calculate the fitness of a given schedule using this lower bound, $f = 655/M$

11.3.5.1 A Sample GA Simulation of the JSS and OSS Problems

To illustrate the way that our GA evolves good schedules, we took our sample 10x10 MT10 problem and simulated it. We first ran a GA when incorporating the precedence constraints, i.e. as a JSS problem. We used a steady state GA with a population size of 300 genotypes. Selection was done according to linear selection probability with a bias of 1.9. The operator used to create new children was LOX (see Section 11.4). The best genotype in the randomly generated initial population had a makespan of 1207 time units (Figure 11.6).

Figure 11.6 shows a gantt chart which describes this solution schedule. The time units taken at each step of the simulation are represented by the X axis. The time in which a machine started and stopped for a particular job is illustrated as a rectangle on the gantt chart. The width of the rectangle, in time units, is given by the processing time matrix t_{jm}. We observe with this random schedule in Figure 11.6 that there are many significant gaps between the machine job blocks for a given machine row. These gaps represent idle time, where a machine is waiting for its scheduled job (which is busy on another machine) to become free. The fitness of this genotype was calculated to be 0.77051 which relates to the ratio of the makespan (1207) to the optimal makespan to this well-studied problem (930). After 6,400 steady state generations the GA reached a makespan of 960 units (Figure 11.7) which is 3% greater than the true optimal makespan for this problem.

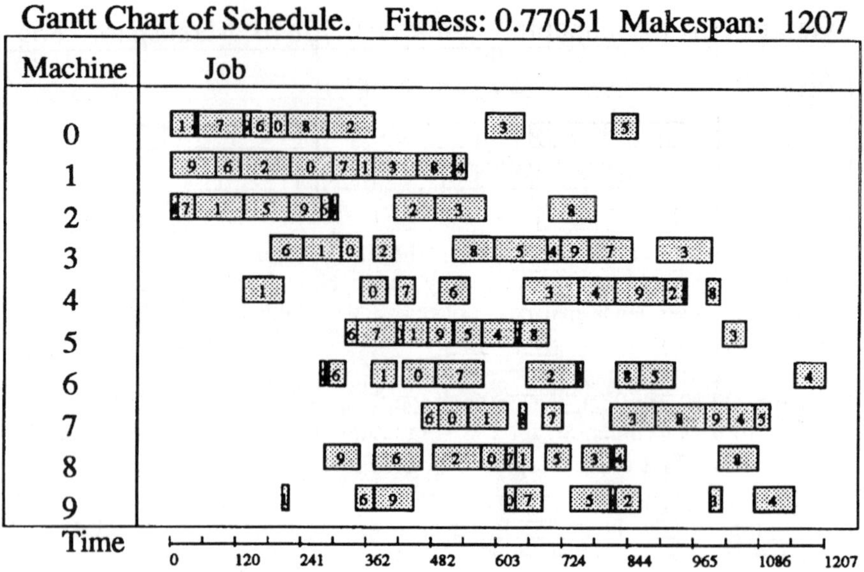

Gantt Chart of Schedule. Fitness: 0.77051 Makespan: 1207

Machine Job

0 1 2 3 4 5 6 7 8 9

Time 0 120 241 362 482 603 724 844 965 1086 1207

Figure 11.6: Best population JSS at generation 0.

We can see from Figure 11.7 that machines 0,1,2, and 3 are in high demand in the earlier phase of the JSS simulation. This early bottlenecking was purposefully designed [MT63] to make the JSS problem especially difficult to solve as it introduces high resource contention into the early part of the schedule.

Gantt Chart of Schedule. Fitness: 0.96875 Makespan: 960

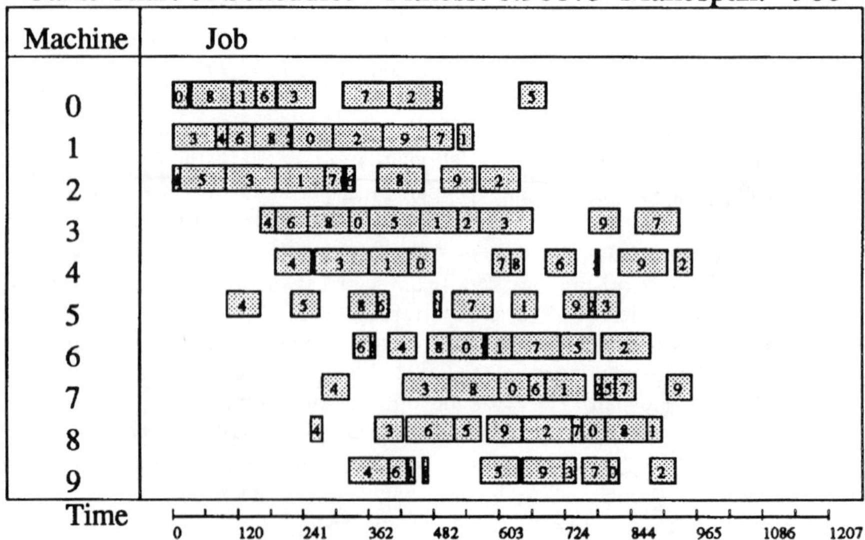

Figure 11.7: Best population JSS at generation 6,400.

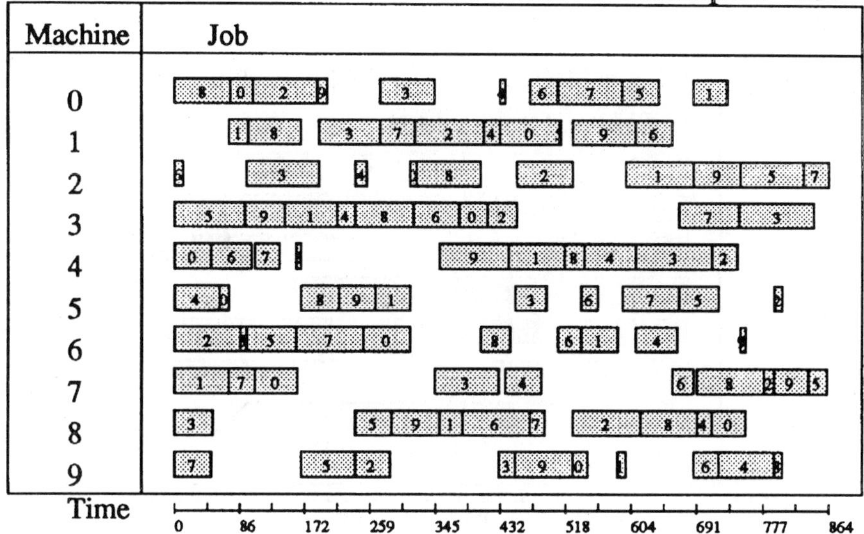

Figure 11.8: Best population OSS at generation 0.

We model a simple OSS problem by removing the precedence constraints in the MT10 problem above. We can see that the best initial randomly generated population member (Figure 11.8) has a makespan of 864 units which is significantly better than that of the JSS problem. In our OSS problem our fitness is relative to the lower bound of makespan (655 units) and has a fitness value of 0.75810. When we run our GA for 160,000 steady state generations a solution to the OSS problem is reached. We also know that this solution is optimal as it forms a schedule with a makespan of 655 units (Figure 11.9) which is the calculated lowest bound makespan (theoretical best). If we follow job id. 3 in Figure 11.9 through in this schedule we can see that there is no idle time introduced between the machine processings of that job. Job id. 3 requires the most aggregate machine processing and is simulated without any delays, which explains why the schedule in Figure 11.9 is optimal.

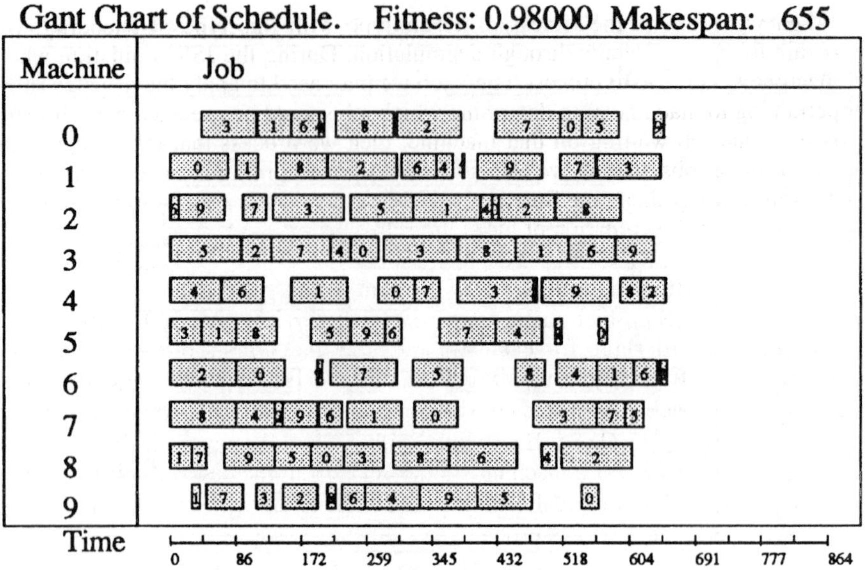

Gant Chart of Schedule. Fitness: 0.98000 Makespan: 655

Machine | Job

0
1
2
3
4
5
6
7
8
9

Time 0 86 172 259 345 432 518 604 691 777 864

Figure 11.9: Best population OSS at generation 160,000.

11.3.6 Encoding a JSS in a Chromosome

For the GA to function efficiently a chromosome must encode a legal JSS schedule. The challenge is to design a genetic representation of a point in the solution space (a schedule) which does not invalidate the precedence constraints which are specified in the JSS problem [FB91]. The chromosome must unambiguously map to a single schedule. The most natural encoding used in job shop scheduling is to partition the chromosome into equally sized blocks of genes which we will call subchromes. One subchrome will be used to encode a preference list for the machine that the subchrome is associated. One subchrome is used for each machine [FB91]. Within a subchrome we encode a list of job preferences (dispatch rules) for the machine the subchrome relates to. Dispatch rules allow us to choose the next job to be processed for machine ra that becomes free, from any number of jobs that are waiting to be processed in the input buffer of machine m. Reading a subchrome from left to right gives us the job to be chosen with the highest priority through to the lowest. For example, the chromosome for a four machine, four job problem might be.

[1 3 2 0|0 1 2 3|1 0 2 3|3 2 1 0]

Note that each subchrome is separated by the '|' symbol. This symbol is included for clarity and is not encoded into the genotype. If we consider the third subchrome [1 0 2 3], we have the job choice preference list for the third machine. This means that if jobs ids. 0, 1, and 2 have stacked up in the input buffer of the third machine, then when the third machine is ready to process a new job we choose job id. 1 to be processed next as it has the highest priority of the waiting jobs.

When we evaluate the fitness of a genotype with this type of encoding, we obtain the timing details through a simulation. During the JSS simulation when machine m is ready to process a new job we may need to apply the dispatch rules pertaining to machine m to determine which job should be executed next. If there is only one job waiting on that machine, then we process that job. If there are two or more jobs waiting we choose the next job according to the dispatch rules. Therefore we evaluate the fitness of a JSS encoded in this way indirectly through a complete mini-simulation of the entire schedule.

11.3.7 Encoding an OSS in a Chromosome

Because the OSS problem has no precedence constraints we can directly encode the machine orderings for each job into the genotype. To encode an OSS schedule into a genotype we divide the chromosome up into j equally sized *subchromes*. Each subchrome consists of m genes (where m is the total number of machines in the OSS). Each gene within the subchrome is then used to specify an identifier for a machine. So the kth subchrome contains a list of genes representing a chronological ordering of machines for job k. For example, in a four machine, four job OSS problem, the subchrome might be.

[1 3 2 0|0 1 2 3|1 0 2 3|3 2 1 0]

If we consider the 3rd subchrome [1 0 2 3], this tells us that for the third job, the order of execution of machines for that job is machine with an id. of 1, followed by machine id. 0, 2, and then 3.

For both the JSS and OSS problems the genetic encoding scheme does *not* specify information relating to the specific time in which any machine is to process a job. The actual schedule, and the associated timings are obtained by simulation of the schedule specified in the genotype.

11.3.8 Implementing the Fitness Function of the JSS and OSS Problems

The first task required in the implernentation of the fitness function of the JSS and the OSS problem is to define a class which has the data items and methods that are relevant to both JSS and OSS problems. The following section details the *sched_base* class which presents such a class.

11.3.8.1 Class Definition of the *sched_base* Class

The *sched_base* is an abstract class and defines data and methods common to both the JSS and OSS problems. The sub-classes *jss and oss* inherit all data and methods from the *sched_base* class and each sub-class introduces additional specialization. We summarize the main data members of the *sched_base* class, and leave the discussion for the defined member functions to subsequent discussion.

i) The data item *job_machine_time,* is a two dimensional array that stores the machining time for a given job id. (first index) and machine id. (second index).

ii) The *job_sequence* array is also another important two dimensional array. For a specified job id. (first index) and a sequence number (second index), this array will give us the machine which is to be processed next. In the JSS the machine orderings are specified as part of the problem definition and the *job_sequence* array will reflect this ordering. When we simulate the OSS problem (Section 11.3.8.5) we populate this array from the machine orderings stored in the input genotype.

iii) The *mach_alloc* data member is an array of a structure of type *mach_alloc_type.* If we index this array by a machine *x* we can determine if *x* is booked *(jobs_hooked),* what time the machine *x* is next free *(free_at),* the id. of any currently running job on the machine *(job_current)* and a list of jobs which are waiting to be processed on machine *x (job_alloc* array). The class definition follows.

```
    class sched_base
protected:

    int job_machine_time [JOBS][MACHINES];
    int job_sequence[JOBS][MACHINES];
    int job_processes_done[JOBS];
    int min_time;
    int current _time, next_completion_time;
    int next_completed_machine, machines, jobs;

    struct mach_alloc_type{
        struct {
            int job;
            int priority;
            char free;
            int queued_at;
        } job_alloc [MAX_ALLOCATIONS];
        int free_at;
        int started_at;
        int jobs_booked;
        int job_current;
    };

    struct mach_alloc_type mach_alloc[MACHINES];

    virtual void scan_genotype(individual *indiv)=0;
    virtual int job_find_next(int m, individual *indiv)=0;
    virtual int job_submit(individual *indiv, int m, int j)=0;
    void print_machine_queue();
    inline int job_remove_current(int m);
    void get_job_mach(char *prob_file);
    inline void machines_process_first_job(individual *indiv);
     inliae int machine_push(int m, int job, individual *indiv, char execute,
       int *jobs_finished);
```

```
    inline int machine_pop(int m, individual *indiv);
    int simulation_run(individual *indiv);

public:
    sched_base(int machs_in, int jobs_in, char *file_name);
    float simulate(individual *indiv, float all_objective_best);
    }; // jss class
```

11.3.8.2 Class Definitions for the *Jss* and *Oss* Derived Classes

We can boil down the differences between the implementation of the *oss* and *jss* classes as differences in the implementation of four member functions. These member functions

i) perform initializations (the constructor)

ii) interpret the meaning of the genes in the input genotype (the *scan_genotype* member function)

iii) choose the next job from a set of waiting jobs at a machine (the *job_find_next* member function) and

iv) determine the way in which we submit a new job is to be processed on a machine (the *job_submit* member function).

Because of these differences between the JSS and OSS problems we have defined the member functions *scan_genotype, jobfind_next, and job_submit as* pure-virtual functions in the base class *(sched_base)* and implement each one differently in the *jss and oss* classes.

The *jss* member function inherits all data and methods from the *sched_base* class. An additional data item is added. The *indiv_priority* data item is a two dimensional array. When the first index is specified as a machine id. (*m*) and the second index is specified as a job (*j*) this array gives us the priority in which the job *j* should be chosen over other jobs which may be awaiting processing on machine *m*. The highest priority corresponds to a value of zero. The class definitions for the *jss* and *oss* classes follow.

```
    class jss: public schod_base {
privats:
    int indiv_priority[MACHINES][JOBS];

    void scan_genotype(individual *indiv);
    int job_find_next(int m, individual *indiv);
    int job_submit(individual *indiv, int m, int j);
public:
    jss(int machs_in, int jobs_in, char *file_name) : sched_base( machs_in,
      jobs_in, file_name) {};
```

```
};

        class oss : public sched_base {
    private:
        void scan_genotype(individual *indiv);
        int job_find_next(int m, individual *indiv);
        int job_submit(individual *indiv, int m, int j);
    public:
        oss(int machs_in, int jobs_in, char *file_name) :
         sched_base( machs_in, jobs_in, file_name) {}
};
```

11.3.8.3 Member Functions Specific to the *sched base* Base Class

The objective function *objfunct* is the high level function called by the GA that is responsible for determining a measure of the fitness of the genotype *indiv* which is passed to it as parameter. The objective function *objfunct* is not a member of any class but has two important responsibilities. The first time that it is called it is responsible for creating a new instance of the *jss* or *oss* class. Once this object has been instantiated, from this point onward, the *objfunct* then calls the simulation member function of either the *jss* or *oss* function which in turn calculates the fitness of the input genotype. This return fitness value is a floating point number between 0 and 1. If we are simulating the JSS (the JSS constant equals 1) then the *sched_prob* pointer is assigned a new instance of the *jss* object. Otherwise, if the JSS parameter is 0, then we are simulating an OSS problem and an *oss* object is created. When creating either of these two simulation objects we pass three parameters to the constructors. We pass *MACHINES,* the number of machines, *JOBS,* the number of jobs, and a string containing the name of the data file which defines the precedence constraints (for JSS) and machine processing times. The simulation is the run by a call to the *simulate* member function of the *jss* or *oss* simulation object.

```
        #inclade "obj.h"
        sched_base #sched_prob;

float objfunct(individual *indiv, int problem_id, float all_objective_best) {

    static first_time=1;
    if (first_time) {
        first_time=0;

        char filename[100]; //got name of problem file which is stored
        extern int get_param(char *name, char *param); //in the GA
        assert(get_param("ob_file", filename));      //parameter file

        if (JSS)
            sched_prob = new jss(MACHINES, JOBS, filename);
        also
            sched_prob = new oss(MACHINES, JOBS, filename);
```

```
    }

    return sched_prob->simulate(indiv, all_objective_best); //run simulation
        }
```

The *jss* or *oss* simulation object is created once for the entire GA run. When this occurs the constructor function is called in the base class *sched_base.* This constructor is responsible for calling the *get_job_mach* member function which in turn reads in the processing times and machine orderings from the problem data file.

```
        sched_base::sched_base(int machs_in, int jobs_in, char *file_nme) {
    machines = machs_in;
    jobs = jobs_in;
    get_job_mach(file_name);
        }
```

This *get_job_mach* function is passed a parameter called *prob_file* which is string containing the name of the problem file name. The function then opens and reads this file and extracts the execution times for each job, on each machine, into the global array called *job_machine_time.* For an example problem with 2 machines and 4 jobs the data file is stored in the following format

```
Problem:                    Simple
Problem Description:        2 machine 4 Job
Problem Size:               2 4
Optimal Time:               0
0 23 3 22 2 34 1 99 1 33 2 92 0 03 3 45
```

In the above example, for a JSS problem each job has a row of machine orderings and corresponding processing times. For the first job, machine 0 is scheduled first followed by machines 3, 2, and then 1. Machine 0 has an execution time of 23 time units, machine 3 has an execution time of 22 time units etc. For the second job the machining sequence is specified as (1, 2, 0, 3). The above data file format is also used to encode OSS problems, however, the order in which the machines appear on a line for a specific job is not important.

In our data file we have a field called *Optimal Time.* If an optimal makespan time has already been calculated for the problem, then this optimal time will appear in this field. If the problem is new, however, and no such optimal time is known, then this is indicated by a zero value in this field. In this case we calculate a theoretical lower bound of the makespan of the best possible schedule. This lower bound is then assigned to class variables called *min_time.* We calculate this value by considering how long each job would take to process given no idle time in the sequence of machine processings. This is to say that we consider that each job is constructed assuming that each machine will be available immediately when required. With this condition in place, we then choose the job that takes the longest to process, and the time of this job forms our lower bound on makespan time. This lower bound l is given by Equation 11 in Section 11.3.4.2.

```
    void sched_base::got_job_mach(char *prob_file) {
 int i,j,total_time, optimal_time;
 min_time = 0;

 cout << "Opening scheduling file: " << prob_file << '\n'; ifstream
 m_j_file(prob_file);
 if (m_j_file.fail()) {
    cout << "'Bad job machine time file\n";
    exit(l);
 }

 char jstr[1000];
 for (j=0; j<3; j++)
    (void) m_j_file.getline(jstr, 1000);
                //skip 3 lines
 (void) m_j_file.get(jstr, 1000, ':');          //go to the colon first
 (void) m_j_file.get(jstr, 2); //move to the optimum value field
 m_j_file >> optimal_time;                      //read in the dmin value
 (void) m_j_file.getline(jstr, 1000);           //go to next line
 int seqn, timen;
 for (j=0; j<jobs; j++) {
    total_time=0;
    for (i=0; i<machines; i++)
        m_j_file >> seqn >> timen; //read in machine sequence and time
        job_sequence[j][i] = seqn;
        job_machine_time[j][seqn]=timen;
        total_time+=job_machine_time[j][seqn];
    } // for i
    if (total_time > min_time)
        min_time = total_time;
 } //for j
 if (optimal_time)
    min_time = optimal_time;

    } //get job_machine_time
```

The *simulate* function first calls the *scan_genotype* member function which reads the genotype information into various class variables. The simulation is then run by a call to the *simulation_run* function. After this function, the *current_time* variable will represent the simulated makespan (in time units) of the schedule encoded in the genotype. We convert this simulation time into a fitness value between 0 and 1 by assigning the *objective_value = min_time/current_time.* The value *rain_time* is a theoretical lowest bound of time that the schedule could run for and is calculated in the *get_job_reach* function. Hence the *objective_value* variable has a maximum value of 1. This objective value is then returned back to the calling function *objfunct.*

```
float sched_base::simulate(individual ,indiv, float all_ objective_ best) {
    scan_genotype(indiv);
```

```
assert(!simulation_run(indiv));
float objective_value= min_time *1.0/ current_time; assert(objective_value
>=0 && objective_value<= 1);

return objective_value;
    ) //objfunct
```

The *simulation_run* function initializes a number of simulation object variables. It sets the status of all machines to free. Next, the first set of jobs are allocated to their respective machines as is determined by the *job_sequence* array. The *machine_start* function then gets the simulation started by processing this first round of jobs; allocating each job to the correct starting machine. The final part of the *simulation_run* loops until all jobs are processed in the simulation. Within a loop, a job j which has finished on a machine m is ejected from machine m with the *machine_pop* function. If the finished job j has not been fully machined it is submitted to its next machine with the *machine_push* function. Once all jobs have been processed by each machine (when *jobs_finished* equals *jobs*) the function ends and the simulation is complete.

```
int sched_base::simulation_run(individual *indiv) {

    //initialize variables
    int i,j, finished_job, jobs_finished=0;
    current_time=0;
    next_completion_time = LARGE_NUMBER
    for(i=0; i<jobs; job_processss_done[i++]=0);
    for(i=0; i<machines; i++) {
        mach_alloc[i].jobs_booked= 0;
        mach_alloc[i].free_at=
        mach_alloc[i].job_current= LARGE_NUMBER; mach_alloc[i].started_at
        = LARGE_NUMBER;
    }
    // set all machine job list to free
    for(i=0; i<machines; i++)
        for(j=0; j<jobs; j++)
            mach_alloc[i].job_alloc[j].free=1;

    // allocate initial jobs
    int unused;
    for (i=0); i<jobs; i++)
        if (machine_push(job_sequence[i][job_processes_done[i]++],i, indiv,
        EX_LOAD, &unused))
            return 1;
    //fire up the machines
    machines_process_first_job(indiv);
    print_machine_queue();

    //loop until all machines processed
    whilo(jobs_finished < jobs) {
```

```
        finished_job = machine_pop(next_completed_machine, indiv);
        if (job_processes_done[finished_job]>= machines) {
            if (REPORT) cout << "Finished Job " << finished_job << '\n';
            ++jobs_finished;
        }
        else
            if (machine_push(job_sequence[finished_job] [job_processes_done
                [finished_job]++], finished_job, indiv, EX_RUN, &jobs_finished))
                 return 1;
    print_machine_queue();
    }// while
    indiv->other = current_time;

    return 0;
        } //simulation_run
```

The *machines_process_first_job* function is called in the early part of the *simulation_run* function. The function is responsible for determining which of the machines initially were allocated for one or more jobs. For each of these machines, the function uses the *job_find_next* function to determine which of any of the queued jobs should become the current job for the machine.

```
inline void sched_base::machines_process_first.job(individual *indiv) {
    for (int m=0; m<machines; m++) {
        int job = job_find_next(m, indiv);
        if (job == NO_JOBS)
            continue;
        mach_alloc[m].job_current = job;
        mach_alloc[m].started_at = 0;
        mach_alloc[m].free_at = current_time+job_machine_time[job][m];
        if (mach_alloc[m].free_at < next_completion_time) {
            next_completion_time = mach.alloc[m].free_at;
            next_completed_machine = m;
        )// if
    }// for m
        }
```

The *machine_push* function is responsible for pushing job *j*, for a specific machine m onto the the input buffer of machine *m*. In the special case that this job *j* happens to have zero processing time for this machine *m*, then we ignore this machine-job pair and advance to the next job in the schedule order determined by the *job_sequence* array. The function then calls the *job_submit* function which registers job *j* as pending (i.e. sitting on the input buffer) of machine *m*.

The function next determines if machine *m* is ready to execute the new job (job *j*). Machine *rn* is ready for job *j*, if *j* it is the first job in the simulation to be processed by that machine, or if the machine is idle. If this is true, then we set the time *t* that our machine *m* is scheduled to be free to the current time plus the processing time of job *j* on machine *m*. We also check to see if time *t* is less than the *next_completion_time* variable and if it is, the next change in state of the simulation will occur when machine *rn* finishes processing job *j*. If so, then we have a new *next_completion_time* value.

```
inline int sched_base::machine_push(int m, int j, individual *indiv, char
                                    execute, int *jobs_finished) {
  int numb_jobs;

  if (REPORT)
    cout << "Pushing machine " << m << ", j " << j;

  if (job_processes_done[j] >= machines+1) {
    (*jobs_finished)++;
    return 0; //if trailing machine has zero procseeing time
    }

  //if zero processing time ignore and advance
  if (!job_machine_time[j][m]) {
       machine_push(job_sequence[j][job_processes_done[j]++], j, indiv,
                                    execute, jobs_finished);
    return 0; //ie if there is no processing req. for this machine
  }//end if

  //submit the job
  if (job.submit(indiv, m, j))
    return 1; //error submitting job

  numb_jobs = mach_alloc[m].jobs_booked;

  //we are scheduling this new job on machine m now?
  if (numb_jobs == 1 && execute==EX_RUN
  ){
    mach_alloc[m].job_current = j;
    mach_alloc[m].started_at = current_time;
    mach_alloc[m].free_at = current_time + job_machine_time[j][m];
    if (mach_alloc[m].free_at < next_completion_time) {
       next_completion_time = mach_alloc[m].free_at;
       next_completed_machine = m;
    } // if
  } // if

  assert (numb_jobs <= MAX_ALLOCATIONS);
    if (REPORT) cout <<" Next Mach Completing" << next_
                                    completed_machine << '\n';
  return 0;
    // machine_push
```

The *machine_pop* function is called when machine *m* becomes free. This function updates the current simulation time by assigning the *free_at* attribute of the *mach_alloc[m]* array to the *current_time* variable. In this way, the *current_time* variable updates to the point in the simulation time when machine *m* becomes free. The function next determines whether there are jobs queued up in the input buffer of machine *m*. If there are, the function determines which job is to run next (with a call to the *job_find_next* function), and then submits that new job as the new running job on machine *m*.

```
    inline int sched_base::machine_pop(int m, individual *indiv) {
int i, job_there, finishing_job;

if (REPORT) cout << "Popping machine " << m << '\n';

// restore current time
current_time = mach_alloc[m].free_at;

finishing_job = mach_alloc[m].job_current;
if (job_remove_current(m))
    return -1; //something wrong

job_there= mach_alloc[m].jobs_booked;

// there are waiting jobs on this machine - run the next job
if (job_there) {
    int next_job;
    next_job = job_find_next(m, indiv);
    if (next_job != NO_JOBS) {
        mach_alloc[m].job_current = next_job;
        mach_alloc[m].free_at = current_time + job_machine_time
                                                    [next_job][m];
        mach_alloc[m].sterted_at = current_time;
    }
}
else mach_alloc[m].free_at = 0;

// now calculate the next time a machine is going to be free

next_completion_time = LARGE_NUMBER;
for (i=0; i< machines; i++)
    if (mach_alloc[i].free_at < next_completion_time &&
        mach_alloc[i].jobe_booked >=1 && mach_alloc[i].job_ current !=
                                                NO_JOBS) {
        next_completion.time = mach_alloc[i].free_at;
        next_completed_machine = i;
    } //if i

return finishing_job;
    }
```

The *job_remove_current* function is called by the *machine_pop* function when a job has finished on a machine *m*. The function removes the finished job (which is *mach_alloc[m].job_current)* from machine *m*.

```
    inline int sched_base::job_remove_current(int m) {
if (mach_alloc[m].job_current == NO_JOBS) return 0;
for (int i=0; i< jobs; i++) {
    if (!mach_alloc[m].job_alloc[i].free && mach_alloc[m].job_ alloc[i].job
                          == mach_alloc[m].job_current) {
    mach_alloc[m].job_alloc[i].free = 1;
    mach_alloc[m].jobs_booked--;
    return 0;
    } // if
} // for
return 1; //error
    }
```

11.3.8.4 Member Functions Specific to the JSS Problem

The *scan_genotype* function is called once at the start of each new simulation from the *simulate* function. The function scans the genotype and populates the *indiv_priority* global array. This global array tells us for a particular machine, what is the priority for each of the jobs which may queue up on the input buffer of that machine. If two or more jobs are queued up on a machine when the machine becomes free, then this array is consulted to determine which job is to be scheduled next on the machine. The lower the value of priority then the higher the precedence of that job.

```
    void jss::scan_genotype(individual *indiv) {
for(int  m=0; m( machines; m++)
    for(int priority_level=0; priority_level < jobs; priority_ level++)
        indiv_priority[m][indiv->value(m*jobs+priority_level)]  =
                                                    priority_level;
if (REPORT) {
    cout << "\nPriorities From Submitted Chrome (y machines, x jobs, v
                                                priority)\n";
    for (i=0; i<machines; i++)
        for (int j=0; j<jobs; j++)
            cout << setw(4) << indiv_priority[i][j];
    cout << '\n';
}
    }
```

The *job_submit* function is responsible for booking a job *j* to be executed on machine *m*. Note that this function only *books* a job *j* on machine *m*, the functions which execute a job on a machine are *machine_pop* and *machine_push*. Note that when we book a job for execution on a machine we also assign the priority in which that job should be selected over any other waiting jobs on that machine. This priority check will be done in the *job_find_next* function by referencing the *indiv_priority* array.

```
    inlino int jss::job_submit(individual *indiv, int m, int j) {
  for (int i=0; i< jobs; i++)
    if (mach_alloc[m].job_alloc[i].free) {
        mach_alloc[m].job_alloc[i].free = 0;
        mach_alloc[m].job_alloc[i].job = j;
        mach_alloc[m].job_alloc[i].priority = indiv_priority[m][j];
        mach_alloc[m].job_alloc[i].queued_at = current_time;
        mach_alloc[m].jobs_booked++;
        return 0;
    } // if
  }// for
  return 1; // error could not submit job
    }
```

The *job_find_next* function is called to decide which job, of the jobs waiting to be processed on machine *m*, is to be next scheduled on *m*. The function searches through the list of waiting jobs on machine *m*, and returns the job found with the highest priority.

Note that this function only chooses a job from those jobs already waiting in the input buffer of our machine *m*. We can call this job of the highest priority j_1. It has been suggested ([CRG93]) that we can look ahead and see if is worth blocking (i.e. machine will be idle), and waiting for a job j_2 to finish on another machine m_2. Blocking is considered worthwhile if the time required to wait for job j_2 to finish processing on m_2 plus the processing time of job j_2 on machine m is less than the processing time of job j_1 on machine m. This look-ahead technique has shown to result in slightly better schedules ([CRG93]) and that most of the code required to implement this would normally appear in the position marked in the *job_find_next* function.

```
    inline int jss::job_find.next(int m, individual *indiv) {
  int highest_priority = jobs, highest_ptr = jobs;
  int seen = 0;

  if (!mach_alloc[m].jobs_booked) return NO_JOBS;
  for(int i=0; i < jobs && seen < mach_alloc[m].jobs_booked; i++) {
    if (!mach_alloc[m].job_alloc[i].free) {
        seen++;
      if (mach_alloc[m].job_alloc[i].priority < highest_priority) {
      highest_priority = mach_alloc[m].job_alloc[i].priority;
      highest_ptr = i;
      } // if
    } // if
  } // for i

  int job_high_waiting = mach_alloc[m].job_alloc[highest_ptr].job;

  // Croce et. al look ahead code would normally appear here
  return job_high_waiting;
```

```
}
```

11.3.8.5 Member Functions Specific to the OSS Problem

When simulating OSS problems we implement the functions *scan_genotype, job_submit,* and *job_find_next* differently to that of the JSS problem. The *scan_genotype* function is called once for each new simulation of the input genotype. The function scans the input genotype and populates the *job_sequence* array. Since there are no mandatory precedent constraints in an open shop scheduling problem, we use the input genotype to encode the machine processing order for each job. In this way, the *scan_genotype* function builds up the *job_sequence* array from the machine orderings that are described in each subchrome in the genotype.

```
    void oss::scan_genotype(individual *indiv) {
for(int j=0; j< jobs; j++)
    for(int seq_no=0; seq_no< machines; seq_no++)
        job_sequence[j][seq_no] = indiv->value(j*machines+seq_no);
    }
```

The *job_submit* member function is responsible for queuing job *j* on machine *m*. The function scan through the *job_alloc* array for machine *ra* and looks for a free booking slot. Once one is found the job is booked in this slot and marks the slot as not free.

```
    inline int oss::job_submit(individual *indiv, int m, int j) {
for (int i=0; i< jobs; i++) {
        if (mach_alloc[m].job_alloc[i].free) {  mach_alloc[m].job_alloc[i].free
        = 0
        mach_alloc[m].job_alloc[i].job = j;
        mach_alloc[m].jobs_booked++;
        return 0;
    } // if
} // for
return 1; //error, could not submit job
    }
```

The *job_find_next* function determines which job will be chosen when a number of jobs have stacked up at the input buffer of a machine and when the machine becomes ready to process a new job. In the JSS problem the GA evolves priority lists to resolve this problem, however, in the OSS problem the GA evolves machine orderings. In the OSS problem, to choose from a set of waiting jobs at machine *m* we apply the simple rule that we always choose that job which first arrived at the input buffer of machine *m*.

```
inline int oss::job_find_next(int m, individual *indiv) {
int earliest_ptr;
int queued_earliest=LARGE_NUMBER;
```

```
int seen = 0;

if (!mach_alloc[m].jobs_booked) return NO_JOBS;
for(int i=0; i < jobs && seen < mach_alloc[m].jobs_booked; i++) {
    if (mach_alloc[m].job_alloc[i].free) {
      seen++;
      if (mach_alloc[m].job_alloc[i].queued.at < queued_earliest) {
            queued_earliest = mach_alloc[m].job.alloc[i].queued_at;
            earliest_ptr = i;
        } // if
    } // if
} // for i

return mach_alloc[m].job_alloc[earliest_ptr].job;
    }
```

11.3.8.6 Reporting

The *print_machine_queue* function is used to display the state of the machine queue as is described by the *mach_alloc* array. The function prints a summary of the job activity current on each of the machines to standard output. For each machine, this information includes; how many jobs are hooked, which is the current job, which job is currently running (if any), which job the machine is waiting for (if it is blocked), which the machine is free (if busy), and when the machine started executing (if busy). For each machine, if there are queued jobs awaiting processing time, each of these waiting jobs are printed along with the priority of each job. (The priority can be ignored when using OSS simulations).

The following sample JSS report shows the current state of the machines at simulation time 29. Machine with id. 4 started the job with id. 3 at time 21. Job id. 3 finishes at time 9.9. We can see from the job waiting list on machine id. 4, that job id. 3 has a priority of three, which is a higher priority list than the other job awaiting processing (job id. 1) which has a priority of 4.

Current Time: 29

Mach	Start	Free	Bkd	Sched	Jobs Waiting
0	25	0	0	4	
1	22	0	0	5	
2	20	0	0	2	
3	29	33	3	2	(5 2) (2 1) (4 5)
4	21	29	2	3	(3 3) (1 4)
s	29	32	1	0	(0 0)

The *print* member function which follows is responsible for generating such a report.

```
    void sched_base::print_machine_queue() {
if (REPORT) {
    int i,j;
    cout << "\nCurront Time:" <<current_time << //print out titles
```

```
"\nMach Start  Free Bkd  Sched  Jobs Waiting\n";

for (i=0; i< machines; i++) {
    int count=0;
    cout << '\n' << setw(4) << i // print out details for machine i
        << setw(6) << mach_alloc[i].started_at
        << setw(6) << mach_alloc[i].free_at
        << setw(4) << mach_alloc[i],jobs_booked
        << setw(4) << mach_alloc[i].job_current
        << "   ";
    for (j=0; count < mach_alloc[i].jobs_booked; j++)
        if (!mach_alloc[i].job_alloc[j],free) {
            count++;
            cout << " (" << mach_alloc[i].job_alloc[j].job <<' '
                << mach_alloc[i].job_alloc[j].priority << " )";
        )//if
    } // for j
} // for i
cout << "\n";
}
} //print
```

11.4 The Linear Order Crossover for JSS and OSS Problems

It is clear from the research into scheduling genetic operators [SMM+91] that all scheduling problems cannot be efficiently solved with a single genetic operator. We must take a careful look at the nature of the scheduling problem and determine what requirements exist for an operator. For both JSS and OSS problems we require a crossover operator that has the following qualities

i) The operator produces genotypes such that highly fit building blocks (strings of adjacent genes) can be transferred from both parents to a new child.

ii) The crossover operator can be applied independently to matching subchrome block between two parents. It is important that subchromes are considered independent of other subchromes in different positions as any attempt to mix gene values between subchromes could easily result in duplicate or absent gene values to appear in the resultant child subchromes.

iii) In the JSS problem, the first gene in a subchrome encodes the job of the highest priority in that subchrome, and the last gene encodes the job of the lowest priority and there is no close relationship between these two extreme gene locations. In the case of the OSS problem the first gene in a subchrome encodes the first machine to be processed by the job *j* that the subchrome maps to. The last gene in the subchrome relates to the last machine to process job *j*. In both cases the subchrome must be considered non-cyclic (unlike the TSP genotype).

iv) In the JSS problem the relative gene positions are important as we are dealing with a list of priorities. This contrasts with TSP-like problems where adjacent gene relationships are more important as these adjacent relationships determine the overall value of the solution.

v) The operator must produce legal children. This means in the JSS problem job ids. must be represented in each subchrome in some order. In the OSS problem each machine id. must be represented in that each subchrome. There should be no duplicate or absent gene values within a child subchrome.

With these requirements in mind Faulkener [FB91] developed a variant on the order-based crossover operator (OX) called Linear Order Crossover (LOX). LOX satisfies all these requirements for the JSS. To implement LOX between two parent subchromes (*ps1* and *ps2*) we:

i) First choose two random crossing points x_1 and x_2 ie

```
ps1 = [1 2 3  0 4 5  6 7]
ps2 = [0 7 1  4 5 2  3 6]
             ^x1    ^x2
```

ii) Replace the jobs in *ps*1 which appear between the crossing points in *ps*2 with 'holes' which are indicated in *ps*1 as asterisks. *ps*1 becomes

```
ps1= [1 * 3 0 * * 6 7]
```

iii) Move the holes int *ps1* into the crossing region by sliding the jobs away from the crossing region toward the extremities of the subchrome to give a new *ps* 1 of

```
ps1= [1 3 0 * * * 6 7]
```

iv) Transfer the jobs from the crossing region of *ps*2 over to the crossing region in *ps*1 to give a new child *cs*1.

```
cs1=[1 3 0  4 5 2  6 7]
```

v) Apply the logic in steps 1-4 to create a new child *cs*2 by reversing the roles *ps*i and *ps*2.

This crossover is guaranteed to generate legal subchromes. The shifting of the jobs away from the crossing region before the transfer ensures that no duplicate and missing jobs are created in the new child. The shifting of jobs preserves as much as possible (despite the new injection of material) the order of the jobs in the parent subchromes. This represents the minimum amount of additional order based disruption occurring when gene block are transferred between parents.

11.4.0.1 Implementing the LOX Operator

The *o_linear_order* function creates a new child genotype *(child)* from two parents *parent1* and *parent2*. The number of different subchromes is denoted in the code by the constant called *sub_chrome&* The number of genes in each subchrome is specified in the constant *sub_chromes_size.* The function either applies the LOX operator to a specific subchrome, or else it will directly copy the subchrome from *parentl* into the *child.* The probability in which crossover is performed (rather than a copy) on a subchrome is specified by the constant called *LOX_PROBABILITY.* We found through experimentation on small sized JSS problems[6] that a useful value for this constant was 0.67. If crossover is to be applied to the *i*th subchrome, then the function selects two points *p*l and *p*2 where both points specify a position within the subchrome. The smallest of *p*l and *p*2 forms the location of the start of the block of genes that will be copied from *parentl* into the *child.* The absolute difference between *p*l and *p*2 plus one, forms the width of this block. The function then calls the *shift* function which performs the LOX operation on the subchrome specified by an offset from the start of the genotype in the *subchrome_offset* parameter.

```
int o_linear_order(individual *parentl, individual *parent2, individual *child) {
    for (int i=0; i< sub_chromes; i++) {

        int subchrom_offset = i*sub_chrome_size; // set offset of subchrom to
                                                   apply LOX
        if (!randobj.flip(LOX_PROBABILITY)) {
            for(int z=0; z< sub_chrome_size; z++)
                child->set(z+subchrome_offset,    parantl->value(z+
                                                   subchrome_offset));
                continue; // this subchrome has been copied, go to next
            }//flip

        int from_pos, to_pos, block_width;

        int pl=0;
        int p2=sub_chrome_size-l;

        while (abs(pl-p2) == sub_chrome_size-l) {
            pl = randobj.random_get(sub_chromo_size); // randomly select 2
                                              positions within subchrome
            p2 = randobj.random_get(sub_chrome_size);
            }

        from_pos = (pl < p2 ? pl : p2); // set the from_pos to smallest position
                                              (pl or p2)
        to_poz = from_pos;
        block_width = abs(pl-p2)+1;
        shift(parentl, parent2, child, from_pos, to_pos, block_ width
                                              ,subchrome_offset);
    } //for loop
    child->op = OT_BLOCK_SHIFT;
    return 0;
```

---

[6] This probability value was obtained by a meta-GA experiment where good GA parameter sets for JSS problems (10 machine/10 job) were bred by a second meta-GA IRon94].

```
}
```

The *shift* function begins by creating an index out of an array called *in_block.* If a gene value *v* in *parentl* is located inside the block of genes to be copied, then the *in_block* array indexed by *v* is set to the value of 1, and 0 otherwise. This gives a fast mechanism to check if a gene value in *parent2* lies in the copy block in *parentl.* The function then loops through all of the gene values *r* of *parent2* and if *in_block[r]* equals zero, then we copy this gene value *r* into the child. This technique will stop any duplicate gene values appearing in the *child.* Inside this loop, when the function gets to the location corresponding to the start of the copy block the block of genes from *parentl* is then copied into the *child.*

```
void shift(individual *parentl, individual *parent2, individual *child, int
                        from_pos, int to_pos, int block_width, int
                        subchrome_offset) {

   // create an index for those values appearing in the block in pl int
   in_block[SUB_CHROME_SIZE];
   for(int i=0; i<sub_chrome_size; i++)
      in_block[i]=0;
   for(i=from_pos; i<from_pos+block_width; i++)
      in_block[parent1->value( i+subchrome_offet)]-l;

   // create an  index for those values appearing in the block in pl
   // now copy across those values in p2 NOT appearing in this in
   // block index, remembering to insert the p1 block at to_pos

   int child_pos_counter=0;

   for (i=0; i< sub_chrome_size; i++) {
      if (i == to_pos) // copy nominated block
         for(int j=0; j<block_width; j++)
            child->set(subchrome_offset+child_pos_counter++, parentl->value
                              (from_ pos+j+subchrome_offset));

      if (!in_block[parent2->value(i+subchrome_offset)])
         child->set(subchrome_offset+child.pos_counter++, parent2->value
         (i+subchrome_offset));
   } // for

    )// shift
```

11.5 Other Genetic Algorithm Scheduling Problems

In Sections 11.2 and 11.3 we examine three hard to solve scheduling problems, namely the Travelling Salesperson Problem, the Job Shop Scheduling Problem and the Open Shop Scheduling problem. However, these problem instances only touch on the current GA applications appearing in the field of routing and scheduling. We present this by discussing six papers in the areas of:

i) VLSI fabrication and the problem of module ordering and placement of varying sized modules with weighted interconnection between modules

ii) VLSI fabrication and the problem of devising the arrangement of standard sized cells together to form a series of rows, and minimizing the connection cost of conductors connecting cells

iii) a school bus routing problem and the problem of genetic-sectoring to determine bus pickup regions and applying heuristics to optimize the TSP tour in the region

iv) a resource scheduling problem employing a direct schedule encoding scheme

v) a train block routing and scheduling problem

vi) a multiprocessor scheduling problem for task graph optimization

11.5.1 The VLSI Floorplan Design Problem

The VLSI Floorplan Design Problem relates to the task of finding an optireal floor-plan arrangement of modules which are to be fabricated onto an integrated circuit. There are N rectangular modules which are to be arranged together to form a larger rectangle with a total area of A. Each module has a required minimum area, and has an upper and lower bound on its allowable aspect ratio. Modules have interconnections determined by a connection matrix. If two modules are connected then the interconnecting conductor contributes to the cost of the final solution according to the Manhattan distance involved in routing the conductor between the two modules. There are two objectives to the floorplan design problem. The first objective of the problem is to find a floor plan arrangement of these N modules which minimizes the area A in the final rectangular perimeter of the region enclosing all of the modules. The second objective is to minimize the sum of the cost of all interconnections between the modules. Cohoon et al. ([CHMR88]) showed that results superior to simulated annealing can be obtained by using a GA. They used a genetic representation of a sequence of module identifiers 1...N. Two slicing operators were defined, the horizontal cut (+) and the vertical cut (*). Thus an ordering of these operators and module identifiers in the list form a Polish slicing expressing which determines how the final rectangle containing all of the modules is to be partitioned. Three different crossover operators, and three mutation operators were presented. The crossover operator ensured that good building blocks of the parent expressions could be inherited from both parents when a new child was produced. The mutation operator concentrated on certain parts of the Polish expression which would result in controlled and useful structural variations, rather than naively swapping two random gene values which would normally result in total destruction of the recursively defined cutting sequence. Both mutation and crossover employed domain specific rules which respected the fundamental problem building blocks. Cohoon et al. used parallel subpopulations which evolved independently. After every x generations (an Epoch) each population exchanges a random selection of population genotypes to neighbouring sub-populations. Coohoon et. al. termed this well known strategy *Genetic*

Algorithm with Punctuated Equilibria (or GAPE). They performed experiments on two problem instances of 16 and 20 modules and a variety of randomly generated problems on a meshcon figured multiprocessor computer. The optimal solutions to both problems were known in advance. The results showed that in some experiments the GA converged on the optimal solution with the GA performing consistently better than simulated annealing.

11.5.2 Standard Cell Placement in VLSI Fabrication

Shahookar and Mazumder in [SM90] described the problem of arranging rows of standard height cells for use in VLSI fabrication. Cells have the same heights, and the width of each cell varies according to its function. The cells are arranged in rows with a number of adjacent cells comprising each row. Interconnecting conductors connect cells and run through the channels at cell boundaries. Because each cell is identical in height a genotype was used to directly encode lists of cell identifiers in the order that each cell appeared in the row. Therefore a gene value represented a cell identifier and the position of the gene value in the genotype represented the position of the cell in the arrangement of cells. The objective of this problem is to find an arrangement of cells that reduces the total length of interconnecting conductor. To create new children three genetic operators were used, crossover, inversion and mutation. Order crossover was found to be the most effective, and mutation was achieved by pairwise interchange of randomly selected gene values. A standard crossover implementation is possible because of the way that this problem maps into a standard ordering encoding scheme. This contrasts the complex Polish expression encoding required to model VLSI floorplans in Section 11.5.1. Inversion was implemented by choosing two gene positions reversing the order of the gene's values between these gene positions. These operators were applied to a new generation with three different probabilities. These three probability values were optimized with a second meta-GA which manipulated these three probability values. In this meta-GA the fitness evaluation of a meta-GA genotype was a complete GA simulation of the cell-placement GA. The meta-GA was applied to three problem instances of 72, 100, and 183 cells an optimized parameter set of three operator probability values was obtained. The resulting optimal parameter values were then used in the cell arrangement GA with the Cycle crossover and the results were compared with a commercial heuristic cell placement program called TimberWolf 3.3. Shahookar and Mazumder concluded that the GA '... examines 19-50 times less configurations compared to TimberWolf for achieving the same or better percentage improvement of wire length'.

11.5.3 School Bus Routing

Thangiah et al. ([TN92]) applied a GA to the school bus routing problem. This problem involves two objectives. First the size of the fleet of buses must be minimized. Second, we wish to find a routing scheme for each bus which minimizes the total length of distance travelled by the fleet of buses. The first task of their GA was to determine the geographic clusters of student where each cluster was serviced by a single bus. To achieve this the genotype encoded a number of seed angles. Each seed angle represented a pie shape sector, relative to the previous sector, radiating from the central location, the school. Each of these sectors defined a cluster which was serviced by a single bus. Once each cluster was defined, the bus routing in that cluster was reduced to a Travelling

Salesperson Problem, and good tours were found using a heuristic technique. The sum of the lengths of each TSP tour in each sector was then fed back as fitness information to the GA. In this way good seed angles were bred to minimize the total distance travelled, and to minimize the degree of overcrowding (and under utilization) of the buses. Thangiah et al. applied a tour improvement heuristic in which the randomly exchanged students between clusters if the exchange led to a further reduction in the cost function. Using the overall technique Thangiah et al. showed the GA produced results superior to the CHOOSE school bus routing system even though the CHOOSE bus routing system is assisted with a high degree of human expert interaction.

11.5.4 Resource Scheduling

Bruns [Bru93] discussed a direct encoding technique for a production scheduling problem. Production scheduling is a problem where we have a number of orders which need to be processed in a production environment. Each order is met by executing a set of operations using a set of resources. The execution of an operation can only occur if the precedence constraints stipulated by a process plan have been met. In the GA described by [Bru93] a direct genetic encoding technique was employed to describe a schedule. In this technique, the ordering of objects (or resources) within the genotype was not important but each operation, for each order was encoded in a fixed position along with the starting and finishing time of that operation. Bruns, encoding scheme was different from most scheduling GAs, as most scheduling GA implementations encode lists of objects which specify the problem routing and the schedule times are obtained through simulation.

Bruns presented a knowledge driven problem specific crossover and mutation operator. His crossover operator inherited a random selection of nondelayed machine processings from the first parent and merged in the remaining information from the second parent. His crossover was designed to comply with the precedence constraints, hence creating legal children. The mutation operator altered existing order/process assignments to alternative, but legal processes, and performed other such alterations to the encoded schedule. Bruns ran his direct encoding on a real life problem with 53 orders with up to 19 machines for each order. Note that the problem often allowed that a machine could be substituted for one or more other machines and his problem allowed at most twelve alternative machines to be used for a process. Bruns compared his direct representation scheme (experiment A) on a real world problem against a scheme which only encoded process orderings (without timing and alternatives) (experiment B). Experiment B was a domain independent scheme where genotypes encoded a list of processing requirements in order blocks. Bruns showed that experiment A produced results approximately 700% better than experiment B for the same test problem. Bruns neglected, however, to give sufficient detail on the domain independent scheme used in experiment B. For instance, he did not adequately explain the cause of the striking pre-convergence of the population in B to a sub-optimal local maximum. Bruns did not discuss if infeasible schedules were generated as a result of the encoding used in experiment B, and if so, how many infeasible schedules were produced, and how this may have affected performance.

Bruns made it clear that experiment B could not make use of alternative machines as the encoding provided no mechanism to do this; a factor more than likely to have contributed to the overall poor results of experiment B. Bruns concluded that the chromosomal structure should contain all information relevant to the optimization problem. He argues that this representation results in considerable performance gains, but at the cost of the loss of generality, as the GA becomes customized for the specific problem (resource scheduling in his case). He concedes, however, that his direct encoding technique is not readily amenable to theoretic analysis such as the application of schemata theory. Bruns concluded that *'Empirical evaluation suggests that a direct representation of schedules seems to be an appropriate representation scheme'*. To support such a conclusion well studied example problems should be chosen in addition to real world problems. Comparisons should be made with a broader range of existing techniques (i.e. preference list encoded chromosomes), and experiments should be designed to focus on the performance of specific problem additions (direct encoded chromosomes or alternative machine choices - but not both). Despite our criticisms, his results look encouraging for his choice of test problem. However, further comparisons are required to conclusively establish the superiority of direct encoding, and evolving timing information relating to the schedule, rather than encoding a set of rules to indirectly obtain the timing for the schedule by simulation.

11.5.5 Train Block Routing and Scheduling

Gabbert et al. in [GBH+89] described a GA that routes and schedules trains in order to shift blocks of train cars from a predetermined set of origins to their specific destinations. Their GA began by establishing a theoretical train timetable in which thousands of trains provide regular service across a train network. The GA was designed to select a small subset of trains from this theoretical schedule. The reduced train schedule is then responsible for efficiently transporting the blocks of train cars. Each scheduled train is given a predefined origin and destination. Each block (grouping of cars) is represented in the genotype by an ordered list of train schedule segments. This list represents the trains which will transfer block A from the origin of A to the destination of block A. The train list for a block is an ordering of train journey segments chosen from the timetable to provide transport of the blocks across contiguous legs of the blocks. The GA then breeds a population of train lists with a single block being mapped to a train list. As an example, if block 4 is mapped to the trains (11, 51, 7, 204), this means that block 4 is shifted by four separate trains on four individual legs of the journey required to shift block 4 from the origin to its destination. Trains are capable of picking up a number of different blocks which might be waiting at an intermediate location. In some situations trains can transport a block express from its origin to its destination.

The crossover used was Syswerda's uniform crossover where children genotypes encoding train lists were constructed by choosing the train in position n randomly from either parent genotype. A greedy tour improvement heuristic was applied occasionally to perform local optimization on a schedule. Occasionally blocks were reassigned to completely different train lists. This reassignment represented a type of mutation.

Gabbert et al. defined a non-linear cost function which related to a number of measurable properties of the resulting simulation of a single genotype. The cost of a schedule related to the rental cost of the train (based on the duration of use of each train in the schedule), a fuel cost, a crew cost, and a locomotive cost which depended on the total weight of the train. They illustrated how a GA can cope with the complexity of such a non-linear cost function. To show this they conducted two experiments (A and B). Experiment A modelled a scenario with high train rental costs. This factor resulted in the GA breeding solution schedules with a fair proportion of express routes (in order to minimize the train rental). Because of the high rental costs solutions were bred in which longer trains resulted. Experiment B had a reduced train rental cost, and the results showed that the GA bred schedules in which more trains were used. This fact resulted in more indirect routes forming in order to increase the accumulation of blocks positioned at intermediate locations. This resulted in increased transit time but lesser crew and other costs. Overall, longer schedules were produced in experiment B as the number of stopovers and indirect travelling was increased. However, both experiments showed that we can tune the nature of final GA schedules by arbitrarily setting the the different weights in the multi-objective cost function. In this way a sensitivity analysis was performed by varying the weights in the economic cost model.

11.5.6 Multiprocessor Scheduling From Task Graphs

Hou et al. [HHA90] described a GA implementation for obtaining good schedules for implementing generalized task graphs on multiple processors. At the heart of this problem we have a task graph that defines the ordering in which sub-tasks must be performed in order to complete the overall task. The sub-tasks form the nodes of the graph, and the precedence relationships between the sub-tasks form the edges. When all sub-tasks have been completed according to the precedence constraints the task is considered complete. We may execute a sub-task on one of the N processors that are available. The goal of the problem is to find an efficient schedule which allocates these sub-tasks to the N physical processors in order that we minimize some cost function such as the overall task execution time.

Hou et al. found a suitable genetic encoding to this problem. Their genotype was divided into variable length partitions or subchromes. A subchrome n encoded a sequential list of tasks for processor n. Crossover worked at the subchrome level. A crossing site at position x in the subchrome was selected and the head of the list of sub-tasks from the first parent at this position x was transferred into the new child. The tail of tasks from the second parent at position x was then added on to the tasks in the child to form the completed child. Hou et al. presented a technique for choosing the crossing region x so that only legal solutions were generated. Thus all new children created with their crossover complied with the precedence constraints of the task graph. The finishing times of each processor were calculated by a simulation algorithm which modelled the schedule represented in the new child genotype. The fitness of each new child was measured as being inversely proportional to finishing time of the last sub-task.

Their algorithm was applied to a task graph which modelled the Newton Euler inverse dynamic equations for a robotic arm manipulator. The task graph contained 88 sub-tasks and was applied to 5 different processor configurations of 3,4,5, and 6 processors. The process time requirements for each of the 88 tasks ranged from 1 ms through to 111 ms. Hou et al. presented results ranging from 12 to 30% greater than the optimal solution time. These results were obtained after approximately 800 new children were produced. These results were promising, however, Hou et al. conceded that further improvements were necessary to improve the genetic representation of the list of processor tasks.

11.5.7 The Future of Scheduling GAs

From the recent interest in scheduling GAs in the research community it is clear that many more novel and powerful GA applications will soon emerge. We can look forward to new, more efficient, standard operators being developed. We will see unusual hybrid schemes which will employ problem specific knowledge at the genetic encoding, crossover, and mutation levels. Complex ways will emerge which will seed an initial GA population with fit and diverse heuristically-generated genotypes. We will see many more examples of GA applications employing multiple-objective non-linear cost functions. GA implementations will be demonstrated which optimize complex real-world problems and will take into account many additional real-world constraints. We can expect the gap between GA performance and the best problem-specific heuristic techniques to narrow, especially when hybrid GA-techniques exploit the best principles that these heuristics rely on. We will see the emergence of powerful new visual tools which will allow the user to interactively construct a model of the scheduling problem with graphical building blocks, visually specify a cost function, and then click on an optimize button. Eventually hybrid GA techniques will be used in many hard and complex routing and scheduling optimization applications.

References

[ABZ88] J. Adams, E. Balas, and D. Zawack. The shifting bottleneck procedure for job shop scheduling. *Management Science,* 34:391401, 1988.

[AC91] D. Applegate and W. Cook. A computational study of the job-shop scheduling problem. *ORSA Journal on Computing,* 3(2):149-56, Spring 1991.

[Bra91] H. Braun. On solving travelling salesman problems by genetic algorithms. In H.P. Schwefel and R. Manner, editors, *Parallel Problem Solving from Nature. Ist Workshop. Proceedings,* pages 129-133. Springer-Verlag, 1991.

[Bru93] R. Bruns. Direct chromosome representation and advanced genetic operators for production scheduling. In S. Forrest, editor, *Proceedings of the Fifth International Conference on Genetic Algorithms,* pages 352-359, San Mateo, California, 1993. Morgan Kaufmann Publishers.

[BW91] K.M. Baumgartner and B.W. Wah. Computer scheduling algorithms: Past, present, and future. *Information Sciences,* 57-58:319-345, Sep-Dec 1991.h.

[CHMR88] J.P. Cohoon, S.U. Hegde, W.N. Martin, and D. Richards. Floorplan design using distributed genetic algorithms. In *IEEE International Conference on Computer-Aided Design,* pages 452-455, Washington, DC, USA, 1988. IEEE Computer Society Press.

[CRG93] F.D. Croce, T. Roberto, and V. Giuseppe. A genetic algorithm for the job shop problem. Technical report, D.A.I Politecnico di Torino. To appear in *Computers and Operations Research,* Italy, 1993.

[Dav91] L. Davis. *Handbook of Genetic Algorithms.* Van Nostrand Reinhold, New York, 1991.

[DP93] U. Dorndorf and E. Pesch. Evolution based learning in a job shop scheduling environment. (to appear in computers & operations research), Faculty of Economics and Business Administration, KE, University of Limburg, P.O. Box 616, NL-6200 MD Maastricht, The Netherlands, 1993.

[EJKe85] E.L. Lawler, J.K. Lenstra, A.H.G. Rinooy Kan, and D.B. Shmoys (ed). *The Traveling Salesman Problem.* John Wiley & Sons, 1985.

[EM92] J.R. Evans and E. Minieka. *Optimization Algorithms for Networks and Graphs. Second Edition.* Marcel Dekker Inc, New York, 1992.

[FB91] E. Falkenauer and S. Bouffouix. A genetic algorithm for job shop. In *Proceedings. 1991 IEEE International Conference on Robotics and Automation,* pages 824-829, 1991.

[Fou92] L.R. Foulds. *Graph Theory Application.* Springer-Verlag, 1992.

[GBH+89] P.S. Gabbert, D.E. Brown, C.L. Huntley, B.P. Markowicz, and D.E. Sappington. A system for learning routes and schedules with genetic algorithms. In J.D. Schafer, editor, *Proceedings of the Third International Conference for Genetic Algorithms,* pages 430-436. Morgan Kaufmann, 1989.

[GMJ79] Garey, R. Michael, and D.S. Johnson. *Computers and Intractability: A Guide to the Theory of NP-Completeness.* W. H. Freeman and Company, San Francisco, CA., 1979.

[Gol89] D.E. Goldberg. *Genetic Algorithms in Search, Optimization, and Machine Learning.* Addison-Wesley, 1989.

[HGL93] A. Homaifar, S. Guan, and G.E. Liepins. A new approach on the traveling salesman problem by genetic algorithms. In S. Forrest, editor, *Proceedings of the Fifth International Conference on Genetic Algorithms,* pages 460-466, San Mateo, California, 1993. Morgan Kaufmann Publishers.

[HHA90] E.S. Hou, R. Hong, and N. Ansari. Efficient multiprocessor scheduling based on genetic algorithms. In *IECON '90. 16th Annual Conference of IEEE Industrial Electronics Society,* pages 1239-1243 vol. 2, New York, NY, USA, 1990. IEEE.

[Hol75] J.H. Holland. *Adaption in Natural and Artificial Systems.* The University of Michigan Press, 1975.

[MT63] J.F. Muth and G.L. Thompson. *Industrial Scheduling.* Prentice Hall, Englewood Cliffs, N J, 1963.

[Rei91] G. Reinelt. Tsplib - a traveling salesman problem library. *ORSA Journal on Computing,* 3(4):376-385, Fall 1991.

(Ron94] S.P. Ronald. The application of genetic algorithms with job shop scheduling capabilities to job shop scheduling problems. Technical report (in preparation), The University of South Australia, Department of Computer and Information Science, 1994.

[RPLH89] J.T. Richardson, M.R. Palmer, G. Liepins, and M. Hilliard. Some guidelines for genetic algorithms with penalty functions. *Proceedings of the Third International Conference on Genetic Algorithms and their Applications,* pages 191-195, 1989.

(SM90] K. Shahookar and P. Mazumder. A genetic approach to standard cell placement using meta-genetic parameter optimization. *IEEE Transactions on Computer-Aided Design,* 9(5):500-511, 1990.

[SMM+91] T. Starkweather, S. McDaniel, K. Mathias, D. Whitley, and C. Whitley. A comparison of genetic sequencing operators. In R.K. Belew and L.B. Booker, editors, *Proceedings of the Fourth International Conference on Genetic Algorithms,* San Mateo, California, 1991. Morgan Kaufmann Publishers.

[TN92] S.R. Thangiah and K.E. Nygard. School bus routing using genetic algorithms. In *Applications of Artificial Intelligence SPIE Conference Proceedings,* pages 387-398, 1992.

[Whi89] D. Whitley. The genitor algorithm and selection pressure: Why rank-based allocation of reproductive trials is best. In J.D. Schaffer, editor, *Proceedings of the Third International Conference for Genetic Algorithms,* pages 116-121. Morgan Kaufmann, 1989. J.

[WSF89] D. Whitley, T. Starkweather, and D. Fuquay. Scheduling problems and traveling salesmen: The genetic edge recombination operator. In J.D. Schaffer, editor, *Proceedings of the Third International Conference for Genetic Algorithms,* pages 133-139. Morgan Kaufmann, 1989.

Chapter 12

Jinwoo Kim
Bernard P. Zeigler
AI and Simulation Group
Department of Electrical and Computer Engineering
University of Arizona
Tucson, AZ 85721

jwkim@helios.ece.arizona.edu

Beneficial Effect of Intentional Noise

0-8493-2519-6/95/$0.00 + $.50

12.1 Introduction

Evolution algorithms which inherit the characteristics of biological evolution often outperform classical optimization methods when applied to complex real-world problems. During the last three decades, there has been much research of these stochastic optimization strategies such as Genetic Algorithm (GA), Evolution Strategy (ES) and Evolution Programming (EP). The application areas of these optimization techniques range from numerical function optimizations to the wide spectrum of engineering design problems [9, 12, 32].

In this chapter, we demonstrate the beneficial effects of noise in Genetic Algorithms. Binary-GA uses a binary encoding scheme for an individual (chromosome) to represent a potential solution in the search space. The size of the binary string depends upon the number of parameters to optimize and the desired precision. This binary encoding scheme enables: (1) the theoretical analysis of GA, such as *schema theorem* and *building block hypothesis* and (2) the elegant genetic operators, such as *crossover* and *mutation.* However, the binary representation scheme, which discretizes the search space, has some drawbacks when applied to high precision numerical function optimization problems. On the other hand, the floating-point GA (FP-GA) which facilitates a real-value vector for an individual sometimes outperforms the binary encoding scheme [32].

In order to enhance the performance of Binary-GA, we intentionally introduce noise into the search variable x_i. The noise enables sampling of the search region that a binary chromosome can not directly access. We also consider noise assignment to the objective function in order to improve the reliability of the global search. In a noisy environment, examining more individuals usually increases the probability of obtaining a better solution with higher resolution [1]. Since GA is a robust optimization algorithm, the introduction of noise (to a certain degree) should not impair the performance of GA. Previous research of GA in a noisy environment [1, 13, 19] has been with regard to the selection of an optimal population size and the number of evaluations performed during the runtime. However, the objective of our experiments with noise is mainly focused on the enhancement of search resolution and global search capability. To study the effects of noise in GA, two different representation schemes of a chromosome, binary string (Binary-GA) and real-value code (FP-GA), are employed in this chapter.

12.1.1 Chromosome Representation Scheme of GAs

As indicated by J. Koza [21], the representation scheme of an individual is a key issue in genetic algorithm research because this scheme can be a serious constraint to the window by which the system observes its world. In the binary implementation (Binary-GA), search parameters for an objective function are coded using the same number of bits. The crossover and mutation operators are functions applied to the binary string to create a new chromosome and the GA search process is associated with the *schema theorem* and *building block hypothesis* of the coding scheme [12].

Suppose we wish to maximize a function of k variables, $f(x_1, x_2, ... x_k): R^k \rightarrow R$ and each variable x_i can take values from a domain $D_i = [a_i, b_i] \subseteq R$ and $f(x_1, x_2, ..., x_k) > 0$ for all $x_i \in D_i$. It is clear that to achieve desired precision, each domain D_i should be cut into $(b_i - a_i) \times 10^n$ equal size ranges, where n is the required precision. Let us denote by m_i, the smallest integer such that $(b_i - a_i) \times 10^n \leq 2^{m_i} - 1$. Then, a representation having each variable x_i's code as a binary string of length m_i clearly satisfies the precision requirement [9].

In the floating point implementation (FP-GA), a chromosome is implemented as a vector of real-value numbers of the same length as the search parameters. The variation of elements is supposed to be within the search range, and the operators are specially designed to preserve this requirement. The precision of each element depends on the underlying machine, but is generally much better than that of the binary encoding scheme. The main objective in modifying the encoding scheme and the genetic operators is to map the genetic algorithms closer to the problem space. Such a transformation forces the operators to be more problem specific by utilizing some specific characteristics of real space [32].

In contrast, the Binary-GA, which samples the discretized value of a problem space, sometimes suffers from the pre-convergence of the entire population to a non-global optimal solution (sometimes called the local optimal solution) and from an inability to perform fine local tuning. In order to alleviate this problem, we devise a noise assignment scheme to the search variables, x_i, in Binary-GA.

12.2 Noise Assignment Scheme in the Binary Representation Chromosome

To provide an initial examination of the convergence properties of a genetic algorithm, consider a state space of possible solutions encoded as a string of k bits {0,1}. Let there be a population, m, of such bit strings and let each possible string have a fitness, f_i, $i = 1, 2, ...2^k$. Let one globally optimum configuration yield the maximum fitness (or equivalently, a minimum loss). For the Binary-GA, the states of an individual can be defined by every possible configuration of a bit string. Therefore, there are 2^k such states for a binary string size of k. In this scheme, the number of search points in the problem space is limited by the number of bits. Increasing the bit size of a chromosome increases the search resolution [9].

In order to find a highly accurate value for the search parameters, we have to employ a large number of bits for each individual, which results in a dramatic increase in the number of search points. If the search space is complex (multi-modal) or unknown, a GA should also have a large population to provide an effective search. There has been a debate over the convergence properties of GA and Evolutionary Programming [10]. Since GA uses a binary encoding of the parameters for evolution, it may prematurelly stagnate at solutions that may not even be locally optimal. Evolutionary Programming, in contrast, employs a real-valued n-dimensional vector, $x = (x_1, x_2, \dots x_n)$, to evaluate an objective function and often claims to outperform GA [10].

In order to alleviate the convergence problem in the Binary-GA, we introduce noise into the search variable xi and evaluate an objective function. Figure 12.1 shows the concept of noise assignment to the search parameters x_i. The parameter of x_i is converted from the bit string of an individual. The noise (random number) is then added to the parameter before the evaluation of an objective function (assume the objective function has 2 parameters x_1, x_2).

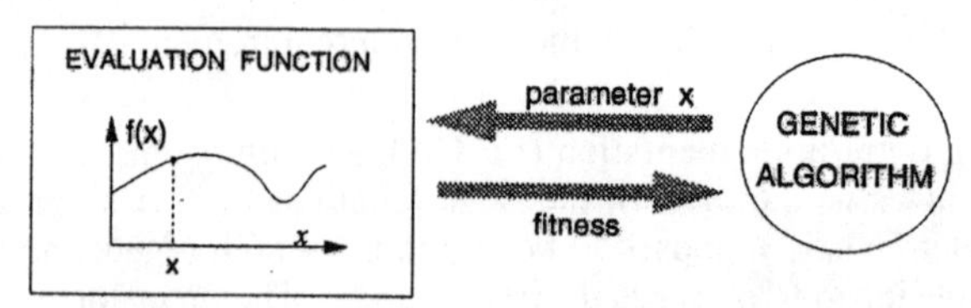

(a) Function evaluation without noise assignment in the search parameters

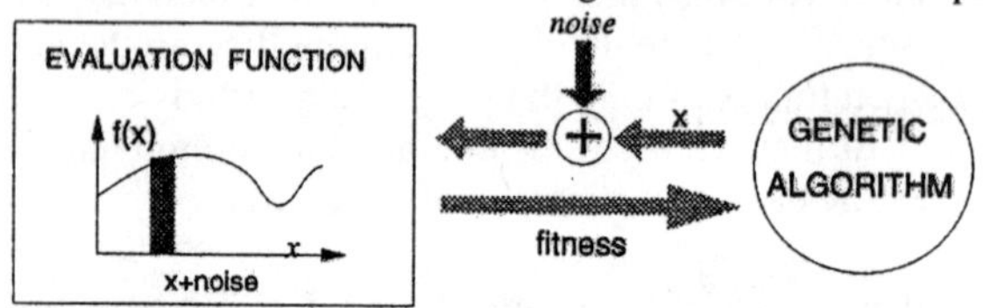

(b) Function evaluation with noise assignment in the search parameters

Figure 12.1: Noise assignment scheme for the search parameters in the function evaluation.

- *Function evaluation without noise assignment:*

$$FITNESS = f(x_1, x_2)$$

- *Function evaluation with noise assignment:*

$$FITNESS = f(x_1 + noise, x_2 + noise)$$

In Figure 12.1a, the output of a function evaluation with arguments passed by the GA becomes a fitness of an individual, while in case (5), the argument contains some noise before evaluation. However, GA does not know whether the returned fitness is evaluated with noise or without noise.

Figure 12.2 shows how the noise (analog sampling point) assignment scheme assists the binary chromosome (digital sampling point) to more fully explore the search ranges. Assume that an individual is 3-bits long and represents one of 8 points in the search range [MIN, MAX]. Then, a GA can only sample eight states from the entire search space. However, instead of employing more bits to increase the precision, we introduce random noise into the parameter x_i. The function consequently analyzes x_i plus some noise. In this experiment, two types of random noise are used: (1) Uniform distributed random noise, U[xi - $D_x/2$, x_i + $D_x/2$] and (2) Gaussian distributed random noise, $m = x_i$, $\sigma = D_x/2$. Here,

$D_x = (MAX - MIN)/(2^n - 1)$ where n = number of bits, is the size of the search space.

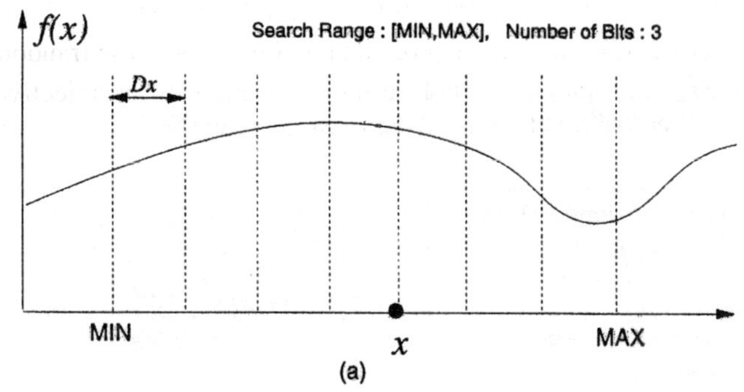

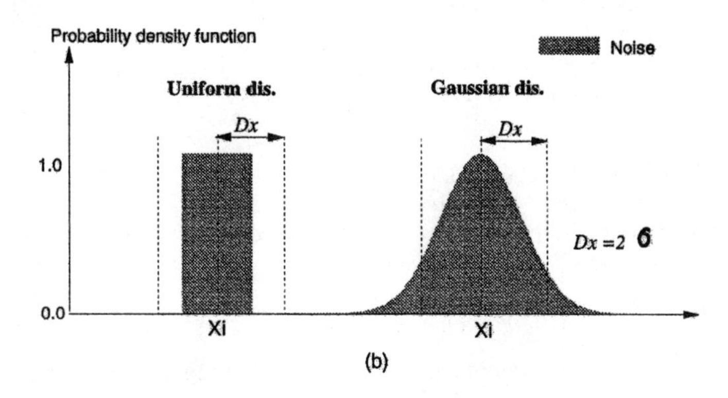

Figure 12.2: Noise assignment to the search variable converted from a binary individual (search range = [MIN,MAX], n = 3, $D_x = (MAX - MIN)/(2^n - 1)$: (a) 8 sampling points in the search ranges, (b) types of the noise assigned to a sampling point.

```
/*-----------------------------------------------------------*/
/* PARA_SIZE : number of parameters of an objective function */
/* gauss_rv(mean,sigma): return Gaussian noise with mean and
sigma */
/* arg[] : real-value vector converted from binary string    */
/* cons[].min, cons[]max : boundary values of parameters     */
/*-----------------------------------------------------------*/
double eval (buffer,ichrom)
int     *buffer;        /* binary string of a chromosome */
int     lchrom;         /* length of chromosome */
{
   int    i,k,start,add;
   double result,chrom_max,bin_number;
   double Dx,arg[PARA_SIZE];

   chrom_max = pow(2.0,(double)ichrom/PARA_SIZE) _ 1.0;
   start = 0;
```

```
    add = (int)(lchrom/Para_SIZE);

    for ( k = 0 ; k < PARA_SIZE ; k++ ) {
       bin_number = 0.0;
       for ( i = start ; i < start + add ; i++ ) {
          bin_number = bin_number +
             pow(2.0,(double)(start+add - i - 1)) • buffer[i];
       }
       start = start + add;
       arg[k] = (cons[k].max - cons[k].min) * bin_number/
             chrom_max + cons[k].min;

       /* add Gaussian noise to the real-value parameters */
       Dx = (cons[k].max - cons[k].min) / chrom_max;
       arg[k] = gauss_rv( arg[k], 0.5*Dx );

       /* prevent noise goes beyond the boundary values */
       if ( arg[k] < cons[k].min )
          arg[k] = cons[k].min;
       else if ( arg[k] > cons[k].max)
          arg[k] = cons[k].max;
    }

    /* evaluate objective function */
    result = objective_function(arg,PARA_SIZE);
    return result;
}
```

Figure 12.3: Example program for assigning noise to the search parameters.

Figure 12.3 shows an example program for assigning noise to the search parameter before evaluation.

With the conventional Binary-GA, the same binary configuration of an individual always has the same fitness value. However, by introducing noise to the search parameters, the same binary configuration of an individual may have different fitness whenever it is evaluated. The smaller the noise, the closer this fitness is to the actual fitness of $f(x_1, x_2)$. Since evolution is performed on the basis of the fitness of individuals, the introduction of too much noise confuses the GA and may lead to a non-optimal direction. We study this problem next.

12.2.1 Study of Noise Assignment in the Search Parameters

We have conducted experiments to study the effect of noise assignment to the search parameters with the numerical test functions given in Table 12.1. FP-GA is also simulated to allow comparison with the performance of Binary-GA. Test functions contain different numbers of search parameters and vary in complexity. In Binary-GA, we perform the search with different search resolution by varying the bit size of a chromosome, and use Gaussian distributed noise.

Number	Function	Constraints
1	$f_1(x_i)=\sum_{i=1}^{n} x_i^2, n=10$	$-5.12 \le x_i \le 5.12$
2	$f_2(x_i)=\dfrac{\sin(2.0(x_1+x_2))}{(1.0+0.0005(x_1^2+x_2^2))}$	$-100.0 \le x_i \le 100.0$
3	$f_3(x_i)=100(x_1^2-x_2^2)^2+(1-x_1)^2$	$-2.048 \le x_i \le 2.048$
4	$f_4(x_i)=\sum_{i=1}^{n} \text{integer}(x_i), n=10$	$-5.12 \le x_i \le 5.12$
5	$f_5(x_i)=\sin(z)/z, z=\sqrt{x_1^1+y_2^2}$	$-10.0 \le x_i \le 25.0$

Table 12.1: Test functions for numerical optimization.

- **Test function 1**

We cannot provide the picture of test function 1 due to the large number of search parameters, but the solution is located at the boundary values and there is no local optimum even though a large number of search parameters are involved. Since the binary configurations of a global solution are either 0's or 1's in the chromosome, the conventional Binary-GA locates the optimum faster than the GA with noise assignment. Reducing the bit size of the chromosome increases the ease on the Binarv-GA in aggregating the schema for the global solutions. Notice that the Binary-GA of bit size 3 outperforms the FP-GA (Table 12.2). Since the test function is simple enough to collect global optimum schema easily, the noise assignment scheme slows down the rate of convergence toward the solution.

Number of bit/para	*without noise*		*with noise*	
	eval.	*fitness*	*eval.*	*fitness*
3	663	262.14400000	2243	262.14400000
6	2842	262.14400000	11305	262.14042833
9	5497	262.14400000	26412	261.98524980
12	8431	262.14400000	29865	261.95878380
FP-GA	*eval.*	740	*Fitness*	262.14400000

Table 12.2: Comparing performance of GA with and without noise assignment in the search variables for different numbers of bit/parameter. Simulation results are for test function 1 (averaged from 100 trials).

- **Test function 2**

Figure 12.4 shows the shape of test function 2 which includes multimodal complexity in a 2-dimensional search space. The global solution exists near the center of the search space. This test function requires the Binary-GA to have enough bits for a high search resolution. If all the possible binary configurations of a chromosome miss a global optimum, the Binary-GA cannot locate the solution regardless of the number of evaluations.

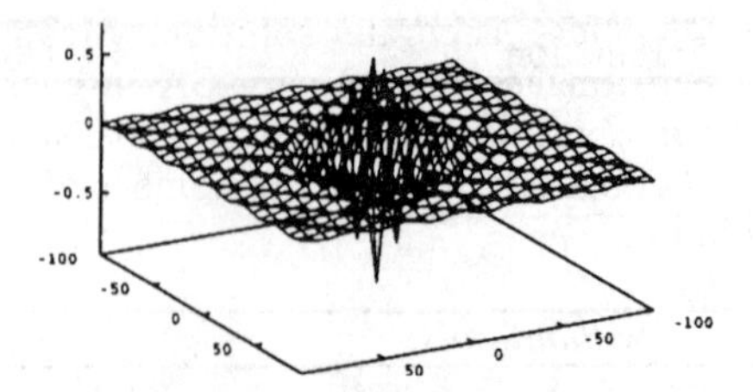

Figure 12.4: Test function 2.

Number of bit/para	*without noise*		*with noise*	
	eval.	*fitness*	*eval.*	*fitness*
5	30000	0.43744994	30000	0.99371483
7	30000	0.05062690	30000	0.99824531
9	30000	0.99512404	30000	0.99765126
11	30000	0.99442124	30000	0.99671854
FP-GA	*eval.*	30000	*Fitness*	0.99829798

Table 12.3: Comparing performance of GA with and without noise assignment in the search variables for different numbers of bit/parameter Simulation results are for test function 2 (averaged from 100 trials).

Table 12.3 shows the simulation results of the experiment. The performance of a Binary-GA without noise is very sensitive to the bit size. If, by any chance, all the discretized search points are located in the valley of the problem space, the Binary-GA displays a severe difficulty in finding a solution, as shown in the case of 7 bits. However, the noise assignment scheme provides consistent accuracy for a variety of different bit sizes. The Binary-GA with noise can sample more information than the conventional Binary-GA. FP-GA also demonstrates good performance with this search problem. Since the individuals in the FP-GA are coded by a real-value n-dimension vector, the FP-GA provides a more efficient way of preserving the information of the problem space.

- **Test function 3**

Figure 12.5 shows test function 3.

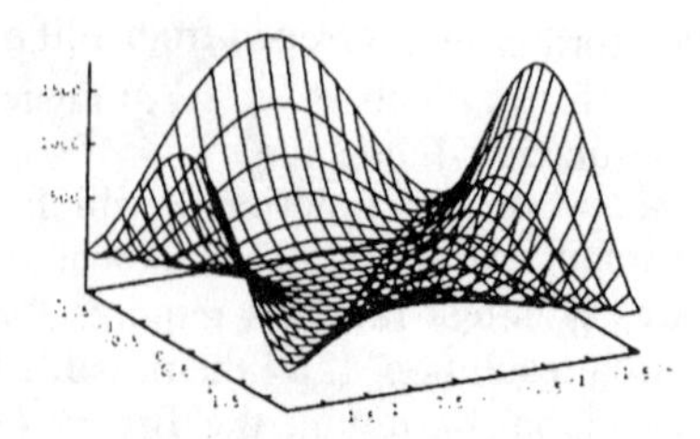

Figure 12.5: Test function 3.

As shown in Table 12.4, Binary-GA with noise assignment provides consistent performance, while Binary-GA without noise assignment evaluates more individuals but finds a less accurate solution. Meanwhile, the FP-GA demonstrates better performance than Binary-GA in every case except one. Binary-GA with a chromosome size of 20 finds a global solution faster than FP-GA. In this example, FP-GA executes a large number of evaluations for local fine-tuning of the solution.

Number of bit/para	*without noise*		*with noise*	
	eval.	*fitness*	*eval.*	*fitness*
3	80000	1697.43676990	46935	1768.50889644
6	80000	1764.84958836	35414	1768.50890202
9	80000	1768.29077166	28245	1768.50890561
20	80000	1768.42698516	9243	1768.50890546
FP-GA	*eval.*	22539	*Fitness*	1768.50890582

Table 12.4: Comparing performance of GA with and without noise assignment in the search variables for different numbers of bit/parameter. Simulation results are for test function 3 (averaged from 100 trials).

The problem space of test function 4 is monotonic increasing like a stepladder. With this function, the Binary-GA with noise assignment shows poor performance and the FP-GA outperforms the Binary-GA as shown in Table 12.5. As with test function 1, since the solution exists at the boundary values, the Binary-GA with small bit size can obtain a solution faster than the FP-GA.

- **Test function 4**

Number of bit/para	*without noise*		*with noise*	
	eval.	*fitness*	*eval.*	*fitness*
2	140	25.0000000	246	25.00000000
3	187	25.0000000	270	25.00000000
6	2144	25.0000000	6877	24.94999999
9	10210	24.8299999	16008	24.73000000
20	20737	24.6700000	19204	24.67000000
FP-GA	*eval.*	146	*Fitness*	25.00000000

Table 12.5: Comparing performance of GA with and without noise assignment in the search variables for different numbers of bit/parameter. Simulation results are for test function 4 (averaged from 100 trials)

- **Test function 5**

In this test function (shown in Figure 12.6), we employ a uniform distributed random noise instead of a Gaussian noise. The degree of noise is Dx ($D_x = (MAX - MIN)/(2^n - 1)$, where n = bit size of a chromosome) in order to investigate the entire problem space.

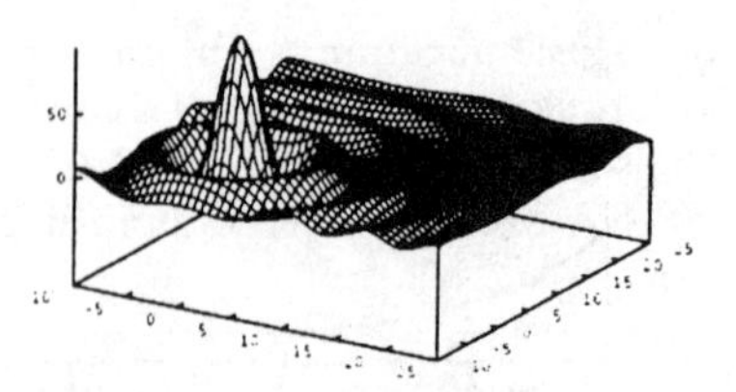

Figure 12.6: Test function 5.

As claimed by D. Fogel [10], the Binary-GA tends to stagnate at nonglobal optima. We have conducted experiments to study the convergence property of our new scheme. The experiments show that the introduction of noise to the parameters overcomes the Binary-GA's convergence problem. This new scheme improves the limited representation scheme of binary strings from discretized sampling to real-value sampling of the search space.

	without noise	*with noise*	*FP-GA*
number of trials	100	100	100
number of find 122.0	0	89	0
avg. error	2.7943e-05	1.33226e-06	4.4843e-8
min. error	2.7943e-05	0.00000	1.5630e-13
max. error	2.7943e-05	6.6160e-5	3.7085e-6

Table 12.6: Compared performance of Binary-GA with and without noise assignment. *error* represents the difference between the exact solution of 122.0 and GA solution.

With the introduction of noise, Binary-GA locates an exact solution (122.0) 89 times out of 100 trials and also outperforms FP-GA as shown in Table 12.6. The Binary-GA without noise assignment scheme consistently stagnated before reading the global solution. Thus, the noise assignment scheme enables the Binary-GA to execute local fine-tuning to the global solution.

Figure 12.7 shows the convergence property of our new scheme which displays the error between the exact solution (122.0) and that found by GA. Notice that the error of Binary-GA with noise assignment shown in the graph comes from 11 cases that didn't find the global solution.

12.2.2 Summarization of the simulation results

Experiments of noise assignment in the search parameters were conducted and the performance of Binary-GA was compared to FP-GA. Noise in the search parameter increases sampling of the problem space and often enables Binary-GA to find a global solution with high precision, even though the chromosome size is small. With the noise assignment, the search resolution of Binary-GA becomes insensitive to the size of a chromosome.

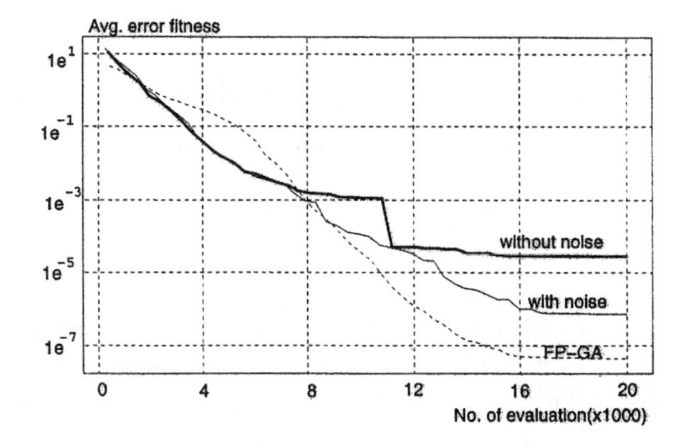

Figure 12.7: Convergence speed of Binary-GA and FP-GA (population size=200, individual size=28, Pc=0.8, Pm=0.015).

In this experiment, FP-GA shows better performance than Binary-GA in such test functions where population can be easily attracted to the global optimum. For the test function 2 case, even though a lot of local optimum exists close to the global solution, the overall fitness increase of the search space leads to the center of search space, where the global solution is located. If the convergence of the population in the search space can be easily identified to a certain direction, FP-GA outperform Binary-GA.

However, if the convergence direction of population cannot be clearly discriminated, such as test functions 3 and 5, the performance of FP-GA degrades. In those cases, certain conditions of Binary-GA outperforms FPGA. As shown in Figure 12.7, the convergence speed of FP-GA to the global optimum is faster than Binary-GA with no noise assignment scheme, but it is unable to locate the exact solution of 122.0. We extend our experiment to a more complex engineering design problem in the next section.

12.3 Noise Assignment of GA for Design of a Control System

Since GA evaluates many parameters of the search space in parallel, it is more likely to converge toward the global solution. GA does not need to assume that the search space is differentiable or continuous and can also iterate several times on each datum received [12, 32]. In this context, the GA-optimizer for a fuzzy logic controller (FLC) affords more reliability in global optimization than does the Adaptive Neural Net approach [5, 7, 20].

In this experiment, we study two types of GA noise assignment on the design of an optimal FLC. The first type is to assign noise to the search parameters as in the previous experiment. The second type is to introduce noise into the training data set. A brief review of the Fuzzy Logic Controller and its integration with a GA-optimizer is presented in the next section.

12.3.1 Brief Review of the Fuzzy Parallel Net Controller

The applications of fuzzy logic to control problems were described in some early works of Zadeh [25]. Mamdani and Assilian [15], King and Mamdani [29] and Kickerr and Mamdani [31] provided the pioneering applications in this domain. Despite the great success in industrial applications of fuzzy logic controllers, a need for an automated design scheme for the optimal FLC still exists [6, 20].

The fundamental principle of the fuzzy logic control lies around the *labeling* process, in which the sensory data is translated into a label as done by a human being. With the expert supplied membership functions for these labels, a reading of a sensor can be *fuzzified* and *defuzzified.* It is important to note that the transition between labels is not abrupt and a given reading might belong to several label regions [6]. However, the fuzzification and defuzzification processing does not need to be sequential. The input signal can be fuzzified/defuzzified simulatenously by matching membership functions. J. Jang [20] developed an ANFIS (Adaptive Network-based Fuzzy Inference Systems) that the fuzzy control processing was realized into the parallel neural network structure where each *neuron* represents functions (fuzzy membership) and each *link* represents the weight of a fuzzy rule.

Figure 12.8 shows the structure of the fuzzy logic processor and its fuzzy rule table [20]. In this example, five membership functions are assigned to the input/output signals and the fuzzy rule table is filled with symmetric fuzzy values. However, unlike ANFIS, the links between layers 3 and 4 in our fuzzy processor are not fixed. These links represent *consequents* of fuzzy rules of which our experiment also tries to find an optimal set. While an earlier Fuzzy Logic Controller [4, 26] was implemented in rulebased form *(if-then),* the FLC in this experiment employs a parallel inferencing network structure. Due to the parallel fuzzification/defuzzification scheme, the FLC can improve real-time performance of the control system for practical applications.

Layer 1 The node i in the first layer has its own node function

$$O_i^1 = \mu_{A_i}(x)$$

O_i^1 is the membership function of A_i (a linguistic label such as *negative, zero* or *positive,* etc.) and it specifies the degree to which quantifier A_i satisfies sensory input x. In our application, we use a bell-shaped function for μ_{A_i} with a maximum of 1 and minimum of 0, such as,

$$\mu_{A_i}(x) = \frac{1}{1 + \left[\left(\frac{x - c_i}{a_i}\right)^2\right]^{b_i}} \qquad a_i\text{: width, } b_i\text{: steepness, } c_i\text{: mean}$$

The basic operation of each layer in the ANFIS is defined as follows:

Layer 2 The node in the second layer performs the logical AND operations with w_i representing the firing strength of rule-i

$$w_i = min\,(\,\mu_{A_i}(x),\ \mu_{B_i}(x)\,) \qquad min\text{:logical AND operation.}$$

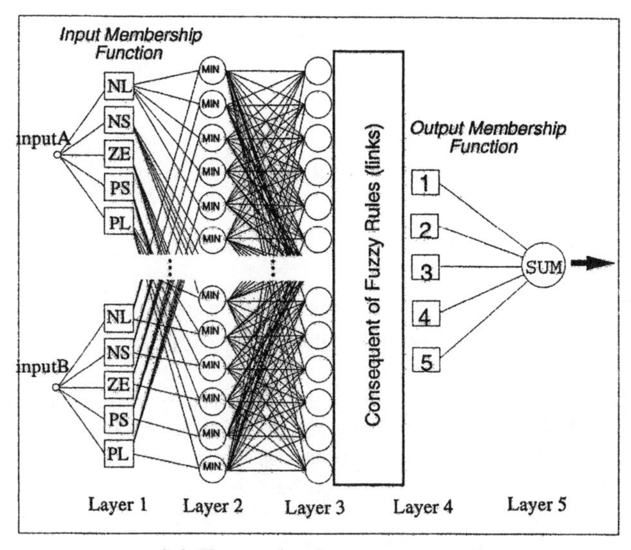

(a) Fuzzy logic processor

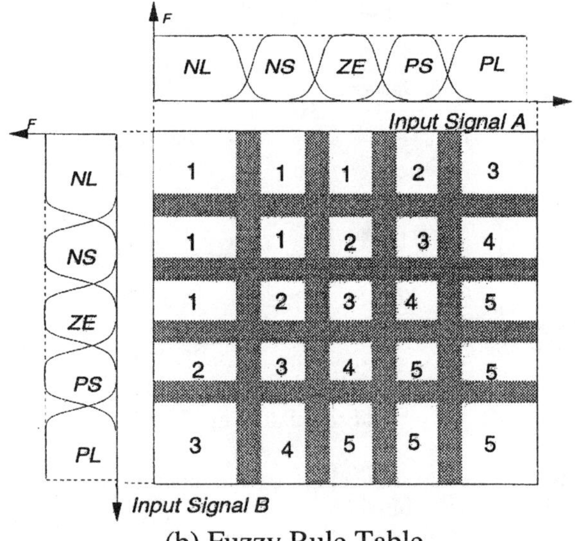

(b) Fuzzy Rule Table

Figure 12.8: (a) Fuzzy inference network and (b) fuzzy subspaces : NL (negative large), NS (negative small), ZE (zero), PS (positive small), PL (positive large).

Layer 3 Computes the ratio of the w_i to the sum of the firing strengths of all the rules.

$$\overline{w}_i = \frac{w_i}{\sum_{k=0}^{n} w_k} \qquad n = \text{number of rules}$$

Layer 4 Computes the defuzzified value of each rule i. We employed the simplified center of mass method for the defuzzification.

$$O_i^4 = \overline{w}_i f_i$$

Layer 5 Computes the overall output as the summation of all incoming signals.

$$O^5 = \sum_{i=0}^{n} O_i^4 = \sum_{i=0}^{n} \overline{w}_i f_i$$

The performance of a FLC is determined by the fuzzy membership functions of layer 1 (input), layer 4 (output), and fuzzy rules (links between layers 3 and 4).

12.3.2 Interaction of the FLC Module with GA Optimizer

Figure 12.9 shows the interaction of the GA-optimizer, FLC module, and simulation model. An individual of a GA-optimizer possesses the parameter information of an FLC, such as membership functions and fuzzy rules. The performance of the FLC operating with a simulation model is the fitness of the trial individual. Various parameter sets for a FLC are generated from the GA-optimizer, and the FLC/simulation model are the objective function module for calculating the fitness of the trial individual.

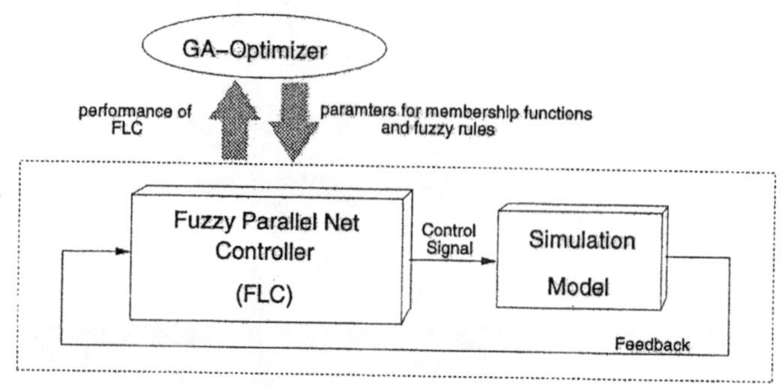

Figure 12.9: Block diagram of the interactions of GA-optimizer, FLC module, and simulation model.

Figure 12.10 provides a flow chart of the operations of the FLC module integrated with the GA-optimizer and a simulation model. An individual from a GA represents one trial set of fuzzy membership functions and rules. If an FLC involves 2 input signals and 1 output signal, each of which uses 3 membership functions (a bell-shaped function which needs parameters a, b, and c), then 27 parameters (9 membership functions x 3) are required to specify the membership functions of the FLC. Nine additional parameters should also be included to describe the fuzzy rules. This results in a total of 36 parameters for each individual to specify a certain FLC. Each parameter is coded by an 8-bit binary number so the total length of an individual is a 288-bit binary code. Using a binary representation scheme for an individual makes it easy to apply the genetic operations (mutation and crossover).

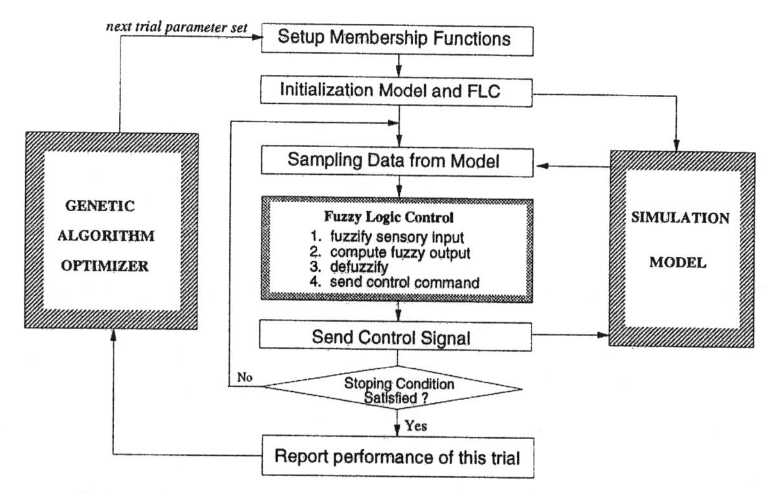

Figure 12.10: Flow chart of the operation of the FLC module.

When the simulation starts, the GA-optimizer sends an individual to setup a certain set of fuzzy membership functions and fuzzy rules for the FLC. The other operational specifications can be pre-set inside the controller. Followed by an FLC setup, the simulation model is reset with its initial conditions. The FLC starts to issue an operational command to the simulation model to control the target plant. The amount of error between the desired trajectory and actual trajectory is considered as the performance of the FLC. A smaller error means a higher fitness of the trial individual.

Figure 12.11 shows the schematic diagram of an inverted pendulum system. The structure of the inverted pendulum system is composed of a rigid pole and a cart onto which the pole is hinged. The cart moves either right or left, depending on the force exerted on the cart. The pole is hinged to the cart through a frictionless free joint so that there exists only one degree of freedom. The control objective is to balance the pole starting from non-zero conditions by supplying the appropriate forces to the cart.

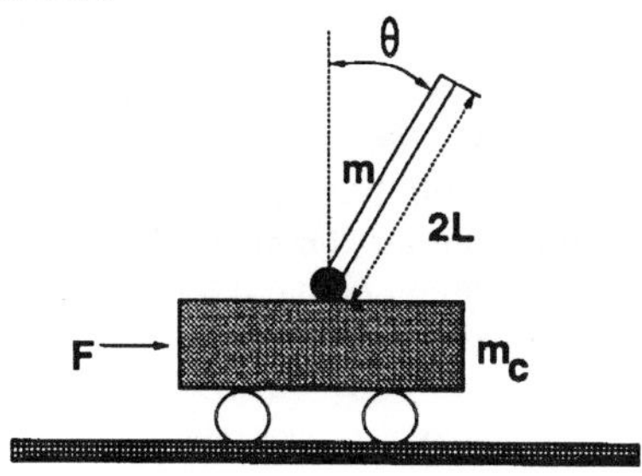

Figure 12.11: The inverted pendulum system.

If we let $x_1(t) = \theta(t)$ and $x_2(t) = \dot{\theta}(t)$, then this system can be defined by the following differential equations [20]:

$$\dot{x}_1 = x_2$$

$$\dot{x}_2 = \frac{g\sin(x_1) + \cos(x_1)\left(\dfrac{-F - mlx_2^2\sin(x_1)}{m_c + m}\right)}{l\left(\dfrac{4}{3} - \dfrac{m\cos^2(x_1)}{m_c + m}\right)}$$

$$= H_2(x_1, x_2, F)$$

where g is 9.8*meter/sec*^{2}, m_C (mass of cart) is 1.0*kg*, m (mass of pole) is 0.1*kg*, l (half length of pole) is 0.5*meter*, and F is the applied force in Newtons. The objective of the GA in this experiment is to find the optimal set of membership functions and fuzzy rules for a FLC. Our fuzzy controller has the following specifications:

- *Input signals of FLC*: angle of pole (degrees), angular velocity of pole (degrees/sec.)
- *Output signal of FLC*: force (Newtons)
- *Type of membership function*: bell-shaped (requires parameters a, b, c for the fuzzy membership function).
- *Fuzzy region*: NE (negative), ZE (zero), PO (positive)
- *Number of fuzzy rules*: 9

The most noticeable problem associated with the characteristics of the performance index under consideration involves differences in the relative fitness between individuals. If the variance of fitness in the population is small, GA undergoes a random search; such behavior may be tolerable during the early life of the population but would be devastating later on [32].

For this design problem, we want to minimize the total amount of error; E = *degree of error* between the pole and 90°+ *Force* spent during control. At every sampling interval (0.01 sec), the degree of error of the pole between the current position and the target position (90°) and the amount of force are considered to compute a fitness. Therefore, a smaller E represents a higher fitness. There are two ways to convert the E to a fitness value of a GA.

- (1) *Fitness = Offset - E.*
- (2) *Fitness = 1/E.*

In real world problems, sometimes it is difficult to select an appropriate *Offset* value. If an inappropriate value is selected, the performance of the GA will be degraded. In order to strengthen the relative fitness differences between individuals in later GA operations, we choose scheme (2) to compute the fitness. The performance index of our GA experiment is

$$\text{Performance Index} = \left(\frac{C_1}{E}\right)^{C_2}$$

where C1 and C2 are heuristically chosen to adjust the fitness difference.

12.3.3 Experimental Results: Noise Assignment to the Search Parameters

The GA searches for an optimal FLC which controls the inverted pendulum with an initial angle of $10deg$, an initial angular velocity of $0deg/sec$, and a length of $0.5m$. We vary the bit size of a chromosome and assign noise (uniform distributed random noise, size = D_X, where $D_x = (MAX - MIN)/(2^n - 1)$, n = bit size of the chromosome) to the parameters. With the FLC optimized by the conditions given above, we also test its ability to control other initial conditions such as:

- *Initial angle (deg):* ±10, ±20, ±30, ±40, ±50, ±60, ±70, ±80

- *Initial angular velocity (deg/sec):* ±10, ±20, ±30, ±40, ±50, ±60, ±70, ±80

- *Pole length* (m): 0.25, 0.5, 1.0

The *success rate* in Table 12.7 represents the number of successes when controlling the balance of the pole when the optimized FLC is applied to different initial conditions (total 768 cases). As explained before, since the E in the table is an amount of degree of error and force during control, the smaller E values mean a higher fitness of the GA. The fitness and success rate of the GA with noise is higher than the GA without noise assignment. However, assigning too much noise degrades the performance of optimization. Notice that the FP-GA has a higher E than the Binary-GA. The simulation results also show that the FP-GA has more chance of locating a local optimum in this design problem. The mutation in FP-GA provides a random number to a real-value chromosome. Since the size of the random number becomes smaller in the later GA process, the FP-GA has higher pressure in the exploration than the Binary-GA. In contrast, the Binary-GA balances the exploration and exploitation in the problem space appropriately.

We have tested the applicability of the optimized FLC to control of the pendulum under different initial conditions. Figures 12.12 and 12.13 show two examples of the performance of the optimized FLCs with an initial angle of 10 degrees. Both of the FLCs show approximately a 90.0% success rate. The FLC of case 1 (Figure 12.12) balances the pole within 3.5 *sec.* in most cases, while the FLC of case 2 (Figure 12.13) shows a high degree of oscillation while balancing the pole. The FLC of case 1 consumes less force than the FLC of case 2. Figure 12.14 shows the optimized fuzzy membership functions of Figure 12.12 and Figure 12.13.

Number of bit/para	*without noise*		*with noise*	
	E	*success rate (%)*	*E*	*success rate (%)*
4	285.65	73.48	294.06	65.76
6	267.10	75.97	265.42	80.55
8	285.03	74.46	263.03	77.78
10	275.15	64.27	271.17	66.69
FP-GA	*E*		414.05	

Table 12.7: Performance of the optimized FLC using different sizes for a chromosome and noise assignment. E is the total sum of *degree of error* and *force* (population size = 300, Pc=0.9, Pm=0.03, training data = 10*deg*, data averaged from 10 trials).

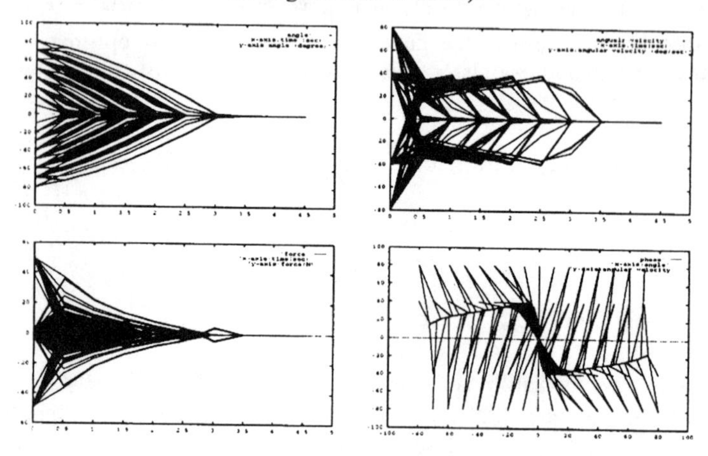

Figure 12.12: Performance of the optimized FLC in controlling the pendulum under different initial conditions : case 1 (total 786 cases).

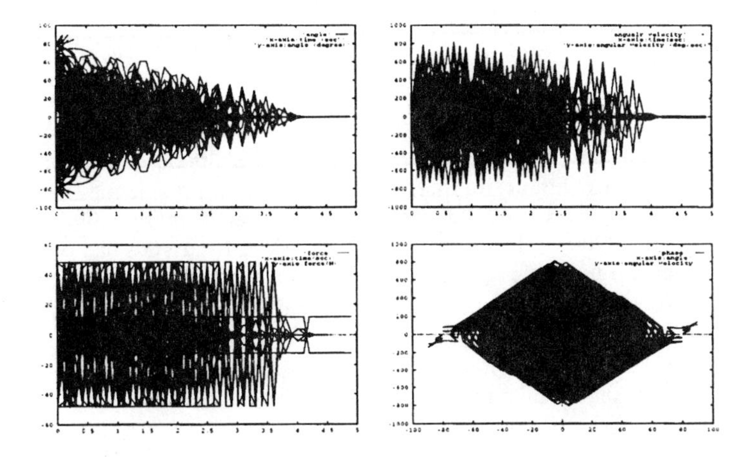

Figure 12.13: Performance of the optimized FLC in controlling the pendulum under different initial conditions : case 2 (total 786 cases).

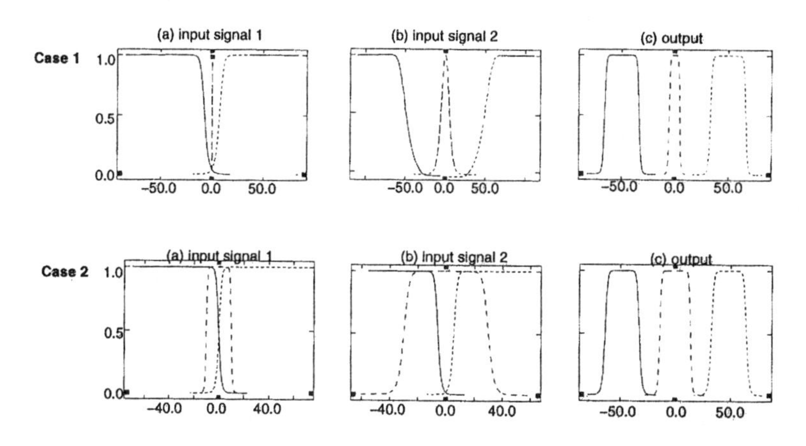

Figure 12.14: Optimized fuzzy membership functions of FLC for cases 1 and 2.

12.3.4 Experimental Results: Noise Assignment to the Training Data Set

We continue the new experiment to investigate other effects of noise assignment. In this experiment, instead of introducing noise to the search parameters, we introduce noise into the training data. During the simulation, the GA is trained to balance the pole with the following initial conditions:

- *Initial angle (deg):* 20.0 + *noise.*
- *Initial angular velocity (deg/sec):* 0.0
- *Pole length (m):* 0.5

As shown in Figure 12.15, we introduced varying amounts of noise to the initial angle so that the FLC is trained with slightly different initial angles. However, GA assumes that the fitness of an individual is still computed based upon a fixed initial condition (20 *deg).* Since GA is a robust search algorithm, the performance is not affected by the small amount of noise.

Figure 12.16 shows the increase of the best fitness individual during the GA search. The best individual of FP-GA stagnates after 30000 evaluations, while that of Binary-GA continues to increase. The fitness improvement of FP-GA is faster than Binary-GA, but the graph ensures that if more individuals are evaluated after 40000 evaluations, the best fitness of BinaryGA can outperform FP-GA. Since we use the performance index as a form of '1/*E*', the fitness that the GA approaches is not assured to be a global optimum. If the performance index is defined as '1/*E*' to minimize E, the global optimum is usually unknown. In order to investigate the fitness as the real performance of a controller, we measure the amount of real error in a five-second simulation.

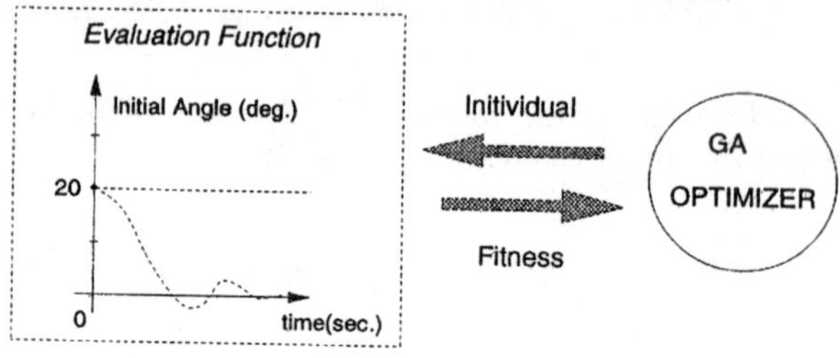

(a) AGA–Optimizer with fixed training set (20 degree)

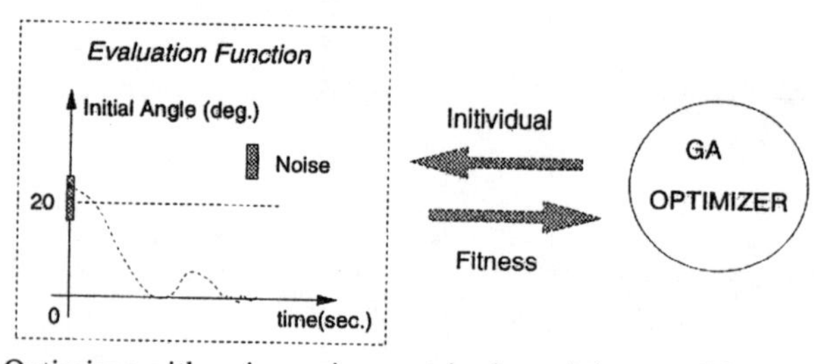

(b) AGA Optimizer with noise assignment in the training set (20 degree + noise)

Figure 12.15: Evaluating an individual with a training data which contains noise.

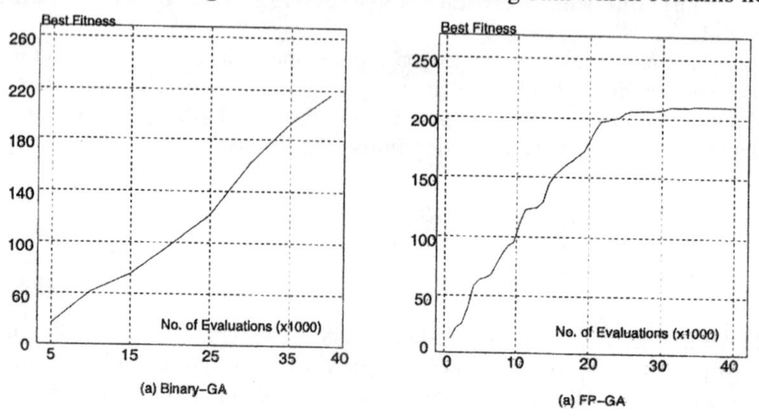

Figure 12.16: The best fitness increase rate of Binary-GA and FP-GA.

Table 12.8 shows the results of Binary-GA and FP-GA when noise is introduced into the training set. As shown in the previous experiments, a small degree of noise does not impair the performance of GA. The small degree of noise rather increases the fitness of Binary-GA. This experiment shows that Binary-GA outperforms FP-GA. However, the performance of GA decrease when too much noise is introduced into the training data.

noise	Binary-GA		FP-GA	
	E	success rate(%)	E	success rate(%)
0.00	1312.76	71.40	1789.10	55.04
0.01	1107.59	66.78	3003.70	53.80
0.02	816.36	79.85	2576.56	46.96
0.03	960.34	73.75	3647.17	55.56
0.04	1553.45	67.49	2236.27	38.15

Table 12.8: Performance of GA for different quantities of noise assigned to the training set (population size = 300, Pc=0.9, Pm=0.03, training data = 20*deg,* data averaged from 10 trials).

12.4 Analysis of Noise Effects in Genetic Algorithms

12.4.1 Noise Assignment to the Search Parameters

In Binary-GA, if the bit size of a chromosome is k, then there are 2^k possible bit configurations, each of which represents a state (diagram) in the search space as shown in Figure 12.17. As the bit size increases, GA can investigate more states. Since a bit configuration can be changed after a crossover and mutation, the state transition diagram shown in Figure 12.17 is fully interconnected to the other states including itself (we eliminated all the lines interconnecting the states from the figure to keep clarity).

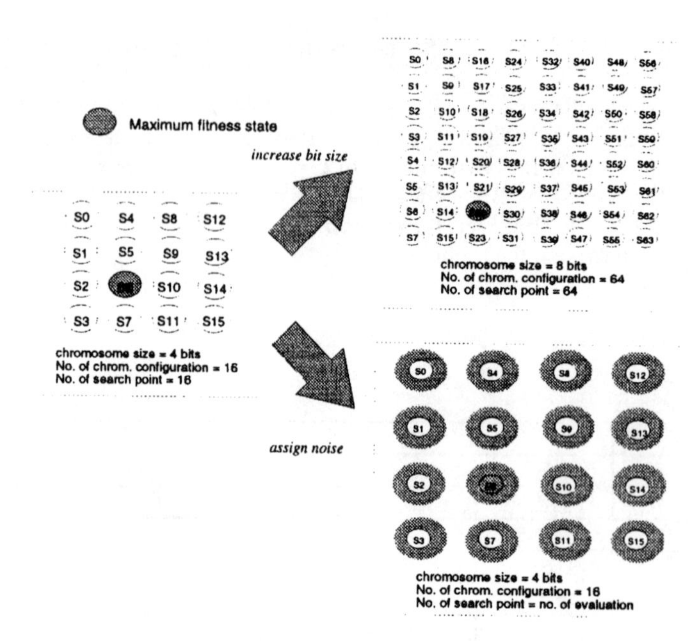

Figure 12.17: State transition diagram of binary representation.

Suppose that there are n individuals in the population, and that a newly evaluated individual updates the gene pool only if its fitness is higher than that of an old individual. Since we always retain the best individuals, there is (at least) one *absorbing state* (possibly maximum fitness), which is a global solution. We can consider the state transition diagram (Figure 12.17) as a discrete-time Markov chain where n tokens travel across the state transition diagram [11]. If at least one of the tokens enters the absorbing state, then GA finds a global solution.

In a traditional Binary-GA, a state (a bit configuration) possesses a single fitness value by which the selection of a token in the state is determined. If more information about the problem space is required to be sampled, we should employ more bits in the chromosome. Even though we expand the state transition diagram, is still not guaranteed that one of the states is the global solution of a real problem. As the size of the state transition diagram grows, the probability of a token (at least) entering the observing state decrease. Therefore when we increase the state transition diagram, we should also employ more tokens (population).

The noise assignment scheme can resolve this problem. Instead of employing more bits to increase the number of states, a certain degree of noise is assigned to each state so that even though a token enters the same state, the fitness may be different. In Binary-GA, a state represents a single value in the search space, while in the noise assignment scheme, a state represent a region of the search space and provides more sampling opportunity.

The fitness of a state affects the transition of tokens. The noise assignment scheme provides more diverse information to a token in determining the next state. However, if we provide too much noise to the state, the transient direction of a token may be misleading.

12.4.2 Noise Assignment to the Training Data

Design tasks using optimization algorithms, such as genetic algorithms and neural nets, require entries of training data. This training data provides real world information to the algorithms. Lowering the involvement of human expertise in current design operations means a greater role played by the training data in the optimization process.

The diverse information about the problem space ensures that the algorithms will be more robust and will find a more global optimum. Apparently, carefully selected multiple entries of training data support the optimization process of algorithms better than a single entry of training data. However, the multiple entries of data require more execution time than a single entry. Therefore, there is a trade-off between the number of training data entries and the program execution time.

Figure 12.18 shows an example of the training data (initial angle of pole) in the GA for the inverted pendulum. For example, if there are three training data entries for a single individual, the performance of each training data is measured and then averaged to compute the fitness of an individual (Figure 12.18a). If the simulation execution time of a model with a single training datum is t, time 3 x t is required to compute a fitness. However, when we use a single training datum with noise assigned, it still takes time t to compute a fitness, since the GA uses a slightly different training datum for every individual. However, adding too much noise to the training data will degrade the GA performance.

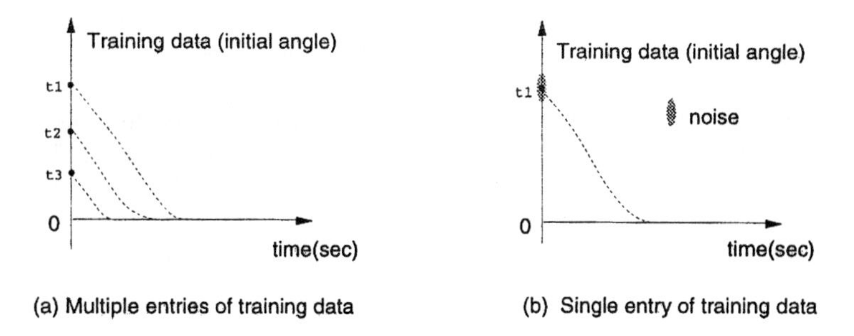

Figure 12.18: Training data for the objective function.

- (a) Three entries of training data *(initial angle = t1, t2, t3)*
Performance $= (P_{t1} + P_{t2} + P_{t3})/3$

- (b) Single entry of training data *(initial angle = t1)*
Performance $= P_{t1+noise}$

where P_{t1} is a performance measurement when given training data (initial angle) is t1.

12.5 Conclusions

We have demonstrated the beneficial effect of noise in GA. The binary representation scheme of a chromosome has drawbacks for high precision numerical problems, where the bit size of a chromosome often gives a significant effect to the search performance. However, by assigning noise (to a certain degree) to the search parameters, we can alleviate the pre-convergence problem of Binary-GA. The noise assignment scheme supports the binary chromosome to more fully explore the search ranges. Binary-GA with noise in the search parameters also provides consistently better performance regardless of the bit size of a chromosome. The new scheme ensures relatively stable performance, even though GA employs a small size of population and chromosome.

The second experiment demonstrates that the noise assigned to the training data set can improve the performance of GA. The noise provides GA more diverse information about a problem space and reduces the evaluation time of an individual. A small degree of noise does not impair the performance of GA, but the performance degrades when too much noise is introduced into the training data.

The performance of Binary-GA is also compared to FP-GA. The FPGA which utilizes the real-value chromosome outperforms the conventional Binary-GA in numerical test functions. However, the noise assignment scheme in the Binary-GA improves the performance of Binary-GA.

For a design problem, Binary-GA shows better performance than FPGA. The simulation result shows that FP-GA has more probability to stagnate at a local optimum in a complicated real-world design experiment. The mutation in FP-GA provides a random number to a real-value chromosome. Since the size of the random number becomes smaller as generation continues, FP-GA has higher pressure in exploration than does the Binary-GA in the later portions of the GA process. The Binary-GA appropriately balances exploration and exploitation during the search.

Bibliography

[1] A.N. Aizawa and B.W. Wah, "Dynamic Control of Genetic Algorithms in a Noisy Environment", *Proc. of 5th International Conference on Genetic Algorithms,* Urbana-Champaign, IL, Morgan Kaufmann, pp. 48-55, 1993.

[2] B.D.O. Anderson, R.P. Bitmead, C.R. Johnson, P.V. Kokotovic, R. L Kosut, I.M.Y. Mareels, L. Praly and B.D. Riedle, *Stability of Adaptive Systems: Passivity and Averaging Analysis,* Cambridge, MA, MIT Press, 1986.

[3] B.P. Zeigler and J.W. Kim, "Asynchronous Genetic Algorithms on Parallel Computers", *Proc. of 5th International Conference on Genetic Algorithms,* Urbana-Champaign, IL, Morgan Kaufmann, pp. 660, 1993.

[4] B.P. Zeigler and J.W. Kim, "Event-based Fuzzy Logic Control System", *Proc. of 8th IEEE International Symposium on Intelligent Control,* Columbus, OH, pp. 611-616, 1993.

[5] C. Karr, "Genetic Algorithms for Fuzzy Controllers", *AI Expert,* vol. 6, no. 2, February, pp. 26-33, 1991.

[6] C.C. Lee, "Fuzzy logic in control systems: fuzzy logic controllerpart 1." *IEEE Trans. on Systems, Man Cybernetics,* vol. 20, no. 2, pp. 404-418, 1990.

[7] C. Karr, "Applying Genetic to Fuzzy Logic", *AI Expert,* March, 1991.

[8] C. Tong and D. Sriram (Ed.), *Artificial Intelligence in Engineering Design,* New York vol. II, III, Academic Press, 1992.

[9] D.B. Fogel, "An Introduction to Simulated Evolutionary Optimization", *IEEE Transactions on Neural Nets,* vol. 5, no. 1, pp. 3-14, 1994.

[10] D.B. Fogel, "Asymptotic Convergence Properties of Genetic Algorithms and Evolutionary Programming:Analysis and Experiments," *Cybernetics and Systems,* in press, 1993.

[11] D.E. Goldberg and J.J. Richardson, "Finite Marchov chain analysis of genetic algorithms", *Proc. of 2nd International Conference on Genetic Algorithm,* Cambridge, MA, Lawrence Erlbaum Associates, July, 1987.

[12] D.E. Goldberg, *Genetic Algorithms in Search, Optimization and Machine Learning*, Reading, MA, Addison-Wesley, 1989.

[13] D.E. Goldberg, and M. Rudnick, "Genetic Algorithms and the Variance of Fitness," *Complex System,* vol. 5, pp. 265-278, 1991.

[14] D.T. Pham (Ed.), *Artificial Intelligence in Design,* Spring-Verlag, New York, 1991.

[15] E.H. Mamdani and S. Assilian, "An Experiment in Linguistic with a Fuzzy Logic Controller," *International Journal of Man-Machine Studies,* no. 7, pp. 1-13, 1975.

[16] E. Rich, *Artificial Intelligence,* New York, McGraw-Hill, 1983.

[17] H.J. Antonisse, "A New Interpretation of Schema Notation that Overtunes the binary encoding constraints", *Proc. of the 3rd International Conference on Genetic Algorithm,* Morgan Kaufmann Publisher, Los Altos, CA 1989.

[18] J.H. Holland, *Adaptation in Natural and Artificial Systems.* Ann Arbor, MI, Univ. of Michigan Press, 1975.

[19] J.M. Fitzpatrick and J. E. Grefenstette, "Genetic Algorithms in Noisy Environments", Machine Learning, vol. 3, no. 2, pp. 101-120, 1998.

[20] J.-S. Jang, "ANFIS:Adaptive-network-based fuzzy inference systems" *IEEE Trans. on Systems, Man Cybernetics,* vol. 23, no. 3, pp. 665-685, 1993.

[21] J.R. Koza, *Genetic Programming: A paradigm for genetically breeding populations of computer programmings to solve problems,* Report No. STAN-CS-90-1314, Stanford University, 1990.

[22] J.W. Kim, Y.K. Moon and B.P. Zeigler, "Designing Fuzzy Neural Net Controllers using GA Optimization," *Proc. of International Joint Symposium IEEE/IFAC on Computer-Aided Control System Design.* Tucson, AZ, 1994.

[23] K. DeJong, *An Analysis of the Behavior of a Class of Genetic Adaptive System,* Ph.D Dissertation, Dept. of Computer and Communication Sciences, Univ. of Michigan, Ann Arbor, 1975.

[24] L. Davis, *Handbook of Genetic Algorithms,* New York, Van Nostrand Reinhold, 1991.

[25] L.A. Zadeh, "Towards a Theory of Fuzzy Systems," *In Aspects of Network and Systems Theory,* R.E. Kalmann and N. DeClaris (eds.), New York, pp. 469-490, 1971.

[26] L. Schooley, B.P. Zeigler, F.E. Cellier, M. Marefat and F. Wang, "High Autonomy Control of Space Resource Processing Plants," *Control Systems Magazine,* June, 1993

[27] M.S. Bazarra and C.M. Sherry, *Nonlinear Programming,* New York, John Wiley, 1979.

[28] N.N. Schraudolph and R.K. Belew, "Dynamic Parameter Encoding for Genetic Algorithms", Tech. Report, LAUR 90-2795, Los Alamos National Laboratory, Los Alamos, NM, 1990

[29] P.J. King and E.H. Mamdani, "The Application of Fuzzy Control Systems to Industrial Process," *Automatica,* no. 13, pp. 235-242, 1977.

[30] T. Back and H.-P. Schwefel, "An Overview of Evolutionary Algorithms for Parameter Optimization," *Evolutionary Computation,* vol. 1, no. 1, MIT Press, 1993.

[31] W.J.M. Kickerr and E.H. Mamdani, "Analysis of a Fuzzy Logic Controller", *Fuzzy Sets and Systems,* vol. 1, pp. 29-44, 1978.

[32] Z. Miachalewicz, *Genetic Algorithm + Data Structure = Evolution Programming,* Springer-Verlag, New York, 1992.

Chapter 13

Paul Harrald
School of Management
PO Box 88 UMIST
Manchester, Lancs.
England

harrald@dir.mcc.ac.uk

Evolving Behaviour in Repeated 2-Player Games

0-8493-2519-6/95/$0.00 + $.50

13.1 Introduction

This chapter provides a brief introduction to a limited survey of evolutionary algorithms designed to model adaptation in repeated *games.* The limited aim here is to demonstrate the use of GAs as an evolutionary dynamic, and within such a dynamic the representation of players with limited memory in these games. The models described here can be thought of as artificial life simulations of a particular structured nature; with a strong foundation in game theory. Examples are pretty much restricted to an area of interest contained in Axelrod and Hamilton [1981], Axelrod [1984, 1987], Axelrod and Dion [1988], in particular Marks [1989a 1989b] and Miller [1989], also Lindgren [1991], Fogel [1993a], Fogel and Harrald [1994], Stanley, Ashlock and Tesfatsion [1994] and elsewhere, which are all, in one form or another, studies of the iterated prisoner's dilemma (described below). What is *not* provided here are sufficiently detailed discussions of game theory or evolutionary game theory. The reader is referred to Luce and Raiffa [1957], Maynard-Smith [1982], Rasmusen [1991], Fundenberg and Tirole [1993], and Malaith [1992]. Credit for suggesting GAs may have a role to play specifically in modeling adaptation in games appears to go to Cohen and Axelrod [1984, p. 40], and is employed in Axelrod [1987]. The first recognisable applications of GA-like algorithms to economics appears in Nelson and Winter's work, see Nelson and Winter [1982]. We begin with formal preliminaries in the shape of a description of game theory, as the GAs are seen as extensions to game theory (as suggested in Aumann [1985], Binmore and Dasgupta [1986] both hint at something along these lines, not specifically GAs but some external dynamic in an evolutionary mould).

13.2 Game Theory

Game theory was developed by von Neumann and Morgenstern [1944] as a means of analysing social interactions in which agents' decisions affect not only their own well-being, but that of other agents too. To introduce an interaction in the formal manner of game theory we begin with the set of *players,* $\mathbf{P} = \{i | i = 1,..., n\}$. We split time into discrete moments, $t = 1,...,T$, at which decisions are made. We do not here consider games in continuous time[7], or with infinite horizons, and many other generalities. At time t each player must choose an *action,* $a_i^t \in A_i^t$. The game ends at time T. Each player has *preferences* over the ways in which the game can be played out – sequences of actions for each player over $t = 1,....T$: a *play of the game,* POG. If a player i does *not* prefer POG β to POG α we write $\alpha \geq i\ \beta$. If both $\alpha \geq i\ \beta$ and $\beta \geq i\ \alpha$ the interpretation is that player i is indifferent between the two POGs, denoted $\alpha =i\ \beta$, whereas if $\alpha \geq i\ \beta$ and $\beta \not\geq i\ \alpha$ we assert that agent i strictly prefers POG α over POG β, denoted $\alpha >i\ \beta$. Let $\mathbf{Q}$ be the set of all possible POGs. We presume $\geq i$ completely preorders $\mathbf{Q}$ $\forall i$. If so, then we can establish the existence of a real-valued function, $U_i : \mathbf{Q} \rightarrow \Re$, such that, for any two POGs, α and β:

$$U_i(\alpha) > U_i(\beta) \text{ iff } \alpha >i\ \beta i$$
$$U_i(\alpha) = U_i(\beta) \text{ iff } \alpha =i\ \beta i$$

[7] Such as the 'war of attrition,' see Maynard-Smith [1982, pp. 28-39].

In game theory this is referred to as agent *i*'s *payoff function.*

To complete a description of a game, we must specify each agent's information when making a decision. The only specification we make is an agent's information over what has gone on prior to his decision at time *t*. Up to time *t* there have been *t* - 1 periods of actions taking by each of *n* players.[8] There is a set of possible sequences of actions over these *t* - 1 periods, each element of this set dubbed a *history.* Even if a player knew nothing about anothers actions, he would be able to narrow down the set of possible histories that have occurred up to time *t* because he can remember some of his own prior actions. Other knowledge of players' actions would make the information finer.[9] We suppose that each player can partition the set of possible histories into *information sets.* Hence, we speak of a player's *information partition.* Each information set contains histories such that if one of them did occur, the player cannot tell which. However, a player can tell which information set contains the actual history of the game. Let player *i*'s information partition at time 3 be $\omega_i^t \in \Omega_i^t$. We do not admit that players can infer the actions of others when speaking of information.

We are now in a position to describe a player's *strategy:* a recipe or rule for playing the game. For player *i* a strategy is a function which associates an action (or 'move') with each information set at each decision time *t*, denoted $s_i = s_i^t : \Omega_i^t, t = 1,...,T$. Denote by S_i the set of all strategies available to player *i*. A *mixed-strategy* is one which maps from information sets into a probability distribution over A_i^t. There are many games in which mixed-strategies are advisable, for example the game of 'rock-scissors-paper' and many popular sports (the 3 and 2 pitch in baseball, the decision to run or pass in football, the choice of serve in tennis, or delivery by a leg-spinner, are all examples in which being predictable might be a disadvantage. The question becomes one of in what manner to be unpredictable (a choice of probability distribution over actions). If a strategy is not mixed it is referred to as a *pure* strategy.

When modeling randomness not caused by players, but extrinsic, we invoke a player, *Nature,* who undertakes actions at known times using a probability distribution known to all players.

A particular collection of strategies, $(s_1,...,s_n)$ is called a *strategy profile.* Denote by $\mathbf{s} \in \mathbf{S}$ the set of all strategy profiles. In the absence of Nature and mixed-strategies, for each strategy profile there is just *one* implied play of the game. Hence, we could just as easily define players' preferences over strategy profiles as plays of the game. When players adopt mixed strategies or Nature is

---

[8] If a player does not have an action to make at time τ, this is represented by A_i^τ being empty.

[9] We do not admit that players can infer the actions of others when speaking of information.

present, there is just one implied probability distribution over POGs. We assume that preferences are still well defined in the previous sense, over the space of possible probability distributions, however.[10] So, when speaking of a player's payoff function now, we mean a function such that, $\forall$ $s^1, s^2 \in \mathbf{S}$:

$$U_i(s^1) > U_i(s^2) \text{ iff } s^1 >_i s^2$$
$$U_i(s^1) = U_i(s^2) \text{ iff } s^1 =_i s^2$$

13.2.1 Solving Games

To predict behaviour in games is to predict a chosen strategy profile (and then forget about actual play, since it is all implied in the profile). We look for strategy profiles that have certain characteristics. One such characteristic often employed is that of a *Nash equilibrium* (Nash [1950]) profile. Let $s = (s_1,...,s_n)$ be a strategy profile, and denote by s_{-i} s with the i'th component removed. Then, s forms a Nash equilibrium if, $\forall i \in \mathbf{P}$:

$$Ui(s_i;s_{-i}) \geq Ui(s'_i;s_{-i}) \ \forall s_i,s'_i \in S_i.$$

Although strategies are not considered observable, a Nash equilibrium has the property that each player makes a *best-response* to the choices of the other players: given their strategy choices, no player can force a preferred POG, or distribution over such, by choosing some other than their own chosen strategy. We can further refine Nash equilibrium, but space does not allow a full discussion. Again, the reader is referred to the references for a full treatment.

13.2.2 Open- and Closed-loop Strategies

If a player adopts a strategy which specifies the same action for all information sets, then that player is said to adopt an *open-loop* strategy. Such a strategy is simply a predetermined sequence of actions, or distribution over actions, that is not contingent upon information revealed during the game. Otherwise, a player is said to adopt a *closed-loop* strategy.

13.2.3 The Repeated Symmetric 2 x 2 Game

We now consider the following case: $A_i^t = A_i^{t+1}, A_i^t = A_j^t \forall i,j \in \mathbf{P}$: players always choose from the same action sets. We also suppose that P = {1,2}, there are two players. At each time interval we suppose, $A_1 = A_2 = \{X,Y\}$. To represent payoffs, we look at what might happen at any one time step. There are four possibilities, $(a1,a2) \in \{X,Y\}^2$. Now consider Figure 13.1 representing in itself a *one-shot* game. Each entry shows the 'payoff' to [*row, column*] depending on the choices made by the two.

	X	Y
X	[e,e]	[g,h]

[10] In fact, under such circumstances an *expected utility function* is used to represent preferences. Such a function is linear in the probabilities of each POG occurring.

Y [h,g] [f,f]

Figure 13.1: The basic 2 x 2 game.

As the repeated game is played out, one of these cells will determine an *increment* in payoffs, and once the game is finished at $t = T$, each player will have accrued a sum of such increments. It is this sum that is used to represent preferences, or its expected value in the presence of mixed strategies.

In iterating this game, there are 4^{t-1} histories up to time t. If a player only remembers his own moves, there will be an information partition containing 2^{t-1} information sets (for each move made by player i there are two possible plays for player j). We refer to Figure 13.1 as the one-shot game, while of course *the* game being played consists of choices from $\{X, Y\}$ over T *iterations*.[11]

Since decisions about strategies are made prior to the game commencing, we might invoke players who discount future increments in payoff in assessing payoff functions, so that when contemplating a strategy profile, implied (expected) payoffs at time t would be calculated with reference to Figure 13.1, and discounted by δ^t, $0 < \delta < 1$. Notice, however, that the matrix presented implies that the players have identical preferences - I could re-label the players without altering the matrix - furnishing us finally with *a symmetric* 2 x 2 game.

Game 1: The Prisoner's Dilemma
Consider the following inequalities:

$$g>e>f>h.$$

A one-shot game such as this is referred to as a *prisoner's dilemma,* PD.[12] Playing this game once only, Nash equilibrium predicts that both players choose Y, since no matter what the other player does this is a best-response. This is true whether a player is informed of the opponent's choice or not. In the PD a common notation is to have what we have labelled as X referred to as 'cooperate' (C), and what we have referred to as Y 'defect' (D). An example is shown in Figure 13.2. This terminology arises because although we predict a play of ($a1$, $a2$) = (D,D), the players could both have achieved a greater payoff (preferred outcome) with a play of (C,C). However, they are unable to support this 'cooperative' agreement because, by definition of the game, they are compelled to choose D. So, played once, we see what economists refer to as a *Pareto inferior*

[11] Often in artificial life simulations, *the* game is a very complex thing to describe, particularly when agents condition behaviour on many aspects of past events. Often, a one-shot game like that depicted is referred to as "the game", whereas it is but a small component of a full description.

[12] Tucker [1950]. Most of the references cited here have a description of the nature of and interest in this game. See also, Rapoport and Chammah [1965].

outcome: it is possible to make both players better-off by constraining them to play *C*.

	C	D
C	[3,3]	[0,4]
D	[4,0]	[1,1]

Figure 13.2: A prisoner's dilemma.

This compelling reason to play *D* does *not* apply to a repeated version of this game, the *iterated prisoner's dilemma,* IPD. For example, consider the case in which players who can observe one another's moves (or their own move and increment in payoff). Imagine playing a player who adopts a strategy which amounts to "play *C* until opponent plays *C*, thereafter play *D*." Against this player it would be unwise to defect in early rounds, since this guarantees (mutually) low payoffs thereafter. This kind of reasoning has led researchers to look into the possibility that players can adopt strategies that support cooperative behaviour.[13] Examples are Kreps *et al* [1982], Harrington [1987], Neyman [1985], Sober [1992], and the reader is referred to these works for an insight into the game-theoretic approach to this issue.

In a series of papers initiated by Axelrod [1978], Axelrod [1981], Axelrod and Hamilton [1981], the issue of whether cooperation emerges in an evolutionary context has arisen, and spawned an active literature, described somewhat in later sections.

We also impose the restriction that:

$$2e>g+h,$$

if this were not true there is the possibility of a more sophisticated kind of cooperation over a cycle of two PDs: players who alternate (D,C), (C,D) obtain higher average payoffs than persistent mutual cooperation.

13.2.4 Modeling Learning and Adaptation in Games

Recently, economists have attempted to promote models of learning and adaptation from artificial intelligence into the game theoretic scenario (Holland and Miller [1991], Arthur [1993], Arthur [1992], Harrald [1994a]). The desire is to model agents in economic contexts who are not necessarily thought of as choosing strategies prior to a game, but who react to experience as the game is played out in what is not thought of as a predetermined manner.[14] The idea is

---

[13] It is important to note that the IPD is not itself a Prisoner's dilemma: it does not have a dominant strategy. However, the word cooperation is understood to mean plays of *C during* the IPD.

[14] Some works look for simulations that reproduce equilibrium notions from economics, Arifovic [1992a, 1992b] and Marimon, McGrattan and Sargent [1990], for example.

not to look for equilibrium strategy choices, but to impose a dynamic on a game that resembles an evolutionary process.[15] There is no presumption that simulations are required (Friedman [1991], Malaith [1992]), but once we allow the use of simulations the range of models that can be analysed is no longer the set that we can 'solve' ourselves (Harrald [1994a]). This external dynamic has underpinnings in the formalisms of evolutionary game theory to which we now turn.

13.3 Evolutionary Game Theory

When evolution operates at certain levels it looks very similar to a GA optimising some objective function. Take wing design in birds for example (see Maynard-Smith [1982, ch. 1]. If we suppose that selection favours birds with more efficient designs, then it is possible to think of evolution as selecting certain parameters describing the physical features of the wing based on an objective function governed by the laws of physics and nothing else. However, there is another type of adaptation that differs in that the fitness of a particular action depends upon the actions of other members of a species. An example is human language. The usefulness of my language depends upon the number of others I meet who use the same language. If everyone speaks the same language, then we can talk of its engineering efficiency in conveying ideas separately from the number of people using it, however, when there are many languages we must separate a languages's engineering efficiency (how long does it take to convey a particular idea), from the strategic aspect (how many people use it). In modeling natural evolution in contexts of inter-player effects of this type[16], Maynard-Smith [1982] used the formal structures of game theory much as we have described, but with new definitions of equilibrium and interpretation of payoffs[17]. This *evolutionary game theory* has since become a fairly popular modeling tool for biologists. We restrict attention to our repeated symmetric 2 x 2 game here, although as the reader is no doubt aware, there are many generalisations.[18]

The first amendment is in the interpretation of payoffs. Each entry, (e,f,g,h) is designed to represent an increment in *Darwinian fitness* to each of two players depending on an expressed behaviour, abstractly X or Y. The meaning of 'fitness' is clearer in a context.

---

[15] Evolutionary economics has its own tradition, from the works of Veblen and Schumpeter to that of Nelson and Winter [1982], which has a close affinity with techniques of Al. Witt (ed.) [1993] contains a collection of older papers, see the references there. Also, Samuelson [1988] considers the evolutionary basis of concepts like Nash equilibrium.

[16] Economists use the evocative term *externalities.*

[17] Hamilton [1967] is the first explicit use of game theory terminology to evolutionary biology. Lewontin [1961] uses game theoretic concepts. Slobodkin and Rapoport [1974] model players as selecting strategies that minimise the probability of extinction. The subject was crystalised with a solution concept, the 'evolutionary stable strategy', in Maynard-Smith and Price [1976].

[18] The version here does not go beyond Maynard-Smith [1982], ch. 2.

13.3.1 The Random-Mixing Population

Our context will be the following. A large population of players (a single species) meet in random pairwise encounters to play a repeated version of the game in Figure 13.1. One can imagine the players guided by Brownian motion on a surface, and each time they encounter another player they have a repeated game. Each player is characterised by a strategy for playing another in the repeated game. *We suppose that previous encounters do not influence behaviour in subsequent encounters,* although this is an interesting issue.[19] We suppose there to be a finite set of strategies that might be in use at any time by a player, S. Note that S contains the set of all strategies that might ever exist, and must be chosen by the modeler. Imposing finiteness again, $S = \{s^1,...,s^m\}$. At any point in time, the population in question can b described by the proportion of all players adopting strategy $j = 1,...$, denoted π_j. When two players meet we denote the payoff to a player adopting s_j against a player adopting s_k as $P(s_j,s_k)$. To calculate this we establish the sequence of actions implied by s_j against s_k, and refer to Figure 13.1 to find the sum of increments in payoffs. If players adopt mixed strategies, or Nature interferes in one way or another, then we can calculate the expected value of the sum in increments in payoffs, which is still referred to as $P(s_j,s_k)$.[20] Now, the actual game being played here is quite complex. First, Nature chooses a random opponent for each player. Then there is a repeated version of the one-shot game, which ends after T iterations. Then Nature moves again to select another opponent, and so on. We are presently limiting the players to have strategies that only account for what transpires during the T iterations with an opponent. Given this, against a random opponent, a player adopting strategy s_j has an expected payoff, E_j, of:

$$E_j = \sum_{k=1}^{m} \pi_j p(s_j, s_k)$$

When we talk of a strategy's or player's *fitness,* it is this that we refer to. Hence, the numbers (e,f,g,h) have significance when embedded in this calculation. The connotation as Darwinian fitness arises since we wish to impose a dynamic on $\pi = (\pi_1,...,\pi_m)$. In particular, we think of π as changing over time such that relatively fitter strategies become relatively more prevalent.[21] Analytical versions of evolutionary game theory would specify difference equations, $\Delta\pi_j$ representing *replicator dynamics,* and solve these (subject to obvious boundary conditions). The actual laws of motion must be decided upon by the modeler The difference equations are also subject to stochastic disturbances due to the effect of

---

[19] Ashlock, Stanley and Tesfatsion [1994] consider a game in which a player's payoff against an opponent influences behaviour should the two meet again. This allows a consideration of reputation issues, and area of active research.

[20] For example, we will describe a game in which players misinterpret what the other player did with a fixed probability.

[21] Since the payoff to a player is independent of that player's identity, we can use the loose phrase 'a strategy's fitness.'

mutation: an arbitrary remapping of s_j into a new strategy, and also effects due to players not meeting a representative sample of the population. The alternative explored here is to model a population as evolving in a stochastic evolutionary process: a GA. To explore the link between such GAs and evolutionary game-theory, we'll briefly look at what we might expect.

13.3.2 The Evolutionary Stable Strategy

The primary solution concept in evolutionary game theory is that of the *evolutionary stable strategy,* ESS. The idea is that when the vast majority of players adopt an ESS, a minuscule minority of mutant invaders cannot achieve a higher payoff than a typical ESS player, and hence cannot thrive. If a strategy is immune to all such invasions, it is an ESS. So, a strategy j is an ESS, if when the proportion $\pi_j \approx 1$ adopt it, if a proportion $\pi_k \approx (1 - p)$ adopt any $s_k \in S$, $E_j \geq E_k$. We assert that this is true if $P(s_j, s_j) > P(s_k, s_j)$, since these two types of encounter dominate activity, or if $P(s_j, s_j) = P(s_k, s_j)$ we take notice of the negligible fraction of the time players play s_k types, and complete the definition of an ESS with $P(s_j, s_k) \geq P(s_k, s_k)$. Notice that the definition of an ESS depends upon S.

13.3.3 Evolutionary Stable Population

When all players accrue the same expected payoff in a population, we would expect their relative proportions to remain the same. Such a configuration is referred to as an *evolutionary stable population,* ESP. Formally, an ESP obtains when

$$E_j = E_k \; \forall s_j, s_k \in S,$$

notice again the dependency on S. In the case of a random mixing population playing the IPD, if we restrict S so that players can only adopt unconditional probabilities of playing C in each iteration of the PD, then the only ESS is for all players to play C with probability 0, which is also the ESP. As we have mentioned, this changes dramatically by allowing players to map from histories into moves.

Game 2: The Hawk-Dove Game

In providing an example, Maynard-Smith [1982, ch. 2] describes a scenario in which two animals are contesting a resource of value V: the increment in fitness associated with the resource is V. The animals have two basic approaches, an aggressive one, like a 'Hawk,' H or a more passive show, like a 'Dove,' d (so as not to confuse with D for defection). It is presumed that two Doves share the resource, experiencing a change in fitness of $V/2$ under such circumstances. A Hawk player will accrue the whole value of the resource when playing against a Dove player, and the Dove player no increment in fitness. Two Hawks will fight, the winner receiving the full value of the resource, the loser nothing. In fights there is a probability of 0.5 of being injured, whether the winner or loser of the fight, and injury carries with it a decrement in fitness of c (not 'C!'). Figure 13.3 illustrates the one-shot game.

	H	d
H	[(V-c)/2,(V-c)/2]	[V,0]
d	[0,V]	[V/2,V/2]

Figure 13.3: The Hawk-Dove Game in general form

The reader can verify that if we restrict players to either a strategy that amounts to unconditionally playing H with some probability, $\pi_i \in S = [0, 1]$, then, if $V > c$ the only ESS is $\pi_i = 1$, and this is the only ESP. If $V < c$ then there is an ESS of $\pi_i = V/c$, and an ESP in which the mean of all players' probabilities of playing H is equal to V/c. Figure 13.4 shows a case for which the mixed-ESS and average probability of playing H in the ESP is $^4/_7$.

	H	d
H	[-3,-3]	[8,0]
d	[0,8]	[4,4]

Figure 13.4: A specific Hawk-Dove Game, $V = 8$, $c = 14$.

Game 3: Simultaneous Hermaphrodites

Simultaneous Hermaphrodites essentially have the ability to choose their sex when copulation is attempted (by choosing which gamete to produce). An example is the sea slug *Navanax Inermis*.[22] Evidence suggests that the act of producing the female gamete is slightly more taxing than that of producing the male gamete. Successful copulation can only occur, of course, if the two slugs produce different gametes. Label the two actions as M and F, then a payoff matrix like Figure 13.5 represents a case like this.

Again, restricting strategies to the open-loop unconditional probability of playing M, there is no pure ESS, a mixed ESS exists with $\pi_i = {}^3/_5$, and a population that plays M on average with probability $^3/_5$ is an ESP.

	M	F
M	[-1,-1]	[9,7]
F	[7,9]	[-3,-3]

Figure 13.5: A simultaneous hermaphrodite game.

---

[22] For a proper treatment of this topic, see Matsuda [1989], Leonard and Lukowiak [1985] and Leonard [1990] on the slug, Fischer [1980, 1981] on the black hamlet (*Hypoplectrus Nigricans*), a coral reef fish. There is a similar vein to the literature on food sharing in the vampire bat *Desmodus Rotundus*, see Wilkinson [1984], for example.

13.3.4 Summary of Open-Loop Equilibria

We are considering a class of players who are memoryless; they do not themselves attempt to adapt to experience, but are completely characterised by an unconditional probability that they play X in any given interaction. Let π_i be player i's strategy, $i = 1,...,n$. An ESP in this scenario consists of a distribution of probabilities, (π_i), such that each player's expected payoff in a random encounter is the same as any other player. It is easy to show that there are three exclusive and exhaustive possible cases (using notation of Figure 13.1):

(a) $(\pi_i) = (0)$

(b) $(\pi_i) = (1)$

(c) $\left\{(\pi_i):\left(\sum_i \pi_i\right)/n = \frac{(g-f)}{(g-f)+(h-e)}\right\}$

A special case of (c) is the strictly-mixed ESS, $(\pi_i) = \frac{(g-f)}{(g-f)+(h-e)}$. We use the notation π^* to indicate the average frequency of playing X in an ESP, so that (a) corresponds to $\pi^* = 0$, (b) to $\pi^* = 1$ and (c) to $\pi^* = \frac{(g-f)}{(g-f)+(h-e)}$. See Maynard-Smith [1982, Appendix B] for proofs of the ESSs. Details are more difficult with three or more moves, see Maynard-Smith [1982, Appendix D].

13.4 Simulating the Random-Mixing Population

We have said little about dynamics, although there is a lot to be said.[23] Our purpose now is to use a GA to simulate evolutionary dynamics. The following pseudocode outlines the essentials of a routine to evolve a population which interacts in the random-mixing fashion we have described:

```
1. for each GENERATION
   1.1 for each PLAYER i
      1.1.1 for each OTHER j not yet encountered
         1.1.1.1 for each ITERATION
            1.1.1.1.1 Play one GAME between i and j
            1.1.1.1.2 update FITNESS[i]
            1.1.1.1.3 update FITNESS[j]
   1.2 create new POPULATION
```

Each member of the population plays each other member in a repeated version of the 2 x 2 game so that players' fitnesses are representative of their expected values meeting randomly chosen opponents. Each play of the one-shot game requires an action by each player, and the one-shot game is played over a series of iterations so that representative outcomes accrue. It may be that the game is played just once, however, depending on the context. Fitnesses are updated based on the increments at each play of the game, although there are alternatives (such

---

23 For examples of analytical approaches see, for example, Maynard-Smith [1982 Appendix D], McCauley *et al* [1991] or Friedman [1991] as samples of this approach in djfferent contexts.

as maintaining a per-game average payoff for each player, and hence avoiding using long integers, but requiring double precision payoffs).

Notice that generations do not overlap, and also that at the beginning of each generation, all player fitnesses are reset to some base level, which is zero for our simulations.

There are two major decisions to be made in specific incarnations of such an algorithm: player representation is one, and the method of generating a new population the other. We begin with a GA which evolves a population of players who adopt open-loop mixed strategies: in each play of the one-shot game each player is characterised by an unconditional probability that they play *X*. Such a scenario has been handled quite readily using some simple analytics of evolutionary game-theory, so that we can check that the GA accords with the basic intuition we associate with this scenario.

13.5 Implementing a GA

The following code, written in QuickBASIC 4.5 implements a GA for our scenario. Each player is encoded as a bit-string of length `leng%` which translates into integers between 0 and $2^{leng\%}$ - 1. Dividing by $2^{leng\%}$ - 1 gives the real-valued probability (performed by the subprogram `decode`).

Each of `numPlayers%` players meets each other player in a `numlts%` played 2 x 2 game, with play governed by the player probabilities (stored in `prb#$(i)` as a bit string, and decoded in `prob#(i)`). The array `pay#(i)` maintains players' cumulative payoffs, and once all the `numPlayers%*(numPlayers%-l)*numlts%` one-shot games have been played, control is passed to the subprogram `newpop` to create a new population.

The new population, in this instance, is created by *tournament selection:* a sample of `numSelect%` parents are chosen with replacement. The one with greatest fitness becomes a parent. This is repeated to obtain a second parent. The children are mutated by bit-flipping with probability `pMut#`. Payoff matrix entries are stored in `xx#, yy#, xy#, yx#`, representing the entries in the payoff matrix in an obvious manner. Other details are commented throughout the code.

```
DECLARE        FUNCTION randPlayer% ()
DECLARE        SUB routate ()
DECLARE        SUB initpop ()
DECLARE        SUB decode ()
DECLARE        SUB stars ()
DECLARE        SUB clean ()
DECLARE        SUB playgame (e%, 1%)
DECLARE        SUB newpop ()
'
'' Note: % = integer, # = double precision.
'
CONST numPlayers% =          30 'Number of players, must be even
CONST xx# = 2# '             e in text matrix
CONST yy# = 1# '             f in text matrix
CONST xy# = O# '             g in text matrix
```

```
CONST yx# = 3# '              h in text matrix
CONST leng% = 15 '            Length of chromosomes
CONST numGens% = 1000 '       Number of generations
CONST numIts% = 8 '   Number of iterations per interaction
CONST pMut# = .008# 'Probability of bit-flipping in mutation
CONST numSelect% = 6 '        Number of players in tournament.
DIM SHARED prbS(numPlayers%) 'The chromosome strings
DIM SHARED pay#(numPlayers%) 'The payoffs
DIM SHARED prob#(numPlayers%)'The numerical probabilities
'
'===============================================================
                         MAIN LOOP START
'===============================================================
CALL initpop ' Get first random generation (of bit-strings)
FOR h% = 1 TO numGens%
   CALL decode ' Translate strings into probabilities
   CALL clean '
   FOR i% = 1 TO numPlayers% - 1
      FOR j% = i% + 1 TO numPlayers%
         FOR k% = 1 TO numIts%
            CALL playgame(i%, j%) 'Play a game and increment
                                                      payoffs
         Next k%
      Next J%
   Next i%
   CALL newpop ' Generates a new population of strings
   CALL mutate ' Mutation of chromosomes
NEXT h%
END
'===============================================================
                         MAIN LOOP END
'===============================================================

'===============================================================
                         START OF clean
'===============================================================
SUB clean
' Often more to do than just this; resetting payoffs.
'
   FOR i% = 1 TO numPlayers%
      pay#(i%) = O#
   NEXT i%
END SUB
'===============================================================
                         END OF clean
'===============================================================

'===============================================================
                         START OF decode
'===============================================================
SUB decode
' Calculates numerical probability associated with bit-string
'
   FOR i% : 1 TO numPlayers%
      s# = O#
```

```
      FOR j% = 0 TO leng% - 1
         s# = s# + VAL(MID$(prb$(i%), j%, 1)) * 2 ^ (j%)
      NEXT j%
      prob#(i%) = s# / (2 ^ leng% - 1)
   NEXT i%
END SUB
'==============================================================
                            END OF decode
'==============================================================


'==============================================================
                           START OF initpop
'==============================================================
'Initial random bit-strings of length leng%.
'
SUB initpop
   FOR i% = 1 TO numPlayers%
      FOR j% = 1 TO leng%
         IF RND < .5 THEN prb$(i%) = prb$(i%) + "0" ELSE prb$(i%)
                                                 = prbS(i%) + "1"
      NEXT j%
   NEXT i%
END SUB
'==============================================================
                            END OF initpop
'==============================================================

'==============================================================
                           START OF mutate
'==============================================================
'Bit-flip with probability pMut#, bit of a pain.
'
SUB mutate
FOR i% = 1 TO numPlayers%
   FOR j% = 1 TO leng%
      IF RND < pMut# THEN
         MIDS(prb$(i%), j%, 1) = STR$(1 - VAL(MID$(prb$(i%), j%,
                                                          1)))
      ENDIF
   NEXT j%
NEXT i%
END SUB
'==============================================================
                            END OF mutate
'==============================================================

'==============================================================
                           START OF newpop
'==============================================================
SUB newpop
'
'  This version uses tournament selection and one-point
```

```
crossover.
'
DIM ch$(numPlayers%)  'Temporary array of children strings
DIM parent%(2)        'Convenient to store index of parent.
'
FOR i% = 1 to numPlayers% / 2
   FOR a% = 1 to 2     'Start of tournament selection.
      parent%(a%) = randPlayer%
      FOR b% = 1 to numSelect% - 1
         x% = randPlayer%
         IF pay#(x%) > pay#(parent%(a%)) THEN parent(a%) = x%
      NEXT b%
   NEXT a%

'              We now have the two parents,
'              indexes in parent%(1) and parent%(2).

   pc% = INT(RND • leng% + i) 'Crossover point
'                             Copy parent portions into children:

   ch$(i%) = prb$(parent%(2))
   ch$(i% + numPlayers% / 2) = prb$(parent%(1))
   MID$(ch$(i%), 1, pc%) = MID$(prb$(parent%(2)), 1, pc%)
   MIDS(ch$(i%+numPlayers%/2),1,pc%)=MID$(prb$(parent%(2)),1,pc%)
NEXT i
'
'
'    Now copy children into parents for next generation:
FOR i% = 1 TO numPlayers%
  prb$(i%) = ch$(i%)
NEXT i%
END SUB
'==============================================================
                          END OF newpop
'==============================================================

'==============================================================
                        START OF playgame
'==============================================================
SUB playgame (e%, f%)
' Plays actual one-shot game, play governed by probabilities.
' Also increments payoffs accordingly.

X%   = 1 'Used to determine who did what.
Y%   = 0 'Many ways to do this, some more compact.

IF RND < prob#(e%) THEN movee% = X% ELSE movee% = Y%
IF RND < prob#(f%) THEN movef% = X% ELSE mover% = Y%

IF movee% + movef% = 2 THEN          'Both played X.
  pay#(e%) = pay#(e%) + xx#
  pay#(f%) = pay#(f%) + xx#

ELSEIF movee% + movef% = 0 THEN      'Both played Y.
  pay#(e%) = pay#(e%) + yy#
```

```
   pay#(f%) = pay#(f%) + yy#

ELSEIF movee% = 1 THEN                    'e% played X, f% played Y.
   pay#(e%) = pay#(e%) + xy#
   pay#(f%) = pay#(f%) + yx#

ELSE                                      'f% played X, e% played Y.
   pay#(e%) = pay#(e%) + yx#
   pay#(f%) = pay#(f%) + xy#
END IF
END SUB

'==================================================================
                          END OF playgame
'==================================================================

FUNCTION randPlayer%
'
'
' Returns a random player
   randPlayer% = INT(RND • numPlayers%) + 1
END FUNCTION
```

Notice that representing players as strings allows a literal implementation of crossover, calling QuickBASIC's `MID$()` function. We could have represented players directly as double-precision probabilities, as in `prob#(i%)` and employed a version of real-valued crossover. One such method is to form two children by making a random convex-combination of the two parents:

```
p# = RND
ch(i%) = p# * prob#(parent%(1)) + (1 - p#) * prob#(parent%(2))
ch(i%+numPlayers%/2)=(1-p#)*prob#(parent%(1))+p#*prob#(parent%(1))
```

This does restrict child probabilities to lie between parent probabilities, at least before mutation, which would usually be accomplished by adding a random variable with mean zero (which may create small probability masses at 1 and 0).

To generate one child from two parents, we could simply take the arithmetic mean of the two parent probabilities (the sum of them divided by two), or the geometric mean (the square-root of the product of the two parent probabilities). Alternative mutation techniques are to add a random disturbance, let the probabilities *creep,* to replace the value with a randomly selected value (with a small probability), or *geometric creep:* to multiply bay a random amount.[24]

13.5.1 Some Results

Running this GA on our three basic games yields results consistently like the ones presented in Figure 13.6. Indicated is the population-wide probability of

---

[24] See Janikow and Michalewicz [1991], Michalewicz and Janikow [1991], Eschelman and Schaffer [1993] and the references there for analyses of floating point representations.

playing C, H, or M depending on the game, over 1000 generations. The GA successfully finds and maintains the ESP, which is encouraging for applications in which the game-theoretic analytics are too difficult to arrive at an ESP readily.

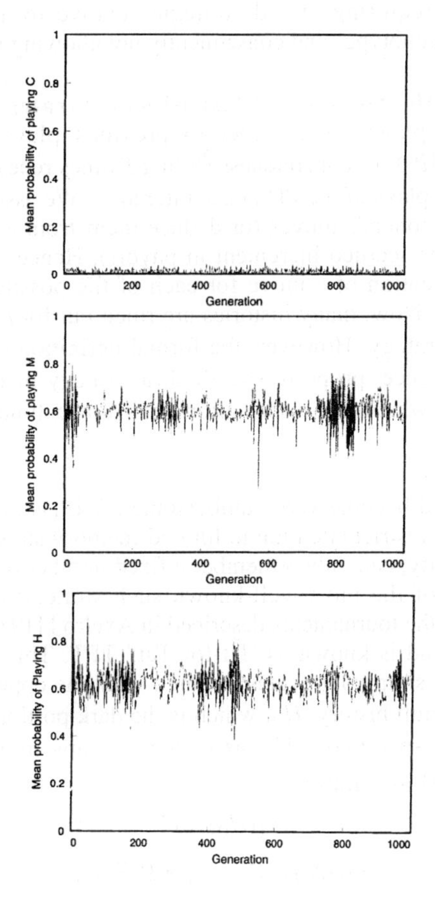

Figure 13.6: Population mean probabilities of playing C, M and H in the three games. In these cases, `numSelect%` = 6, `numPlayers%` = 30, `numIts%` = 5, `pMut#` = 0.008, `leng` = 15.

Moreover, and this is interesting, the GA finds the ESS, not just any ESP. For example the first 18 chromosomes in the final generation of the simultaneous hermaphrodite game were:

```
110010011010101   110010111010101   110010011010001
110010011010101   110010011010101   110010011010001
110010011010001   110010011010101   110010011000001
110010011010101   110010011010101   111010011010101
110010011000001   110010011010001   110010011010101
110010011010001   110010011010001   110010011010101
```

showing a remarkable uniformity, presumably due to the crossover operator. The chromosome 110010011010101 decodes to a probability of playing M of 0.6674703207, while the mixed-ESS is 0.6.[25] The GA predicts then, that players in games requiring mixed strategies evolve to a uniform type, not sustaining various genotypes and consequently not allowing various phenotypes.

13.6 Players with Memory: Closed-Loop Strategies

Considering now players who react to previous plays of the game, we concentrate on the IPD for concreteness.[26] In a T times repeated PD, players can generate 4 possible plays of the PD at each iteration. We assume that the players can observe the opponent's moves (or deduce them from a knowledge of their own move and their accrued increment in payoff). Hence, a strategy for these players is a specification of a move for each of the possible histories at each iteration, $t = 1,...,T$. Now, many histories are ruled out for $t = 2,...,T$ because of the player's own strategy. However, the formal definition of a strategy retains this redundancy. Hence, properly specified, a strategy maps each of the 4^{t-1} possible histories (which are information sets) up to time t into $\{C, D\}$, the move for $t1$.

This mapping would become very cumbersome for IPDs of a few iterations or more. We therefore restrict attention to limited-memory strategies: players who, by choice or necessity, can only remember a fixed number of previous iterations of the game. One of the most well-known such strategies was submitted by Anatol Rapoport in the tournaments described in Axelrod [1984] and Axelrod and Hamilton [1988], and is known as 'Tit-for-Tat,' TFT. This strategy begins by playing C, and then simply replicates at time t what the opponent did at time t - 1. We conjecture a null history, H_0, which is the dark portion of time before the game begins. Fully specified, TFT, as player 1, maps from histories H_0 and $\left(a_1^{t-1}, a_2^{t-1}\right)$ into $\{C, D\}$ as follows:

$$s_1(H_0) = C$$

$$s_1(\bullet,\mathrm{C}) = \mathrm{C} \qquad s_1(\bullet,D) = D,$$

and is an example of a "memory-1" player: recalling just 1 previous encounter. A player with a memory-2 strategy will consist of an action for the start of the IPD, an action specified for each of 4 possible histories at $t = 2$, and is completed by a specification of an action for each of the 16 possible histories of length 2

---

25 You need to read the chromosome backwards as a binary number, that's the way they are decoded.

26 This is certainly *not* the only interesting repeated game. The simultaneous hermaphrodites we mentioned previously seem to adopt strategies that involve an active switching of roles--alternating in a coordinated fashion the male and female roles, thus suggesting that these players adopt closed-loop strategies of a type mapping at least from the opponent's last move, and the players' last move into a conditional probability over moves in the next iteration of the one-shot game.

iterations. This generalises in an obvious manner to the case of an memory-m player.

13.6.1 Player Representation

We would like to encode players so that the traditional operators of a GA are applicable. Attention is restricted at present to pure strategies. An example encoding, used in Axelrod [1987] and Lindgren [1991] is to denote the action C by 1 and D by 0. A player of memory-m can be represented by a bit-string of length

$$\sum_{i=1}^{m} 4^{i-1}$$

For instance, the first bit represents the move in 'response' to H_O, the next 4 represent actions for each of the four possible histories at $t = 1$, the next 16 bits moves for each of the possible histories for $t = 2$ and so on.[27] For purposes of this paper we turn to an alternative characterisation, however.

13.6.2 Finite Automata as Game Players

The use of *deterministic finite automata* (DFA) for these games is suggested and analysed in Miller [1989], Fogel [1993], Rubinstein [1986], Binmore and Samuelson [1992], and the evolution of DFAs can be found in different contexts in Fogel [1962a, 1962b], Fogel, Owens and Walsh [1966]. Formally, an DFA consists of 5 mathematical objects.

$$M = \{Q, \alpha, \beta, s, o\}$$

where Q is a set of states, α is a set of input symbols, β is a set of output symbols, s: $Q \times \alpha \rightarrow Q$ is a next-state function, and o: $Q \times \alpha \rightarrow \beta$ is an output function. For our purposes, the inputs to the DFA will be opponent's moves from the previous iteration, and the output in each state the DFA's move for the

[27] Neither Axelrod [1987] nor Lindgren [1991] in fact use an encoding exactly like this. and Lindgen [1991] does not explicitly employ an evolutionary algorithm. The players are thought of as playing the infinitely repeated PD, and at each iteration there is a fixed probability that a player mistakenly makes the opposite to that suggested by the strategy. Since the game is played infinitely with mistakes, moves need not be specified for the m - 1 initial rounds, since payoffs accrued during this period are inconsequential. Player interactions form a Markov process: given the m-length history up to t - 1, the probabilities that the m-length history at t is any one of its possibilities can be deduced from the players' strategies, and the fixed probability of mis-employing an action. Hence, the expected payoff to players against one another can be calculated. Thanks to Kristian for clarifying this in personal communication.

next.[28] TFT described thus has two states, $Q = \{0, 1\}$, accepts input from $\alpha = \{0, 1)$, and generates output from the same set, $\beta = \alpha$. Completing this description yields:

$$s(0,1)=0 \qquad s(0,0)=1 \qquad s(1,1)=0 \qquad s(1,0)=1$$

$$o(0) = 1 \qquad o(1)=0$$

A transition diagram for this DFA is provided in Figure 13.7. Another example is shown in Figure 13.8. This latter DFA ("punish-twice-and-wait") (PTAW) punishes *all* defections by an opponent with two defections, but will then return to cooperation should the opponent cooperate.[29] Note that changing just *one* transition makes TFT a pathological cooperator.[30]

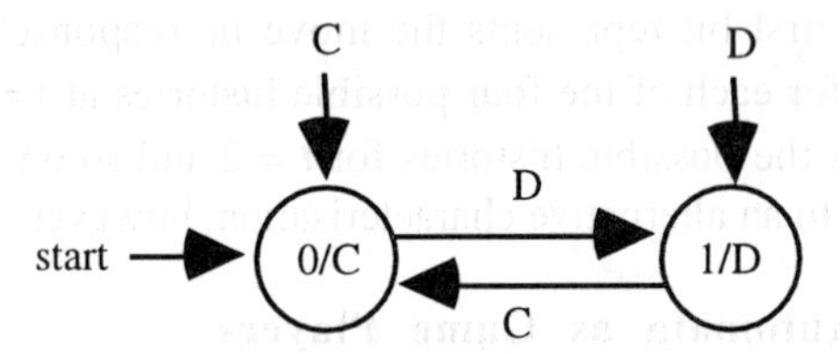

Figure 13.7: Transition diagram for the DFA which plays TFT. Arrows indicate transitions in response to the opponent's last move, inside the nodes are the state number (left of the virgule) and move when in that state. The DFA begins in state 0.

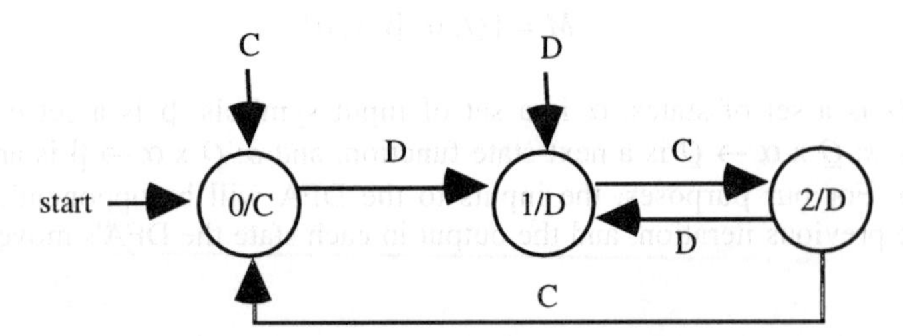

Figure 13.8: Transition diagram for "punish-twice-and-wait."

[28] To be finicky, technically the automata begin the game with an e-transition to state 0, a 'starting' state. 2. From that point on the players act as deterministic DFA's. Also, since output is associated with entry to a state, these DFA are *Moore machines,* Moore [1957]. See Hopcroft and Ullman [1979], an excellent introduction to automata theory.

[29] I would like to thank Michael Spencer for pointing out an error in a previous incarnation of this transition diagram.

[30] This kind of mutation lies at the heart of the instability of small cooperating populations of TFT-like players when invaded by opportunistic defectors.

When DFAs play a repeated game, they must, given enough iterations, begin to cycle through the same sequence of states, and hence actions. Rubinstein [1986] uses average payoff over this cycle to consider a game in which two players have as strategy sets the set of all DFAs. Imposing a cost for the number of states in a chosen DFA, restrictions on Nash equilibria are found. See also Binmore and Samuelson [1992] for an evolutionary analysis of games played by DFAs.

13.7 Encoding the DFAs

There is a fairly natural encoding for these DFAs which generalises naturally to more complex scenarios. For example, all two-state DFAs, like TFT, can be represented by a string #1#2#3#4#5#6. The interpretations are as follows:

$$\#1 = o(0), \#4 = o(1)$$
$$\#2 = s(0, 1), \#5 = s(1, 0) \quad \#3 = s(0, 0), \#6 = s(1, 1)$$

This yields $2^6 = 64$ possible encodings for 2-state DFAs. Many of these represent observationally equivalent players, however. For example, the schema 1## 1## pathologically cooperate, as do 100###. There are, in fact, 14 distinct strategies representable by 2-state DFA's. TFT is encoded as 101001. Such encodings illustrate nicely the property of *epistasis,* describing the case in which some genes inhibit the effect of other genes (by preventing entry to one state or another. Maynard-Smith [1982, ch. 4] contains an interesting discussion of the notions of ESS at the genotype level, after Lloyd [1977]. See also Sober [1992, Appendix].

For DFAs of more states, a similar interpretation is placed on characters in the string, although it is most convenient to move away from bit-strings to accommodate the greater number of states. In a hopefully clear way, for example, PTAW could be encoded as 101 012 002, where the states have been separated for clarity. In general, an s-state DFA for playing the IPD consists of $3s$ characters, the first three representing state 0, the second 3 state 1, and so on. Note that state i is described by characters at positions $3i + 1$, $3i + 2$, $3i + 3$, where the character at $3i + 1$ is $o(i)$, character $3i + 2$ is $s(i, 1)$, and character $3i + 3$ is $s(i, 0)$. Characterised as strings in this manner, we can now implement a GA to evolve a population of DFA players.

13.8 A GA for DFAs in the IPD

We need new routines for `initpop`, `mutate` and `playgame(e%,f%)`. We need a new routine for `initpop` and `mutate` since the strings are no longer bit-strings but include digits to represent states labelled 2 and above.[31] The subprogram

---

[31] We could, actually, restrict players to a number of states of $2^I - 1$ where I is an integer, so that a state consist of $(2I + 1)$ bits, the first representing a move,

included in what follows mutates integer arrays that represent DFAs of numStates% states by checking whether the current position in the string represents a move or a transition. A move is mutated with probability pmutM# while a character representing a state transition is chosen for mutation with probability pmutT#. In fact, a state transition may not be mutated with probability 1/numStates% even if chosen for mutation, as it may simply be assigned to its original value.

The sub-program playgame(e%,f%) requires us to first determine moves for the current iteration, and then determine the next state for each player. Payoffs are updated in the same manner as previously. The DFAs are, in fact, stored in an array of integers, which makes crossover a little messy, involving looping through the player's chromosomes instead of calls to QuickBASIC's MID$() function. Since the IPD matrix we have been dealing with has non-negative values, we could implement roulette-wheel selection, but retain tournament selection.

```
DECLARE FUNCTION randPlayer% ()
DECLARE SUB mutate ()
DECLARE SUB stars ()
DECLARE SUB newpop ()
DECLARE SUB clean ()
DECLARE SUB initpop ()
DECLARE SUB playgame (a%, b%)
CONST xx# = 3#
CONST yy# = 1#
CONST xy# = 0#
CONST yx# = 4#
CONST numPlayers% = 30
CONST numGens% = 5000
CONST numStates% = 8
CONST numIts% = 150
CONST pmutT# = .003#
CONST pmutM# = .001#
DIM SHARED fsm%(numPlayers%, numStates% * 3)
DIM SHARED pay#(numPlayers%)
DIM SHARED state%(2)
CALL initpop
'=================================================================
                         MAIN LOOP START
'=================================================================
FOR g% = 1 to numGens%
   CALL clean
   FOR i% = 1 TO numPlayers% - 1
      FOR j% = iX + 1 TO numPlayers%
         state%(1) = 0 'Always start here, but could make this
         state%(2) = 0 'player-dependent, and mutable.
         FOR k% = 1 TO numIts%
```

---

the next l the state entered if the opponent cooperated, the tail l bits of a state the state entered if the opponent defected. In this way bit-flipping always produces well-formed strings.

```
            CALL playgame(i%, j%)
         NEXT k%
      NEXT j%
   NEXT i%
   CALL newpop
   CALL routate
NEXT g%
'===============================================================
                          MAIN LOOP END
'===============================================================




'===============================================================
                          START initpop
'===============================================================
SUB initpop
' Check whether creating a random move or a random transition.
'
FOR i% = 1 TO numPlayers%
   FOR j% = 1 TO numStates% * 3
      IF (j% - 1) / 3 = INT((j% - 1) / 3) THEN  ' A move.
         fsm%(i%, j%) = INT(RND * 2)
      ELSE fsm%(i%, j%) = INT(RND * numStates%) ' A transition.
      END IF
   NEXT j%
NEXT i%
END SUB
'===============================================================
                          END initpop
'===============================================================

'===============================================================
                          START newpop
'===============================================================
SUB newpop
DIM ch%(numPlayers% , numStates% * 3)
DIM parent%(2)
                 .
                 .
                 .
FOR i% = 1 to numPlayers% / 2
                 .
                 .
                 .
'We now have the two parents, indexed by parent(1) and
parent(2).

   pc% = INT(RND * numStates% * 3+1) 'Crossover point, and do it:
   FOR j% = 1 TO pc%
      ch%(i%, j%) = fsm%(parent%(i), j%)
      ch%(i% + numPlayers% / 2, j%) = fsm%(parent%(2), j%)
   NEXT j%
   FOR j% = pc% to numStates%*3
```

```
      ch%(i%, j%) = fsm%(parent%(2), j%)
      ch%(i% + numPlayers% / 2, j%) = fsm%(parent%(1), j%)
   NEXT j%
NEXT i%
'
' Copy children into next population:
'
FOR i% = i TO numPlayers%
   FOR j% = 1 TO numStates% * 3
      fsm%(i%, j%) = ch%(i%, j%)
   NEXT j%
NEXT i%
END SUB
'=====================================================================
                              END newpop
'=====================================================================

'=====================================================================
                             START playgame
'=====================================================================
'
' Just careful to pick off correct components of DFA chromosome.

SUB playgame (e%, f%)
   movee% = fsm%(a, state%(1) * 3 + 1)            ' Pick-off moves.
   movef% = fsm%(b, state%(2) * 3 + 1)
'                                             Do state transitions:
   state%(2) = fsm%(b, state%(2) * 3 + 3 - movee%)
   state%(1) = fsm%(a state%(1) * 3 + 3 - movef%)
                        .
                        .
                        .
END SUB
'=====================================================================
                              END playgame
'=====================================================================

'=====================================================================
                              START mutate
'=====================================================================
'
' Again, need to check if muteting a move or a transition.

SUB mutate
   FOR i% = 1 TO numPlayers%
      FOR j% = 1 TO numStates% * 3
         IF (j% - 1) / 3 = INT((j% - 1) / 3) THEN
            ' Mutate a move gene:
            IF RND < pmutM# THEN
               fsm(i%, j%) = 1 - fsm(i%, j%)
            END IF
         ' Or a transition gene:
         ELSEIF RND<pmutT# THEN
            fsm(i%, j%) = INT(RND * numStates%)
         END IF
```

```
      NEXT j%
   NEXT i%
END SUB
END mutate
```

13.8.1 Results

An important factor is the total number of games played, since the emergence of TFT-like players is aided if they can compensate for the loss against a defector in the first iteration of the game by a long period of mutual cooperation. Once TFT players (or somewhat similar players) emerge by mutation, they rapidly dominate the population. However, should a mutation cause such a player to indiscriminately cooperate, an opportunistic defector can gain a significant advantage.[32] It might be argued that a fair bit of parameterisation in these algorithms has been undertaken to overtly generate cooperation, which certainly does emerge in many circumstances. In fact, Nachbar [1988] argues that such parameterisation questions the robustness of results. With this in mind, we note that under certain circumstances, the population undergoes a sequence of *punctuated equilibria,* alternating between apparently stable populations of one kind or another. Prolonged periods of cooperation *can* be generated, see the references for examples, but must be searched for. Rather than show these, Figure 13.9 shows a punctuated equilibrium example, notice the small encounter lengths and relatively small tournament size. Further parameterisations can be found in the literature, the reader might like to try other examples. As the encounter length increases, prolonged periods of cooperation become more likely, but are *not* stable. Very short encounter lengths will result in almost perpetual defection. In the middle ground are these interesting dynamics. Notice that it is improper to introduce some *a priori* definition of convergence to study these algorithms.

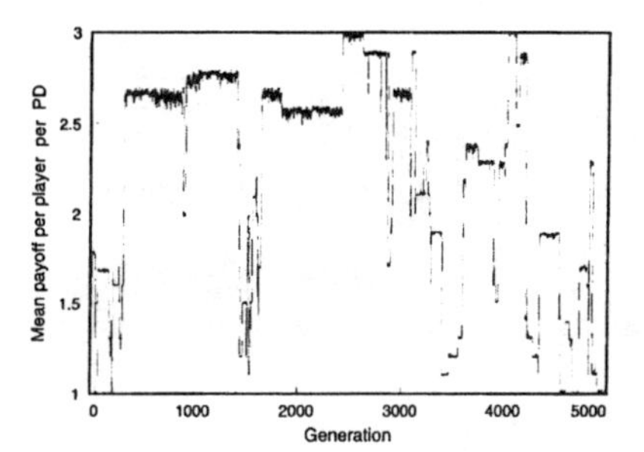

[32] Note that Boyd and Lorderbaum [1987] show that no *pure* strategy is evolutionary stable in the IPD. We have not yet considered DFAs that implement a mixed strategy, but is is clear how tiffs might work: a chromosome is the same except for a floating point entry representing the probability of playing C in each iteration. The crossover of these portions can be implemented as the random convex combination we have already described.

Figure 13.9: Sample results for a population of 200 20-state DFA's, with 6 used as tournament size and encounter length 13. Show is the mean payoff per PD per player, with 3 representing mutual cooperation.

13.9 Variations on the Basic Algorithm

We must allow Nachbar [1988] his point, but variations on the basic algorithm are usually undertaken with an eye to modeling the evolution of cooperation, not the evolution of defection. It's also impossible to describe all the variations on the basic simulations presented here, but here's a few briefly described.

13.9.1 Mistakes

There is some interest in the emergence of cooperation in the presence of "noise," for example Miller [1988], Nalebuff [1987], Boyd [1989]. There are two types of noise we might consider. The first is to have the players misimplement their intended move with a fixed probability (Lindgren [1991]), say `pMisImp#`. Having this at 0.5 makes all play random, having it at 1 redefined the allele for cooperation to 0 and defection to 1, other value represents genuine noise. It makes sense, then, to have such a value lie between 0 and 0.5. Coding for this:

```
SUB playgame (e%, f%)
   movee% = fsm%(a, state%(1) * 3 + 1)
   movef% = fsm%(b, state%(2) * 3 + 1)
   IF RND < pMisImp# THEN movee% = i - movee%
   IF RND < pMisImp# THEN movef% = i - movef%
END SUB
```

Alternatively, we can have the mistake lie in interpreting what the other player did: with probability `probMisint#` a player makes the incorrect transition, once payoffs have been incremented:

```
IF RND < pMisImp# THEN movee% = 1 - movee%
IF RND < pMisImp# THEN movef% = 1 - mover%
state%(2) = fsm%(b, state%(2) * 3 + 3 - movee%)
state%(1) = fsm%(a, state%(1) * 3 + 3 - movef%)
```

Again, a probability of 0.5 makes transitions purely random, and a probability of one is the same as a change of gene interpretation for transitions.

13.9.2 Continuous Action Spaces

Fogel and Harrald [1994] consider a variant of the PD in which agents choose from the interval, $A_i = [-1, 1]$. In any one iteration, if a player 1 chooses $a_1 \in [-1, 1]$ and a player 2 $a_2 \in [-1, 1]$, then the payoff to A is given by:

$$P(a_1,a_2) = 2.25 + 1.75a_2 - 0.75a_1.$$

This one-shot payoff function retains the basic structure of the discrete PD. A play of -1 is 'complete' defection, and mutual complete defection is the only Nash equilibrium when this game is played once. A play of +1 is complete cooperation, and mutual complete cooperation maximises joint payoffs.

The players in this game are represented by artificial *neural networks,* NNs.[33] An example architecture is shown in Figure 13.10.

The NNs are activated by six values: their own actions for the previous three interactions, shown as SELF[t - i], and the actions of their opponent over same period, OPP[t - i], i = 1,...,3 (the *activation values).* Each node in the hidden layer takes a weighted sum of the inputs, offset by a bias. If we label the activation values (α_1, α_2, α_3, α_4, α_5, α_6), then a hidden node j receives the term

$$x_j = \left(\sum_{i=1}^{6} w_{1,i} \alpha_i \right) + b_j$$

which is processed via the non-linear sigmoid filter

$$f(x_j) = 2[(1 + \exp(-x_j))^{-1} - 0.5]$$

before being passed as part of an offset weighted sum to the output node. Hence, an NN with two hidden nodes passes the following term to the output node:

---

[33] Harrald [1994b] considers the use of NNs in the setting of dynamic oligopoly in economics. This paper, work in progress, uses *evolution strategies,* see Back, Rudolph and Schwefel [1993] for example, to model a monopolist learning about a demand curve, and duopolists predicting one another's behaviour. The ES proceeds by generating a random NN, and using it to decide on production levels. intermittently, a player generates another NN by mutating the existing one. If the mutation *would have* done better than the existing one did, it is adopted, otherwise it is discarded. With many players, a GA is implemented much like the one described here.

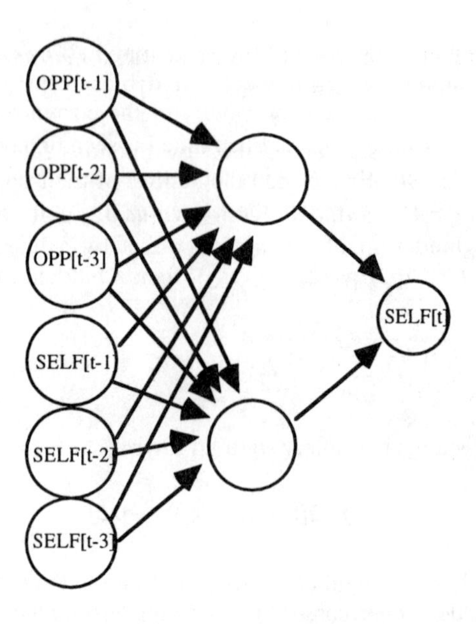

Figure 13.10: Feedforward perceptron used in Fogel and Harrald [1994] for the continuous PD.

$$y = W_1 f(x_1) + W_2 f(x_2) + b_3,$$

yielding a final output of

$$\mathrm{SELF}[t] = 2[(1 + \exp(-y))^{-1} - 0.5],$$

which always lies in the range (-1, 1). Hence, an NN player with 2 hidden nodes is described by 17 parameters, 14 weights and 3 bias terms:

$$\mathrm{NN} = (w_{1,1},\ldots,w_{6,2}, W_1, W_2, b_1, b_2, b_3)$$

Fogel and Harrald [1994] did *not* use a GA to evolve this population, rather an *evolutionary program,* EP, was used. EPs were developed around the same time as GAs and evolution strategies by L.J. Fogel and colleagues to address much the same purpose. Early work operated on population of DFAs to evolve predicting mechanisms.[34] The main difference between a GA and an EP is that while the crossover operator is thought to be essential in the former, there is no such operator applied in an EP. It is uncommon to see roulette-wheel selection used in EPs also. Successive populations are generated most often by tournament selection, or something similar. For example, Fogel and Harrald [1994] use a technique whereby the worst performing half of the population are deleted, and

[34] See Fogel, Owens and Walsh [1966], and Fogel [1992, 1993b, 1993c] for overviews, and an example of the power of these algorithms can be found in Fogel [1994] and the references there.

replaced by mutating each member of the remaining half by adding a Gaussian (Normal) random deviate to each weight and bias term (creeping). Other techniques include matching each member *i* of the current population with a sequence of randomly selected others. Each time *i* is seen to have a better fitness, it scores one 'win.' The number of wins then serves as a basis for selection in some manner. The reader is referred to the references for further details.[35]

However, should a GA be applied, the list of weights and biases form floating-point representation of each player, which can each be recombined in a fashion described previously (Section 13.5).

13.9.3 Players as Prologue and Lisp Code

In some fascinating related work, Danielson [1994] and Fujiki and Dickinson [1987] represent players as source code, the former in Prolog, the latter in Lisp. In more recent work, the techniques are those, therefore, of {\it genetic programming}, after Koza [1992]. This allows some interesting strategies to emerge. For example, Danielson [1994] considers players who can perform the equivalent of the `UNIX DIFF` on one another, and have an action specified in the PD depending on the result. *Many* strategies are considered, for example it is also possible to have the player make a call to another's code. The reader is referred to these works for further details.

13.9.4 Spatially Distributed Agents

Axelrod [1984], Nowak and May [1992] and Routledge [1993] consider agents who are distributed on some space. This space is a grid, with agents interacting only with immediate neighbours in an IPD. These simulations tend to have the same interesting spatio-temporal patterns that many cellular automata do, such as the well known "game of life".[36] Populations in Routledge [1993] for example, consist of players encoded directly as bit-strings as described in Section 13.6.1, who play their four neighbours in the IPD. The worst performing few players are replaced each generation using crossover, the other players remain the same. Depending on parameterisation various patterns emerge, including frozen areas of defection and cooperation, cyclical patterns, frozen evolution and more.[37]

13.9.5 Social Dilemmas

---

[35] The advantage of EPs is that since they rely on mutation only, there can be very general formulations of the structures that represent players. For example, it is conceivable to construct populations of DFAs for whjch a mutation included adding or deleting a state (as was done in early work by L.J. Fogel), and its not clear how crossover can be pelionned on two DFAs of differing sizes.

[36] Bak, Chen and Creutz [1989].

[37] I would like to thank Bryan for Ms discussion and demonstration of this simulation.

In Turbo Pascal, Bryan makes his code available. Currently (until the end of 1994 I should think) reachable at the Faculty of Commerce, the University of British Columbia, ROUT@phdlab. commerce. ubc. ca.

The 2 player IPD has been generalised to a multi-player (the MPD) context several times. Schelling [1976] describes the clearest generalisation, one such that there are 2 actions, *C* and *D*, but *n* players. A player's payoff depends on his own action, and the number of cooperators and defectors. Defection is a dominant move in the one-shot game.[38] The multi-player PD has just one Nash equilibrium played once, for all players to defect.[39] The game is then completed by specifying the size of a subset of defectors required such that if they all changed to cooperation, it would increase payoffs to all of that subset (and everyone else). It's hard to imagine an evolutionary process in which all players can do is play *C* or *D* that could foster cooperation. In the 2 player game, TFT-like players gain a *relative* advantage over defectors since they only get 'ripped-off' once by defectors, but enjoy many rounds of cooperation with one another (isolated from the rest of the population). In the MPD, however, every act of cooperation increases all player's fitnesses, including defectors. While cooperation many benefit the group, our simulations have fitness described at the individual level (in the process of selection). Nevertheless, studies of what are loosely termed 'social dilemmas' do exhibit cooperation.[40] If, however, agents have the ability to ostracise defectors, then it is possible for cooperation to emerge.[41]

The issue of player representation in multi-player games is fairly straightforward. For DFAs we could have transitions based on the fraction of players adopting one of the available moves falling into some set that is part of a partition of [0, 1] (see Marks [1989]). For NNs, just the raw input, lagged, would suffice. The output layer would resemble those used for character recognition, specifying an element of A_i.

13.9.6 Reputation Effects in the IPD

One form of ostracism considered in Stanley, Ashlock and Tesfatsion [1994] is for players to be able to refuse encounters with players they don't wish to play with.[42] Stanley, Ashlock and Tesfatsion [1994] have players remember specific interactions between themselves and the others, and on this basis update

---

[38] Games that can be expressed this way have been discussed in various contexts in Harrald [1993]

[39] This being the *tragedy of the commons,* after the well known Hardin [1968]. More general games involving many players often have many equilibria played once, and there is a consequent coordination problem among players. For example, these games may have equllibria like *Nx* players do *x, Ny* players do y. The problem is that in the absence of focal points or conventions, the players can't tell *who* should play x and who y. On many occasions social institutions emerge to solve these coordination problems (public lotteries), and sometimes, as with the adoption of language, it is an evolutionary process.

[40] Fader and Hauser [1988], Marks [1989a], Glance and Huberman [1993a, 1993b, 1994] for example.

[41] Hirshleifer and Rasmusen [1989].

[42] See also Kitchef [1992] and Dugatkin [1991]

expectations about payoffs against them again. There are given payoffs to refusal, which are compared to expected payoffs when interacting with others. Players refuse to play when the expected payoff when playing is less than that of refusal.

One way to endogenise the refusal payoff in such a simulation, is to limit the total number of PDs each player can play per generation. For example, consider players randomly moving on a grid, one move per tick of a clock. If players are adjacent they can engage in an IPD, one PD per tick of the clock, and remain stationary. There are a fixed number of ticks of the clock per generation. When adjacent and playing a IPD it is possible that a third player becomes adjacent to either engaged in the IPD. Players are characterised by a DFA for play in the IPD, and a criterion for a) offering to play the IPD, b) disengaging from an IPD, c) a special case of b) being whether to accept an offer initially, and whether to maintain an ongoing IPD when another offer comes along. To make these decisions each player is endowed with a 'reputation,' something along the lines of the percentage of the time they have played *C* in the past for example. Even if all the payoffs in the PD are positive, it still may pay to refuse some offers, even though time is running out, if players aren't too sparse. This because there is a chance of meeting a cooperator (either to go along with or to rip-off). The problem with such a design is the resetting of reputation of each generation, as it is undefined initially for example. Further research is ongoing on this matter.

13.9.7 The Evolution of Coordination

There is some work in progress that looks at players interacting in games that require coordination, like that in Figure 13.11.

	X	Y
X	[0,0]	[g,h]
Y	[h,g]	[0,0]

Figure 13.11: A coordination game, $h,g > 0$

Of course, space does not allow a discussion of signalling in the biological literature. In the repeated game context, Harrald [1994c] considers players evolving a simple language to announce their intentions when meeting to play this game once against opponents. The players have a two word vocabulary, say {fnord, foo} and when they meet Nature chooses one of them to speak. When chosen to speak, a player utters "fnord" with probability π_i. Players are characterised by four other probabilities: if they hear each of the two words, the probability they play *X*, and if they say each of the two words, the probability they play *X*. Evolved in a GA (using random convex-combination), language quickly allows the required coordination as words, interpretations and actions become correlated. This is similar to correlated equilibrium in games whereby players coordinate activity by using the same randomising device in playing mixed-strategies. Current research now focuses on a game like Figure 13.12, which allows the potential for deception in language (there are many variants on this particular game that do the job-encourage lying).

	X	Y
X	[4,0]	[2,2]
Y	[2,2]	[0,4]

Figure 13.12: Another coordination game.

References

Arthur, W.B. "On Designing Economic Agents that Behave Like Human Beings," *Journal of Evolutionary Economics* 3, pp. 1-22, 1993.

Arthur, W.B. "On Learning and Adaptation in the Economy," Working Paper 92-07-038, Santa Fe Institute Economics Research Program, 1992.

Arifovic, J. "Genetic Algorithm in the Overlapping Generations Economy" *mimeo.,* McGill University 1992a.

Arifovic, J. "Genetic Algorithm in the Cobweb Model" *mimeo.,* McGill University, 1992b.

Aumann, R. "Repeated Games," in G.R. Feiwel (ed.) *Issues in Contemporary Microeconomics and Welfare,* London: Macmillan, 1985.

Axelrod, R. "Artificial Intelligence and the Repeated Prisoner's Dilemma," Discussion Paper 120, Institute for Public Policy Studies, University of Michigan, 1978.

Axelrod, R. "The Emergence of Cooperation Among Egoists," *American Political Science Review,* 75: 306-318, 1981.

Axelrod, R. *The Evolution of Cooperation.* New York: Basic Books, 1984.

Axelrod, R. In *Genetic Algorithms and Simulated Annealing,* edited by D. Davis, pp. 32-42, London: Pitman, 1987.

Axelrod, R. and E. Dion "The Further Evolution of Cooperation," *Science* 242, pp. 1385-1390, 1988.

Axelrod, R. and W.D. Hamilton "The Evolution of Cooperation" *Science* 211, pp. 1390-1396, 1981.

Back, T., G. Rudolph and H-P. Schwefel "Evolutionary Programming and Evolution Strategies: Similarities and Differences," in *Proceedings of the Second Annual Conference on Evolutionary Programming* D.B. Fogel and W. Atmar (editors), La Jolla CA: Evolutionary Programming Society, 1993.

Bak, P., K. Chen and M. Creutz "Self-Organised Criticality in the 'Game of Life,'" *Nature* 342, pp. 780-782, 1989.

Boyd, R. "Mistakes Allow Evolutionary Stability in the Repeated Prisoner's Dilemma" *Journal of Theoretical Biology,* 136, 47-56, 1989.

Binmore, K.G. and P. Dasgupta "Game Theory: A Survey," in K.G. Binmore and P. Dasgupta (eds.) *Economic Organisations as Games,* Oxford: Basil Blackwell, 1986.

Binmore, K.G. and L. Samuelson "Evolutionary Stability in Repeated Games Played by Finite Automata," *Journal of Economic Theory* 57, pp. 278-305, 1992.

Boyd, R. and J .R. Lorderbaum "No Pure Strategy is Evolutionary Stable in the Repeated Prisoner's Dilemma Game," *Nature* 327, pp. 58-59, 1987.

Cohen, M.D and R. Axelrod "Coping with Complexity: the Adaptive Value of Changing Utility," *American Economic Review* 74, pp. 30-42, 1984.

Danielson, P. *Artificial Morality: Virtuous Robots for Virtual Games,* New York: Routledge, 1992.

Danielson, P. "Artificial Morality" mimeo, University of British Columbia, 1994.

Dugatkin, L.A. "Rover: A Strategy for Exploiting Cooperators in a Patchy Environment," *The American Naturalist* 138, pp. 687-693, 1991.

Eshelman, L.J. and J.D. Schaffer "Real-Coded Genetic Algorithms and Interval Schemata," in L.D. Whitley (ed.) *Foundations of Genetic Algorithms, 2,* Morgan Kaufmann, 1993.

Fader, P. and J. Hauser "Implicit Coalitions in a Generalised Prisoner's Dilemma," *Journal of Conflict Resolution* 32, pp. 553-582, 1988.

Fischer, E. "The Relationship Between Mating System and Simultaneous Hermaphrodism in Coral Reef Fish *Hypolectrus Nigricans* (Serranidae)," *Animal Behavior* 28, pp. 620-633, 1981.

Fischer, E. "Sexual Allocation in the Simultaneously Hermaphroditic Coral Reef Fish," *American Naturalist* 117, pp. 620-633, 1981.

Fogel, D.B "Evolving Behaviours in the Iterated Prisoner's Dilemma," *Evolutionary Computation* 1(1), pp. 77-97, 1993a.

Fogel, D.B. "Applying Evolutionary Programming to Selected Control Problems," *Computers Mathematical Applications* 12(11), pp. 89-104.

Fogel, D.B. "On the Philosophical Differences Between Evolutionary Algorithms and Evolutionary Programming," in *Proceedings of the Second*

Annual Conference on Evolutionary Programming D.B. Fogel and W. Atmar (editors), La Jolla CA: Evolutionary Programming Society, 1993b.

Fogel, D.B. "An Introduction to Simulated Evolutionary Optimisation," in *Proceedings of the Second Annual Conference on Evolutionary Programming* D.B. Fogel and W. Atmar (editors), La Jolla CA: Evolutionary Programming Society, 1993c.

Fogel, D.B. *Evolving Artificial Intelligence,* Doctoral Dissertation, University of California, San Diego, 1992.

Fogel, D.B. and P. Harrald "Evolving Continuous Behaviours in the Iterated Prisoner's Dilemma," in A.V. Sebald and L.J. Fogel (eds.) *Proceedings of the Third Annual Conference on Evolutionary Programming,* River Edge N J: World Science Publishers, 1994.

Fogel, L.J. "Autonomous Automata," *Industrial Research,* 4, pp. 1419, 1962a.

Fogel, L.J. "Toward Inductive Inference Automata," *Proceedings of the International Federation for Information Processing Congress,* Munich, 1962b.

Fogel, L.J., A.J. Owens and M.J. Walsh *Artificial Intelligence Through Simulated Evolution,* New York: John Wiley, 1966.

Friedman, D., "Evolutionary Games in Economics," *Econometrica* 59 (4), pp. 637-666, 1991.

Fujiki, C. and J. Dickinson "Using the Genetic Algorithm to Generate Lisp Source Code to Solve the Prisoner's Dilemma," in J.J. Grefensette (ed.) *Genetic Algorithms and Their Applications,* Hillsdale NJ: Lawrence Erlbaum Assoc, 1987.

Fundenberg, D. and J. Tirole *Game Theory,* Cambridge MA: MIT Press, 1993.

Glance, N. and B. Huberman "Organisational Fluidity and Sustainable Cooperation," in K. Carley and M. Prietula (eds.) *Computational Organisation Theory,* Lawrence Erlbaum Associates, 1993a.

Glance, N. and B. Huberman "The Outbreak of Cooperation," *Journal of Mathematical Sociology,* 17, 1993b.

Glance, N. and B. Huberman "Dynamics of Social Dilemmas," *Scientific American* March 1994.

Hamilton, W.D. "Extraordinary Sex Ratios," *Science, Wash.* 156, pp. 477-488, 1967.

Hardin, G. "The Tragedy of the Commons," *Science* 162, pp. 1243-1248, 1968.

Harrald, P. *Numbers Externalities: n-Player Anonymous Symmetric Games,* unpublished Ph.D. Dissertation, Simon Fraser University, 1993.

Harrald, P. "Artificial Intelligence for Modeling Adaptation in Economics: Observations and Speculations," in A.V. Sebald and L.J. Fogel (eds.) *Proceedings of the Third Annual Conference on Evolutionary Programming,* River Edge NJ: World Science Publishers, 1994a.

Harrald, P. "Learning in Dynamic Oligopoly by Artificial Neural Networks," paper presented at the meetings of the Canadian Economic Association, Calgary, 1994b.

Harrald, P. "A Simple Exposition of the Evolution of Coordination Devices" *mimeo,* 1994c.

Harrington, J.E. '"Finite Rationalisability and Cooperation in the Finitely Repeated Prisoner's Dilemma," *Economic Letters* 23, pp. 233 237, 1987.

Hirshleifer, J. and E. Rasmusen *"Cooperation* in a repeated Prisoner's Dilemma with Ostracism," *Journal of Economic Behaviour and Organisation* 12, pp. 87-106, 1989.

Holland, J. and J. Miller "Artificial Adaptive Agents in Economic Theory," *American Economic Review* 81, Papers and Proceedings, pp. 365-370, 1991.

Hopcroft, J. and R. Ullman *Introduction to Automata Theory, Languages and Computation,* Reading MA: MIT Press, 1979.

Janikow, C.Z. and Z. Michalewicz "An Experimental Comparison of Binary and Floating Point Representations in Genetic Algorithms," in R.K. Belew and L.B. Booker (eds.) *Proceedings of the Fourth International Conference on Genetic Algorithms,* Morgan Kaufmann, 1991.

Kitchef, P. "Evolution or Altruism in Repeated Optional Games," Working Paper, Department of Philosophy, UCSD, July 1992.

Koza, J. *Genetic Programming,* Cambridge MA: MIT Press, 1992.

Kreps, D., P. Milgroin, J. Roberts and J. Wilson "Rational Cooperation in the Finitely Repeated Prisoner's Dilemma," *Journal of Economic Theory* 27(2), pp. 326-337, 1982.

Leonard, J. "The Hermaphrodite's Dilemma," *Journal of Theoretical Biology* 147, pp. 361-372, 1990.

Leonard, J. and K. Lukowiak "Courtship, Copulation, and Sperm Trading in the Sea Slug *Navanax Inermis* (Opisthaobranchia: Cephalaspidea)," *Canadian Journal of Zoology* 63, pp. 2719-2719, 1985.

Lewontin, R.C. "Evolution and the Theory of Games," *Journal of Theoretical Biology* 1, pp. 382-403, 1961.

Lindgren, K. "Evolutionary Phenomena in Simple Dynamics," in *Artificial Life H: SFI Studies in the Sciences of Complexity, Vol X,* edited by C.G. Langton, C. Taylor, J.D. Farmer, & S. Rasmussen, pp. 295-312, Addison-Wesley 1991.

Lloyd, D.G. "Genotypic and Phenotypic Models of Natural Selection," *Journal of Theoretical Biology* 69, pp. 543-560, 1977.

Luce, R. and H. Raiffa *Games and Decisions,* New York: Wiley, 1957.

McCauley, E., W. Wilson and A. De Roos "Dynamics of Age Structured and Spatially Structured Predator-Prey Relationships: Individual Based Models and Population-Level Formulations," *The American Naturalist* 142 (3), pp. 412-442, 1991.

Malaith, G.J. "Introduction: Symposium on Evolutionary Game Theory," *Journal of Economic Theory,* 57. pp. 259-277.

Marimon, R., E. McGrattan and T. Sargent "Money as a Medium of Exchange in an Economy with Artificially Intelligent Agents," *Journal of Economic Dynamics and Control* 14, pp. 329-373, 1990.

Marks, R. "Breeding Hybrid Strategies: Optimal Behaviour for Oligopolists" in J.D. Schaffer (ed.) *Proceedings of the Third International Conference on Genetic Algorithms,* George Mason University, 1989a.

Marks, R. "Niche Strategies: The Prisoner's Dilemma Computer Tournaments Revisited," Australian Graduate School of Business, Working Paper 89-009, 1989b.

Matsuda, H. "A Game Analysis of Reciprocal Cooperation: Sequential Food Sharing and Sex Role Alternation," *Journal of Ethology* 7, pp. 198-207, 1989.

Maynard-Smith, J. *Evolution and the Theory of Games,* Cambridge: CUP, 1982.

Maynard-Smith, J. and G.R Price "The Logic of Animal Conflict," *Nature, London,* 246, pp. 15-18, 1976.

Michalewicz, Z. and C.Z. Janikow "Handling Constraints in Genetic Algorithms," in R.K. Belew and L.B. Booker (eds.) *Proceedings of the Fourth International Conference on Genetic Algorithms,* Morgan Kaufmann, 1991.

Miller, J.H. "The Evolution of Automata in the Repeated Prisoner's Dilemma," *mimeo,* Department of Economics, University of Michigan, August 1988.

Miller, J.H. "The Coevolution of Automata in the Repeated Prisoner's Dilemma" Santa Fe Institute Working Paper 89-103, 1989.

Moore, E.F. "Gedanken-Experiments on Sequential Machines," *Annals of Mathematical Studies* 34, pp. 129-153, 1957.

Nachbar, J.H. "The Evolution of Cooperation Revisited," *mimeo.,* Santa Monica: RAND Corp., June 1988.

Nash, J. "Equilibrium Points in n-Person Games," *Proceedings of the National Academy of Sciences, USA* 36, pp. 286-295, 1950.

Nalebuff, B. "Economic Puzzles: Noisy Prisoners, Manhattan Locations and More," *Journal of Economic Perspectives* 1, pp. 185-191, 1987.

Nelson, and S. Winter *An Evolutionary Theory of Economic Change,* London: Macmillan, 1982.

Neyman, A. "Bounded Complexity Justifies Cooperation in the Finitely Repeated Prisoner's Dilemma," *Economic Letters* 19, pp. 227-229, 1985.

Nowak, M. and R. May "Evolutionary Games and Spatial Chaos" *Nature* 349, pp. 826-829, 1992.

Rapoport, A. and A. Chammah *Prisoner's Dilemma,* Ann Arbor: University of Michigan Press, 1965.

Rasmusen, E. *Games and Information,* Oxford: Basil Blackwell, 1991.

Routledge, B. "Co-Evolution and Spatial Interaction," *mimeo,* University of British Columbia, 1993.

Rubinstein, A. "Finite Automata Play the Repeated Prisoner's Dilemma," *Journal of Economic Theory* 39, pp. 83-96, 1986.

Samuelson, L. "Evolutionary Foundations of Solution Concepts for Finite, Two-Player, Normal Form Games," *mimeo,* Department of Economics, Penn State University, 1988.

Schelling, T. *Micromotives and Macrobehaviour,* New York: Norton, 1976.

Slobodkin, L.B. and A. Rapoport "An Optimal Strategy of Evolution," *Quarterly Review of Biology* 49, pp. 181-200, 1974.

Sober, E. "The Evolution of Altruism: Correlation, Cost, and Benefit," *Biology and Philosophy* 7, pp. 177-187, 1992.

Stanley, E., D. Ashlock and L. Tesfatsion "Iterated Prisoner's Dilemma with Choice and refusal of Partners," in C. Langton (ed.) *Artificial Life 111,* Addison-Wesley, 1994.

Tucker, A. "A Two-Person Dilemma" *mimeo.* Stanford University. Reprinted in Philip Straffin "The Prisoner's Dilemma" *UMAP Journal 1,* pp. 101-103, 1950.

von Neumann, J. and O. Morgenstern *The Theory of Games and Economic Behaviour,* Princeton: Princeton University Press, 1944.

Wilkinson, G.S. "Reciprocal Food Sharing in the Vampire Bat," *Nature* 308, pp. 181-184, 1984.

Witt, U. (ed.) *Evolutionary Economics,* Cheltenham: Edward Elgar Publishers, 1993.

Appendix A: ga-test.cfg

```
# User data tile
#   This information is not used by the GA, however, it is a
#   convenient way to input a data file name or other information to
#   your application.
#
# user_data datafile
#
# Seed for random number generator
#
# Usage: rand_seed my_pid
#   rand_seed number

#   my_pid   = use system pid as random seed
#   number   = seed for random number generator, a positive integer

# DEFAULT: rand_seed i

# rand_seed my_pid
# rand_seed

# The data type of the allele

# Usage: datatype [bit l int l int_perm l real]

#   bit       = bit string
#   int       = integers
#   int_perm  = permutation of integers
#   real      = real numbers

# DEFAULT: int_perm

# datatype   bit
# datatype   int
# datatype   int_perm
# datatype   real


# How to initialize the pool

# Usage: initpool [random l from_file filename l interactive]

#   random           = generate at random based on
#        datatype, chrom_len, & pool_size
#   from_file        = read from a file
#   filename         = the name of the file to read from
#   interactive      = read from stdin
```

```
# DEFAULT:  initpool  random

# initpool     random
# initpool     from_file    initpool.dat
# initpool     interactive

# Chromosome length, needed when "initpool random" selected

# Usage: chrom_len length

#  length = chromosome length, a positive integer

# DEFAULT: chrom_len 10

# chrom_len 25

# Pool size, needed when "initpool random" selected

# Usage: pool_size size

#  size = pool size, a positive integer

# DEFAULT; 100

# pool_size 200

(*** edit here

# When to stop the GA

#    Convergence means when the variance = O, or equivalently, when all
#    the fitness values in the pool are identical.

#    Iterations means the number of generations for the generational model
#    and the number of trials for the steady state model. Numbers must
#    be given as positive integers. It takes roughly pool_size/2
#    iterations of the steady state model to equal one iteration of the
#    generational model.

# Usage: stop_after convergence
#   stop_after number [use_convergence | ignore_convergence]

# convergence                  - stop when the GA converges
# number                       - stop after specified number of iterations
# use_convergence            - will stop early if GA converges (default)
# ignore_convergence    - WILL NOT stop early even if GA converges

# DEFAULT: stop_after convergence
```

```
# stop_after  convergence
# stop_after  500
# stop_after  5O0 use_convergence
# stop_after  5O0 ignore_convergence

# GA  Type:
#
# Usage: ga [generational I steady_state]

#       generational = generational GA
#       steady_state = steady-state GA
#
# WARNING: This directire has the following side effects:

#               GA  type          Directives set as a side effect
#               generational      selection          roulette
#                                 replacement        append
#                                 rp_interval        1

#               steady-state      selection          rank_biased
#                                 replacement        by_rank
#                                 rp_interval        100

# DEFAULT:  ga  generational

#    ga generational            # most commonly used
#    ga  steady_state              # used by Genitor


# Generation  gap:

#  The generation gap represents a percentage of the population to copy
#  (clone) to the new pool at each generation. This only makes sense in
#  a GA with two pools as in the generational model. A gap of 0.0 is the
#  traditional generational algorithm. As the gap increases, it becomes
#  more like a steady-state algorithm. A gap of 1.0 essentially disables
#  crossover since only reproduction occurs.

# Usage: gap number

#  number = generation gap, valid range = [0.0 .. 1.0]

# DEFAULT: gap 0.0

#  gap  0.3

# Selection  method:
```

```
# Usage: selection [roulette | rank_biased | uniform_random]

#  roulette        = Roulette wheel
#  rank_biased     = Ranked, biased selection as in Genitor
#  uniform_random       = Pick one at random

# DEFAULT: selection roulette

# selection roulette          # use with generational GA
# selection rank_biased       # use with steady-state GA
# selection uniform_random    # experimental


# Selection bias

# Usage: bias number

#  number = selection bias, valid range = [1.0 .. 2.0]
#             only used for rank-biased selection

# DEFAULT: bias 1.8

# bias 1.1


#   Crossover   method:
#
# Usage: crossover  [simple | uniform | order1 order2 position | cycle |
#                                  pmx |uox | rox | asexual]

#  simple    =  children get alternate "halves" of parents
#  uniform   =  alleles snapped uniformly
#  order1    =  order based
#  order2    =  order based
#  position  =  order based
#  cycle  =  order based
#  pmx       =  order based
#  uox       =  uniform order
#  rox       =  relative order
#          asexual         =    snap  two  alleles

# DEFAULT: crossover order1


#crossover   simple
#crossover   uniform
#crossover   order1  use only with integer permutations
#crossover   order2  use only with integer permutations
```

```
#crossover   position   use only with integer permutations
#crossover   cycle   use only with integer permutations
#crossover   pmx     use only with integer permutations
#crossover   uox     use only with integer permutations
#crossover   rox           use only with integer permutations
#crossover   asexual

# Crossover Rate

# Usage: x_rate number

# number = crossover rate (percentage), valid range = [0.0..1.0]
#          A crossover rate of 0.0 disables crossover

# DEFAULT:  x_rate 1.0

# x_rate 0.6
```

 # Mutation method:

```
# Usage:  mutation  [simple_invert | simple_random | swap]

#  simple-invert = invert a bit
#  simple-random = random bit value
#  swap          = swap two alleles

# DEFAULT:  mutation  swap

# mutation   simple_invert      #use only with bits
# mutation   simple_random      #use only with bits
# mutation swap                 #use with any datatype
```

 # Mutation Rate

```
# Usage: mu_rate number

#  number = mutation rate (percentage), valid range = [0.0 .. 1.0]
#  A mutation rate of 0.0 disables mutation

# DEFAULT: mu_rate 0.0

# mu_rate O.1
```

```
# Replacement method:

# Usage: replacement [append | by_rank | first_weaker | weakest]

# append        = append to new pool, as in generational GA
# by_rank       = insert in sorted order, as in Genitor
# first_weaker  = replace first weaker found in linear scan of pool
# weakest       = replace weakest member of the pool

# DEFAULT: replacement append

# replacement    append        # use with roulette (generational GA)
# replacement    by_rank       # use with rank-biased (steady-state GA)
# replacement    first_weaker  # experimental
# replacement    weakest       # experimental

# Objective 0f GA:

# Usage: objective [minimize | maximize]

#   minimize = minimize evaluation function
#   maximize = maximize evaluation function

# DEFAULT:  objective  minimize

# objective      minimize
# objective      maximize



# Elitism

#       Elitism has two actions. For a generational GA, elitism makes two
#       copies of the best performer in the old pool and places them in the
#       new pool, thus ensuring the most fit chromosome survives. The other
#       action works with both models. In this case, elitism picks the best
#       two chromosomes from the parents and children. Thus, if a child is
#       not as fit as either parent, it will not be placed in the new pool.
#       Selecting elitism in LibGA performs both actions.

# Usage:elitism [true | false]

#      true   = ensure best members survive until next generation
#      false  =  no guarantee best will survive

# DEFAULT:  elitism  true

#   elitism  true
```

```
#   elitism  false

# Report type

# Usage: rp_type [none | minimal | short | long]

#   none       output nothing
#   minimal    output configuration and final result
#   short      output minimal + statistics only
#   long       output   short + dump pool

# DEFAULT:  rp_type  short

# rp_type  none
# rp_type  minimal
# rp_type  short
# rp_type  long

# Report  interval

# Usage:  rp_interval  number

#   number = interval between reports, a positive integer

# DEFAULT: rp_interval 1

#  rp_interval  10

# Output report filename

# Usage: rp_file file_name [file_mode]

#   file_name    = name of report file
#   file_mode    = optional file mode for fopen()
#      a       = append (DEFAULT)
#      w       =  overwrite

# DEFAULT: (write to stdout)

# rp_file  ga.out
# rp_file  ga.out  a
# rp_file  ga.out  w
```

Appendix B: Crossover Code

1 // random.hpp

```
// Routines that help in the selection of random numbers.
//

#ifndef RANDOM_HPP_INCLUDED
#define RANDOM_HPP_INCLUDED 1

#if defined (_AIX)
   extern "C" unsigned long random(void);
#endif

inline
unsigned long int
random(int x)
   // Return a random number from 0 to x - 1.
 {
   assert ( x > 0 );

   if ( x == 1 )
     {
    return 0;
     }
   else
     {
      return (random() % x);
     }
 }

double
random_real(void);
  // Return a random number in the interval [0,1).


extern
void
random_choose_n_fast(const int nb, const int max, int chosen[]);
  // Chose n random integers in the interval [0,max-1] which
  // are no duplicated.
  //
  // This algorithm uses a lot of memory but is fast when
  // nb is close to max.

void
random_choose_n_small(const int nb, const int max, int chosen[]);
  // Chose n random integers in the interval [0,max-1] which
  // are no duplicated.
  //
```

```
  // This algorithm has the potential to be very slow.  The
  // algorithm uses verly little memory, and should be
  // used when nb is much smaller than max.


inline
void
random_choose_n(const int nb, const int max, int chosen[])
   // Use a heuristic to choose the best method of
   // picking random numbers.
  {
   if ( (nb < 4) || (nb < max/10 ) )
     {
    random_choose_n_small(nb, max, chosen);
     }
   else
     {
    random_choose_n_fast(nb, max, chosen);
     }
  }


74 #endif
```

79 // random.cpp

```
//
// Routines that help in the selection of random numbers.

#ifdef _AIX
   #pragma alloca
#endif

#include <assert.h>
#include <malloc.h>
#include <stdlib.h>

#include "random.hpp"

#ifdef __BCC_
  #include <time.h>
  #define FAR far
#elif defined (_AIX)

  #include <sys/time.h>

  #define FAR

  extern "C"
  void
  *lfind (void* Key, void* Base, size_t* NumberOfElementsPointer,
         size_t Width,
         int (*ComparisonPointer)(const void*, const void*));


  extern "C"
  int
  gettimer(int Timer_type, timestruc_t* TimePointer);

  extern "C"
  int
  srandom(int);

  void
  randomize()
   // Set the random number seed, to a value based on the clock.
  {
    timestruc_t time_now;
    gettimer(TIMEOFDAY, &time_now);
    srandom(time_now.tv_nsec % RAND_MAX);
    unsigned long int r = random(); // used in debugging to
                        // ensure random is being set.
  }

#endif
```

```
static
class Startup
  // A simple class that ensures that the random number
  // generator is randomized.
{
public:
  Startup()
    {
     randomize();
    }
} startup;


double
random_real(void)
  // Return a random number in the interval [0,1).
{
  unsigned long int r = random();
  r = r << 16;
  return ( double(r % RAND_MAX) / double(RAND_MAX) );
}


void
random_choose_n_fast(const int nb, const int max, int chosen[])
  // Chose n random integers in the interval [0,max-1] with
  // no duplicated numbers chosen.
  //
  // This algorithm uses a lot of memory but is fast when
  // nb is close to max.
{
  assert(max >= 0);
  assert(nb >= 0);
  assert(max >= nb);

  if ( nb == 0 )
    {
     return;
    }

170  int *to_do = (int *) alloca(max * sizeof(int));
171  assert(to_do != 0);
172
173  for (int i = 0; i < max; i++)
174    {
175      to_do[i] = i;
176    }

  for ( i=0; i < nb; i++)
```

```
    {
180      int j = random(max - i); // LINE randomlookreduce
181      chosen[i] = to_do[j]; // LINE randomreducestart
182      to_do[j] = to_do[max - i - 1]; // LINE randomreduceend
    }
}

static
int FAR compare(const void *xv, const void *yv)
{
   const int *x = (int *)xv;
   const int *y = (int *)yv;
   return( *x - *y );
}

void
random_choose_n_small(const int nb, const int max, int chosen[])
  // Chose n random integers in the interval [0,max-1] with
  // no duplicated numbers chosen.
  //
  // This algorithm has the potential to be very slow.  The
  // algorithm uses verly little memory, and should be
  // used when nb is much smaller than max.
{
  assert(max >= 0);
  assert(nb >= 0);
  assert(max >= nb);

  if ( nb == 0 )
    {
      return;
    }

  chosen[0] = random(max);

  for (int i = 1; i < nb; i++)
    {
      int j;
      size_t nelem;
      do
        {
          nelem = i;
221          j = random(max);
        }
223      while ( lfind(&j, chosen, &nelem, sizeof(int), compare) );
      chosen[i] = j;
    }
}
```

229 // array.hpp

```
//
// This class implements an array of objects that can
// be easily copied.
//
// The objects are owned by the array, and if the objects are
// changed, then they must be "put" into the array.
//
// When an array is copied, just a pointer to the original
// array is made, but if any copy of an array is modified, then the contents
// of the pointer are replaced with a new array, with the appropriate
// values. This all is hidden from the user, but should provide
// space/time savings.
//
// The bad part of this class, is that all accesses take an
// extra indirection operation.

#ifndef ARRAY_HPP_INCLUDED
#define ARRAY_HPP_INCLUDED 1


#include <assert.h>


 template <class C> class Array;

 // class Array_Storage should be a nested class of Array
 // but the Borland C++ 3.0 compiler can not handle it.

 template <class C>
 class Array_Storage
 {

 protected:

  Array_Storage(const int new_size, const C c[]) :
   nb_references(1)
  {
   size = new_size;
   contents = new C [size];
   for (int i = 0; i < size; i++)
     {
      contents[i] = c[i];
     }
  }

  Array_Storage(const int new_size = 0) :
   nb_references(1)
  {
   size = new_size;
```

```
    contents = new C [size];
    assert(contents != 0);
   }

   ~Array_Storage()
   {
    delete[] contents;
   }


  const Array_Storage<C>&
  operator=(const Array_Storage<C>&)
    // This function should never be called. This defintion
    // prevents the compiler from creating it.
   {
    assert(0);
    return *this;
   }

  protected:

   C* contents;
   int nb_references;
   int size;

  friend class Array<C>;
  };


template<class C>
class Array
{
public:

  Array(const int size = 0)
   // Create an array with size elements.
  {
   store = new Array_Storage<C> (size);
   assert(store);
  }

  Array(const int size, const C c[])
   // Create an array with the contents of C being copied.
  {
   store = new Array_Storage<C> (size, c);
   assert(store);
  }

330  Array(const Array<C>& array) // LINE arraycopystart
331  {
```

```
332   store = array.store;
333   if (store)
334     {
335      (store->nb_references)++;
336     }
337 } // LINE arraycopyend


340 ~Array() // LINE arraydestructorstart
341   {
342     if (store)
343    {
344        (store->nb_references)--;
345        if ( store->nb_references <= 0 )
346           {
347            delete store;
348           }
349       }
350   } // LINE arraydestructorend


353 Array<C>& // LINE arrayassignstart
354 operator= (const Array<C>& array)
355 {
356   // No longer point to the contents of array.
357   if (store)
358     {
359      store->nb_references--; // LINE arraydecrementfreestart
360      if ( store->nb_references <= 0 )
361         {
362          delete store;
363     }
364    } // LINE arraydecrementfreeend
365
366   // Point to the new contents.
367   store = array.store; // LINE arrayreferencesetincrementstart
368   if (store)
369     {
370
371    } // LINE arrayreferencesetincrementend
372
373   return *this;
374 } // LINE arrayassignend


377 const C // LINE arraygetputstart
378 get(const int i) const
379   // Return the i-th element of the array.
380 {
381   assert(i >= 0);
```

```
382   assert(store);
383   assert(i < store->size);
384
385   return store->contents[i];
386  }
387
388
389  void
390  put(const int i, const C& c)
391   // Replace the i-th element of the array with c.
392  {
393   assert(i >= 0);
394   assert(store);
395   assert(i < store->size);
396
397   if ( store->nb_references != 1 ) // LINE arraygetcheckref
398     {
399      // Make a non-shared copy of the storage.
400      Array_Storage<C>* old_store = store; // LINE arrayputcopystart
401      store = new Array_Storage<C> (old_store->size, old_store->contents);
// LINE arrayputcopyend
402     old_store->nb_references--; // LINE arrayputdecrementref
403     }
404
405   store->contents[i] = c;
406  } // LINE arraygetputend

  const int
  size() const
   // Return the size of the array.
  {
   return store?store->size:0;
  }

  operator const C*() const
  {
   return store?store->contents:0;
  }

protected:
421   Array_Storage<C>* store; // LINE arraystore

};

425 #endif
```

430 // chromo.hpp

```
//
// This file defines class Chromosome<class Genome>

#ifndef CHROMO_HPP_INCLUDED
#define CHROMO_HPP_INCLUDED 1


#include <math.h>
#include <iostream.h>

#include "array.hpp"
#include "ordered.hpp"

template<class Genome>
class Chromosome
{
  public:

    Chromosome() :
     gene_data(0)
    { }

    Chromosome(const int len) :
       gene_data(len)
      // Create a Chromosome of length len, with the
      // contents unset.
      //
    { }

    Chromosome(const Chromosome<Genome>& chrom)
      : gene_data(chrom.gene_data)
    { }

    ~Chromosome()
    { }

    Chromosome<Genome>&
    operator=(const Chromosome<Genome>& other)
    {
     this->gene_data = other.gene_data;
      return *this;
    }


    virtual
    double
    distance(const Chromosome<Genome>& other)
     // A measure of how close the two chromsomes are
     // to each other.
```

```
 {
// For binary alphabets this is the Hamming distance,
// for other alphabets this is the taxicab distance.
assert(nb_genes() == other.nb_genes());

double result = 0.0;
const int length = nb_genes();

for (int i = 0; i < length; i++)
  {
   result += abs(  gene_data.get(i)
            - other.gene_data.get(i) );
  }
return result;
 }


Array<Genome>
genes() const
{
 return gene_data;
}

virtual
void
set_genes(const Array<Genome>& genes)
{
 gene_data = genes;
}


int
nb_genes() const
{
 return gene_data.size();
}


virtual
void
randomize() = 0;
 // Set the values of the genes to random values.

protected:

525   Array<Genome> gene_data; // LINE chromogene

private:

 Genome
```

```
    abs(Genome g)
      {
     return (g>=0)?g:-g;
      }


};



template<class Genome>
class Chromosome_Ordered :
    public Chromosome<Genome>,
   public Ordered< Chromosome_Ordered<Genome> >
  // A class of Chromosomes, such that the strength of
  // the chromosome is ordered.
{
  public:

   Chromosome_Ordered() :
     Chromosome<Genome>(0)
   {
   }

   Chromosome_Ordered(const int len) :
       Chromosome<Genome>(len)
     // Create a Chromosome of length len, with the
     // contents unset.
   { }

   Chromosome_Ordered(const Chromosome_Ordered<Genome>& chrom) :
     Chromosome<Genome>(chrom)
   {
   }

   Chromosome_Ordered()
   {
   }

   Chromosome_Ordered<Genome>&
   operator=(const Chromosome_Ordered<Genome>& other)
   {
     this->Chromosome<Genome>::operator=(other);
      return *this;
   }
  protected:

};
582 #endif
```

```
587 // ordered.hpp
//
// This file defines a fully ordered relationship.

#ifndef ORDERED_HPP_INCLUDED
#define ORDERED_HPP_INCLUDED 1

#include <iostream.h>


template <class C>
class Ordered
  // Definition of a fully ordered relationship
{
public:

  virtual
  int
  compare(const C& other) const
    // Return negative if this is less than other.
    //             0 if this an other are equal.
    //        positive if this is greater than other.
  {
    cerr << "This function should never be called.\n";
    assert(0);
    return 0;
  }
};


617 #endif
```

```
618 #ifndef XOVER_HPP_INCLUDED
#define XOVER_HPP_INCLUDED 1

//  xover.hpp
//
// This file defines various crossover methods.

#include <string.h>
#include "chromo.hpp"
#include "random.hpp"
#include "table.hpp"

static
int
ascending_order(const int size, const int array[])
  // Check that the array is in ascending order.
  //
  // return 1 if they are in ascending order.
{
  assert(size >= 0);

  for (int i = 1; i < size; i++)
    {
     if ( array[i] < array[i-1] )
       {
        return 0;
       }
    }

  return 1;
}

static
int
compare_int(const void* xx, const void* yy)
  // Compare the integers pointed to by xx and yy.
{
  int* x = (int*) xx;
  int* y = (int*) yy;
  return (*x - *y);
}


668 template<class Genome> // LINE copygenesstart
669 static
670 void
671 copy_genes(const Chromosome<Genome>& parent0,
672           const Chromosome<Genome>& parent1,
673           Chromosome<Genome>* child0,
```

```
674         Chromosome<Genome>* child1,
675         const int start,
676         const int stop)
677  // Copy genes from parent0 to child0, and parent1 to
678  // child1 in the range start to stop inclusive.
679 {
680  const Array<Genome> p0 = parent0.genes();
681  const Array<Genome> p1 = parent1.genes();
682  Array<Genome> c0 = child0->genes();
683  Array<Genome> c1 = child1->genes();
684
685  for ( int i = start; i <= stop; i++)
686   {
687    Genome gene = p0.get(i);
688    c0.put(i, gene);
689    gene = p1.get(i);
690    c1.put(i, gene);
691   }
692  child0->set_genes(c0);
693  child1->set_genes(c1);
694 } // LINE copygenesend


698 template<class Genome> // LINE crossovernpointhelperstart
699 void
700 crossover_n_point(
701   const Chromosome<Genome>& parent0,
702   const Chromosome<Genome>& parent1,
703   Chromosome<Genome>* child0,
704   Chromosome<Genome>* child1,
705   const int nb_crossover_points,
706   const int crossover_points[])
707  // Perform n-point crossover, producing two children.
708  // The crossover points are defined, as the parents donating
709  // material to the child switches after a gene in the
710  // crossover_points has been copied.
711  //
712  // This algorithm requires that the crossover points
713  // are in ascending order, and there are two extra
714  // crossover points (not counted in nb_crossover_points).
715  // The first entry (crossover_points[0] must be -1,
716  // and the last entry (crossover_points[nb_crossover_points + 1]
717  // must be parent0.length - 1.
718 {
719  assert( parent0.nb_genes() == parent1.nb_genes() );
720  assert(nb_crossover_points > 0);
721  assert( nb_crossover_points < parent0.nb_genes() );
722  assert(ascending_order(nb_crossover_points, crossover_points));
732  assert(crossover_points[0] == -1);
```

```
724 assert(crossover_points[nb_crossover_points + 1] ==
725       parent0.nb_genes() - 1);
726
727 // Copy the genes, switching after a crossover point.
728 int i = 0; // LINE crossoveralternatecopystart
729 while ( i <= nb_crossover_points )
730   {
731    copy_genes(parent0, parent1, child0, child1,
732        crossover_points[i] + 1, crossover_points[i+1]);
733     i++;
734    if ( i <= nb_crossover_points )
735      {
736       copy_genes(parent1, parent0, child0, child1,
737         crossover_points[i] + 1, crossover_points[i+1]);
738        i++;
739      }
740   } // LINE crossoveralternatecopyend
741 } // LINE crossovernpointhelperend


744 template<class Genome> // LINE crossovernpointstart
745 void
746 crossover_n_point(
747     const Chromosome<Genome>& parent0,
748    const Chromosome<Genome>& parent1,
749    Chromosome<Genome>* child0,
750    Chromosome<Genome>* child1,
751    const int nb_crossover_points)
752 // Perform n-point crossover, producing two children.
753 // with the position of the crossover_points randomly chosen.
754 {
755 assert( parent0.nb_genes() == parent1.nb_genes() ); //LINE
      crossoverassertstart
756 assert( parent0.nb_genes() == child0->nb_genes() );
757 assert( parent0.nb_genes() == child1->nb_genes() );
758 assert( nb_crossover_points > 0);
759 assert( nb_crossover_points < parent0.nb_genes() ); //LINE
      crossoverassertend
760
761 const int length = parent0.nb_genes();
762
763 // Find the crossover points.
764 int* xover_points =
765    (int*) alloca(sizeof(int) * (nb_crossover_points + 2) );
766
767 // Choose nb_crossover_points between 0, and length - 2,
768 // where a crossover point means that a switch in copying
769 // will occur after the gene is copied. (Remember in C++,
770 // the genes are numbered from 0 to l-1.)
```

```
771 random_choose_n(nb_crossover_points, length-1, //LINE
crossovernpointchoosestart
772      xover_points + 1);
773 qsort(xover_points + 1, nb_crossover_points, sizeof(int), //LINE
      crossovernpointaddstart
774      compare_int);
775
776 // Add two extra crossover points at the start and the end
777 // of the gene, so no special code is needed in the copying
778 // loop.
779 xover_points[0] = -1;
780 xover_points[nb_crossover_points + 1] = parent0.nb_genes() - 1; //LINE
      crossovernpointchooseend
781
782 crossover_n_point(parent0, parent1, child0, child1, //LINE
      crossovernpointcallstart
783      nb_crossover_points, xover_points); // LINE crossovernpointcallend
784 } //LINE crossovernpointend
```

```
template<class Genome>
static
void
copy_shuffled_genes(
   const Array<Genome>& parent0,
   const Array<Genome>& parent1,
   Array<Genome>* child0,
   Array<Genome>* child1,
   const int start,
   const int stop,
   const int permutation[])
 // Copy genes from parent0 to children[0], and parent1 to
 // children[1] according to the genes in the permutation
 // table in the range start to stop inclusive.
{
 for ( int i = start; i <= stop; i++)
   {
    int j = permutation[i];
    child0->put(j, parent0.get(j));
    child1->put(j, parent1.get(j));
   }
}
```

```
template <class Genome> // LINE crossovernpointshuffleslowstart
void
crossover_n_point_shuffle_slow(
   const Chromosome<Genome>& parent0,
   const Chromosome<Genome>& parent1,
```

```
    Chromosome<Genome>* child0,
    Chromosome<Genome>* child1,
    const int nb_crossover_points)
  // Perform n-point shuffle crossover, producing two children.
  //
  // This algorithm is the classic way, and is slower, than the
  // alternative that is also provided. The advantage of this
  // method is that no costly initial work is needed, and
  // tables do not need to be stored.
  //
  // This algorithm is conceptually the same as tagging each
  // gene with a position, performing crossover, and resorting
  // the chromosome according to the tagged positions.
{
  assert(parent0.nb_genes() == parent1.nb_genes());
  assert(nb_crossover_points > 0);
  assert(nb_crossover_points < parent1.nb_genes());

  int length = parent0.nb_genes();

  Array<Genome> c0(length), c1(length);
  const Array<Genome> p0 = parent0.genes();
  const Array<Genome> p1 = parent1.genes();

  int* order = (int *)alloca( length * sizeof(int) );
  random_choose_n(length, length, order);

  // Find the crossover points.
  int* xover_points = (int *)
    alloca(sizeof(int) * (nb_crossover_points + 2) );
  assert(xover_points);

  // Choose nb_crossover_points between 0, and length - 2,
  // where a crossover point means that a switch in copying
  // will occur after the gene is copied. (Remember in C++,
  // the genes are numbered from 0 to l-1.)
853 random_choose_n(nb_crossover_points, length - 1, //LINE
       crossoverslowchoosefragmentsstart
854     xover_points + 1);
855 qsort(xover_points + 1, nb_crossover_points, sizeof(int),
856     compare_int);
857
858 // Add two extra crossover points at the start and the end
859 // of the gene, so no special code is needed in the copying
860 // loop.
861 xover_points[0] = -1;
862 xover_points[nb_crossover_points+1] = length - 1; //LINE
       crossoverslowchoosefragmentsend

  // Copy the genes, switching after a crossover point.
```

```
865  int i = 0; // LINE crossoverslowcopystart
866  while ( i <= nb_crossover_points )
867    {
868      copy_shuffled_genes(p0, p1, &c0, &c1,
869          xover_points[i] + 1, xover_points[i+1], order);
870       i++;
871      if ( i <= nb_crossover_points )
872        {
873          copy_shuffled_genes(p1, p0, &c0, &c1,
874              xover_points[i] + 1, xover_points[i+1], order);
875           i++;
876        }
877    } // LINE crossoverslowcopyend

  child0->set_genes(c0);
  child1->set_genes(c1);
} // LINE crossovernpointshuffleslowend


template <class Genome>
static
void
crossover_shuffle_copy_c_genes(
    const Chromosome<Genome>& parent0,
    const Chromosome<Genome>& parent1,
    Chromosome<Genome>* child0,
    Chromosome<Genome>* child1,
    const int nb_genes_to_copy)
  // Perform a shuffle crossover, producing two children.
  //
  // This method copies nb_genes_to_copy genes chosen independently
  // from the first parent, and length - nb_genes_to_copy from
  // the second parent, to child0, and similarly the reverse
  // child1.
{
  assert(parent0.nb_genes() == parent1.nb_genes());

  int length = parent0.nb_genes();
  int* order = (int *)alloca( length * sizeof(int) );
  assert(order);

  random_choose_n(nb_genes_to_copy, length, order);

  // The first child is a copy of the first parent, except
  // fot the genes transferred from the second parent. Similarly,
  // the second child is a copy of the second parent, except
  // for the genes transferred from the first parent.
  Array<Genome> pgenes0 = parent0.genes();
  Array<Genome> pgenes1 = parent1.genes();
  Array<Genome> cgenes0 = pgenes0;
```

```
  Array<Genome> cgenes1 = pgenes1;

  for (int i = 0; i < nb_genes_to_copy; i++)
    {
      int j = order[i];
      cgenes0.put(j, pgcnes1.get(j));
      cgenes1.put(j, pgenes0.get(j));
    }

  child0->set_genes(cgenes0);
  child0->set_genes(cgenes0);
}


template<class Genome>
void
crossover_n_point_shuffle(
    const Chromosome<Genome>& parent0,
    const Chromosome<Genome>& parent1,
    Chromosome<Genome>* child0,
    Chromosome<Genome>* child1,
    const int nb_crossover_points)
  // Perform n-point shuffle crossover, producing two children.
{
  assert(parent0.nb_genes() == parent1.nb_genes());
  assert(parent0.nb_genes() > 0);
  assert(nb_crossover_points < parent0.nb_genes());

  int length = parent0.nb_genes();
  double* table;
945  table_crossover_probability(length, nb_crossover_points, &table); // LINE
crossovergettable

  double prob = random_real();

949  int low = 1;  // LINE crossoverbsearchstart
950  int high = length;
951  int mid = (low + high) / 2;
952
953  while (   (table[mid - 1] > prob) || (prob > table[mid]) )
954    {
955      assert(low < high);
956      assert(table[low-1] <= prob);
957      assert(prob < table[high]);
958      assert(low <= mid);
959      assert(mid <= high);
960
961      if ( table[mid-1] > prob )
962      {
963        high = mid - 1;
```

```
964    }
965     else if ( prob > table[mid] )
966    {
967      low = mid + 1;
968    }
969     mid = (low + high) / 2;
970   } // LINE crossoverbsearchend

  if ( mid <= length-mid )
   {
    crossover_shuffle_copy_c_genes(parent0, parent1, child0, child1,
        mid);
   }
  else
   {
    crossover_shuffle_copy_c_genes(parent0, parent1, child0, child1,
        length-mid);
   }
}


template<class Genome>
void
crossover_uniform(
  const Chromosome<Genome>& parent0,
  const Chromosome<Genome>& parent1,
  Chromosome<Genome>* child0,
  Chromosome<Genome>* child1)
   // Perform uniform crossover, producing two children.
   //
   // Uniform crossover is performed by copying half of
   // the genes from the first parent, and half of the
   // genes by the second child for each parent.  The
   // genes are chosen independantly.
  {
   assert(parent0.nb_genes() == parent1.nb_genes());

   int length = parent0.nb_genes();
   const int half = length / 2;
   crossover_shuffle_copy_c_genes(parent0, parent1,
    child0, child1, half);
  }


template<class Genome>
void
crossover_uniform_modified(
   const Chromosome<Genome>& parent0,
   const Chromosome<Genome>& parent1,
   Chromosome<Genome>* child0,
```

```
  Chromosome<Genome>* child1)
// Perform modified crossover, producing two children.
{
  assert(parent0.nb_genes() == parent1.nb_genes());
  assert(parent0.nb_genes() > 1);
  const int length = parent0.nb_genes();

  Array<Genome> p0 = parent0.genes();
  Array<Genome> p1 = parent1.genes();
  Array<Genome> c0(length);
  Array<Genome> c1(length);

  // to_do holds the genes that are to be copied.
  int* to_do = (int*) alloca(sizeof(int) * length);
  assert(to_do);

  // Look for genes that are the same between the
  // two parents.
  int j = 0;
  for (int i = 0; i < length; i++)
   {
   Genome g = p0.get(i);
1036  if ( g == p1.get(i) ) // LINE copysamestart
1037    {
1038      c0.put(i, g);
1039      c1.put(i, g);
1040    } // LINE copysameend
   else
    {
      to_do[j++] = i;
    }
   }

1047   const int half = j/2;  // LINE copyfirstpartstart
1048   for ( i = 0; i < half; i++ )
1049     {
1050  int k = random(j - i);
1051  int l = to_do[k];
1052  c0.put(l, p1.get(l));
1053  c1.put(l, p2.get(l));
1054  to_do[l] = to_do[j--];
1055     } // LINE copyfirstpartend

  while ( j >= 0 )
   {
   int l = to_do[j--];
   c1.put(l, p1.get(l));
   c0.put(l, p0.get(l));
   }
```

```
  child0.set_genes(c0);
  child1.set_genes(c1);
 }



1070 static  // LINE decodecrossoverstart
1071 void
1072 decode_crossover(
1073  const Array<char>& genes,
1074  int* shuffle,
1075  int* nb_xover_pts,
1076  const int offset,
1077  const int size)
1078   // Set shuffle and nb_xover_points, for the encoding in genes.
1079   // shuffle indicates whether shuffle crossover should be
1080   // used.
1081   //
1082   // This implematation will only work for binary encodings.
1083  {
1084   *shuffle = (genes.get(offset) == '1'); // LINE decodeshuffle
1085
1086   // Find the number of crossover points.
1087   char* buffer = (char*) alloca(sizeof(char) * size);
1088   assert(buffer);
1089
1090   for (int i = 1; i < size; i++) // LINE convertstart
1091     {
1092      buffer[i-1] = genes.get(offset+i);
1093     }
1094   buffer[size-1] = '\0';
1095   char* end_pointer;
1096   *nb_xover_pts =
1097      int( double(strtoul(buffer, &end_pointer, 2)) / pow(2, size) *
1098          double(genes.size() - 2) ) + 1; // LINE convertend
1099
1100   // Use heursitic that for non-shuffled crossover a low
1101   // number of crossover points is good, and that for
1102   // shuffled crossover a high number of crossover points
1103   // is good.
1104   if ( *shuffle ) // LINE heuristicstart
1105     {
1106      *nb_xover_pts = genes.size() - *nb_xover_pts;
1107     } // LINE heuristicend
1108
1109   assert(*nb_xover_pts > 0);
1110   assert(*nb_xover_pts < genes.size());
1111  } // LINE decodecrossoverend
```

```
template<class Chrom> // LINE variablestart
void
crossover_variable(
  const Chrom& parent0,
  const Chrom& parent1,
  Chrom* child0,
  Chrom* child1,
  const int offset,
  const int size)
   // Perform variable shuffled crossover, producing two children.
   //
   // Chrom must inherit from chromosome.
   //
   // The parameters offset and size, indicate the
   // location, and the number of genomes that are used
   // for the encoding of the crossover method.
   //
   // This crossover is different from either shuffle crossover,
   // or n-point crossover in that the children are not
   // necessarily the reverse of each other, in terms in of
   // gene origin.  i.e. both children can have the i-th
   // gene come from the same parent.
  {
   assert(offset >= 0);
   assert(size > 2);
   assert(parent0.nb_genes() == parent1.nb_genes());
   assert(parent0.nb_genes() >= offset + size - 1);

   int length = parent0.nb_genes();

   int shuffle, nb_xover_pts;
   Chrom dummy;

   // Create the first child using the first parents encoding scheme.
   decode_crossover(parent0.genes(),  // LINE decodestart
     &shuffle, &nb_xover_pts, offset, size); // LINE decodeend

   if ( shuffle )
     {
      crossover_n_point_shuffle(parent0, parent1,
          child0, &dummy, nb_xover_pts);
     }
   else
     {
      crossover_n_point(parent0, parent1,
          child0, &dummy, nb_xover_pts);
     }

   // Create the second child using the second parents encoding scheme.
   decode_crossover(parent1.genes(),
```

```
      &shuffle, &nb_xover_pts, offset, size);

    if ( shuffle )
      {
       crossover_n_point_shuffle(parent0, parent1,
           &dummy, child1, nb_xover_pts);
      }
    else
      {
       crossover_n_point(parent0, parent1,
           &dummy, child1, nb_xover_pts);
      }
  } // LINE variableend
1183 #endif
```

```
1188 // table.hpp
//
// Definition of crossover probabilities.

#ifndef TABLE_HPP_INCLUDED
#define TABLE_HPP_INCLUDED 1

void
table_crossover_probability(
    const int chrom_length,
    const int nb_crossover_points,
    double** prob);
  // Return in *prob the table of probabilities for genes copied.
  // (*prob)[i] is the probability that i or less genes will
  // be copied from a parent using n-point shuffle crossover.
  // Beware the size of prob must be chrom_length + 1.
1206 #endif


#ifndef XOVER_HPP_INCLUDED
#define XOVER_HPP_INCLUDED 1
//
```

1210 table.c

```
#include <assert.h>
#include <malloc.h>

#include "table.hpp"

static
double
product(const int a, const int b)
   // Return the product of a x (a+1) x (a+2) x ... x (b-1) x b
{
 assert(a <= b);

 double result = 1.0;
 for (int i = a; i <=b ; i++)
   {
    result = result * i;
   }
 return result;
}


static
double
permutations(const int a, const int b)
   //   / a \
   //   |   |  = a! /(b!(a-b)!)
   //   \ b /
{
 double result;

 if ( (a < 0) || (b < 0) || (b > a) )
   {
    result = 0.0;
   }
 else if ( ( b == 0 ) || ( a == b ) )
   {
    result = 1.0;
   }
 else if ( b > (a-b) )
   {
    result = product(b+1, a) / product(1,a-b);
   }
 else
   {
    result = product(a-b+1, a) / product(1,b);
   }
 return result;
}
```

```
static
non_decreasing(
  const double table[],
  const int size)
    // Return 0 if table is strictly non-decreasing.
    // Else return the index of the element that is larger
    // than its predecessor.
{
  for (int i=1; i < size; i++)
    {
     if (table[i] < table[i-1])
       {
         return i;
       }
    }
  return 0;
}


static
void
calc_crossover_probability(
    const int chrom_length,
    const int nb_crossover_points,
    double prob[])
  // Place in prob[i] the probability of a chromosome copying,
  // more than i genes to one child, and less than
  // chrom_length-i genes to the other child.
  //
  // Beware prob must be of length chrom_length + 1.
{
  const int k = (nb_crossover_points + 2) / 2;
  const int mid = ( chrom_length + 1) / 2;
    // Notice the implicit floor in the division.

  // Calculate the probabilities of copying exactly
  // c genes.
  const double denom =
      2 *
      permutations(chrom_length - 1,
           chrom_length - nb_crossover_points - 1);
  for (int c = k; c <= mid; c++)
    {
     double nom11 = permutations(c-1, c-k);
     double nom12 = permutations(chrom_length - c -1,
                 nb_crossover_points - k);
     double nom21 = permutations(chrom_length - c - 1, k - 1);
     double nom22 = permutations(c - 1, nb_crossover_points - k);
```

```
    prob[c] = ( nom11 * nom12 + nom21 * nom22 ) / denom;
    prob[chrom_length - c] = prob[c];
   }

  // Calculate the cumulative proabilities
  for (c=0; c<k;c++)
   {
    prob[chrom_length-c] = 1.0;
    prob[c] = 0.0;
   }
  for (c = k; c <= chrom_length - k; c++)
   {
    prob[c] = prob[c] + prob[c-1];
   }

  assert(non_decreasing(prob, chrom_length + 1) == 0);
}

static int max_table_length = 0;
static double*** tables = 0;
 // A two dimension table indexed by chromosome size - 1,
 // and number of crossover points - 1, that contains the
 // probability table if calculated.

void
table_crossover_probability(
   const int chrom_length,
   const int nb_crossover_points,
   double** prob)
   // Return in prob the table of probabilities for genes copied.
   // prob[i] is the probability that i or less genes will
   // be copied from a parent using n-point shuffle crossover.
{
 assert(chrom_length > 0);
 assert(nb_crossover_points < chrom_length);
 assert(nb_crossover_points > 0);
 if (chrom_length > max_table_length)
   {
    tables = (double***)
         realloc(tables, sizeof(double***) * chrom_length);
    assert(tables);
    for (int chrom_len = max_table_length + 1;
      chrom_len <= chrom_length; chrom_len++)
    {
     tables[chrom_len - 1] = 0;
    }
    max_table_length = chrom_length;
   }

 if ( tables[chrom_length - 1] == 0 )
```

```
    {
     tables[chrom_length - 1] = (double**)
      malloc(sizeof(double**) * (chrom_length - 1));
     assert(tables[chrom_length - 1]);

     for (int xover = 1; xover < chrom_length; xover++)
     {
      tables[chrom_length - 1][xover - 1] = 0;
     }
    }

  if ( tables[chrom_length - 1][nb_crossover_points - 1] == 0 )
    {
     tables[chrom_length - 1][nb_crossover_points - 1] =
    (double*) malloc(sizeof(double) * (chrom_length + 1));
     assert(tables[chrom_length-1][nb_crossover_points-1]);

     calc_crossover_probability(chrom_length,
      nb_crossover_points,
      tables[chrom_length-1][nb_crossover_points-1]);
    }
  *prob = tables[chrom_length-1][nb_crossover_points-1];
1383 }
```

Appendix C: GenAlg Code

The following code, in HyperCards HyperTalk, is a very simple implementation of the pseudo-code contained in Goldberg's book. I would hope that the code is self explanitory since HyperTalk is a very English like programming language.

Each code segment is broadly commented upon and it should therefore be a simple task to isolate code of particular interest or to convert this simple implementation into any other langyage of choice.

SCRIPTS FROM STACK: GenAlg

The function 'flip' is employed to set a flag to true or false, dependent upon a random number generated as tested against a passed value (i.e. develops a weighted random test).

```
••••• BACKGROUND No.1 genalg
function flip prob
  put random(1000) into p
  if p<=prob*1000 then
    return true
  else
    return false
  end if
end flip
```

The function 'objfunc' determines the fitness of the passed decoded chromosome. The objective function varies dependent upon the objective being sought. In the model the objective function is the sum of the absolute differences of the desired ending factor values and the modelled ending factor values.

```
function objfunc x
  -- ************************************************************
  -- Fitness function for this run is f(x) = x**10
  -- The objective function needs to change for different problems
  -- ************************************************************
  put 1073741824 into coef
  return (x/coef)
end objfunc
```

The function 'decode' takes a passed chromosome and the number of bits (allelles) in the chromosome and converts it to a value that can be tested by the objfunc function. In this instance we are converting the binary string (chromosome) into a value by direct substitution of a 1 bit (true or on bit) to its power of 2 equivalent.

```
function decode chrom,ibits
  -- ************************************************************
  -- The decode function needs to change for different problems
  -- Here we are decoding the binary string into powers of 2 based
  -- upon their position in the string.
  -- ************************************************************
  put 0 into accum
```

```
  put 1 into powerof2
  repeat with i = ibits down to 1
    if char i of chrom = 1 then
      put powerof2+accum into accum
    end if
    put powerof2*2 into powerof2
  end repeat
  return accum
end decode
```

The 'statistics' procedure determines the maximum, minimum and average fitness of a population of chromosomes. The population size can be altered by the user. The size of the population is important in ensuring sufficient diversity in the population from which crossover (breeding) and mutation occur.

```
on statistics
  global popsize, lmax, lavg, lmin, sumfitness, pop, fitness

  put item 1 of fitness into sumfitness
  put sumfitness into lmin
  put sumfitness into lmax

  repeat with j = 2 to popsize
    add item j of fitness to sumfitness
    if item j of fitness < lmin then put item j of fitness into lmin
    if item j of fitness > lmax then put item j of fitness into lmax
  end repeat

  put sumfitness/popsize into lavg

end statistics
```

The 'report'' procedure writes results to the screen.

```
on report
  global reportnum, popsize, oldpop, newpop, lmax, lmin
  global sumfitness, nmutation, ncross, fitness, oldfitness
  add 1 to reportnum

  -- Write out the stats and pop results
  put empty into card field "Results"
  put "Generation Report "&reportnum into line 1 of card field results

  put return after card field "results"

  repeat with j = 1 to popsize
    put return & line j of oldpop&& line j of oldfitness after card field results
  end repeat

  put return after card field "results"

  repeat with j = 1 to popsize
```

```
  put return& line j of newpop&& item j of fitness after card field results
 end repeat

 put return&"Max = "&lmax after card field results
 put return&"Min = "&lmin after card field results
 put return&"Sumfitness = "&sumfitness after card field results
 put return&"Number of Mutations = "&nmutation after card field results
 put return&"Number of Crossovers = "&ncross after card field results
end report
```

The 'select' function selects chromosomes for crossover based upon their fitness.

```
function select
 global fitness, popsize, sumfitness
 put 0 into partsum
 put 0 into j

 put random(10000)/10000 * sumfitness into rand
 repeat until (partsum >= rand) or (j = popsize)
  add 1 to j
  put partsum + item j of fitness into partsum
 end repeat

 return j
end select
```

The 'mutation' function selects an allele randomly for mutation. A mutation is the flipping, of an allelle, of a 1 to a 0 or a 0 to a 1.

```
function mutation allele
 global nmutation, pmutation
 put flip(pmutation) into mutate
 if mutate is true then
  add 1 to nmutation
  if allele = 1 then
   put 0 into mutation
  else
   put 1 into mutation
  end if
 else
  put allele into mutation
 end if

 return mutation
end mutation
```

The 'crossover' procedure selects two locations along the chromosome to execute an allele swap. This procedure also places the mutated allele into the chromosome if a mutation occurs.

```
on crossover
 global newpop, lchrom, ncross, mate1, mate2
 global jcross, j, pcross, oldpop
```

```
  if flip(pcross) is true then
    put random(lchrom-1) into jcross
    if jcross = 0 then put 1 into jcross
    add 1 to ncross
  else
    put lchrom into jcross
  end if

  repeat with ji = 1 to jcross
    put mutation(char ji of line mate1 of oldpop) into char ji of line j of newpop
      put mutation(char ji of line mate2 of oldpop) into char ji of line j+1 of
newpop
  end repeat

  if jcross<>lchrom then
    repeat with ji = jcross to lchrom
      put mutation(char ji of line mate2 of oldpop) into char ji of line j of newpop
        put mutation(char ji of line mate1 of oldpop) into char ji of line j+1 of
newpop
    end repeat
  end if

end crossover
```

The 'generation' procedure selects two parent chromosomes for mating, performs the crossover and decodes the resultant children to determine their fitness.
I have inserted a test to ensure that any children generated have a fitness greater than their parents, if they do not then the children do not replace their parents in the seed population. This was found to increase the speed at which fitness of the population increases, however; it performs this function with an attendant loss of genetic diversity (which has its own attendant problems). I determined that, since we are only seeking movement directions in cross- and self-impact values and not optimality, that speed was preferred to genetic diversity (which is vital to the search for optimality).

```
on generation
  global popsize, oldpop, lchrom, newpop, sumfitness, fitness, j
  global mate1, mate2, oldfitness
  put 1 into j
  repeat until j>popsize

    put select() into mate1
    put select() into mate2

    crossover

    put decode(line j of newpop,lchrom) into x
    put objfunc(x) into line j of fitness
    if line j of fitness < line j of oldfitness then
      put line j of oldpop into line j of newpop
```

```
    put line j of oldfitness into line j of fitness
   end if

   put decode(line j+1 of newpop,lchrom) into x
   put objfunc(x) into line j+1 of fitness
   if line j+1 of fitness < line j+1 of oldfitness then
     put line j+1 of oldpop into line j+1 of newpop
     put line j+1 of oldfitness into line j+1 of fitness
   end if

   add 2 to j

 end repeat
end generation
```

The 'execute' procedure is the starting point for the algorithm. This is the first procedure executed. A number of variables are set up and the seed population initialized. This procedure also controls the number of generations that are tested and exits when the number of generations developed meets the number requested by the user.

```
on execute
  global maxgen
  global oldpop, newpop, reportnum, fitness, oldfitness
  put 1 into reportnum
  put 1 into gen
  initialize
  repeat until gen >= maxgen
    add 1 to gen
    generation
    statistics
    report
    put newpop into oldpop
    put fitness into oldfitness
  end repeat
end execute
```

The 'initdata' procedure reads values for constants that are set by the user. These constants are the size of the seed population, the length of the chromosome, the maximum number of generations to test, the probability of a crossover and the probability of a mutation.

```
on initdata
  global popsize, lchrom, maxgen, pcross, pmutation, nmutation, ncross
  put 0 into ncross
  put 0 into nmutation

  put card field fpopsize into popsize
  put card field flchrom into lchrom
  put card field fmaxgen into maxgen
  put card field fpcross into pcross
  put card field fpmutation into pmutation
```

```
end initdata
```

Initpop creates the seed population by randomly creating 0 or 1 alleles and tests the fitness of each chromosome generated.

```
on initpop
  global popsize, oldpop, lchrom, chrom, fitness, newpop
  repeat with j = 1 to popsize
    repeat with j1 = 1 to lchrom
      put flip(0.5) into flag
      if flag = true then
        put 1 into char j1 of chrom
      else
        put 0 into char j1 of chrom
      end if
    end repeat

    put chrom into line j of oldpop
    put chrom into line j of newpop

    put decode(chrom,lchrom) into x
    put objfunc(x) into line j of fitness

  end repeat
end initpop
```

Perform the initialization procedures.

```
on initialize
  initdata
  initpop
  statistics
  initreport
end initialize
```

A unique first generation report is printed. This procedure generates that report.

```
on initreport
  global popsize, lchrom, maxgen, pcross, pmutation
  global lmax, lavg, lmin, sumfitness, newpop

  put "Population size (popsize)              = "&popsize into line 1 of card field
results
  put "Chromasome Length (LChrom)      = "&lchrom into line 2 of card field
results
  put "Maximum Number of Generations (Maxgen) = "&maxgen into line 3 of
card field results
  put "Crossover Probability (PCross)      = "&pcross into line 4 of card field
results
  put "Mutation Probability (PMutation) = "&pmutation into line 5 of card field
results

  put "Initial Generation Statistics" into line 7 of card field results
```

```
  put "Initial Population Maximum Fitness = "&lmax into line 9 of card field
results
  put "Initial Population Average Fitness = "&lavg into line 10 of card field
results
  put "Initial Population Minimum Fitness = "&lmin into line 11 of card field
results
  put "Initial Population Sum of Fitness = "&sumfitness into line 12 of card
field results
  put return after card field "results"

  repeat with i = 1 to popsize
    put return&line i of newpop after card field "results"
  end repeat
end initreport
```

Below is an example of the output from the twentieth generation run. The row of 0's and 1's is a particular chromosome and the decimal value at the end of each line is the calculated fitness for that chromosome. Two generations are displayed, the results of the 19th generation followed by the 20th. Measures of the maximum, minimum and sum of fitnesses for the 20th generation are given as are the number of crossovers and mutation to date. Note that there has been an improvement in only one chromosome (the last) in this generation run. This slow improvement tends to occur as the average fitness approaches unity.

```
••••• CARD No.1  FIELD No.6  results Field Contents
Generation Report 19
1111110001011010000011000110001 0.985749
1111111111011110010011110101100 0.999486
1111111011011010010010111011100 0.995518
1111110111001001110011011101010 0.991361
1111110111010110001011111011110 0.991549
1111100111100111000011111110010 0.976182
1111111011100111001101011101000 0.995715
1111011001011010010011000101100 0.962315
1111111000000000011101111011010 0.992195
1111100100001011100000111001010 0.972832
Generation Report 20
1111110001011010000011000110001 0.985749
1111111111011110010011110101100 0.999486
1111111011011010010010111011100 0.995518
1111110111001001110011011101010 0.991361
1111110111010110001011111011110 0.991549
1111100111100111000011111110010 0.976182
1111111011100111001101011101000 0.995715
1111011001011010010011000101100 0.962315
1111111000000000011101111011010 0.992195
1111110100001011100000111001010 0.988741
Max = 0.999486
Min = 0.962315
Sumfitness = 9.878811
```

NMutation = 1916
NCross = 95

This is the end of the HyperTalk source code.

Appendix D: Disk contents

The disk that comes with this book contains working applications and code (in ASCII). The disk is formatted as a PC disk. However; the code segments are ONLY for Macintosh. The code segments can, however, be used and copied to any type of machine as long as you can read the contents of the disk.

For a Macintosh user to read the disk there are a number of possibilities, use a commercial product (such as PC Access), use System 7.5 or some other mechanism to read the floppy.

Appendix E: CONTRIBUTOR AGREEMENT

THIS AGREEMENT made and entered into this _____day of _________, 19__, by and between CRC PRESS, INC. (hereinafter referred to as the 'Publisher'), and _________________________________ .(hereinafter referred to as the 'Contributor'). The Publisher and the Contributor hereby agree as follows:

FIRST: (a) The Contributor agrees to write for the Publisher the textual material, including all references, figures, and tables (hereinafter the 'Manuscript') and to furnish all artwork and photographs (hereinafter the 'Illustrations') (the Manuscript and the Illustrations shall be referred to jointly hereinafter as the

'Contribution') for a chapter on the subject of ____________________________

____

for inclusion in a work tentatively entitled: ______________________________

____

(hereinafter referred to as the 'Work'). It is expressly agreed that the Contribution shall be considered a work made for hire for the Publisher.

(b) The Contributor agrees to deliver the Manuscript in form and content acceptable to the Publisher and the Editor of the Work; the textual content of the Manuscript shall be approximately 20 - 60 pages in length. Variation in the length of the Manuscript requires the approval of the Editor of the Work.

SECOND: (a) The Contributor agrees to deliver to the Editor the original, plus a copy, of the typewritten completed Manuscript or a computer disk containing the completed Manuscript, together with the Illustrations, on or before the __________ day of ____________ ,19__

(b) When delivered to the Publisher, the Contribution shall be typed double spaced on 8-1/2 x 11 inch paper, or in electronic form approved by the Publisher on a computer disk, prepared in accordance with the Publisher's *Instruction Guide for Authors,* a copy of which has been provided by the Publisher, and the Illustrations shall be in camera-ready condition ready for reproduction in the form specified by the Publisher.

(c) If the Contributor has not delivered the Contribution within thirty (30) days of the date specified in paragraph SECOND (a) above, or if the Publisher shall not accept the Manuscript or the Illustrations submitted as being in form and content acceptable for publication in the Work, the Editor or the Publisher may request that the Contributor revise the Contribution or may terminate this Agreement and shall return the Contribution to the Contributor.

(d) In the event the Work is not published, or the Contribution is not published in the Work by Publisher for any reason, the Publisher and the Editor shall have no liability to the Contributor except to return the Contribution to the Contributor.

THIRD: The Contributor acknowledges that the Contribution is being specifically ordered and commissioned by the Publisher for publication by the Publisher and that the Publisher is the owner of the copyright in the Contribution and, in any event, the Contributor hereby grants, transfers, and assigns to the Publisher all of the Contributor's rights, including the copyright, in and to the Contribution. The Contributor may, however, draw on and refer to material contained in the contribution in preparing articles for publication in professional journals, for teaching purposes, and for delivery at professional meetings and symposia, provided appropriate credit is given to the Publisher and the Work. Other uses require written permission of the Publisher.

FOURTH: (a) The Contributor represents and warrants to the Publisher that the Contributor has full power and authority to enter into this Agreement; that the Contribution is original except for the material in the public domain and such excerpts from other works as may be included with the prior written permission of the copyright owners; that the Contributor has not assigned, transferred, or otherwise encumbered the Publisher's rights to the Contribution; that the Contribution contains no libelous or unlawful statements; that the Contribution shall not infringe upon or violate any rights of third parties or any copyright, trademark, or any other proprietary or privacy right of others; the Contribution contains no material which, in the best judgment of the Contributor, is inaccurate; and the use of any instruction, material, or formula contained in the Contribution will not result in any injury to person or property.

(b) Should any matter be submitted for publication in the Work which in the opinion of the Publisher is libelous or otherwise actionable, the Publisher shall have the exclusive right to refuse to include such matter in the Work. Nothing contained herein shall be deemed to impose upon the Publisher any duty of independent investigation or relieve the Contributor of the obligations assumed by the Contributor hereunder.

(c) The Contributor agrees to indemnify the Publisher and hold the Publisher and the Editor of the Work harmless from any and all losses, damages, liabilities, costs, charges, and expenses, including reasonable attorneys' fees arising out of any breach or alleged breach of any of the Contributor's representations and warranties contained in this Agreement or third party claims relating to the matters covered by these representations and warranties.

FIFTH: (a) The Contributor shall identify all material in the Contribution belonging to others and shall obtain, at the Contributor's expense, from each owner of copyrighted material appearing in the Contribution, a letter granting permission to the Publisher to reproduce such material in the Work. Each permission which the Contributor is obligated to obtain under this Agreement must be consistent with the rights of the Publisher in the Contribution pursuant to this Agreement so that they will cover all uses to which such material may eventually be put. The Contributor, where necessary, shall have all Illustrations prepared for him or her as a work for hire, and shall execute and deliver to the

Publisher all documents as the Publisher may require transferring all right, title, and interest, including the copyright, in and to the Illustrations to the Publisher. The Contributor shall deliver to the Publisher with the completed Contribution an executed copy of all such permissions or documents.

(b) The Contributor shall also identify all material in the Contribution taken from documents prepared and published by the United States government, and therefore not subject to copyright, when such material constitutes a significant portion of the Contribution.

<u>SIXTH:</u> The Publisher and Editor shall make, or cause to be made, all corrections, including deletions, additions, and other revisions as may be considered desirable, and in the form and style reasonably required by the Publisher.

<u>SEVENTH:</u> The Publisher reserves the right to determine who shall revise the Contribution in any revised edition of the Work. If, in the Publisher's sole discretion, significant portions of the Contribution are used by a reviser who is not the Contributor, the Contributor shall be given credit with respect to that revision as the original author of the Contribution, and the reviser shall receive credit as the reviser of the Contribution. No additional compensation shall be payable to the Contributor with respect to the publication of any reprints or revisions of the Contribution unless the Contributor is requested to prepare the revision, in which event compensation shall be as specified in a separate agreement mutually satisfactory to the Publisher and the Contributor.

<u>EIGHTH:</u> This Agreement may not be assigned by the Contributor without the prior written consent of the Publisher and is binding upon the parties hereto, their heirs, personal representatives, and assignees.

<u>NINTH:</u> The Publisher shall have the right to use the Contributor's name, likeness, and relevant biographical data in, and in connection with, the sale and promotion of the Work, the Contribution, and any work or other publication in which the Contribution appears.

<u>TENTH:</u> This Agreement, regardless of its place of physical execution, shall in all respects, be governed by and construed in accordance with the internal law, and not the law pertaining to conflicts or choice of law, of the State of Florida.

<u>ELEVENTH:</u> Each of the parties to this Agreement hereby expressly and irrevocably agrees and consents that any suit, action, or proceeding arising out of or relating to this Agreement, shall be instituted exclusively and only in a state or federal court sitting in Miami, Florida, and, by execution of this Agreement, each of the parties hereto expressly waives any objection that it may have now or hereafter to the laying of venue or to the jurisdiction of any such suit, action, or proceeding in Miami, Florida, and each of the parties to this Agreement further irrevocably, exclusively, and unconditionally submits to the personal jurisdiction of any state or federal court sitting in Miami, Florida, in connection with any such suit, action, or proceeding.

IN WITNESS WHEREOF, the parties hereto have duly executed this Agreement the __________ day of ______________ , 19_____

CONTRIBUTOR: ____

Social Security No.: ________________________________

Place of Birth: ____

Business address: ____

____

(include department, building, room number)

Telephone No.: ______________________________________
(include international codes: if applicable)

FAX No.: ____
(include international codes: if applicable)

Permanent Domicile: ____

____

____

Date: ________________________________

Citizenship: ____________________________________

Signature: ____________________________________

PUBLISHER: Lance Chambers
Date: 27th June, 1994
Title of Company Officer: EDITOR

Signature: ____________________________________

Please return the signed contract and your contribution on a 3.5" floppy (MS WORD, WORD PERFECT, or POSTSCRIPT for Macintosh or IBM PC. Note: Please also include a complete hard-copy of the contribution as well.) to:

Lance Chambers
140 Treasure Road
Queens Park
Perth, West Australia 6107

If you wish to make a contribution to future volumes of the GA Handbook series please complete this contract and send off your work.

INDEX

E

F

G

H

I

J

K

L

M

N

O

P

Q

R

S

T

U

V

W

X

Z